Maximum Principles and Sharp Constants for Solutions of Elliptic and Parabolic Systems

Mathematical Surveys and Monographs

Volume 183

Maximum Principles and Sharp Constants for Solutions of Elliptic and Parabolic Systems

Gershon Kresin
Vladimir Maz'ya

American Mathematical Society
Providence, Rhode Island

2010 *Mathematics Subject Classification.* Primary 35A23, 35B50, 35J47, 35K40; Secondary 31B10, 35J30, 35Q35, 35Q74.

The text was translated from Russian by Tatiana O. Shaposhnikova.

For additional information and updates on this book, visit
www.ams.org/bookpages/surv-183

Library of Congress Cataloging-in-Publication Data

Kresin, Gershon, author.
Maximum principles and sharp constants for solutions of elliptic and parabolic systems / Gershon Kresin, Vladimir Maz′ya
p. cm. — (Mathematical surveys and monographs ; volume 183)
Includes bibliographical references and index.
ISBN 978-0-8218-8981-7 (alk. paper)
1. Inequalities (Mathematics) 2. Maximum principles (Mathematics) I. Maz′ya, V. G., author. II. Title.

QA295.K85 2012
515′.983–dc23 2012020950

∞ The paper used in this book is acid-free and falls within the guidelines
established to ensure permanence and durability.
Visit the AMS home page at http://www.ams.org/

10 9 8 7 6 5 4 3 2 1 17 16 15 14 13 12

Contents

Introduction

Background. The classical maximum modulus principles are fundamental properties of solutions to partial differential equations. They have numerous important applications to various problems in the theory of these equations, both linear and nonlinear.

Let Ω be a bounded domain in the Euclidean space $\mathbb{R}^n$. For the elliptic equation

$$(1) \qquad \sum_{j,k=1}^{n} a_{jk}(x)\frac{\partial^2 u}{\partial x_j \partial x_k} - \sum_{j=1}^{n} a_j(x)\frac{\partial u}{\partial x_j} - a_0(x)u = 0\,, \quad x \in \Omega,$$

with bounded coefficients, positive-definite matrix $((a_{jk}(x)))$, and with $a_0(x) \geq 0$, two classical facts are basic: the weak and strong maximum modulus principles. By the weak one, a solution $u \in \mathrm{C}^2(\Omega) \cap \mathrm{C}(\overline{\Omega})$ to equation (1) satisfies

$$\max_{\overline{\Omega}} |u| = \max_{\partial\Omega} |u|\,.$$

According to the strong principle the maximum modulus of a non-constant solution u is never attained inside Ω. Note that the last property is not dealt with in the present book.

The weak and strong maximum modulus principles hold also for the parabolic equation

$$(2) \qquad \frac{\partial u}{\partial t} - \sum_{j,k=1}^{n} a_{jk}(x,t)\frac{\partial^2 u}{\partial x_j \partial x_k} + \sum_{j=1}^{n} a_j(x,t)\frac{\partial u}{\partial x_j} + a_0(x,t)u = 0$$

with bounded coefficients, positive-definite matrix $((a_{jk}(x,t)))$, and with $a_0(x,t) \geq 0$ in a cylinder $Q_T = \Omega \times (0,T]$. By the weak maximum modulus principle, if $u \in \mathrm{C}^{(2,1)}(Q_T) \cap \mathrm{C}(\overline{Q}_T)$ is a solution of (2), then

$$\max_{\overline{Q}_T} |u| = \max_{\overline{\Gamma}_T} |u|\,,$$

where $\Gamma_T = \{(x,t) \in \partial Q_T :\ 0 \leq t < T\}$.

The maximum principles just mentioned were obtained for different classes of solutions and under various assumptions about the coefficients, but are valid almost exclusively for scalar equations of the second order (with weakly coupled systems as the only exception).

The not so sharp but incomparably more general property of the same nature, which holds for equations and systems of arbitrary order in smooth domains, is the so-called Miranda-Agmon maximum principle. In particular, for a homogeneous elliptic equation of order 2ℓ, this principle is the estimate

$$(3) \qquad \max_{\overline{\Omega}} |\nabla_{\ell-1} u| \leq c(\Omega) \max_{\partial\Omega} |\nabla_{\ell-1} u|.$$

A weaker variant of the Miranda-Agmon maximum principle runs as follows:

$$\max_{\overline{\Omega}} |\nabla_{\ell-1} u| \leq K \max_{\partial\Omega} |\nabla_{\ell-1} u| + C\|u\|_{L^1(\Omega)}. \tag{4}$$

Motivation, subject, and method. Various maximum principles for scalar equations of the second order are treated in numerous papers and in the monographs by Miranda [**Mir**], Protter and Weinberger [**PW**], Sperb [**Sp**], Fraenkel [**Fr**] , Pucci and Serrin[**PS**], López-Gómez [**LG**]. Less is written about systems and higher order equations and, in particular, there are no books dealing with this subject whatsoever.

The main goal of the present monograph, which has almost nothing in common both in results and methods with the books just mentioned, is to partially fill in this gap. Here we deal with different maximum modulus principles for elliptic and parabolic systems of arbitrary order. In particular, we present necessary and sufficient conditions for validity of the classical maximum modulus principles for systems of the second order. We obtain sharp constants in inequalities (3) and (4) and in many other inequalities of a similar nature. Somewhat related to this topic are explicit formulae for the norms and the essential norms of integral operators of potential type.

Our proofs are based on a unified approach, using on one hand, representations of the norms of matrix integral operators whose target spaces are linear and finite dimensional, and, on the other hand, on solving certain finite dimensional optimization problems.

The book reflects results obtained by the authors during the last three decades and published in [**Kr1**]-[**KM10**], [**KM12**], [**KM13**], [**MK**]. Some new material is added as well.

Structure of the book. Let us sketch briefly the main contents of the present volume which consists of two parts. Part I is mostly devoted to linear elliptic equations and systems.

Chapter 1 is auxiliary. It also contains general facts used throughout the book. We establish representations for the norms of linear integral operators acting from the space of n-component continuous or L^p-functions into the Euclidean space $\mathbb{R}^m$ or into the unitary space $\mathbb{C}^m$. Continuous vector-valued functions are defined on a locally compact Hausdorff space. Vector-valued functions in L^p are considered on spaces with a measure.

Chapter 2 concerns necessary and sufficient conditions for validity of the classical maximum modulus principle for strongly elliptic systems of the second order. First we consider the systems with constant coefficients and the second order derivatives only. It is shown that the classical maximum modulus principle holds for such a model system if and only if the differential operator of the system is the product of a positive-definite matrix and a scalar elliptic operator, which essentially means that the operator of the system is scalar.

Further we turn to strongly elliptic systems of the second order with variable coefficients and obtain, in particular, algebraic sufficient conditions for validity of the classical modulus principle. These conditions prove to be also necessary if the maximum modulus principle holds for every subdomain of Ω. We give an example showing that the scalarity of the principal part of the system with variable coefficients may not be necessary for validity of the classical maximum modulus principle.

In Chapter 3 we find representations for the best constants in the Miranda-Agmon maximum principle for solutions of homogeneous strongly elliptic systems of the second order with constant coefficients in a half-space. We also obtain explicit formulas for sharp constants in inequalities of the Miranda-Agmon type for solutions of the Stokes and Lamé systems as well as a system of viscoelasticity in a half-space. For instance, we prove that the velocity vector $\boldsymbol{u}$ satisfying the homogeneous Stokes system in $\mathbb{R}^n_+ = \{x = (x_1, \dots, x_n) : x_n > 0\}$ admits the estimate

$$|\boldsymbol{u}(x)| \leq K_n \sup\{|\boldsymbol{u}(x', 0)| : x' \in \partial\mathbb{R}^n_+\}$$

with the sharp constant

$$K_n = \frac{2}{\sqrt{\pi}} \frac{\Gamma(\frac{n}{2}+1)}{\Gamma(\frac{n+1}{2})}.$$

In particular, $K_2 = 4/\pi$ and $K_3 = 3/2$.

It is proved that under certain algebraic conditions imposed on the Dirichlet boundary data the constant in an inequality of the Miranda-Agmon type is equal to one, i.e., the classical maximum modulus principle holds.

Chapter 4 concerns linear elliptic systems of the second order with constant coefficients without lower order terms. We find a representation of the sharp coefficient in a pointwise estimate for the modulus of a solution of the system in a half-space by the norm of the Dirichlet data in L^p. Solving a certain optimization problem on the unit sphere in a finite dimensional space, we derive explicitly the coefficients just mentioned for the Lamé and Stokes systems.

Unlike previous chapters, in Chapters 5 and 6 we turn to pointwise estimates for derivatives of solutions to elliptic equations. In Chapter 5 we are concerned with the best coefficients in the Miranda-Agmon type estimates (3) and (4). For example, we obtain the C. Miranda inequality for biharmonic functions with the sharp constant

$$\sup_{\mathbb{R}^n_+} |\nabla u| \leq K \sup_{\partial\mathbb{R}^n_+} |\nabla u|.$$

In particular, $K = 4/\pi$ for $n = 2, K = 1/2 + 2\pi\sqrt{3}/9$ for $n = 3$ and $K = 2$ for $n = 4$.

Chapter 6 contains inequalities with sharp coefficients for the modulus of directional derivatives of harmonic functions in a half-space and in a ball. The boundary values of harmonic functions are assumed to be bounded, semibounded or L^p-integrable. Simultaneously with deriving explicit formulae for the best coefficients in pointwise estimates of the modulus of the gradient of a harmonic function, we find extremal directions for harmonic fields in a half-space or in a ball. This leads to a solution of the Khavinson type extremal problems for harmonic functions. In particular, we obtain the inequality with the sharp constant

$$|\nabla u(x)| \leq \frac{4(n-1)^{(n-1)/2}}{n^{n/2}} \frac{\omega_{n-1}}{\omega_n x_n} \sup_{y \in \mathbb{R}^n_+} |u(y)|,$$

where ω_n is the area of the unit sphere in $\mathbb{R}^n$. The estimates obtained in this chapter can be interpreted as multidimensional analogues of certain "sharp real-part theorems" going back to the famous Hadamard theorem in complex function theory.

In Chapter 7, which is the last chapter of Part I, we consider vector-valued potentials of the double layer type and their corresponding matrix-valued integral operators in spaces of continuous vector-valued functions on the boundaries of irregular domains. The potentials and operators in question arise when the Dirichlet problem for elliptic systems is reduced to regular boundary integral equations. Unlike the case of smooth boundaries, the boundary integral operators under consideration are not compact but only bounded. As a measure of non-compactness of a bounded operator L one uses the essential norm $\mathrm{ess}\|L\|$ which appears in the estimate $R(L) \geq \big(\mathrm{ess}\|L\|\big)^{-1}$ for the Fredholm radius $R(L)$ of L. The notions of the essential norm and of the Fredholm radius were introduced and applied by Radon who developed a logarithmic potential theory on curves with bounded rotation. We find representations for the norms and for the essential norms of matrix-valued integral operators of double layer potential type in spaces of continuous vector-valued functions. Representations for the essential norm are concretized for boundary integral operators of the elasticity theory (Lamé system) and hydrodynamics (Stokes system) both in the planar case for a domain with angular points, and in the three-dimensional case for a domain with conic points or edges.

The second part of the book concerns equations and systems of parabolic type.

In Chapter 8 we consider systems of partial differential equations of the first order in t and of order 2ℓ in x variables, which are uniformly parabolic in the sense of Petrovskiĭ. We show that the classical maximum modulus principle is not valid in $\mathbb{R}^{n+1}_T = \mathbb{R}^n \times (0, T]$ for $\ell \geq 2$. For the system

$$\frac{\partial \boldsymbol{u}}{\partial t} - \sum_{j,k=1}^{n} \mathcal{A}_{jk}(x,t) \frac{\partial^2 \boldsymbol{u}}{\partial x_j \partial x_k} + \sum_{j=1}^{n} \mathcal{A}_j(x,t) \frac{\partial \boldsymbol{u}}{\partial x_j} + \mathcal{A}_0(x,t)\boldsymbol{u} = \mathbf{0} \tag{5}$$

we obtain necessary and, separately, sufficient conditions for the classical maximum modulus principle to hold in the layer $\mathbb{R}^{n+1}_T$ and in the cylinder $Q_T = \Omega \times (0, T]$, where Ω is a bounded subdomain of $\mathbb{R}^n$. If the coefficients of the system do not depend on t, these conditions coincide. A necessary and sufficient condition in this case is that the principal part of the system is scalar, i.e., $\mathcal{A}_{jk}(x) = a_{jk}(x)I$, $1 \leq j,\ k \leq n$, and that the coefficients of the system satisfy the algebraic inequality

$$\sum_{j,k=1}^{n} a_{jk}(x)(\boldsymbol{\xi}_j, \boldsymbol{\xi}_k) + \sum_{j=1}^{n} (\mathcal{A}_j(x)\boldsymbol{\xi}_j, \boldsymbol{\zeta}) + (\mathcal{A}_0(x)\boldsymbol{\zeta}, \boldsymbol{\zeta}) \geq 0$$

for all $x \in \Omega$ ($x \in \mathbb{R}^n$) and for any $\boldsymbol{\xi}_j$, $\boldsymbol{\zeta} \in \mathbb{R}^m$, $j = 1, \ldots, n$, with $(\boldsymbol{\xi}_j, \boldsymbol{\zeta}) = 0$. Here $((a_{jk}))$ is a positive-definite $(n \times n)$-matrix-valued function, and I is the identity matrix of order m.

The topic of Chapter 9 is related to one of the main assertions in the preceding chapter. We consider solutions to the initial boundary value problem in the cylinder $\Omega \times (0, T]$ with zero Dirichlet data for a linear second order system, which is strongly parabolic, with smooth $(m \times m)$-matrix-valued coefficients depending only on x. It is assumed that Ω is a bounded subdomain of $\mathbb{R}^n$ with smooth boundary. It is shown that the criterion for validity of the classical maximum modulus principle, obtained in the previous chapter for the case of the parabolic second order system with coefficients independent of t remains necessary if one assumes a priori that the boundary data are zero. We note that this result can be interpreted as a criterion for the $[\mathrm{C}(\overline{\Omega})]^m$-contractivity of a semigroup generated by the operator $\mathfrak{A}(x, D_x)$ of the strongly parabolic system

$$\frac{\partial \boldsymbol{u}}{\partial t} - \mathfrak{A}(x, D_x)\boldsymbol{u} = \boldsymbol{0}.$$

Unlike the rest of Part II, in Chapters 10 and 11 we deal with a *maximum norm principle*, understanding the norm in a generalized sense, i.e., as the Minkowski functional of a compact convex body in $\mathbb{R}^m$ containing the origin in its interior.

In Chapter 10 we consider systems of partial differential equations, which contain only second derivatives in the x variables and which are uniformly parabolic in the sense of Petrovskiĭ. For such systems we obtain necessary and, separately, sufficient conditions for the maximum norm principle to hold in the layer $\mathbb{R}^n \times (0, T]$ and in the cylinder $\Omega \times (0, T]$, where Ω is a bounded subdomain of $\mathbb{R}^n$. The necessary and sufficient conditions coincide if the coefficients of the system do not depend on t. The above mentioned criterion for validity of the maximum norm principle is formulated as a number of equivalent algebraic conditions describing the relation between the geometry of the unit sphere of the given norm and coefficients of the system under consideration. Simpler formulated criteria ensuring the maximum norm principle are given for certain classes of norms: for differentiable norms, p-norms $(1 \le p \le \infty)$ in $\mathbb{R}^m$, as well as for norms whose unit balls are m-pyramids, m-bipyramids, cylindrical bodies, m-parallelepipeds.

In Chapter 11, the question of validity of the maximum norm principle is considered for linear strongly coupled systems of partial differential equations of the first order in t and of the second order in the x variables, which are uniformly parabolic in the sense of Petrovskiĭ. It is supposed that the norm is twice continuously differentiable. For such systems necessary and, separately, sufficient conditions for the maximum norm principle are obtained. These necessary and sufficient conditions coincide if the coefficients of the system do not depend on t. The criterion for validity of the maximum norm principle in this case is that the principal part of the system is scalar and that the coefficients satisfy a certain algebraic inequality, describing a relation between the system and the geometry of the sphere of the given norm. Simpler formulated criteria are given for certain classes of norms, in particular, for p-norms $(2 < p < \infty)$ in $\mathbb{R}^m$.

More detailed information on the material in this book can be found in the introductions to the chapters. We collect historical and bibliographical notes in comments to each chapter.

Readership. The volume is addressed to mathematicians working in partial differential equations, in particular, in elasticity theory and hydrodynamics.

Prerequisites for reading this book are undergraduate courses in linear algebra, theory of partial differential equations, and functional analysis.

Acknowledgments. The research of G. Kresin was supported by the KAMEA program of the Ministry of Absorption, State of Israel, and by the Ariel University Center of Samaria. The work of V. Maz'ya was supported by the Linköping University, University of Liverpool, and the Ohio State University. The authors express their thanks to these institutions.

Ariel, Israel — *Gershon Kresin*
Linköping, Sweden — *Vladimir Maz'ya*

Part 1

Elliptic Equations and Systems

CHAPTER 1

Prerequisites on Operators Acting into Finite Dimensional Spaces

1.1. Introduction

The present chapter is auxiliary. The material collected here will be applied to special integral operators throughout the book.

Section 1.2 is devoted to functionals and operators acting on n-component continuous vector-valued functions which are given on a locally compact Hausdorff space X and vanish at infinity. The space of such functions, endowed with the norm $\|\boldsymbol{f}\| = \sup\{|\boldsymbol{f}(x)| : x \in X\}$, is denoted by $[\mathrm{C_v}(X)]^n$ in the case of real vector-valued functions and by $[\mathbf{C_v}(X)]^n$ in the case of complex vector-valued functions. From the Riesz theorem it follows that any linear bounded functional L on $[\mathrm{C_v}(X)]^n$ admits the representation

$$L(\boldsymbol{f}) = \int_X d\boldsymbol{\mu}(x)\boldsymbol{f}(x) = \sum_{j=1}^{n} \int_X f_j(x)d\mu_j(x),$$

where $\boldsymbol{\mu} = (\mu_1, \dots, \mu_n)$ is a vector-valued measure whose elements are signed finite regular measures on the σ-algebra $\mathcal{B}_X$ of Borel subsets of X and $\boldsymbol{f} = (f_1, \dots, f_n)$. A similar representation is valid for a linear bounded functional on $[\mathbf{C_v}(X)]^n$ with $\boldsymbol{\mu} = (\mu_1, \dots, \mu_n)$ being complex regular measures on $\mathcal{B}_X$. We show that

$$\|L\| = |\boldsymbol{\mu}|,$$

where $|\cdot|$ is the variation of the measure. We also consider a linear bounded operator T mapping $[\mathrm{C_v}(X)]^n$ into the m-dimensional Euclidean space $\mathbb{R}^m$ (and $[\mathbf{C_v}(X)]^n$ into the unitary space $\mathbb{C}^m$) which is defined by

$$T(\boldsymbol{f}) = \int_X dM(x)\boldsymbol{f}(x),$$

where $M = ((\mu_{ij}))$ is a $(m \times n)$-matrix whose elements are finite signed (complex) regular measures on $\mathcal{B}_X$. We derive representations for the norms of the operator T and the functional $\mathcal{T}_{\boldsymbol{z}} = (\boldsymbol{z}, T(\cdot))$ in the real and complex cases, i.e. for $T : [\mathrm{C_v}(X)]^n \to \mathbb{R}^m$, $\boldsymbol{z} \in \mathbb{R}^m$ and $T : [\mathbf{C_v}(X)]^n \to \mathbb{C}^m$, $\boldsymbol{z} \in \mathbb{C}^m$. These representations remain valid if the operators and functionals are defined on the spaces $[\mathrm{C_b}(X)]^n$ and $[\mathbf{C_b}(X)]^n$ of continuous and bounded n-component vector-valued functions on X supplied with the norm $\|\boldsymbol{f}\| = \sup\{|\boldsymbol{f}(x)| : x \in X\}$.

In Section 1.3 we consider the operator

$$S(\boldsymbol{f}) = \int_{\mathcal{X}} G(x)\boldsymbol{f}(x)d\mu(x),$$

acting from the space $[L^p(\mathcal{X}, \mathcal{A}, \mu)]^n$ ($[\boldsymbol{L}^p(\mathcal{X}, \mathcal{A}, \mu)]^n$) of real (complex) vector-valued n-component functions into $\mathbb{R}^m$ ($\mathbb{C}^m$). Here $(\mathcal{X}, \mathcal{A}, \mu)$ is the space with a measure, $G = ((g_{ij}))$ is an $(m \times n)$-matrix-valued function with real (complex) components $g_{ij} \in L^q(\mathcal{X}, \mathcal{A}, \mu)$ ($g_{ij} \in \boldsymbol{L}^q(\mathcal{X}, \mathcal{A}, \mu)$), $1 \le p \le \infty$, $1/p + 1/q = 1$ and the norm in $[L^p(\mathcal{X}, \mathcal{A}, \mu)]^n$ ($[\boldsymbol{L}^p(\mathcal{X}, \mathcal{A}, \mu)]^n$) is defined by

$$\|\boldsymbol{f}\|_p = \left\{ \int_{\mathcal{X}} |\boldsymbol{f}(x)|^p d\mu(x) \right\}^{1/p}$$

for $1 \le p < \infty$ and $\|\boldsymbol{f}\|_\infty = \text{ess sup}\{|\boldsymbol{f}(x)| : x \in \mathcal{X}\}$, where $|\cdot|$ is the length of a vector in $\mathbb{R}^m$ ($\mathbb{C}^m$). We derive a representation for the norm

$$\|S\|_p = \sup_{|\boldsymbol{z}|=1} \|G^* \boldsymbol{z}\|_q,$$

where G^* stands for the transposed (adjoint) matrix of G and $\boldsymbol{z} \in \mathbb{R}^m$ ($\boldsymbol{z} \in \mathbb{C}^m$).

1.2. Linear bounded operators defined on spaces of continuous vector-valued functions and acting into $\mathbb{R}^m$ or $\mathbb{C}^m$

Let X be a locally compact Hausdorff space. By $|\cdot|$ and $(\cdot,\cdot)$ we denote the length of a vector and the inner product in the Euclidean space $\mathbb{R}^m$ and in the unitary finite dimensional space $\mathbb{C}^m$, i.e. for $\boldsymbol{z} = (z_1, \dots, z_m)$ and $\boldsymbol{\zeta} = (\zeta_1, \dots, \zeta_m) \in \mathbb{C}^m$ we put $(\boldsymbol{z}, \boldsymbol{\zeta}) = \overline{z}_1\zeta_1 + \cdots + \overline{z}_m\zeta_m$ and $|\boldsymbol{z}| = (\boldsymbol{z}, \boldsymbol{z})^{1/2}$. By $[\mathrm{C_v}(X)]^n$ we mean the space of continuous real vector-valued functions on X with n components vanishing at infinity. This space is endowed with the norm $\|\boldsymbol{f}\| = \sup\{|\boldsymbol{f}(x)| : x \in X\}$. In a similar way, we introduce the space $[\mathrm{C_b}(X)]^n$ (respectively, $[\mathrm{B_b}(X)]^n$) of real vector-valued functions with n components which are continuous (Borel) and bounded on X. This space is endowed with the norm $\|\boldsymbol{f}\| = \sup\{|\boldsymbol{f}(x)| : x \in X\}$. We use the notation $[\mathbf{C_v}(X)]^n, [\mathbf{C_b}(X)]^n, [\mathbf{B_b}(X)]^n$ for similar spaces of complex vector-valued functions with n components supplied with the same norm. We omit the upper index n in notation of spaces in the case $n = 1$.

Let $\mathcal{B}_X$ be the Borel σ-algebra of the space X. By $|\cdot|$ we mean the variation of the scalar or vector-valued measure. We denote by $\mathfrak{M}_\mathbb{R}(\mathcal{B}_X)$ (respectively, $\mathfrak{M}_\mathbb{C}(\mathcal{B}_X)$) the vector space of all finite signed (complex) regular Borel measures λ with the total variation norm $\|\lambda\| = |\lambda|(X)$.

The following assertion contains a relation between the spaces of functions and measures $\mathrm{C_v}(X)$ and $\mathfrak{M}_\mathbb{R}(\mathcal{B}_X)$, $\mathbf{C_v}(X)$ and $\mathfrak{M}_\mathbb{C}(\mathcal{B}_X)$ (see, for example, Cohn [**Co**]).

THEOREM 1.1. *The map that takes the finite signed (complex) regular measure λ on $\mathcal{B}_X$ to the linear continuous functional*

$$f \mapsto \int_X f(x) d\lambda(x)$$

on $\mathrm{C_v}(X)$ ($\mathbf{C_v}(X)$), is an isometric isomorphism of the Banach space $\mathfrak{M}_\mathbb{R}(\mathcal{B}_X)$ ($\mathfrak{M}_\mathbb{C}(\mathcal{B}_X)$) onto the dual of the Banach space $\mathrm{C_v}(X)$ ($\mathbf{C_v}(X)$).

Let μ_{ij} be finite signed (complex) measures on $\mathcal{B}_X$, $1 \le i \le m, 1 \le j \le n$. We define an $(m \times n)$-matrix-valued measure $M = ((\mu_{ij}))$ on $\mathcal{B}_X$. Suppose $\boldsymbol{f} = (f_1, \dots, f_n)$ belongs to one of the spaces $[\mathrm{C_v}(X)]^n, [\mathrm{C_b}(X)]^n, [\mathrm{B_b}(X)]^n$ or $[\mathbf{C_v}(X)]^n, [\mathbf{C_b}(X)]^n$, $[\mathbf{B_b}(X)]^n$.

We introduce the integral of the real (complex) vector-valued function f with respect to the matrix-valued measure M with real (complex) elements

$$\int_X dM(x)\boldsymbol{f}(x) = \sum_{j=1}^{n} \begin{pmatrix} \int_X f_j(x)d\mu_{1j}(x) \\ \cdots\cdots\cdots\cdots\cdots \\ \cdots\cdots\cdots\cdots\cdots \\ \int_X f_j(x)d\mu_{mj}(x) \end{pmatrix}. \tag{1.2.1}$$

The next assertion follows directly from the above theorem on the representation of linear functionals on the spaces $\mathrm{C_v}(X)$, $\mathbf{C_v}(X)$.

COROLLARY 1.1. *Any linear bounded operator*

$$T : [\mathrm{C_v}(X)]^n \to \mathbb{R}^m \ (T : [\mathbf{C_v}(X)]^n \to \mathbb{C}^m)$$

admits the representation

$$T(\boldsymbol{f}) = \int_X dM(x)\boldsymbol{f}(x),$$

where $M = ((\mu_{ij}))$ *is the* $(m \times n)$*-matrix-valued measure with elements* $\mu_{ij} \in \mathfrak{M}_{\mathbb{R}}(\mathcal{B}_X)$ *(respectively,* $\mu_{ij} \in \mathfrak{M}_{\mathbb{C}}(\mathcal{B}_X)$*).*

PROOF. We consider the space of real-valued functions. The case of complex-valued functions is treated in the same way. Let T be an arbitrary linear bounded operator $[\mathrm{C_v}(X)]^n \to \mathbb{R}^m$,

$$T(\boldsymbol{f}) = \begin{pmatrix} T_1(\boldsymbol{f}) \\ \cdots\cdots \\ \cdots\cdots \\ T_m(\boldsymbol{f}) \end{pmatrix}, \tag{1.2.2}$$

where $T_1, \ldots, T_m$ are functionals on $[\mathrm{C_v}(X)]^n$. Clearly, the functionals $T_1, \ldots, T_m$ are linear and, in view of

$$|T_i(\boldsymbol{f})| \le \left[\sum_{i=1}^{m} |T_i(\boldsymbol{f})|^2\right]^{1/2} = |T(\boldsymbol{f})| \le \|T\| \, \|\boldsymbol{f}\|,$$

they are bounded.

Further, for any function $\boldsymbol{f} = (f_1, \ldots, f_n) \in [\mathrm{C_v}(X)]^n$ there holds

$$T_i(\boldsymbol{f}) = T_i(f_1, 0, \ldots, 0) + T_i(0, f_2, \ldots, 0) + \cdots + T_i(0, \ldots, 0, f_n). \tag{1.2.3}$$

We introduce the notation

$$T_{i1}(f_1) = T_i(f_1, 0, \ldots, 0), \ldots\ldots\ldots, T_{in}(f_n) = T_i(0, \ldots, 0, f_n). \tag{1.2.4}$$

Clearly, each of the functionals $T_{i1}, \ldots, T_{in}$ is linear and bounded on $\mathrm{C_v}(X)$. Hence, by Theorem 1.1, $T_{ij}(f_j)$ admits the representation

$$T_{ij}(f_j) = \int_X f_j(x)d\mu_{ij}(x), \tag{1.2.5}$$

$1 \le i \le m$, $1 \le j \le n$, where $\mu_{ij} \in \mathfrak{M}_{\mathbb{R}}(\mathcal{B}_X)$. Thus, (1.2.3)-(1.2.5) imply

$$T_i(\boldsymbol{f}) = \sum_{j=1}^{n} T_{ij}(f_j) = \sum_{j=1}^{n} \int_X f_j(x)d\mu_{ij}(x). \tag{1.2.6}$$

Combining (1.2.2), (1.2.6) and the notation $M = ((\mu_{ij}))$, $1 \le i \le m$, $1 \le j \le n$, we complete the proof. ☐

Let $(\mathcal{X}, \mathcal{A})$ be a measurable space, and let $\mu_1, \ldots, \mu_n$ be finite signed (complex) measures. Let, further $\boldsymbol{\mu} = (\mu_1, \ldots, \mu_n)$. The variation $|\boldsymbol{\mu}|(A)$ on $A \in \mathcal{A}$ of the vector-valued measure $\boldsymbol{\mu} : \mathcal{A} \to \mathbb{R}^n$ ($\boldsymbol{\mu} : \mathcal{A} \to \mathbb{C}^n$) is the supremum of the numbers

$$\sum_{j=1}^{k} |\boldsymbol{\mu}(A_j)|,$$

where $\{A_j\}_{j=1}^k$ ranges over all finite partitions of A into $\mathcal{A}$-measurables sets (cf. Diestel and Uhl, Jr. [**DU**], Dunford and Schwartz [**DS**])). The variation $|\boldsymbol{\mu}|$ of the vector-valued measure $\boldsymbol{\mu}$ is a finite positive measure on $(\mathcal{X}, \mathcal{A})$; the finiteness and countable additivity of $|\boldsymbol{\mu}|$ result from the inequality $|\boldsymbol{\mu}|(A) \leq |\mu_1|(A) + \cdots + |\mu_n|(A)$ and corresponding properties of measures $|\mu_i|, 1 \leq i \leq n$ (cf. Cohn [**Co**]).

The following assertion contains a representation of the norm of an arbitrary linear continuous functional on the space of continuous (complex) vector-valued functions with n components which are continuous on X and vanish at infinity.

PROPOSITION 1.1. *Let $\boldsymbol{\mu} = (\mu_1, \ldots, \mu_n)$ be a vector-valued measure on $\mathcal{B}_X$ with $\mu_i \in \mathfrak{M}_{\mathbb{R}}(\mathcal{B}_X) (\mu_i \in \mathfrak{M}_{\mathbb{C}}(\mathcal{B}_X))$. The norm of the linear continuous functional L*

$$L(\boldsymbol{f}) = \int_X d\boldsymbol{\mu}(x) \boldsymbol{f}(x) \tag{1.2.7}$$

defined on $[\mathrm{C_v}(X)]^n$ ($[\mathbf{C_v}(X)]^n$) is equal to

$$\|L\| = |\boldsymbol{\mu}|(X). \tag{1.2.8}$$

In particular, if

$$\boldsymbol{\mu}(A) = \int_A \boldsymbol{\mathcal{N}}(x) d\lambda(x), \tag{1.2.9}$$

where $A \in \mathcal{B}_X$, $\boldsymbol{\mathcal{N}} \in [\mathrm{B_b}(X)]^n$ (or $\boldsymbol{\mathcal{N}} \in [\mathbf{B_b}(X)]^n$) and λ is a finite signed (complex) regular Borel measure, then

$$\|L\| = \int_X |\boldsymbol{\mathcal{N}}(x)| d|\lambda|(x). \tag{1.2.10}$$

PROOF. We shall deduce (1.2.8) and (1.2.10) for the functional (1.2.7) on the space $[\mathbf{C}_0(X)]^n$. The case of real-valued functionals is treated in the same way. Let $\boldsymbol{f}$ be a vector-valued function with n components on X with values $\boldsymbol{z}_1, \ldots, \boldsymbol{z}_k \in \mathbb{C}^n$ on the sets $A_1, \ldots, A_k \in \mathcal{B}_X$, respectively. Then

$$|L(\boldsymbol{f})| = \left| \int_X d\boldsymbol{\mu}(x) \boldsymbol{f}(x) \right| = \left| \sum_{j=1}^{k} (\overline{\boldsymbol{z}}_j, \boldsymbol{\mu}(A_j)) \right| \leq \sum_{j=1}^{k} |\boldsymbol{z}_j| |\boldsymbol{\mu}(A_j)| \leq \|\boldsymbol{f}\| \sum_{j=1}^{k} |\boldsymbol{\mu}(A_j)|,$$

and therefore,

$$|L(\boldsymbol{f})| \leq \|\boldsymbol{f}\| | \boldsymbol{\mu}|(X). \tag{1.2.11}$$

Since every vector-valued function from $[\mathbf{B_b}(X)]^n$ is a uniform limit of a sequence of simple Borel vector-valued functions with n components (a function is called simple if it has only finitely many values), it follows that (1.2.11) holds for any $\boldsymbol{f} \in [\mathbf{B_b}(X)]^n$ and, in particular, for $\boldsymbol{f} \in [\mathbf{C_v}(X)]^n$. Thus,

$$\|L\| \leq |\boldsymbol{\mu}|(X). \tag{1.2.12}$$

Next, we show, that $\|L\| \geq |\boldsymbol{\mu}|(X)$. Let ε be a positive number. By the definition of $|\boldsymbol{\mu}|$ there exists a finite partition $A_1, \ldots, A_k \in \mathcal{B}_X$ of the space X such that

$$\sum_{j=1}^{k} |\boldsymbol{\mu}(A_j)| > |\boldsymbol{\mu}|(X) - \varepsilon. \tag{1.2.13}$$

It is well known (see Cohn [**Co**], Lemma 7.3.4) that for any measure $\mu_i \in \mathfrak{M}_{\mathbb{C}}(\mathcal{B}_X)$ and for any set A_j there exists a compact set $K_{ij} \subset A_j$ such that

$$|\mu_i(A_j \backslash K_{ij})| = |\mu_i(A_j) - \mu_i(K_{ij})| < \frac{\varepsilon}{nk}, \tag{1.2.14}$$

$1 \leq i \leq n, 1 \leq j \leq k$. We introduce the notation

$$K_j = \bigcup_{i=1}^{n} K_{ij}, \qquad K = \bigcup_{j=1}^{k} K_j.$$

In view of (1.2.14) and the just cited Lemma,

$$|\mu_i(A_j \backslash K_j)| = |\mu_i(A_j) - \mu_i(K_j)| < \frac{\varepsilon}{nk}.$$

Hence

$$\begin{aligned} \sum_{j=1}^{k} |\boldsymbol{\mu}(A_j)| &= \sum_{j=1}^{k} |\boldsymbol{\mu}(A_j \backslash K_j) + \boldsymbol{\mu}(K_j)| \leq \sum_{j=1}^{k} |\boldsymbol{\mu}(A_j \backslash K_j)| + \sum_{j=1}^{k} |\boldsymbol{\mu}(K_j)| \\ &\leq \sum_{j=1}^{k} \sum_{i=1}^{n} |\mu_i(A_j \backslash K_j)| + \sum_{j=1}^{k} |\boldsymbol{\mu}(K_j)| < \sum_{j=1}^{k} |\boldsymbol{\mu}(K_j)| + \varepsilon, \end{aligned}$$

which together with (1.2.13) implies

$$|\boldsymbol{\mu}|(X) - 2\varepsilon < \sum_{j=1}^{k} |\boldsymbol{\mu}(K_j)| \leq \sum_{j=1}^{k} |\boldsymbol{\mu}|(K_j). \tag{1.2.15}$$

Given nonoverlapping compact sets $K_1, \ldots, K_k$, one can construct nonoverlapping open sets $U_1 \supset K_1, \ldots, U_k \supset K_k$ and functions $\alpha_j \in \mathrm{C_b}(X)$ with compact supports such that $\operatorname{supp} \alpha_j \in U_j, \chi_{K_j} \leq \alpha_j \leq \chi_{U_j}, 1 \leq j \leq k$ (see Cohn [**Co**], Prop. 7.1.8, 7.1.11). We introduce the vector-valued function with n components

$$\boldsymbol{g}(x) = \sum_{j=1}^{k} \alpha_j(x) \boldsymbol{\gamma}_j, \tag{1.2.16}$$

where

$$\boldsymbol{\gamma}_j = \begin{cases} \dfrac{\overline{\boldsymbol{\mu}(K_j)}}{|\boldsymbol{\mu}(K_j)|} & \text{for} \quad |\boldsymbol{\mu}(K_j)| \neq 0, \\[2ex] \mathbf{0} & \text{for} \quad |\boldsymbol{\mu}(K_j)| = 0. \end{cases} \tag{1.2.17}$$

Note that $\boldsymbol{g} \in [\mathbf{C}_0(X)]^n$ and $|\boldsymbol{g}(x)| \leq 1$ for all $x \in X$. Using (1.2.16), (1.2.17), we find

$$\int_K d\boldsymbol{\mu}(x) \boldsymbol{g}(x) = \sum_{j=1}^{k} (\overline{\boldsymbol{\gamma}}_j, \boldsymbol{\mu}(K_j)) = \sum_{j=1}^{k} |\boldsymbol{\mu}(K_j)|.$$

This and (1.2.15) imply the estimate

$$\int_K d\boldsymbol{\mu}(x)\boldsymbol{g}(x) > |\boldsymbol{\mu}|(X) - 2\varepsilon. \tag{1.2.18}$$

Using an estimate similar to (1.2.11) as well as (1.2.15) and the additivity of $|\boldsymbol{\mu}|$, we obtain

$$\begin{aligned}\left|\int_{K^c} d\boldsymbol{\mu}(x)\boldsymbol{g}(x)\right| &\leq \|\boldsymbol{g}\|\, |\boldsymbol{\mu}|(K^c) \leq |\boldsymbol{\mu}|(K^c) = |\boldsymbol{\mu}|\left(X \setminus \bigcup_{j=1}^{k} K_j\right) \\ &= |\boldsymbol{\mu}|(X) - \sum_{j=1}^{k} |\boldsymbol{\mu}|(K_j) < 2\varepsilon.\end{aligned}$$

Combining this with (1.2.18), we find

$$\begin{aligned}|L(\boldsymbol{g})| &= \left|\int_X d\boldsymbol{\mu}(x)\boldsymbol{g}(x)\right| = \left|\int_K d\boldsymbol{\mu}(x)\boldsymbol{g}(x) + \int_{K^c} d\boldsymbol{\mu}(x)\boldsymbol{g}(x)\right| \\ &\geq \left|\int_K d\boldsymbol{\mu}(x)\boldsymbol{g}(x)\right| - \left|\int_{K^c} d\boldsymbol{\mu}(x)\boldsymbol{g}(x)\right| > |\boldsymbol{\mu}|(X) - 4\varepsilon,\end{aligned}$$

which together with $|\boldsymbol{g}(x)| \leq 1$ implies the estimate $\|L\| > |\boldsymbol{\mu}|(X) - 4\varepsilon$. Using the arbitrariness of $\varepsilon > 0$ and taking into account (1.2.12), we arrive at (1.2.18).

Suppose now that μ is defined by (1.2.9), where $A \in \mathcal{B}_X$, $\boldsymbol{\mathcal{N}} \in [\mathbf{B_b}(X)]^n$ and λ is a complex regular Borel measure. Each component μ_i of measure (1.2.9) belongs to $\mathfrak{M}_{\mathbb{C}}(\mathcal{B}_X)$ (see Cohn [**Co**]). Hence, using (1.2.8) and the well known formula for the variation of measure (1.2.9)

$$|\boldsymbol{\mu}|(X) = \int_X |\boldsymbol{\mathcal{N}}(x)| d|\lambda|(x), \tag{1.2.19}$$

(see Dunford and Schwartz [**DS**]), we arrive at (1.2.10). □

The next assertion contains representations for the norms of the operator T and the functional $\mathcal{T}_z(\cdot) = (\boldsymbol{z}, T(\cdot))$ in the real and complex cases, i.e. for $T : [\mathrm{C_v}(X)]^n \to \mathbb{R}^m$, $\boldsymbol{z} \in \mathbb{R}^m$ and $T : [\mathbf{C_v}(X)]^n \to \mathbb{C}^m$, $\boldsymbol{z} \in \mathbb{C}^m$.

COROLLARY 1.2. *Let $M = ((\mu_{ij}))$ be an $(m \times n)$-matrix-valued measure on $\mathcal{B}_X$ with elements $\mu_{ij} \in \mathfrak{M}_{\mathbb{R}}(\mathcal{B}_X)$ ($\mu_{ij} \in \mathfrak{M}_{\mathbb{C}}(\mathcal{B}_X)$) and let the operator $T : [\mathrm{C_v}(X)]^n \to \mathbb{R}^m$ ($T : [\mathbf{C_v}(X)]^n \to \mathbb{C}^m$) be given by*

$$T(\boldsymbol{f}) = \int_X dM(x)\boldsymbol{f}(x). \tag{1.2.20}$$

Then

$$\|\mathcal{T}_{\boldsymbol{z}}\| = |M^*\boldsymbol{z}|(X), \tag{1.2.21}$$

$$\|T\| = \sup_{|\boldsymbol{z}|=1} |M^*\boldsymbol{z}|(X), \tag{1.2.22}$$

where M^* stands for the transposed (adjoint) matrix of M and $\boldsymbol{z} \in \mathbb{R}^m$ ($\boldsymbol{z} \in \mathbb{C}^m$).

Let, in particular,

$$M(A) = \int_A \mathcal{M}(x)d\lambda(x), \tag{1.2.23}$$

where $A \in \mathcal{B}_X$, $\mathcal{M}$ is an $(m \times n)$-matrix-valued function on X with bounded real (complex) Borel functions as its elements, and λ is a finite signed (complex) regular Borel measure. Then

$$\|\mathcal{T}_{\boldsymbol{z}}\| = \int_X |\mathcal{M}^*(x)\boldsymbol{z}|d|\lambda|(x), \tag{1.2.24}$$

$$\|T\| = \sup_{|\boldsymbol{z}|=1} \int_X |\mathcal{M}^*(x)\boldsymbol{z}|d|\lambda|(x). \tag{1.2.25}$$

PROOF. We consider the operator $T : [\mathbf{C}_{\mathrm{v}}(X)]^n \to \mathbb{C}^m$. The case of T acting between real spaces is treated in the same way. Let $\boldsymbol{z} = (z_1, \dots, z_m) \in \mathbb{C}^m$. Then

$$\begin{aligned} \mathcal{T}_{\boldsymbol{z}}(\boldsymbol{f}) &= \left(\boldsymbol{z}, \int_X dM(x)\boldsymbol{f}(x)\right) = \sum_{i=1}^m \overline{z}_i \left(\sum_{j=1}^n \int_X f_j(x) d\mu_{ij}(x)\right) \\ &= \sum_{j=1}^n \left(\sum_{i=1}^m \int_X f_j(x)\overline{z}_i d\mu_{ij}(x)\right) = \sum_{j=1}^n \int_X f_j(x) d\left(\sum_{i=1}^m \mu_{ij}\overline{z}_i\right)(x). \end{aligned} \tag{1.2.26}$$

The vector-valued measure

$$\left(\sum_{i=1}^m \mu_{i1}\overline{z}_i, \dots, \sum_{i=1}^m \mu_{in}\overline{z}_i\right) = \left(\sum_{i=1}^m \overline{\overline{\mu_{i1}} z_i}, \dots, \sum_{i=1}^m \overline{\overline{\mu_{in}} z_i}\right)$$

can be written in the form $\overline{M^*\boldsymbol{z}}$. Then (1.2.26) implies

$$\mathcal{T}_{\boldsymbol{z}}(\boldsymbol{f}) = \int_X d(\overline{M^*\boldsymbol{z}})(x)\boldsymbol{f}(x). \tag{1.2.27}$$

Applying Proposition 1.1 to this functional, we obtain the representation $\|\mathcal{T}_{\boldsymbol{z}}\| = |\overline{M^*\boldsymbol{z}}|(X)$ which is equivalent to (1.2.21) owing to the definition of the variation of the vector-valued measure and the equality $|\overline{M^*(A)\boldsymbol{z}}| = |M^*(A)\boldsymbol{z}|$.

Since the elements of measure (1.2.23) are regular Borel measures, it follows from (1.2.27), (1.2.23) and Proposition 1.1 that (1.2.24) holds.

Further, using

$$\|T\| = \sup_{\|\boldsymbol{f}\|\le 1} |T(\boldsymbol{f})| = \sup_{\|\boldsymbol{f}\|\le 1} \sup_{|\boldsymbol{z}|=1} |(\boldsymbol{z}, T(\boldsymbol{f}))| = \sup_{|\boldsymbol{z}|=1} \sup_{\|\boldsymbol{f}\|\le 1} |\mathcal{T}_{\boldsymbol{z}}(\boldsymbol{f})| = \sup_{|\boldsymbol{z}|=1} \|\mathcal{T}_{\boldsymbol{z}}\|,$$

together with (1.2.21), (1.2.24), we arrive at (1.2.22) and (1.2.25). □

REMARK 1.1. If the measure $\boldsymbol{\mu}$ is given by (1.2.9), functional (1.2.7) can be written in the equivalent form

$$L(\boldsymbol{f}) = \int_X \left(\overline{\boldsymbol{\mathcal{N}}(x)}, \boldsymbol{f}(x)\right) d\lambda(x) \tag{1.2.28}$$

because of the statement on the "change of measure" (see Dunford and Schwartz [**DS**], Ch. 3).

Similarly, if the $(m \times n)$-matrix-valued measure M is given by (1.2.23), then the operator (1.2.20) can be written as

$$T(\boldsymbol{f}) = \int_X \mathcal{M}(x)\boldsymbol{f}(x)d\lambda(x). \tag{1.2.29}$$

The converse holds as well, namely, the statement on the change of measure guarantees the transition from (1.2.28), (1.2.29) to (1.2.7) and (1.2.20), respectively.

REMARK 1.2. Since

$$[\mathrm{C_v}(X)]^n \subset [\mathrm{C_b}(X)]^n \subset [\mathrm{B_b}(X)]^n \ ([\mathbf{C_v}(X)]^n \subset [\mathbf{C_b}(X)]^n \subset [\mathbf{B_b}(X)]^n)$$

and since the upper estimate (1.2.12) for $\|L\|$ on $[\mathrm{B_b}(X)]^n$ ($[\mathbf{B_b}(X)]^n$) coincides with the lower estimate for $\|L\|$ on $[\mathrm{C_v}(X)]^n$ ($[\mathbf{C_v}(X)]^n$), it follows that Proposition 1.1 remains valid for L defined on $[\mathrm{C_b}(X)]^n$ ($[\mathbf{C_b}(X)]^n$). Therefore, Corollary 1.2 holds also for the integral operator T on $[\mathrm{C_b}(X)]^n$ ($[\mathbf{C_b}(X)]^n$).

Sometimes, one needs to know the norms of the functional

$$\mathcal{T}_{\boldsymbol{z}}^{\langle \Re \rangle}(\cdot) = \Re(\boldsymbol{z}, T^{\langle R \rangle}(\cdot))$$

and the operator $T^{\langle R \rangle}$, where $T^{\langle R \rangle}$ stands for the restriction of $T : [\mathbf{C_v}(X)]^n \to \mathbb{C}^m$ to $[\mathrm{C_v}(X)]^n$ and $\boldsymbol{z} \in \mathbf{S}^{m-1} = \{\boldsymbol{z} \in \mathbb{C}^m : |\boldsymbol{z}| = 1\}$. We shall use the same notation for the restriction of $T : [\mathbf{C_b}(X)]^n \to \mathbb{C}^m$ to $[\mathrm{C_b}(X)]^n$.

In what follows, we represent a vector $\boldsymbol{z} \in \mathbb{C}^m$ as $\boldsymbol{z} = \Re\boldsymbol{z} + i\Im\boldsymbol{z}$, where $\Re\boldsymbol{z}, \Im\boldsymbol{z} \in \mathbb{R}^m$, and we use a similar notation for matrix-valued functions and operators.

COROLLARY 1.3. *Let $M = ((\mu_{ij}))$ be an $(m \times n)$-matrix-valued measure on $\mathcal{B}_X$ with elements $\mu_{ij} \in \mathfrak{M}_{\mathbb{C}}(\mathcal{B}_X)$ and let the operator $T : [\mathbf{C_v}(X)]^n \to \mathbb{C}^m$ be given by* (1.2.20)*. Then*

$$\|\mathcal{T}_{\boldsymbol{z}}^{\langle \Re \rangle}\| = |\Re\{M^*\boldsymbol{z}\}|(X), \tag{1.2.30}$$

$$\|T^{\langle R \rangle}\| = \sup_{|\boldsymbol{z}|=1} |\Re\{M^*\boldsymbol{z}\}|(X), \tag{1.2.31}$$

where M^* stands for the adjoint matrix of M and $\boldsymbol{z} \in \mathbb{C}^m$.

Let, in particular, M be given by (1.2.23)*, where $A \in \mathcal{B}_X$, $\mathcal{M}$ is an $(m \times n)$-matrix-valued function on X whose elements are bounded complex Borel functions and λ is a finite signed regular Borel measure. Then*

$$\|\mathcal{T}_{\boldsymbol{z}}^{\langle \Re \rangle}\| = \int_X |\Re\{\mathcal{M}^*(x)\boldsymbol{z}\}| d|\lambda|(x), \tag{1.2.32}$$

$$\|T^{\langle R \rangle}\| = \sup_{|\boldsymbol{z}|=1} \int_X |\Re\{\mathcal{M}^*(x)\boldsymbol{z}\}| d|\lambda|(x). \tag{1.2.33}$$

PROOF. Using the same arguments as in the proof of Corollary 1.2, we find

$$\mathcal{T}_{\boldsymbol{z}}^{\langle \Re \rangle}(\boldsymbol{f}) = \int_X d\left(\Re\left\{\overline{M^*\boldsymbol{z}}\right\}\right)(x)\boldsymbol{f}(x) = \int_X d(\Re\left\{M^*\boldsymbol{z}\right\})(x)\boldsymbol{f}(x), \tag{1.2.34}$$

which together with Proposition 1.1 proves (1.2.30). The measure M given by (1.2.23) with $\lambda \in \mathfrak{M}_{\mathbb{R}}(\mathcal{B}_X)$ satisfies

$$\Re\left\{M^*(A)\boldsymbol{z}\right\} = \int_A \Re\left\{\mathcal{M}^*(x)\boldsymbol{z}\right\} d\lambda(x).$$

Hence, by (1.2.34) and Proposition 1.1, we arrive at (1.2.32).

Let $\boldsymbol{z}, \boldsymbol{\zeta} \in \mathbb{C}^m$. From

$$\sup_{|\boldsymbol{z}|=1} |\Re(\boldsymbol{z}, \boldsymbol{\zeta})| = \sup_{|\boldsymbol{z}|=1} |(\Re\boldsymbol{z}, \Re\boldsymbol{\zeta}) + (\Im\boldsymbol{z}, \Im\boldsymbol{\zeta})| = |\boldsymbol{\zeta}|, \tag{1.2.35}$$

it follows that

$$\|T^{\langle R\rangle}\| = \sup_{\|\boldsymbol{f}\|\le 1} |T^{\langle R\rangle}(\boldsymbol{f})| = \sup_{\|\boldsymbol{f}\|\le 1}\sup_{|\boldsymbol{z}|=1} |\Re(\boldsymbol{z}, T^{\langle R\rangle}(\boldsymbol{f}))|$$

$$= \sup_{|\boldsymbol{z}|=1}\sup_{\|\boldsymbol{f}\|\le 1} |\mathcal{T}_{\boldsymbol{z}}^{\langle\Re\rangle}(\boldsymbol{f})| = \sup_{|\boldsymbol{z}|=1} \|\mathcal{T}_{\boldsymbol{z}}^{\langle\Re\rangle}\|. \tag{1.2.36}$$

Combining the last equality with (1.2.30) and (1.2.32), we obtain (1.2.31) and (1.2.33). □

REMARK 1.3. By (1.2.35),

$$|\Re\,\boldsymbol{\zeta}| = \sup\{\, |\Re(\boldsymbol{z},\boldsymbol{\zeta})| : |\boldsymbol{z}| = 1,\ \Im \boldsymbol{z} = \boldsymbol{0}\},\quad |\Im\,\boldsymbol{\zeta}| = \sup\{\, |\Re(\boldsymbol{z},\boldsymbol{\zeta})| : |\boldsymbol{z}| = 1,\ \Re \boldsymbol{z} = \boldsymbol{0}\}.$$

Setting here $\boldsymbol{\zeta} = T^{\langle R\rangle}(\boldsymbol{f})$ and taking the supremum over $\boldsymbol{f} \in [\mathrm{C_v}(X)]^n, \|\boldsymbol{f}\| \le 1$, we find

$$\|\Re\, T^{\langle R\rangle}\| = \sup\{\, \|\mathcal{T}_{\boldsymbol{z}}^{\langle\Re\rangle}\| : |\boldsymbol{z}| = 1,\ \Im \boldsymbol{z} = \boldsymbol{0}\}, \tag{1.2.37}$$

$$\|\Im\, T^{\langle R\rangle}\| = \sup\{\, \|\mathcal{T}_{\boldsymbol{z}}^{\langle\Re\rangle}\| : |\boldsymbol{z}| = 1,\ \Re \boldsymbol{z} = \boldsymbol{0}\}. \tag{1.2.38}$$

Thus, one arrives at representations for $\|\Re\, T^{\langle R\rangle}\|$ and $\|\Im\, T^{\langle R\rangle}\|$ after combining (1.2.37), (1.2.38) with (1.2.30) and (1.2.32).

REMARK 1.4. Using the same arguments as in Remark 1.2, one can replace the space $[\mathbf{C_v}(X)]^n$ by $[\mathbf{C_b}(X)]^n$ in Corollary 1.3.

1.3. Linear bounded operators defined on Lebesgue spaces of vector-valued functions and acting into $\mathbb{R}^m$ or $\mathbb{C}^m$

Let $(\mathcal{X}, \mathcal{A}, \mu)$ be a measure space, let $1 \le p \le \infty$, and let q be defined by $1/p + 1/q = 1$.

It is known (see, e.g., Cohn [**Co**], Proposition 3.5.2 and Ex. 2, p. 153), that each element $g \in L^q(\mathcal{X}, \mathcal{A}, \mu)$ induces a bounded linear functional Φ_g on $L^p(\mathcal{X}, \mathcal{A}, \mu)$ by means of the formula

$$\Phi_g(f) = \int_{\mathcal{X}} f(x)g(x)d\mu(x),$$

and that the operator Φ mapping g to Φ_g is an isometry of $L^q(\mathcal{X}, \mathcal{A}, \mu)$ into $(L^p(\mathcal{X}, \mathcal{A}, \mu))^*$. A similar assertion holds for functionals on $\boldsymbol{L}^p(\mathcal{X}, \mathcal{A}, \mu)$ (see Sect. 3.3, 3.5 in Cohn [**Co**]).

Moreover, the following statement is known (Cohn [**Co**], Theorem 4.5.1).

THEOREM 1.2. *If $p = 1$ and μ is σ-finite, or if $1 < p < \infty$ and μ is arbitrary, then the operator $\Phi : L^q(\mathcal{X}, \mathcal{A}, \mu) \to (L^p(\mathcal{X}, \mathcal{A}, \mu))^*$ in the real case, and the operator $\Phi : \boldsymbol{L}^q(\mathcal{X}, \mathcal{A}, \mu) \to (\boldsymbol{L}^p(\mathcal{X}, \mathcal{A}, \mu))^*$ in the complex case, defined above is an isometric isomorphism.*

We introduce the spaces $[L^p(\mathcal{X}, \mathcal{A}, \mu)]^n$ and $[\boldsymbol{L}^p(\mathcal{X}, \mathcal{A}, \mu)]^n$ of real and complex vector-valued functions $\boldsymbol{f} = (f_1, \dots, f_n)$ with components in $L^p(\mathcal{X}, \mathcal{A}, \mu)$ and $\boldsymbol{L}^p(\mathcal{X}, \mathcal{A}, \mu)$, respectively, endowed with the norm

$$\|\boldsymbol{f}\|_p = \left\{ \int_{\mathcal{X}} |\boldsymbol{f}(x)|^p d\mu(x) \right\}^{1/p} \tag{1.3.1}$$

for $1 \le p < \infty$, and

$$\|\boldsymbol{f}\|_\infty = \text{ess sup}\{|\boldsymbol{f}(x)| : x \in \mathcal{X}\}\ .$$

The next assertion follows directly from the above theorem on the representation of a linear functional on the spaces $L^p(\mathcal{X}, \mathcal{A}, \mu)$ and $\boldsymbol{L}^p(\mathcal{X}, \mathcal{A}, \mu)$.

COROLLARY 1.4. *If $p = 1$ and μ is σ-finite, or if $1 < p < \infty$ and μ is arbitrary, then any linear bounded operator $S : [L^p(\mathcal{X}, \mathcal{A}, \mu)]^n \to \mathbb{R}^m$ ($S : [\boldsymbol{L}^p(\mathcal{X}, \mathcal{A}, \mu)]^n \to \mathbb{C}^m$) admits the representation*

$$S(\boldsymbol{f}) = \int_{\mathcal{X}} G(x)\boldsymbol{f}(x)d\mu(x),$$

where $G = ((g_{ij}))$ is the $(m \times n)$-matrix-valued function with elements $g_{ij} \in L^q(\mathcal{X}, \mathcal{A}, \mu)$ (respectively, $g_{ij} \in \boldsymbol{L}^q(\mathcal{X}, \mathcal{A}, \mu)$).

PROOF. We consider the space of real-valued functions. The case of complex-valued functions is treated in the same way. Let S be an arbitrary linear bounded operator $[L^p(\mathcal{X}, \mathcal{A}, \mu)]^n \to \mathbb{R}^m$,

$$S(\boldsymbol{f}) = \begin{pmatrix} S_1(\boldsymbol{f}) \\ \cdots\cdots \\ \cdots\cdots \\ S_m(\boldsymbol{f}) \end{pmatrix}, \tag{1.3.2}$$

where $S_1, \ldots, S_m$ are functionals on $[L^p(\mathcal{X}, \mathcal{A}, \mu)]^n$. Clearly, these functionals $S_1, \ldots, S_m$ are linear and, in view of

$$|S_i(\boldsymbol{f})| \le \left[\sum_{i=1}^m |S_i(\boldsymbol{f})|^2\right]^{1/2} = |S(\boldsymbol{f})| \le \|S\|_p\ \|\boldsymbol{f}\|_p,$$

they are bounded, where $\|S\|_p$ is the norm of the operator $S : [L^p(\mathcal{X}, \mathcal{A}, \mu)]^n \to \mathbb{R}^m$.

Further, for any function $\boldsymbol{f} = (f_1, \ldots, f_n) \in [L^p(\mathcal{X}, \mathcal{A}, \mu)]^n$ there holds

$$S_i(\boldsymbol{f}) = S_i(f_1, 0, \ldots, 0) + S_i(0, f_2, \ldots, 0) + \cdots + S_i(0, \ldots, 0, f_n). \tag{1.3.3}$$

We introduce the notation

$$S_{i1}(f_1) = S_i(f_1, 0, \ldots, 0), \ldots\ldots\ldots, S_{in}(f_n) = S_i(0, \ldots, 0, f_n). \tag{1.3.4}$$

Clearly, each of the functionals $S_{i1}, \ldots, S_{in}$ is linear and bounded on $L^p(\mathcal{X}, \mathcal{A}, \mu)$. Hence, by Theorem 1.2, $S_{ij}(f_j)$ admits the representation

$$S_{ij}(f_j) = \int_{\mathcal{X}} f_j(x)g_{ij}(x)d\mu(x), \tag{1.3.5}$$

$1 \le i \le m$, $1 \le j \le n$, where $g_{ij} \in L^q(\mathcal{X}, \mathcal{A}, \mu)$. Thus, (1.3.3)-(1.3.5) imply

$$S_i(\boldsymbol{f}) = \sum_{j=1}^n S_{ij}(f_j) = \sum_{j=1}^n \int_{\mathcal{X}} f_j(x)g_{ij}(x)d\mu(x). \tag{1.3.6}$$

Combining (1.3.2), (1.3.6) and the notation $G = ((g_{ij}))$, $1 \le i \le m$, $1 \le j \le n$, we complete the proof. □

The following assertion contains a representation of the norm $\|S\|_p$ of the integral operator S defined on $[L^p(\mathcal{X}, \mathcal{A}, \mu)]^n$ ($[\boldsymbol{L}^p(\mathcal{X}, \mathcal{A}, \mu)]^n$), $1 \le p \le \infty$, and acting into $\mathbb{R}^m$ ($\mathbb{C}^m$).

PROPOSITION 1.2. *Let $(\mathcal{X}, \mathcal{A}, \mu)$ be a measure space, let p satisfy $1 \le p \le \infty$, and let q be defined by $1/p + 1/q = 1$. Let, further, $G = ((g_{ij}))$ be an $(m \times n)$-matrix-valued function with the elements $g_{ij} \in L^q(\mathcal{X}, \mathcal{A}, \mu)$ $(g_{ij} \in \boldsymbol{L}^q(\mathcal{X}, \mathcal{A}, \mu))$ whose values g_{ij} $(|g_{ij}|)$ are everywhere finite. The norm of the linear continuous operator $S : [L^p(\mathcal{X}, \mathcal{A}, \mu)]^n \to \mathbb{R}^m$ $(S : [\boldsymbol{L}_p(\mathcal{X}, \mathcal{A}, \mu)]^n \to \mathbb{C}^m)$ defined by*

$$S(\boldsymbol{f}) = \int_{\mathcal{X}} G(x)\boldsymbol{f}(x)d\mu(x) \tag{1.3.7}$$

is equal to

$$\|S\|_p = \sup_{|\boldsymbol{z}|=1} \|G^*\boldsymbol{z}\|_q, \tag{1.3.8}$$

where G^* stands for the transposed (adjoint) matrix of G and $\boldsymbol{z} \in \mathbb{R}^m$ $(\boldsymbol{z} \in \mathbb{C}^m)$.

PROOF. We deduce (1.3.8) for the operator $S : [L^p(\mathcal{X}, \mathcal{A}, \mu)]^n \to \mathbb{R}^m$ defined by (1.3.7). The case of the operator $S : [\boldsymbol{L}^p(\mathcal{X}, \mathcal{A}, \mu)]^n \to \mathbb{C}^m$ is treated in the same way.

1. Upper estimate for $\|S\|_p$. For any vector $\boldsymbol{z} \in \mathbb{R}^m$,

$$(S(\boldsymbol{f}), \boldsymbol{z}) = \int_{\mathcal{X}} (G(x)\boldsymbol{f}(x), \boldsymbol{z})d\mu(x) = \int_{\mathcal{X}} (\boldsymbol{f}(x), G^*(x)\boldsymbol{z})d\mu(x), \tag{1.3.9}$$

where we denote by G^* the transposed matrix of G. Hence by Hölder's inequality

$$|(S(\boldsymbol{f}), \boldsymbol{z})| \le \int_{\mathcal{X}} |(\boldsymbol{f}(x), G^*(x)\boldsymbol{z})|d\mu(x) \le \int_{\mathcal{X}} |G^*(x)\boldsymbol{z}||\boldsymbol{f}(x)|d\mu(x) \le \|G^*\boldsymbol{z}\|_q\|\boldsymbol{f}\|_p.$$

Therefore, taking into account that $|S(\boldsymbol{f})| = \sup\{|(S(\boldsymbol{f}), \boldsymbol{z})| : |\boldsymbol{z}| = 1\}$ we arrive at the estimate

$$\|S\|_p \le \sup_{|\boldsymbol{z}|=1} \|G^*\boldsymbol{z}\|_q. \tag{1.3.10}$$

2. Lower estimate for $\|S\|_p$. Let us fix $\boldsymbol{z} \in \mathbb{S}^{m-1} = \{\boldsymbol{z} \in \mathbb{R}^m : |\boldsymbol{z}| = 1\}$. We introduce the vector-valued function with n components

$$\boldsymbol{h}_{\boldsymbol{z}}(x) = \boldsymbol{g}_{\boldsymbol{z}}(x)h(x), \tag{1.3.11}$$

where $h \in L_p(\mathcal{X}, \mathcal{A}, \mu)$, $\|h\|_p \le 1$, and

$$\boldsymbol{g}_{\boldsymbol{z}}(x) = \begin{cases} G^*(x)\boldsymbol{z}|G^*(x)\boldsymbol{z}|^{-1} & \text{for} \quad |G^*(x)\boldsymbol{z}| \ne 0, \\ \boldsymbol{0} & \text{for} \quad |G^*(x)\boldsymbol{z}| = 0. \end{cases}$$

Note that $\boldsymbol{h}_{\boldsymbol{z}} \in [L^p(\mathcal{X}, \mathcal{A}, \mu)]^n$ and $\|\boldsymbol{h}_{\boldsymbol{z}}\|_p \le 1$. Setting (1.3.11) as $\boldsymbol{f}$ in (1.3.9) we find

$$(S(\boldsymbol{h}_{\boldsymbol{z}}), \boldsymbol{z}) = (S(\boldsymbol{g}_{\boldsymbol{z}}h), \boldsymbol{z}) = \int_{\mathcal{X}} (\boldsymbol{g}_{\boldsymbol{z}}(x), G^*(x)\boldsymbol{z})h(x)d\mu(x) = \int_{\mathcal{X}} |G^*(x)\boldsymbol{z}|h(x)d\mu(x).$$

Hence

$$\begin{aligned} \|S\|_p &= \sup_{\|\boldsymbol{f}\|_p \le 1} |S(\boldsymbol{f})| \ge \sup_{\|h\|_p \le 1} |S(\boldsymbol{g}_{\boldsymbol{z}}h)| \ge \sup_{\|h\|_p \le 1} |(S(\boldsymbol{g}_{\boldsymbol{z}}h), \boldsymbol{z})| \\ &= \sup_{\|h\|_p \le 1} \left| \int_{\mathcal{X}} |G^*(x)\boldsymbol{z}|h(x)d\mu(x) \right| = \|G^*\boldsymbol{z}\|_q. \end{aligned}$$

By the arbitrariness of $\boldsymbol{z} \in \mathbb{S}^{m-1}$,

$$\|S\|_p \geq \sup_{|\boldsymbol{z}|=1} \|G^* \boldsymbol{z}\|_q, \tag{1.3.12}$$

which together with (1.3.10) leads to (1.3.8). □

REMARK 1.5. Let $1 < p < \infty$. Estimate (1.3.12) can be derived with the help of the function

$$\boldsymbol{h_z}(x) = \begin{cases} \dfrac{G^*(x)\boldsymbol{z}|G^*(x)\boldsymbol{z}|^{q-2}}{\|G^* z\|_q^{q/p}} & \text{for} \quad |G^*(x)\boldsymbol{z}| \neq 0, \\ \mathbf{0} & \text{for} \quad |G^*(x)\boldsymbol{z}| = 0. \end{cases} \tag{1.3.13}$$

Indeed, since $q/(q-1) = p$, it follows that

$$\|\boldsymbol{h}_z\|_p = \|G^* z\|_q^{-q/p} \left\{ \int_{\mathcal{X}} |G^*(x)\boldsymbol{z}|^{(q-1)p} d\mu(x) \right\}^{1/p} = \|G^* z\|_q^{-q/p} \|G^* z\|_q^{q/p} = 1.$$

Using (1.3.9) and (1.3.13), we obtain

$$\begin{aligned} \|S\|_p &= \sup_{\|\boldsymbol{f}\|_p \leq 1} |S(\boldsymbol{f})| \geq (S(\boldsymbol{h_z}), \boldsymbol{z}) = \|G^* z\|_q^{-q/p} \int_{\mathcal{X}} |G^*(x)\boldsymbol{z}|^q d\mu(x) \\ &= \|G^* z\|_q^{-q/p} \|G^* \boldsymbol{z}\|_q^q = \|G^* \boldsymbol{z}\|_q, \end{aligned}$$

which implies (1.3.12), because $\boldsymbol{z}$ is arbitrary. The case $p = \infty$ is treated in a similar way.

1.4. Comments to Chapter 1

The first representation theorem in the form of Stieltjes' integral for an arbitrary linear functional in the space C[0, 1] was proved by F. Riesz [**Ri1**]. This theorem was generalized to the case of functionals on $C(K)$, where K is a compact subset of $\mathbb{R}^n$ (Radon, 1913) and to the case of a compact metric space (Banach, 1937). Kakutani [**Ka**] proved an analogue of the Riesz representation theorem for a compact Hausdorff space K. The first attempt to generalize this result to noncompact topological spaces is due to Markov [**Ma**].

The general form of a linear functional on the space $L^p(a, b)$ with $1 < p < \infty$ is due to F. Riesz [**Ri2**]. The space $L^1(a, b)$ was treated by Steinhaus [**Ste**]. Nikodým [**Nikd**] obtained a generalization to abstract spaces with finite measure. The material of Section 1.3 was published in [**KM10**].

General facts on representation of functionals in various function spaces, as well as information of historical character can be found in the books by Aliprantis and Border [**AB**], Cohn [**Co**], Dunford and Schwartz [**DS**], Edwards [**Ed**], Kantorovich and Akilov [**KA**], Zaanen [**Zaa**], Maz'ya and Shaposhnikova [**MSh**], Section 12.5.

CHAPTER 2

Maximum Modulus Principle for Second Order Strongly Elliptic Systems

2.1. Introduction

This chapter contains criteria for validity of the maximum modulus principle for second order strongly elliptic systems.

Section 2.2 is concerned with the question of validity of the classical maximum modulus principle for strongly elliptic second order systems with constant coefficients of the form

$$\sum_{j,k=1}^{n} \mathcal{A}_{jk} \frac{\partial^2 \boldsymbol{u}}{\partial x_j \partial x_k} = \mathbf{0}, \tag{2.1.1}$$

where the $\mathcal{A}_{jk} = \mathcal{A}_{kj}$ are real or complex constant $(m \times m)$-matrices, and $\boldsymbol{u}$ is an m-component vector-valued function. We obtain a representation for the exact constant $\mathcal{K}(\mathbb{R}^n_+)$ in the Miranda-Agmon inequality

$$|\boldsymbol{u}(x)| \le \mathcal{K} \sup \{|\boldsymbol{u}(x')| : x' \in \partial\mathbb{R}^n_+\},$$

where $x \in \mathbb{R}^n_+ = \{x = (x_1, \dots, x_n) : x_n > 0\}$, and $\boldsymbol{u}$ is a solution of the strongly elliptic system (2.1.1) in the class $[\mathbf{C_b}(\overline{\mathbb{R}^n_+})]^m \bigcap [\mathbf{C}^2(\mathbb{R}^n_+)]^m$ for complex coefficients and in the class $[\mathrm{C_b}(\overline{\mathbb{R}^n_+})]^m \bigcap [\mathrm{C}^2(\mathbb{R}^n_+)]^m$ for real coefficients. It is shown that an analogous constant for a domain with compact closure and smooth boundary majorizes $\mathcal{K}(\mathbb{R}^n_+)$. With the help of these two results it is established that the systems satisfying classical maximum modulus principle are exhausted by a scalar equation with real coefficients repeated m times. More precisely, the maximum modulus principle implies the identity

$$\sum_{j,k=1}^{n} \mathcal{A}_{jk} \frac{\partial^2}{\partial x_j \partial x_k} = \mathcal{A} \sum_{j,k=1}^{n} a_{jk} \frac{\partial^2}{\partial x_j \partial x_k},$$

where $\mathcal{A}$ is an $(m \times m)$-matrix such that $\Re(\mathcal{A}\boldsymbol{\zeta}, \boldsymbol{\zeta}) > 0$ for all $\boldsymbol{\zeta} \in \mathbb{C}^m \backslash \{\mathbf{0}\}$ and $((a_{jk}))$ is a positive-definite real $(n \times n)$-matrix.

Let $\mathcal{A}$ be a real non-necessarily symmetric matrix of order m. We call $\mathcal{A}$ positive-definite if $(\mathcal{A}\boldsymbol{\zeta}, \boldsymbol{\zeta}) > 0$ for all $\boldsymbol{\zeta} \in \mathbb{R}^m \backslash \{\mathbf{0}\}$.

In Section 2.3 we give criteria for validity of the maximum modulus principle for solutions of the strongly elliptic system

$$\sum_{j,k=1}^{n} \mathcal{A}_{jk}(x) \frac{\partial^2 \boldsymbol{u}}{\partial x_j \partial x_k} - \sum_{j=1}^{n} \mathcal{A}_j(x) \frac{\partial \boldsymbol{u}}{\partial x_j} - \mathcal{A}_0(x)\boldsymbol{u} = \mathbf{0} \tag{2.1.2}$$

with real or complex coefficients defined in a domain $\Omega \subset \mathbb{R}^n$ with compact closure $\overline{\Omega}$ and with the boundary $\partial\Omega$ of the class $\mathrm{C}^{2,\alpha}, 0 < \alpha \le 1$. Here $\boldsymbol{u}$ is an m-component vector-valued function, $\mathcal{A}_{jk} = \mathcal{A}_{kj}$, and $\mathcal{A}_{jk}, \mathcal{A}_j, \mathcal{A}_0$ are $(m \times m)$-matrix valued functions with elements in the spaces $\mathrm{C}^{2,\alpha}(\overline{\Omega}), \mathrm{C}^{1,\alpha}(\overline{\Omega}), \mathrm{C}^{\alpha}(\overline{\Omega})$ for real coefficients and in the spaces $\mathbf{C}^{2,\alpha}(\overline{\Omega}), \mathbf{C}^{1,\alpha}(\overline{\Omega}), \mathbf{C}^{\alpha}(\overline{\Omega})$ for complex coefficients, respectively.

Henceforth, using spaces of continuous real or complex-valued functions given on compact subsets of $\mathbb{R}^n$, we use the notation C or $\mathbf{C}$ instead of $\mathrm{C_b}$ or $\mathbf{C_b}$.

We list basic results obtained here for system (2.1.2) in the case of real coefficients. We obtain necessary and, separately, sufficient conditions for the classical maximum modulus principle to hold for solutions of (2.1.2) which belong to the class $[\mathrm{C}(\overline{\Omega})]^m \bigcap [\mathrm{C}^2(\Omega)]^m$. As a corollary of these conditions we prove the following assertion.

THEOREM 1. *The classical maximum modulus principle*

$$\max_{\overline{\omega}} |\boldsymbol{u}| \le \max_{\partial\omega} |\boldsymbol{u}|$$

is valid for solutions of system (2.1.2) *in an arbitrary domain* $\omega \subset \Omega$ *with boundary of the class* $\mathrm{C}^{2,\alpha}$ *if and only if:*

(i) *for all* $x \in \overline{\Omega}$ *the equalities*

$$\mathcal{A}_{jk}(x) = \mathcal{A}(x) a_{jk}(x), \quad 1 \le j,\ k \le n,$$

hold, where $\mathcal{A}$ *and* $((a_{jk}))$ *are positive-definite matrices in* $\overline{\Omega}$ *of orders* m *and* n, *respectively;*

(ii) *for all* $x \in \Omega$ *and any* $\boldsymbol{\xi}_j,\ \boldsymbol{\zeta} \in \mathbb{R}^m$, $j = 1, \dots, n$, *with* $(\boldsymbol{\xi}_j, \boldsymbol{\zeta}) = 0$ *the inequality*

$$\sum_{j,k=1}^{n} a_{jk}(x)(\boldsymbol{\xi}_j, \boldsymbol{\xi}_k) + \sum_{j=1}^{n} (\mathcal{A}^{-1}(x)\mathcal{A}_j(x)\boldsymbol{\xi}_j, \boldsymbol{\zeta}) + (\mathcal{A}^{-1}(x)\mathcal{A}_0(x)\boldsymbol{\zeta}, \boldsymbol{\zeta}) \ge 0$$

is valid.

The next assertion immediately follows from this theorem.

COROLLARY 1. *The classical maximum modulus principle holds for solutions of system* (2.1.2) *in an arbitrary domain* $\omega \subset \Omega$ *with boundary of the class* $\mathrm{C}^{2,\alpha}$ *if and only if condition* (i) *of Theorem* 1 *is satisfied and*

(i) *for all* $x \in \Omega$ *and for any* $\boldsymbol{\zeta} \in \mathbb{R}^m \backslash \{\mathbf{0}\}$ *the inequality*

$$|\boldsymbol{\zeta}|^{-2} \sum_{j,k=1}^{n} b_{jk}(x)(\mathcal{A}^{-1}(x)\mathcal{A}_j(x)\boldsymbol{\zeta}, \boldsymbol{\zeta})(\mathcal{A}^{-1}(x)\mathcal{A}_k(x)\boldsymbol{\zeta}, \boldsymbol{\zeta})$$

$$- \sum_{j,k=1}^{n} b_{jk}(x)(\mathcal{A}_j^*(x)(\mathcal{A}^*(x))^{-1}\boldsymbol{\zeta},\ \mathcal{A}_k^*(x)(\mathcal{A}^*(x))^{-1}\boldsymbol{\zeta}) + 4(\mathcal{A}^{-1}(x)\mathcal{A}_0(x)\boldsymbol{\zeta}, \boldsymbol{\zeta}) \ge 0$$

is valid. Here $((b_{jk}))$ *is the inverse matrix of* $((a_{jk}))$ *and* * *means passage to the transposed matrix.*

We demonstrate by an example that the possibility to represent the principal part of system (2.1.2) in the form

$$\mathcal{A}(x) \sum_{j,k=1}^{n} a_{jk}(x) \frac{\partial^2 \boldsymbol{u}}{\partial x_j \partial x_k} \tag{2.1.3}$$

everywhere in $\overline{\Omega}$, when the coefficients $\mathcal{A}_{jk}$ depend on x, is not necessary for validity of the classical modulus principle

$$\max_{\overline{\omega}} |\boldsymbol{u}| \le \max_{\partial\omega} |\boldsymbol{u}|,$$

where $\boldsymbol{u}$ is the solution of (2.1.2) in Ω which belongs to the class $[\mathrm{C}(\overline{\Omega})]^m \bigcap [\mathrm{C}^2(\Omega)]^m$.

We also prove a theorem on conditions ensuring that the modulus of the solution of (2.1.2) with a special principal part (2.1.3) has no points of the strict local maximum in Ω.

THEOREM 2. *Let system* (2.1.2) *be of the form*

$$\mathcal{A}(x) \sum_{j,k=1}^{n} a_{jk}(x) \frac{\partial^2 \boldsymbol{u}}{\partial x_j \partial x_k} - \sum_{j=1}^{n} \mathcal{A}_j(x) \frac{\partial \boldsymbol{u}}{\partial x_j} - \mathcal{A}_0(x)\boldsymbol{u} = \mathbf{0}, \tag{2.1.4}$$

where $\mathcal{A}$ and $((a_{jk}))$ are positive-definite $(m \times m)$- and $(n \times n)$-matrix-valued functions in $\overline{\Omega}$ with elements in the space $\mathrm{C}^{2,\alpha}(\overline{\Omega})$.

The system (2.1.4) *has no solution $\boldsymbol{u} \in [\mathrm{C}(\overline{\Omega})]^m \bigcap [\mathrm{C}^2(\Omega)]^m$ with modulus which attains its strict local maximum at some point $x \in \Omega$ if and only if for all $x \in \Omega$ and any $\boldsymbol{\xi}_j$, $\boldsymbol{\zeta} \in \mathbb{R}^m$, $j = 1, \dots, n$, with $(\boldsymbol{\xi}_j, \boldsymbol{\zeta}) = 0$ the inequality*

$$\sum_{j,k=1}^{n} a_{jk}(x)(\boldsymbol{\xi}_j, \boldsymbol{\xi}_k) + \sum_{j=1}^{n} (\mathcal{A}^{-1}(x)\mathcal{A}_j(x)\boldsymbol{\xi}_j, \boldsymbol{\zeta}) + (\mathcal{A}^{-1}(x)\mathcal{A}_0(x)\boldsymbol{\zeta}, \boldsymbol{\zeta}) \geq 0$$

is valid.

Theorem 2 can be also formulated with the last condition replaced by condition (i) of Corollary 1.

It is shown that all the facts concerning the maximum modulus principle for solutions of systems with complex coefficients are corollaries of corresponding assertions for systems with real coefficients.

2.2. Systems with constant coefficients without lower order terms

2.2.1. The case of real coefficients

2.2.1.1. *The maximum principle in a half-space.* We introduce the operator

$$\mathfrak{A}_0(D_x) = \sum_{j,k=1}^{n} \mathcal{A}_{jk} \frac{\partial^2}{\partial x_j \partial x_k},$$

where $D_x = (\partial/\partial x_1, \dots, \partial/\partial x_n)$ and $\mathcal{A}_{jk} = \mathcal{A}_{kj}$ are constant real $(m \times m)$-matrices. Assume that the operator $\mathfrak{A}_0$ is strongly elliptic, i.e. that for all $\boldsymbol{\zeta} = (\zeta_1, \dots, \zeta_m) \in \mathbb{R}^m$ and $\boldsymbol{\sigma} = (\sigma_1, \dots, \sigma_n) \in \mathbb{R}^n$, with $\boldsymbol{\zeta}$, $\boldsymbol{\sigma} \neq \mathbf{0}$, we have the inequality

$$\left(\sum_{j,k=1}^{n} \mathcal{A}_{jk} \sigma_j \sigma_k \boldsymbol{\zeta},\ \boldsymbol{\zeta} \right) > 0.$$

Let Ω be a domain in $\mathbb{R}^n$ with boundary $\partial\Omega$ and closure $\overline{\Omega}$. Let $[\mathrm{C_b}(\overline{\Omega})]^m$ denote the space of bounded m-component vector-valued functions which are continuous in $\overline{\Omega}$. The norm on $[\mathrm{C_b}(\overline{\Omega})]^m$ is $\|\boldsymbol{u}\| = \sup\{|\boldsymbol{u}(x)| : x \in \overline{\Omega}\}$. The notation $[\mathrm{C_b}(\partial\Omega)]^m$ has a similar meaning. By $[\mathrm{C}^2(\Omega)]^m$ we denote the space of m-component vector-valued functions with continuous derivatives up to the second order in Ω.

Let

$$\mathcal{K}(\Omega) = \sup \frac{\|\boldsymbol{u}\|_{[\mathrm{C_b}(\overline{\Omega})]^m}}{\|\boldsymbol{u}\|_{[\mathrm{C_b}(\partial\Omega)]^m}}, \tag{2.2.1}$$

where the supremum is taken over all vector-valued functions in the class $[\mathrm{C_b}(\overline{\Omega})]^m \cap [\mathrm{C}^2(\Omega)]^m$ satisfying the system $\mathfrak{A}_0(D_x)\boldsymbol{u} = \mathbf{0}$.

Clearly, $\mathcal{K}(\Omega)$ is the best constant in the inequality

$$|\boldsymbol{u}(x)| \leq \mathcal{K}(\Omega) \sup \{|\boldsymbol{u}(y)| : y \in \partial\Omega\},$$

where $x \in \Omega$ and $\boldsymbol{u}$ is a solution of the system $\mathfrak{A}_0(D_x)\boldsymbol{u} = 0$ in the class $[\mathrm{C_b}(\overline{\Omega})]^m \cap [\mathrm{C}^2(\Omega)]^m$. If $\mathcal{K}(\Omega) = 1$, then the maximum modulus principle holds for the system $\mathfrak{A}_0(D_x)\boldsymbol{u} = \boldsymbol{0}$.

According to Agmon, Douglis and Nirenberg [**ADN2**], Lopatinskiĭ [**Lo1**], Shapiro [**Shap**], Solonnikov [**So1**] there exists a bounded solution of the problem

$$\mathfrak{A}_0(D_x)\boldsymbol{u} = \boldsymbol{0} \text{ in } \mathbb{R}^n_+, \quad \boldsymbol{u} = \boldsymbol{f} \text{ on } \partial\mathbb{R}^n_+, \tag{2.2.2}$$

with $\boldsymbol{f} \in [\mathrm{C_b}(\partial\mathbb{R}^n_+)]^m$, such that $\boldsymbol{u}$ is continuous up to $\partial\mathbb{R}^n_+$, and can be represented in the form

$$\boldsymbol{u}(x) = \int_{\partial\mathbb{R}^n_+} \mathcal{M}\left(\frac{y-x}{|y-x|}\right) \frac{x_n}{|y-x|^n} \boldsymbol{f}(y')dy'. \tag{2.2.3}$$

Here $y = (y', 0)$, $y' = (y_1, \dots, y_{n-1})$, and $\mathcal{M}$ is a continuous $(m \times m)$-matrix-valued function on the closure of the hemisphere $\mathbb{S}^{n-1}_- = \big\{x \in \mathbb{R}^n : \ |x| = 1, x_n < 0\big\}$ such that the integral

$$\int_{\mathbb{S}^{n-1}_-} \mathcal{M}(\sigma)d\sigma$$

is the identity matrix.

We note that equality (2.2.3) can be represented in the form

$$\boldsymbol{u}(x) = \int_{\partial\mathbb{R}^n_+} \mathcal{M}\left(\frac{y-x}{|y-x|}\right) \boldsymbol{f}(y')\, \omega(x, dy'), \tag{2.2.4}$$

where

$$\omega(x, \mathcal{B}) = \int_{\mathcal{B}} \frac{x_n}{|y-x|^n} dy'$$

is the solid angle at which a Borel set $\mathcal{B} \subset \partial\mathbb{R}^n_+$ is seen from the point $x \in \mathbb{R}^n_+$. The solid angle $\omega(x, \cdot)$ is a finite regular Borel measure on $\partial\mathbb{R}^n_+$ for any fixed $x \in \mathbb{R}^n_+$, and $\omega(x, \mathcal{B}) \geq 0$.

REMARK 2.1. The uniqueness of the solution of the Dirichlet problem (2.2.2) in the class $[\mathrm{C_b}(\overline{\mathbb{R}^n_+})]^m \cap [\mathrm{C}^2(\mathbb{R}^n_+)]^m$ can be derived by means of standard arguments from (2.2.3) and from local estimates of the derivatives of solutions to elliptic systems (see Agmon, Douglis and Nirenberg [**ADN2**], Solonnikov [**So1**]).

The following statement immediately follows from (2.2.4), Corollary 1.2 and Remark 1.2.

THEOREM 2.1. *The formula*

$$\mathcal{K}(\mathbb{R}^n_+) = \sup_{|\boldsymbol{z}|=1} \int_{\mathbb{S}^{n-1}_-} |\mathcal{M}^*(\sigma)\boldsymbol{z}| d\sigma \tag{2.2.5}$$

is valid, where the * *denotes passage to the transposed matrix.*

THEOREM 2.2. *The equality* $\mathcal{K}(\mathbb{R}^n_+) = 1$ *is valid if and only if*

$$\mathfrak{A}_0(D_x) = \mathcal{A} \sum_{j,k=1}^{n} a_{jk} \frac{\partial^2}{\partial x_j \partial x_k}, \tag{2.2.6}$$

where $\mathcal{A}$ *and* $((a_{jk}))$ *are positive-definite constant matrices of orders* m *and* n*, respectively.*

PROOF. *1. The structure of the matrix* $\mathcal{M}$. From (2.2.5) it follows that

$$\mathcal{K}(\mathbb{R}^n_+) = \sup_{|\boldsymbol{z}|=1} \sup_{|\boldsymbol{f}(y')|\le 1} \int_{\partial\mathbb{R}^n_+} \left(\mathcal{M}^*\left(\frac{y-x}{|y-x|}\right)\boldsymbol{z},\ \boldsymbol{f}(y')\right) \frac{x_n}{|y-x|^n} dy', \tag{2.2.7}$$

where x is an arbitrary point in $\mathbb{R}^n_+$. Denote by $\Gamma_x(\boldsymbol{z})$ the subset of points $y \in \partial\mathbb{R}^n_+$ on which $\mathcal{M}^*((y-x)/|y-x|)\boldsymbol{z} = 0$.

Assume that $\mathcal{K}(\mathbb{R}^n_+) = 1$ and that there exists a unit m-dimensional vector $\boldsymbol{z}_0$ such that

$$\boldsymbol{z}_0 \neq \left|\mathcal{M}^*\left(\frac{y-x}{|y-x|}\right)\boldsymbol{z}_0\right|^{-1} \mathcal{M}^*\left(\frac{y-x}{|y-x|}\right)\boldsymbol{z}_0$$

for all $y \in \partial\mathbb{R}^n_+ \backslash \Gamma_x(\boldsymbol{z}_0)$, where x is an arbitrary but fixed point of $\mathbb{R}^n_+$. Then, using (2.2.7), we get that

$$\begin{aligned}
1 &= \mathcal{K}(\mathbb{R}^n_+) \ge \sup_{|\boldsymbol{f}(y')|\le 1} \int_{\partial\mathbb{R}^n_+} \left(\mathcal{M}^*\left(\frac{y-x}{|y-x|}\right)\boldsymbol{z}_0,\ \boldsymbol{f}(y')\right) \frac{x_n}{|y-x|^n} dy' \\
&= \sup_{|\boldsymbol{f}(y')|\le 1} \int_{\partial\mathbb{R}^n_+ \backslash \Gamma_x(\boldsymbol{z}_0)} \left(\mathcal{M}^*\left(\frac{y-x}{|y-x|}\right)\boldsymbol{z}_0,\ \boldsymbol{f}(y')\right) \frac{x_n}{|y-x|^n} dy' \\
&= \int_{\partial\mathbb{R}^n_+ \backslash \Gamma_x(\boldsymbol{z}_0)} \left(\mathcal{M}^*\left(\frac{y-x}{|y-x|}\right)\boldsymbol{z}_0,\ \frac{\mathcal{M}^*((y-x)/|y-x|)\boldsymbol{z}_0}{|\mathcal{M}^*((y-x)/|y-x|)\boldsymbol{z}_0|}\right) \frac{x_n}{|y-x|^n} dy' \\
&> \int_{\partial\mathbb{R}^n_+ \backslash \Gamma_x(\boldsymbol{z}_0)} \left(\mathcal{M}^*\left(\frac{y-x}{|y-x|}\right)\boldsymbol{z}_0,\ \boldsymbol{z}_0\right) \frac{x_n}{|y-x|^n} dy' \\
&= \int_{\partial\mathbb{R}^n_+} \left(\mathcal{M}^*\left(\frac{y-x}{|y-x|}\right)\boldsymbol{z}_0,\ \boldsymbol{z}_0\right) \frac{x_n}{|y-x|^n} dy' = \left(\int_{\mathbb{S}^{n-1}_-} \mathcal{M}^*(\sigma)\boldsymbol{z}_0\, d\sigma,\ \boldsymbol{z}_0\right) = 1.
\end{aligned}$$

Consequently, if $\mathcal{K}(\mathbb{R}^n_+) = 1$, then for all $\boldsymbol{z} \in \mathbb{R}^m_+$, $|\boldsymbol{z}| = 1$, and all $y \in \partial\mathbb{R}^n_+ \backslash \Gamma_x(\boldsymbol{z})$

$$\boldsymbol{z} = \left|\mathcal{M}^*\left(\frac{y-x}{|y-x|}\right)\boldsymbol{z}\right|^{-1} \mathcal{M}^*\left(\frac{y-x}{|y-x|}\right)\boldsymbol{z}. \tag{2.2.8}$$

Let μ_{jk}, where $j, k = 1, \dots, m$, denote the elements of the matrix $\mathcal{M}$. Setting $\boldsymbol{z}_1 = (1, \dots, 0), \dots, \boldsymbol{z}_m = (0, \dots, 1)$ successively in (2.2.8), we get that $\mu_{jk}(\sigma) = 0$ for $j \neq k$ and all $\sigma \in \mathbb{S}^{n-1}_-$. In (2.2.8) let $\boldsymbol{z} = m^{-1/2}(1, 1, \dots, 1)$. Then for any $j = 1, \dots, m$

$$(\mu^2_{11}(\sigma) + \cdots + \mu^2_{mm}(\sigma))^{1/2} = \mu_{jj}(\sigma)m^{1/2}.$$

Hence, $\mu_{11}(\sigma) = \cdots = \mu_{mm}(\sigma)$ on $\mathbb{S}^{n-1}_-$. Let $\mu(\sigma) = \mu_{jj}(\sigma)$, $1 \le j \le m$. Then

$$\boldsymbol{u}(x) = \int_{\partial\mathbb{R}^n_+} \mu\left(\frac{y-x}{|y-x|}\right) \frac{x_n}{|y-x|^n} \boldsymbol{f}(y')dy'. \tag{2.2.9}$$

2. The structure of the operator $\mathfrak{A}_0$. Let f_0 be a scalar function that is continuous and bounded on $\partial\mathbb{R}^n_+$, and let

$$u_0(x) = \int_{\partial\mathbb{R}^n_+} \mu\left(\frac{y-x}{|y-x|}\right) \frac{x_n}{|y-x|^n} f_0(y')dy'. \tag{2.2.10}$$

According to (2.2.9), the vector-valued function $\boldsymbol{h_c} = \boldsymbol{c}u_0$, where $\boldsymbol{c} \in \mathbb{R}^m$, is a solution of the boundary value problem $\mathfrak{A}_0(D_x)\boldsymbol{h_c} = 0$ in $\mathbb{R}^n_+$, $\boldsymbol{h}_c = \boldsymbol{c}f_0$ on $\partial\mathbb{R}^n_+$.

Setting $\boldsymbol{c}_1 = (1, \dots, 0), \dots, \boldsymbol{c}_m = (0, \dots, 1)$ successively in $\mathfrak{A}_0(D_x)\boldsymbol{h}_{\boldsymbol{c}} = 0$, we get the m^2 boundary value problems

$$\mathfrak{A}_{ps}(D_x)u_0 = 0 \text{ in } \mathbb{R}^n_+, \quad u_0 = f_0 \text{ on } \partial\mathbb{R}^n_+, \quad p, s = 1, 2 \dots, m.$$

Here $\mathfrak{A}_{ps}(D_x)$ is a scalar differential operator

$$\sum_{j,k=1}^{n} \mathcal{A}^{(ps)}_{jk} \frac{\partial^2}{\partial x_j \partial x_k},$$

where $\mathcal{A}^{(ps)}_{jk}$ is the element of the matrix $\mathcal{A}_{jk}$ situated at the intersection of the p-th row and the s-th column.

We consider the two scalar equations

$$\mathfrak{A}_{pp}(D_x)u_0 = 0 \text{ and } \mathfrak{A}_{ps}(D_x)u_0 = 0 \text{ in } \mathbb{R}^n_+$$

with the boundary condition $u_0 = f_0$ on $\partial\mathbb{R}^n_+$, where p and s are fixed elements of the set $\{1, \dots, m\}$ and $p \neq s$.

By the original assumption, the operator $\mathfrak{A}_0(D_x)$ is strongly elliptic, so the operator $\mathfrak{A}_{pp}(D_x)$ is elliptic.

Without loss of generality it can be assumed that $\mathcal{A}^{(pp)}_{nn} > 0$. Setting

$$x_n = \sqrt{\mathcal{A}^{(pp)}_{nn}}\, y_n,$$

we perform a linear change of variables that takes the operator $\mathfrak{A}_{pp}(D_x)$ to the canonical form

$$\tilde{\mathfrak{A}}_{pp}(D_y) = \sum_{i=1}^{n} \frac{\partial^2}{\partial y_i^2}.$$

Assume that the function f_0 in (2.2.10) has compact support. If we apply the Fourier transform with respect to the variables $y_1, \dots, y_{n-1}$ to the equation $\tilde{\mathfrak{A}}_{pp}(D_y)\tilde{u}_0(y) = 0$ then we get that

$$\frac{d^2 F[\tilde{u}_0]}{dy_n^2} - |\boldsymbol{\xi}'|^2 F[\tilde{u}_0] = 0,$$

where $\boldsymbol{\xi}' = (\xi_1, \dots, \xi_{n-1})$ and $|\boldsymbol{\xi}'| = (\xi_1^2 + \dots + \xi_{n-1}^2)^{1/2}$. The last equation implies that

$$F[\tilde{u}_0](\boldsymbol{\xi}', y_n) = F[\tilde{u}_0](\boldsymbol{\xi}', 0)\exp(-|\boldsymbol{\xi}'|y_n) = F[\tilde{f}_0](\boldsymbol{\xi}')\exp(-|\boldsymbol{\xi}'|y_n).$$

At the same time we also transform the equation $\mathfrak{A}_{ps}(D_x)u_0 = 0$ to the variables $y_1, \dots, y_n$, and then we apply to it the Fourier transform with respect to $y_1, \dots, y_{n-1}$. As a result,

$$\tilde{\mathcal{A}}^{(ps)}_{nn} \frac{d^2 F[\tilde{u}_0]}{dy_n^2} - 2i \frac{dF[\tilde{u}_0]}{dy_n} \sum_{j=1}^{n-1} \tilde{\mathcal{A}}^{(ps)}_{jn} \xi_j - F[\tilde{u}_0] \sum_{j,k=1}^{n-1} \tilde{\mathcal{A}}^{(ps)}_{jk} \xi_j \xi_k = 0. \tag{2.2.11}$$

From the last equation and the equality $F[\tilde{u}_0](\boldsymbol{\xi}', y_n) = F[\tilde{f}_0](\boldsymbol{\xi}')\exp(-|\boldsymbol{\xi}'|y_n)$ we conclude that

$$\sum_{s=1}^{n-1} \tilde{\mathcal{A}}^{(ps)}_{jn} \xi_j = 0,$$

i.e., $\tilde{\mathcal{A}}_{jn}^{(ps)} = 0$ for all $j = 1, \dots, n-1$. Therefore, differentiating $F[\tilde{u}_0](\boldsymbol{\xi}', y_n)$ with respect to y_n and substituting in (2.2.11), we find that

$$\tilde{\mathcal{A}}_{nn}^{(ps)} |\boldsymbol{\xi}'|^2 - \sum_{j,k=1}^{n-1} \tilde{\mathcal{A}}_{jk}^{(ps)} \xi_j \xi_k = 0.$$

Hence, $\tilde{\mathcal{A}}_{jk}^{(ps)} = \delta_{jk} \tilde{\mathcal{A}}_{nn}^{(ps)}$, $1 \le j, k \le n-1$. Thus, the operator $\tilde{\mathfrak{A}}_{ps}(D_y)$ turns out to be the Laplacian (up to a constant factor) and, consequently,

$$\tilde{\mathfrak{A}}_0(D_y) = \mathcal{A} \sum_{i=1}^{n} \frac{\partial^2}{\partial y_i^2}, \tag{2.2.12}$$

where $\mathcal{A}$ is a constant $(m \times m)$-matrix. Since the operator $\mathfrak{A}_0(D_x)$ is strongly elliptic, the matrix $\mathcal{A}$ must be positive-definite. The inverse transformation of variables $y \to x$ in (2.2.12) gives us (2.2.6).

The fact that the representation (2.2.6) suffices for the equality $\mathcal{K}(\mathbb{R}^n_+) = 1$ to hold follows from the positivity of the conormal derivative $\partial G / \partial N$ of Green's function of the Dirichlet problem for an elliptic equation with real coefficients in the half-space and from the Poisson formula

$$\boldsymbol{u}(x) = \int_{\partial \mathbb{R}^n_+} \frac{\partial G(x, y')}{\partial N_{y'}} \boldsymbol{f}(y')\, dy'$$

for the solution of the system

$$\mathcal{A} \sum_{j,k=1}^{n} a_{jk} \frac{\partial^2 \boldsymbol{u}}{\partial x_j \partial x_k} = \boldsymbol{0}.$$

□

2.2.1.2. *The maximum principle in a bounded domain.* Now we establish an analogue of Theorem 2.2 for a domain with compact closure and C^1-boundary. The proof is based on the following theorem.

THEOREM 2.3. *Let Ω be a domain in $\mathbb{R}^n$ with compact closure and C^1-boundary. Then*

$$\mathcal{K}(\Omega) \ge \sup\{\mathcal{K}(\mathbb{R}^n_+(\boldsymbol{\nu})) : \boldsymbol{\nu} \in \mathbb{S}^{n-1}\},$$

where $\mathbb{R}^n_+(\boldsymbol{\nu})$ is a half-space with inward normal $\boldsymbol{\nu}$ and $\mathbb{S}^{n-1} = \{x \in \mathbb{R}^n : |x| = 1\}$.

PROOF. Let l be an arbitrary straight line in $\mathbb{R}^n$, and let $\mathcal{H}(l)$ stand for one of the two hyperplanes that support Ω and are orthogonal to l. Let $\mathcal{O}$ denote some point of the set $\partial\Omega \bigcap \mathcal{H}(l)$, and $\boldsymbol{\nu}$ the inward normal to $\partial\Omega$ at the point $\mathcal{O}$. Assume that $\mathbb{R}^n_+(\boldsymbol{\nu}) \supset \Omega$ and $\partial \mathbb{R}^n_+(\boldsymbol{\nu}) = \mathcal{H}(l)$. We introduce a Cartesian coordinate system with origin at $\mathcal{O}$, with the axis $\mathcal{O}x_n$ directed along the normal, and we set $\mathbb{B}'_r = \{x' = (x_1, \dots, x_{n-1}) : |x'| < r\}$ and $\mathbb{B}_r = \{x \in \mathbb{R}^n : |x| < r\}$.

Let $\boldsymbol{f} \in [\mathrm{C_b}(\partial \mathbb{R}^n_+(\boldsymbol{\nu}))]^m$ and supp $\boldsymbol{f} \subset \mathbb{B}'_r$. We fix a point $x \in \mathbb{R}^n_+(\boldsymbol{\nu})$ and find the norm $||\boldsymbol{u}(x)||$ of the mapping

$$[\mathrm{C_b}(\partial \mathbb{R}^n_+(\boldsymbol{\nu}))]^m \ni \boldsymbol{f} \to \boldsymbol{u}(x) = \int_{\mathbb{B}'_r} \mathcal{M}_{\boldsymbol{\nu}} \left(\frac{y-x}{|y-x|} \right) \boldsymbol{f}(y') \frac{x_n}{|y-x|^n} dy' \in \mathbb{R}^m.$$

Using Corollary 1.2, we arrive at the equality

$$||\boldsymbol{u}(x)|| = \sup_{|\boldsymbol{z}|=1} \int_{\mathbb{B}'_r} \left| \mathcal{M}^*_{\boldsymbol{\nu}} \left(\frac{y-x}{|y-x|} \right) \boldsymbol{z} \right| \frac{x_n}{|y-x|^n} dy'. \tag{2.2.13}$$

From this we get

$$||\boldsymbol{u}(x)|| = \sup_{|\boldsymbol{z}|=1} \int_{\mathbb{S}^{n-1}_-(x)\cap V_r(x)} \left| \mathcal{M}^*_{\boldsymbol{\nu}} \left(\frac{y-x}{|y-x|} \right) \boldsymbol{z} \right| d\sigma(y), \tag{2.2.14}$$

where $V_r(x)$ is the collection of rays emanating from the point x and intersecting the set $\mathbb{B}'_r$.

It follows from equality (2.2.14) and Theorem 2.1 that there exist a point $\xi = (0, 0, \dots, \xi_n)$ and a vector-valued function $\boldsymbol{\varphi}$ continuous on $\partial\mathbb{R}^n_+(\boldsymbol{\nu})$ and with $\operatorname{supp} \boldsymbol{\varphi} \subset \mathbb{B}'_1$ and $||\boldsymbol{\varphi}|| \le 1$ such that

$$|\boldsymbol{u}(\xi)| = \left| \int_{\mathbb{B}'_1} \mathcal{M}_{\boldsymbol{\nu}} \left(\frac{\zeta-\xi}{|\zeta-\xi|} \right) \boldsymbol{\varphi}(\zeta') \frac{\xi_n}{|\zeta-\xi|^n} d\zeta' \right| \ge \mathcal{K}(\mathbb{R}^n_+(\boldsymbol{\nu})) - \mu,$$

where μ is a prescribed arbitrarily small constant.

Replacing ξ and ζ by $w = \varepsilon\xi$ and $y = \varepsilon\zeta$ here and setting $\boldsymbol{u}_\varepsilon(w) = \boldsymbol{u}(w/\varepsilon)$ and $\boldsymbol{\varphi}_\varepsilon(y') = \boldsymbol{\varphi}(y'/\varepsilon)$, we represent the last inequality in the form

$$|\boldsymbol{u}_\varepsilon(w)| = \left| \int_{\mathbb{B}'_\varepsilon} \mathcal{M}_{\boldsymbol{\nu}} \left(\frac{y-w}{|y-w|} \right) \boldsymbol{\varphi}_\varepsilon(y') \frac{w_n}{|y-w|^n} dy' \right| \ge \mathcal{K}(\mathbb{R}^n_+(\boldsymbol{\nu})) - \mu. \tag{2.2.15}$$

The vector-valued function

$$\boldsymbol{u}_\varepsilon(x) = \int_{\mathbb{B}'_\varepsilon} \mathcal{M}_{\boldsymbol{\nu}} \left(\frac{y-x}{|y-x|} \right) \boldsymbol{\varphi}_\varepsilon(y') \frac{w_n}{|y-x|^n} dy', \quad x \in \mathbb{R}^n_+(\boldsymbol{\nu}),$$

is a solution of the system $\mathfrak{A}_0(D_x)\boldsymbol{u}_\varepsilon = \boldsymbol{0}$ in $\mathbb{R}^n_+(\boldsymbol{\nu})$ and, in particular, in Ω.

The assumption that the boundary is smooth provides the existence of $\varepsilon_0 > 0$ such that $\varepsilon\xi = w \in \Omega$ for all $\varepsilon \in (0, \varepsilon_0]$.

Suppose that $\varepsilon^{1/2} < \delta < 1$ and $\varepsilon \le \varepsilon_0$. We estimate $|\boldsymbol{u}_\varepsilon(x)|$ under the condition that x belongs to one of the sets $\partial\Omega \backslash \mathbb{B}_{\varepsilon/\delta}$ or

$$T_{\varepsilon/\delta} = \partial\Omega \bigcap \left(\mathbb{B}'_{\varepsilon/\delta} \times [0, \delta) \right).$$

Using (2.2.13), we get that

$$\begin{aligned} ||\boldsymbol{u}_\varepsilon(x)|| &= \sup_{|\boldsymbol{z}|=1} \int_{\mathbb{B}'_\varepsilon} \left| \mathcal{M}^*_{\boldsymbol{\nu}} \left(\frac{y-x}{|y-x|} \right) \boldsymbol{z} \right| \frac{x_n}{|y-x|^n} dy' \\ &\le \sup\{ ||\mathcal{M}^*_{\boldsymbol{\nu}}(\sigma)||_s : \sigma \in \mathbb{S}^{n-1}_- \} \int_{\mathbb{B}'_\varepsilon} \frac{x_n}{|y-x|^n} dy', \end{aligned}$$

where $\|\cdot\|_s$ is the spectral norm of a matrix. Denote the constant in front of the last integral by c. For $x \in \partial\Omega\backslash\mathbb{B}_{\varepsilon/\delta}$

$$\begin{aligned}
|\boldsymbol{u}_\varepsilon(x)| &\le \sup\{\|\boldsymbol{u}_\varepsilon(x)\| : x \in \partial\Omega\backslash\mathbb{B}_{\varepsilon/\delta}\} \le c \sup\left\{\int_{\mathbb{B}'_\varepsilon} \frac{x_n}{|y-x|^n} dy' : x \in \partial\Omega\backslash\mathbb{B}_{\varepsilon/\delta}\right\} \\
&\le c\, v_{n-1}\varepsilon^{n-1} \sup\left\{\frac{x_n}{|y-x|^n} : x \in \partial\Omega\backslash\mathbb{B}_{\varepsilon/\delta},\ y' \in \mathbb{B}'_\varepsilon,\ y_n = 0\right\} \\
&= c\, v_{n-1}\varepsilon^{n-1} \sup\left\{\frac{x_n}{(|y'-x'|^2+x_n^2)^{1/2}} \frac{1}{(|y'-x'|^2+x_n^2)^{(n-1)/2}} :\right. \\
&\qquad \left. x \in \partial\Omega\backslash\mathbb{B}_{\varepsilon/\delta},\ y' \in \mathbb{B}'_\varepsilon\right\} \le c_1\varepsilon^{n-1}((\varepsilon/\delta)-\varepsilon)^{1-n} \\
&= c_1\delta^{n-1}(1-\delta)^{1-n}, \qquad (2.2.16)
\end{aligned}$$

where $c_1 = cv_{n-1}$, and v_{n-1} being the volume of the $(n-1)$-dimensional unit ball.

We proceed to the case of $x \in T_{\varepsilon/\delta}$. Suppose that the variables x and χ are connected by the equality $x = \varepsilon\chi$. Consider $|\boldsymbol{u}_\varepsilon(x)|$ for $x' \in \mathbb{B}'_{\varepsilon/\delta}$, $x_n = \mathcal{F}(x')$, where $\mathcal{F} \in \mathrm{C}^1(\mathbb{B}'_{\varepsilon/\delta})$, or, which is the same, the quantity

$$|\boldsymbol{u}(\chi)| = \left|\int_{\mathbb{B}'_1} \mathcal{M}_{\boldsymbol{\nu}}\left(\frac{\zeta-\chi}{|\zeta-\chi|}\right)\boldsymbol{\varphi}(\zeta')\frac{\chi_n}{|\zeta-\chi|^n}d\zeta'\right|$$

for $\chi' \in \mathbb{B}'_{1/\delta}$, $\chi_n = \varepsilon^{-1}\mathcal{F}(\varepsilon\chi')$.

Since $\mathcal{F}(0) = 0$ and $(\nabla\mathcal{F})(0) = 0$, it follows that $\chi_n = \varepsilon^{-1}o(\varepsilon\chi')$ as $\varepsilon \to 0$. Consequently, $\chi_n = \varepsilon^{-1}\mathcal{F}(\varepsilon\chi') \to 0$ as $\varepsilon \to 0$ uniformly for all $\chi' \in \mathbb{B}'_{1/\delta}$. Therefore, for all $x \in T_{\varepsilon/\delta}$

$$|\boldsymbol{u}_\varepsilon(x)| \to |\boldsymbol{u}(\chi', 0)| = |\boldsymbol{\varphi}(\chi')| \qquad (2.2.17)$$

as $\varepsilon \to 0$. Since $\|\boldsymbol{\varphi}\| \le 1$, it follows from (2.2.15)-(2.2.17) that there is $\varepsilon > 0$ such that $\|\boldsymbol{u}_\varepsilon\|_{[C(\partial\Omega)]^m} \le 1 + \mu$ for any prescribed $\mu > 0$. From this and (2.2.15), using the definition of the constant $\mathcal{K}(\Omega)$, we obtain

$$\mathcal{K}(\Omega) \ge (\mathcal{K}(\mathbb{R}^n_+(\boldsymbol{\nu})) - \mu)/(1+\mu),$$

which, by the arbitrariness of the constant μ and the choice of the line l, proves the assertion of Theorem. □

It follows from Theorems 2.2 and 2.3 that the equality $\mathcal{K}(\Omega) = 1$ can be satisfied only if the operator $\mathfrak{A}_0(D_x)$ has the form (2.2.6). Conversely, if $\mathfrak{A}_0(D_x)$ is representable in the form (2.2.6), then the classical maximum principle holds (see Bitsadze [**Bi1**]). Thus, the following assertion is obtained.

THEOREM 2.4. *Let Ω be a domain in $\mathbb{R}^n$ with compact closure and C^1-boundary. The equality $\mathcal{K}(\Omega) = 1$ holds if and only if the operator $\mathfrak{A}_0(D_x)$ has the form* (2.2.6).

REMARK 2.2. Does the quantity $\mathcal{K}(\Omega)$ depends on the domain Ω in general? And can the inequality sign be replaced by equality in the estimate given in Theorem 2.3?

We answer the first of these questions in the positive and the second in the negative by considering the Lamé system

$$\mu\Delta\boldsymbol{u} + (\lambda+\mu)\text{grad div }\boldsymbol{u} = \boldsymbol{0} \qquad (2.2.18)$$

in a three-dimensional ball $\mathbb{B}_\rho$, where $\mu > 0$ and $\lambda + \mu > 0$. This system is strongly elliptic (cf. Remark 7.4).

Following Polya [**Po**], who showed that $\mathcal{K}(\mathbb{B}_\rho) > 1$ for solutions of the Lamé system, we introduce the vector-valued function $\boldsymbol{w}$ with components

$$w_1(x) = \alpha|x|^2/2 - \beta^3, \; w_2(x) = -\gamma x_1 x_2, \; w_3(x) = -\gamma x_1 x_3,$$

where $\alpha > 0, \beta > 0$ and γ are constants. We have

$$\begin{aligned} |\boldsymbol{w}(x)|^2 &= \beta^6 - \alpha\beta^3|x|^2 + \alpha^2|x|^4/4 + \gamma^2 x_1^2(x_2^2 + x_3^2) \\ &= \beta^6 - \alpha\beta^3|x|^2 + \alpha^2|x|^4/4 + \gamma^2 x_1^2\big(|x|^2 - x_1^2\big). \end{aligned} \tag{2.2.19}$$

The function $\boldsymbol{w}$ is a solution of the Lamé system under the condition that $(\lambda + 2\mu)\alpha = 2(\lambda + \mu)\gamma$. Substituting the expression of γ in terms of α from the last equality in (2.2.19), we get

$$|\boldsymbol{w}(x)|^2 = \beta^6 - \alpha\beta^3|x|^2 + \alpha^2|x|^4/4 + \alpha^2\eta^2 x_1^2\big(|x|^2 - x_1^2\big),$$

where $\eta = (\lambda + 2\mu)(\lambda + \mu)^{-1}/2$. It is clear that

$$\begin{aligned} &\max\{|\boldsymbol{w}(x)|^2 : |x| = \rho\} = \beta^6 - \alpha\beta^3\rho^2 + \alpha^2\rho^4(1 + \eta^2)/4, \\ &\max\{|\boldsymbol{w}(x)|^2 : |x| \le \rho\} \ge |\boldsymbol{w}(0)| = \beta^6. \end{aligned}$$

Consequently,

$$\mathcal{K}(\mathbb{B}_\rho) \ge \beta^3[\beta^6 - \alpha\beta^3\rho^2 + \alpha^2\rho^4(1+\eta^2)/4]^{-1/2} = [1 - t + (1+\eta^2)t^2/4]^{-1/2},$$

where $t = \alpha\rho^2/\beta^3$. Since $\inf\{1 - t + (1+\eta^2)t^2/4 : 0 < t < \infty\} = \eta^2(1+\eta^2)^{-1}$, it follows that

$$\mathcal{K}(\mathbb{B}_\rho) \ge \sqrt{1 + \eta^{-2}} = \left[1 + 4\left(\frac{\lambda + \mu}{\lambda + 2\mu}\right)^2\right]^{1/2}. \tag{2.2.20}$$

Further, it is shown in Corollary 3.4 (Section 3.3) that

$$\mathcal{K}(\mathbb{R}^3_+) = \frac{1}{2}\left[1 + 2\kappa + \frac{(1-\kappa)^2}{\sqrt{3\kappa(\kappa+2)}}\log\frac{1 + 2\kappa + \sqrt{3\kappa(\kappa+2)}}{1-\kappa}\right],$$

where $\kappa = (\lambda + \mu)(\lambda + 3\mu)^{-1}$.

If $\lambda/\mu \to \infty$, then $\mathcal{K}(\mathbb{R}^3_+) \to 3/2$ and $[1 + 4(\lambda+\mu)^2(\lambda+2\mu)^{-2}]^{1/2} \to \sqrt{5}$. Therefore,

$$\liminf_{\lambda/\mu \to \infty} \mathcal{K}(\mathbb{B}_\rho)/\mathcal{K}(\mathbb{R}^3_+) \ge 2\sqrt{5}/3.$$

Hence, the best constant $\mathcal{K}(\mathcal{D})$ depends on the domain and the inequality sign cannot be replaced by equality for the ball.

2.2.2. The case of complex coefficients

We introduce the operator

$$\mathfrak{C}_0(D_x) = \sum_{j,k=1}^{n} \mathcal{C}_{jk}\frac{\partial^2}{\partial x_j \partial x_k},$$

where $\mathcal{C}_{jk} = \mathcal{C}_{kj}$ are constant complex $(m \times m)$-matrices. Assume that the operator $\mathfrak{C}_0$ is strongly elliptic, i.e. that

$$\Re\left(\sum_{j,k=1}^{n} \mathcal{C}_{jk}\sigma_j\sigma_k\boldsymbol{\zeta},\ \boldsymbol{\zeta}\right) > 0$$

for all $\boldsymbol{\zeta} = (\zeta_1, \dots, \zeta_m) \in \mathbb{C}^m$ and $\boldsymbol{\sigma} = (\sigma_1, \dots, \sigma_n) \in \mathbb{R}^n$, with $\boldsymbol{\zeta},\ \boldsymbol{\sigma} \neq \mathbf{0}$. Here $\mathbb{C}^m$ is a complex linear m-dimensional space with the elements $\boldsymbol{a}+i\boldsymbol{b}$, where $\boldsymbol{a}, \boldsymbol{b} \in \mathbb{R}^m$. The inner product in $\mathbb{C}^m$ is $(\boldsymbol{c}, \boldsymbol{d}) = c_1\overline{d}_1 + \cdots + c_m\overline{d}_m$, $\boldsymbol{c} = (c_1, \dots, c_m)$, $\boldsymbol{d} = (d_1, \dots, d_m)$. The length of the vector $\boldsymbol{d}$ in $\mathbb{C}^m$ is $|\boldsymbol{d}| = (\boldsymbol{d}, \boldsymbol{d})^{1/2}$.

Let $\mathcal{R}_{jk}$ and $\mathcal{H}_{jk}$ be constant real $(m\times m)$-matrices such that $\mathcal{C}_{jk} = \mathcal{R}_{jk}+i\mathcal{H}_{jk}$. We define

$$\mathfrak{R}_0(D_x) = \sum_{j,k=1}^{n} \mathcal{R}_{jk}\frac{\partial^2}{\partial x_j \partial x_k}, \qquad \mathfrak{H}_0(D_x) = \sum_{j,k=1}^{n} \mathcal{H}_{jk}\frac{\partial^2}{\partial x_j \partial x_k}.$$

Separating the real and imaginary parts of the system $\mathfrak{C}_0(D_x)\boldsymbol{u} = 0$, where $\boldsymbol{u} = \boldsymbol{v} + i\boldsymbol{w}$, we get a system with real coefficients

$$\mathfrak{R}_0(D_x)\boldsymbol{v} - \mathfrak{H}_0(D_x)\boldsymbol{w} = \mathbf{0}, \quad \mathfrak{H}_0(D_x)\boldsymbol{v} + \mathfrak{R}_0(D_x)\boldsymbol{w} = \mathbf{0},$$

which, like the original system, is strongly elliptic.

We introduce the matrix

$$\mathfrak{K}_0(D_x) = \begin{pmatrix} \mathfrak{R}_0(D_x) & -\mathfrak{H}_0(D_x) \\ \mathfrak{H}_0(D_x) & \mathfrak{R}_0(D_x) \end{pmatrix}.$$

Let $[\mathbf{C_b}(\Omega)]^m$ be the space of m-component complex vector-valued functions $\boldsymbol{u} = \boldsymbol{v}+i\boldsymbol{w}$ which are bounded and continuous on $\Omega \subset \mathbb{R}^n$. The norm on $[\mathbf{C_b}(\Omega)]^m$ is $\|\boldsymbol{u}\| = \sup\{(|\boldsymbol{v}(x)|^2 + |\boldsymbol{w}(x)|^2)^{1/2} \ :\ x \in \Omega\}$. The notation $[\mathbf{C_b}(\partial\Omega)]^m$ has a similar meaning. By $[\mathbf{C}^2(\Omega)]^m$ we denote the space of m-component complex vector-valued functions with continuous derivatives up to the second order in Ω.

By analogy with the definition of $\mathcal{K}(\Omega)$, let

$$\mathcal{K}'(\Omega) = \sup \frac{\|\boldsymbol{u}\|_{[\mathbf{C_b}(\overline{\Omega})]^m}}{\|\boldsymbol{u}\|_{[\mathbf{C_b}(\partial\Omega)]^m}},$$

where the supremum is over all vector-valued functions in the class $[\mathbf{C}^2(\Omega)]^m \cap [\mathbf{C_b}(\overline{\Omega})]^m$ satisfying the system $\mathfrak{C}_0(D_x)\boldsymbol{u} = 0$ in Ω.

It is clear that the constant $\mathcal{K}'(\Omega)$ for the system $\mathfrak{C}_0(D_x)\boldsymbol{u} = 0$ with complex coefficients coincides with the constant $\mathcal{K}(\Omega)$ for the system $\mathfrak{K}_0(D_x)\{\boldsymbol{v}, \boldsymbol{w}\} = 0$ with real coefficients if we replace m by $2m$, $\mathfrak{A}_0(D_x)$ by $\mathfrak{K}_0(D_x)$, and $\boldsymbol{u}$ by $\{\boldsymbol{v}, \boldsymbol{w}\}$ in definition (2.2.1). Therefore, all assertions about $\mathcal{K}'(\Omega)$ are direct consequences of the analogous assertions about $\mathcal{K}(\Omega)$. Using this fact, from Theorems 2.1-2.4 we get the assertions.

THEOREM 2.5. *The formula*

$$\mathcal{K}'(\mathbb{R}^n_+) = \sup_{|\boldsymbol{z}|=1} \int_{\mathbb{S}^{n-1}_-} |\mathcal{U}^*(\sigma)\boldsymbol{z}| d\sigma$$

is valid, where $\mathcal{U}$ *is the* $(2m \times 2m)$*-matrix-valued function on* $\mathbb{S}^{n-1}_-$ *appearing in the integral representation for a solution of the Dirichlet problem in* $\mathbb{R}^n_+$ *for the system* $\mathfrak{K}_0(D_x)\{\boldsymbol{v}, \boldsymbol{w}\} = 0$ *(analogous to representation* (2.2.3)*) and* $\boldsymbol{z} \in \mathbb{R}^{2m}$.

THEOREM 2.6. *The equality* $\mathcal{K}'(\mathbb{R}^n_+) = 1$ *is valid if and only if*

$$\mathfrak{C}_0(D_x) = \mathcal{C} \sum_{j,k=1}^{n} a_{jk} \frac{\partial^2}{\partial x_j \partial x_k}, \tag{2.2.21}$$

where $\mathcal{C}$ *is a constant complex-valued* $(m \times m)$*-matrix such that* $\Re(\mathcal{C}\boldsymbol{\zeta}, \boldsymbol{\zeta}) > 0$ *for all* $\boldsymbol{\zeta} \in \mathbb{C}^m$, $\boldsymbol{\zeta} \neq \mathbf{0}$, *and* $((a_{jk}))$ *is a real positive-definite* $(n \times n)$*-matrix.*

THEOREM 2.7. *Let* Ω *be a domain in* $\mathbb{R}^n$ *with compact closure and* C^1*-boundary. Then*

$$\mathcal{K}'(\Omega) \geq \sup\{\mathcal{K}'(\mathbb{R}^n_+(\boldsymbol{\nu})) : \boldsymbol{\nu} \in \mathbb{S}^{n-1}\},$$

where $\mathbb{R}^n_+(\boldsymbol{\nu})$ *is a half-space with inward normal* $\boldsymbol{\nu}$.

THEOREM 2.8. *Let* Ω *be a domain in* $\mathbb{R}^n$ *with compact closure and* C^1*-boundary. The equality* $\mathcal{K}'(\Omega) = 1$ *holds if and only if the operator* $\mathfrak{C}_0(D_x)$ *has the form* (2.2.21).

REMARK 2.3. Let $\mathfrak{A}_0(D_x)\boldsymbol{u} = \mathbf{0}$ be a system with real coefficients. Using the notation of the present section, we regard it as a system with complex coefficients. Let $\boldsymbol{u}$ be a vector-valued function in the class $[\mathbf{C}^{(2)}(\Omega)]^m \cap [\mathbf{C_b}(\overline{\Omega})]^m$. We have that

$$\mathcal{U}(\sigma) = \begin{pmatrix} \mathcal{M}(\sigma) & 0 \\ 0 & \mathcal{M}(\sigma) \end{pmatrix}$$

and, by Theorem 2.1,

$$\mathcal{K}'(\mathbb{R}^n_+) = \sup_{|\boldsymbol{z}_1|^2 + |\boldsymbol{z}_2|^2 = 1} \int_{\mathbb{S}^{n-1}_-} \left(|\mathcal{M}^*(\sigma)\boldsymbol{z}_1|^2 + |\mathcal{M}^*(\sigma)\boldsymbol{z}_2|^2 \right)^{1/2} d\sigma, \tag{2.2.22}$$

where $\boldsymbol{z}_1, \boldsymbol{z}_2 \in \mathbb{S}^{m-1}$. Consequently, $\mathcal{K}'(\mathbb{R}^n_+) \geq \mathcal{K}(\mathbb{R}^n_+)$ for any strongly elliptic system with real coefficients. We give an example showing that the inequality sign cannot be replaced here by equality. To do this we consider the Lamé system in $\mathbb{R}^2_+$ under the assumption that the ratio λ/μ is sufficiently large.

It is shown in Section 3.3 that

$$|\mathcal{M}^*(\sigma)\boldsymbol{z}| = \pi^{-1} \left[(1-\kappa)^2 + 4\kappa(\boldsymbol{e}_\sigma, \boldsymbol{z})^2 \right]^{1/2}$$

for the Lamé system in the two-dimensional case, where $\kappa = (\lambda + \mu)(\lambda + 3\mu)^{-1}$. Here $\boldsymbol{e}_\sigma$ is the unit vector joining the origin with a point $\sigma \in \mathbb{S}^1$. By (2.2.22),

$$\begin{aligned} \mathcal{K}'(\mathbb{R}^2_+) &= \frac{1}{\pi} \sup_{|\boldsymbol{z}_1|^2 + |\boldsymbol{z}_2|^2 = 1} \int_{S^1_-} \left[2(1-\kappa)^2 + 4\kappa(\boldsymbol{e}_\sigma, \boldsymbol{z}_1)^2 + 4\kappa(\boldsymbol{e}_\sigma, \boldsymbol{z}_2)^2 \right]^{1/2} d\sigma \\ &= \frac{1}{2\pi} \sup_{|\boldsymbol{z}_1|^2 + |\boldsymbol{z}_2|^2 = 1} \int_{S^1} \left[2(1-\kappa)^2 + 4\kappa(\boldsymbol{e}_\sigma, \boldsymbol{z}_1)^2 + 4\kappa(\boldsymbol{e}_\sigma, \boldsymbol{z}_2)^2 \right]^{1/2} d\sigma. \end{aligned}$$

Setting $\boldsymbol{z}_1 = (0, \sqrt{2}/2)$ and $\boldsymbol{z}_2 = (\sqrt{2}/2, 0)$ here and taking account that $0 < \kappa < 1$, we find that

$$\mathcal{K}'(\mathbb{R}^2_+) \geq \frac{1}{2\pi} \int_0^{2\pi} \left[2(1-\kappa)^2 + 2\kappa(\sin^2\varphi + \cos^2\varphi) \right]^{1/2} d\varphi \geq \kappa\sqrt{2}. \tag{2.2.23}$$

From the formula

$$\mathcal{K}(\mathbb{R}^2_+) = \frac{2}{\pi}(1+\kappa)\mathbf{E}\left(\frac{2\sqrt{\kappa}}{1+\kappa}\right) = 1 + \frac{1}{2^2}\kappa^2 + \frac{1}{2^2 4^2}\kappa^4 + \cdots + \left[\frac{(2m-3)!!}{2^m m!}\right]^2 \kappa^{2k} + \ldots,$$

obtained in Section 3.3 (Corollary 3.4), and from the inequality $0 < \kappa < 1$ it follows that $\mathcal{K}(\mathbb{R}^2_+) < 4/\pi$. Therefore, by (2.2.23), $\mathcal{K}'(\mathbb{R}^2_+) > \mathcal{K}(\mathbb{R}^2_+)$ for $\kappa > 2\sqrt{2}/\pi$.

2.3. General second order strongly elliptic systems

2.3.1. The case of real coefficients

2.3.1.1. *Some notations.* Let Ω be a domain in $\mathbb{R}^n$ with compact closure $\overline{\Omega}$ and with boundary $\partial\Omega$ of the class $\mathrm{C}^{2,\alpha}, 0 < \alpha \leq 1$. Below we use the following notation. Let $[\mathrm{C}(\overline{\Omega})]^m$ be the space of continuous m-component vector-valued functions on $\overline{\Omega}$ with the norm $\|\boldsymbol{u}\| = \max\{|\boldsymbol{u}(y)| : y \in \overline{\Omega}\}$. The notation $[\mathrm{C}(\partial\Omega)]^m$ has a similar meaning. By $[\mathrm{C}^k(\Omega)]^m$ we denote the space of m-component vector-valued functions with continuous derivatives up to order k in Ω. The space of $(m \times m)$-matrix-valued functions whose elements have continuous derivatives up to order k and satisfy the Hölder condition with exponent $\alpha, 0 < \alpha \leq 1$, on $\overline{\Omega}$ is denoted by $[\mathrm{C}^{k,\alpha}(\overline{\Omega})]^{m\times m}$. We omit the upper indices m and $m \times m$ in notation of spaces in the case $m = 1$. Let G be an open set in $\mathbb{R}^n$. Henceforth, by $\mathrm{C}_0^k(G)$ we mean the space of real functions with continuous derivatives up to order k in G and with compact support in G. We remove k in similar notation in the case $k = 0$.

In what follows by $[W_p^l(\Omega)]^m$ we mean the Sobolev space of m-component vector-valued functions $\boldsymbol{u} = (u_1, \ldots, u_m)$ on Ω with each component in $W_p^l(\Omega)$ and endowed with the norm

$$\|\boldsymbol{u}\|_{[W_p^l(\Omega)]^m} = \left(\sum_{k=1}^m \|u_k\|^2_{W_p^l(\Omega)}\right)^{1/2}.$$

The notation $[\mathring{W}_p^l(\Omega)]^m$ has a similar meaning.

We introduce the differential operator

$$\mathfrak{A}(x, D_x) = \sum_{j,k=1}^n \mathcal{A}_{jk}(x)\frac{\partial^2}{\partial x_j \partial x_k} - \sum_{j=1}^n \mathcal{A}_j(x)\frac{\partial}{\partial x_j} - \mathcal{A}_0(x),$$

where $\mathcal{A}_{jk} = \mathcal{A}_{kj}, \mathcal{A}_j, \mathcal{A}_0$ are real $(m \times m)$-matrix-valued functions in the spaces

$$[\mathrm{C}^{2,\alpha}(\overline{\Omega})]^{m\times m}, \qquad [\mathrm{C}^{1,\alpha}(\overline{\Omega})]^{m\times m}, \qquad [\mathrm{C}^{\alpha}(\overline{\Omega})]^{m\times m},$$

respectively. When the coefficients of the operator $\mathfrak{A}(x, D_x)$ do not depend on x we use the notation $\mathfrak{A}(D_x)$. Let the principal homogeneous part of the operator $\mathfrak{A}(x, D_x)$ be denoted by $\mathfrak{A}_0(x, D_x)$.

We assume that $\mathfrak{A}(x, D_x)$ is strongly elliptic in $\overline{\Omega}$ that is for all $x \in \overline{\Omega}$, $\boldsymbol{\zeta} = (\zeta_1, \ldots, \zeta_m) \in \mathbb{R}^m$, $\boldsymbol{\sigma} = (\sigma_1, \ldots, \sigma_n) \in \mathbb{R}^n$, with $\boldsymbol{\zeta}$, $\boldsymbol{\sigma} \neq \mathbf{0}$, the inequality

$$\left(\sum_{j,k=1}^n \mathcal{A}_{jk}(x)\sigma_j\sigma_k\boldsymbol{\zeta},\ \boldsymbol{\zeta}\right) > 0 \tag{2.3.1}$$

is valid.

2.3.1.2. *Necessary conditions. The structure of the principal part of the system.* The following assertion concerns the form of the principal part of the operator $\mathfrak{A}(y, D_x)$ for $y \in \partial\Omega$ under the assumption that the maximum modulus principle holds for solutions of the system $\mathfrak{A}(x, D_x)\boldsymbol{u} = \mathbf{0}$ in Ω.

In what follows, when proving the theorems, we use the shift of the origin to a point $\mathcal{O}$ at the boundary or inside Ω, i.e. the shift transformation $x' = x - x_{\mathcal{O}}$. To simplify the representation we retain the same notation after this transform.

THEOREM 2.9. *If the maximum modulus principle*

$$\max_{\overline{\Omega}} |\boldsymbol{u}| \le \max_{\partial\Omega} |\boldsymbol{u}|, \tag{2.3.2}$$

holds for solutions $[\mathrm{C}^2(\Omega)]^m \cap [\mathrm{C}(\overline{\Omega})]^m$ *of the system* $\mathfrak{A}(x, D_x)\boldsymbol{u} = \boldsymbol{0}$, *then for any* $y \in \partial\Omega$ *the equality*

$$\mathfrak{A}_0(y, D_x) = \mathcal{A}(y) \sum_{j,k=1}^{n} a_{jk}(y) \frac{\partial^2}{\partial x_j \partial x_k} \tag{2.3.3}$$

holds, where $\mathcal{A}$ *and* $((a_{jk}))$ *are positive-definite* $(m \times m)$*- and* $(n \times n)$*-matrices, respectively.*

PROOF. We place the origin at a point $y \in \partial\Omega$ and denote by $\boldsymbol{\nu}$ the inward normal to $\partial\Omega$ at y. Further, let $(\cdot, \cdot)$ be the scalar product in $\mathbb{R}^n$, $\boldsymbol{r}_{yx} = x - y$, and let $\mathbb{R}^n_+(y)$ denote the half-space $\{x \in \mathbb{R}^n : (\boldsymbol{r}_{yx}, \boldsymbol{\nu}) > 0\}, \mathbb{B}_r = \{x \in \mathbb{R}^n : |x| < r\}$. We put $\mathbb{B} = \mathbb{B}_1$.

In this proof by $\boldsymbol{v}$ we mean an m-component vector-valued function on $\overline{\mathbb{R}^n_+(y)}$ which is continuous on $\overline{\mathbb{R}^n_+(y)}$, vanishes at infinity and satisfies the system

$$\mathfrak{A}_0(0, D_x)\boldsymbol{v} = \boldsymbol{0}.$$

We show, that under the assumption of the theorem, for any point $y \in \partial\Omega$ the maximum modulus principle is valid

$$\sup_{\overline{\mathbb{R}^n_+(y)}} |\boldsymbol{v}| \le \sup_{\partial\mathbb{R}^n_+(y)} |\boldsymbol{v}|. \tag{2.3.4}$$

Without loss of generality we assume a part of the boundary $\partial\Omega$ contained in $\mathbb{B}$ with center at y to be plane. The corresponding mapping can be constructed in such a way that the operator $\mathfrak{A}_0(0, D_x)$ does not change.

It suffices to verify (2.3.4) for solutions $\boldsymbol{v}$ subjected to the conditions:

$$\boldsymbol{v} \in [\mathrm{C}^\infty(\overline{\mathbb{R}^n_+(y)})]^m, \quad \operatorname{supp} \boldsymbol{v}\big|_{\partial\mathbb{R}^n_+(y)} \subset \mathbb{B}.$$

Put $\boldsymbol{v}_\varepsilon(x) = \boldsymbol{v}(x/\varepsilon)$, where ε is a small positive number and introduce the cut-off function $\eta \in \mathrm{C}_0^\infty(\mathbb{B}), \eta(x) = 1$ for $|x| < 1/2, 0 \le \eta(x) \le 1$. Let $\boldsymbol{u}_\varepsilon = \eta \boldsymbol{v}_\varepsilon$ on $\mathbb{B}$ and let $\boldsymbol{u}_\varepsilon = \boldsymbol{0}$ outside $\mathbb{B}$.

We have

$$\mathfrak{A}(x, D_x)\boldsymbol{u}_\varepsilon = \boldsymbol{F}_\varepsilon \quad \text{on} \quad \Omega, \tag{2.3.5}$$

where

$$\boldsymbol{F}_\varepsilon = (\mathfrak{A}(x, D_x) - \mathfrak{A}_0(0, D_x))\, \boldsymbol{u}_\varepsilon + [\, \mathfrak{A}_0(0, D_x), \eta]\, \boldsymbol{v}_\varepsilon,$$

where $[A, B]$ stands for the commutator $AB - BA$ of the operators A and B. Since all the coefficients of the operator $\mathfrak{A}$ satisfy the Hölder condition with exponent α, then

$$|\boldsymbol{F}_\varepsilon| \le c(r^\alpha \eta |\nabla_2 \boldsymbol{v}_\varepsilon| + |\nabla(\zeta \boldsymbol{v}_\varepsilon)| + |\zeta \boldsymbol{v}_\varepsilon|), \tag{2.3.6}$$

where $r = |x|, \zeta \in \mathrm{C}_0^\infty(\mathbb{B}), \zeta\eta = \eta$ and ∇, ∇_2 are the first and the second gradients, respectively.

By $\boldsymbol{w}_\varepsilon$ we denote a solution of the Dirichlet problem

$$\mathfrak{A}(x, D_x)\boldsymbol{w}_\varepsilon = \boldsymbol{F}_\varepsilon \text{ on } \Omega, \quad \boldsymbol{w}_\varepsilon\big|_{\partial\Omega} = \boldsymbol{0}. \tag{2.3.7}$$

Since $\mathfrak{A}$ is strongly elliptic, then the mapping

$$\mathfrak{A}(x, D_x) : [W_p^2(\Omega)]^m \cap [\mathring{W}_p^1(\Omega)]^m \to [L^p(\Omega)]^m$$

is Fredholm and has zero index for any $p \in (1, \infty)$. For $2p > n$, by the imbedding $W_p^2(\Omega) \subset \mathrm{C}(\overline{\Omega})$ and by the maximum modulus principle (2.3.2), the problem (2.3.7) has at most one solution in $[W_p^2(\Omega)]^m$. Consequently, this problem is solvable in $[W_p^2(\Omega)]^m$ for any right-hand side and the estimate

$$\|\boldsymbol{w}_\varepsilon\|_{[W_p^2(\Omega)]^m} \le c\|\boldsymbol{F}_\varepsilon\|_{[L_p(\Omega)]^m}$$

holds. This and (2.3.6) imply

$$\|\boldsymbol{w}_\varepsilon\|_{[\mathrm{C}(\overline{\Omega})]^m} \le c\left(\|r^\alpha \eta \nabla_2 \boldsymbol{v}_\varepsilon\|_{[L_p(\Omega)]^m} + \|\zeta \boldsymbol{v}_\varepsilon\|_{[W_p^1(\Omega)]^m}\right),$$

which is equivalent to the estimate

$$\|\boldsymbol{w}_\varepsilon\|_{[\mathrm{C}(\overline{\Omega})]^m} \le c\left(\|r^\alpha \eta \nabla_2 \boldsymbol{v}_\varepsilon\|_{[L_p(\mathbb{R}^n_+(y))]^m} + \|\zeta \boldsymbol{v}_\varepsilon\|_{[W_p^1(\mathbb{R}^n_+(y))]^m}\right). \tag{2.3.8}$$

From the representation of the function $\boldsymbol{v}$ by its Dirichlet data on $\partial\mathbb{R}^n_+(y)$ one concludes that

$$\sum_{|\beta|\le 2} (1+|x|)^{|\beta|} \left|\frac{\partial^\beta \boldsymbol{v}}{\partial x^\beta}\right| \le c\,(1+|x|)^{1-n},$$

where $\beta = (\beta_1, \ldots, \beta_n)$, $|\beta| = \beta_1 + \cdots + \beta_n$, or, which is the same,

$$\sum_{|\beta|\le 2} (\varepsilon+|x|)^{|\beta|} \left|\frac{\partial^\beta \boldsymbol{v}_\varepsilon}{\partial x^\beta}\right| \le c\frac{\varepsilon^{n-1}}{(\varepsilon+|x|)^{n-1}}. \tag{2.3.9}$$

Therefore, the first norm on the right-hand side of (2.3.8) does not exceed

$$c\,\varepsilon^{n-1}\left\|\frac{r^\alpha}{(\varepsilon+r)^{n+1}}\right\|_{L_p(\mathbb{R}^n_+(y))} = c_1\varepsilon^{\alpha-2+n/p},$$

and the second is not greater than

$$c\,\varepsilon^{n-1}\left(\|(\varepsilon+r)^{-n}\|_{L_p(\mathbb{R}^n_+(y))} + \|(\varepsilon+r)^{1-n}\|_{L_p(\mathbb{R}^n_+(y))}\right) \le c_2\varepsilon^{(n-p)/p}.$$

Thus

$$\|\boldsymbol{w}_\varepsilon\|_{[\mathrm{C}(\overline{\Omega})]^m} \le c\,\varepsilon^{\alpha-2+n/p}. \tag{2.3.10}$$

Let p be any number in the $(n/2, (n/(2-\alpha))$.

By (2.3.5) and (2.3.7)

$$\mathfrak{A}(x, D_x)(\boldsymbol{u}_\varepsilon - \boldsymbol{w}_\varepsilon) = \boldsymbol{0} \text{ on } \Omega$$

and the maximum modulus principle (2.3.2) is applicable to $\boldsymbol{u}_\varepsilon - \boldsymbol{w}_\varepsilon$:

$$\|\boldsymbol{u}_\varepsilon - \boldsymbol{w}_\varepsilon\|_{[\mathrm{C}(\overline{\Omega})]^m} \le \|\boldsymbol{u}_\varepsilon\big|_{\partial\Omega}\|_{[\mathrm{C}(\partial\Omega)]^m}.$$

From this and (2.3.10) we obtain

$$\|\boldsymbol{u}_\varepsilon\|_{[\mathrm{C}(\overline{\Omega})]^m} \le \max_{\partial\mathbb{R}^n_+(y)} |\boldsymbol{v}| + c\,\varepsilon^{\alpha-2+n/p}.$$

Further,

$$\max_{\mathbb{R}^n_+(y)} |\boldsymbol{v}_\varepsilon| \le \max_{\partial\mathbb{R}^n_+(y)} |\boldsymbol{v}| + \max_{\mathbb{R}^n_+(y)} |(1-\eta)\boldsymbol{v}_\epsilon| + c\,\varepsilon^{\alpha-2+n/p}.$$

According to (2.3.9) the inequality $|\boldsymbol{v}_\varepsilon(x)| \le c\,\varepsilon^{n-1}$ hods on the support of the function $1-\eta$. Hence

$$\max_{\mathbb{R}^n_+(y)} |\boldsymbol{v}| \le \max_{\partial\mathbb{R}^n_+(y)} |\boldsymbol{v}| + c\left(\varepsilon^{n-1} + \varepsilon^{\alpha-2+n/p}\right).$$

Passing here to the limit as $\varepsilon \to 0$ we get (2.3.4). So, by Theorem 2.2 the representation (2.3.3) takes place. □

COROLLARY 2.1. *If for any $\omega \subset \Omega$ with the boundary in the class $\mathrm{C}^{2,\alpha}$ all solutions of the system $\mathfrak{A}(x, D_x)\boldsymbol{u} = 0$ from the class $[\mathrm{C}^2(\omega)]^m \cap [\mathrm{C}(\overline{\omega})]^m$ satisfy the inequality*

$$\|\boldsymbol{u}\|_{[\mathrm{C}(\overline{\omega})]^m} \le \|\boldsymbol{u}\big|_{\partial\omega}\|_{[\mathrm{C}(\partial\omega)]^m},$$

then $\mathcal{A}_{jk}(x) = \mathcal{A}(x)a_{jk}(x), 1 \le j,k \le n$, where $\mathcal{A}$ and $((a_{jk}))$ are positive-definite in $\overline{\Omega}$ $(m \times m)-$ and $(n \times n)-$matrix-valued functions. (One can essentially restrict the class of domains assuming that ω is an arbitrary ball in Ω). Moreover, one can choose $\mathcal{A}$ and a_{jk} in such a way that elements of $\mathcal{A}$ and $((a_{jk}))$ belong to $\mathrm{C}^{2,\alpha}(\overline{\Omega})$.

PROOF. The first part of this assertion immediately follows from Theorem 2.9.

Since $\mathcal{A}(x)$ is a positive definite matrix in $\overline{\Omega}$ and $\mathcal{A}_{11} \in [\mathrm{C}^{2,\alpha}(\overline{\Omega})]^{m\times m}$, then $\mathcal{A}_{11}^{-1} \in [\mathrm{C}^{2,\alpha}(\overline{\Omega})]^{m\times m}$. Suppose that $\mathcal{A}(x) = \mathcal{A}_{11}(x)$. By using $\mathcal{A}_{jk} \in [\mathrm{C}^{2,\alpha}(\overline{\Omega})]^{m\times m}$ and the equality

$$\mathcal{A}_{jk}(x) = a_{jk}(x)\mathcal{A}(x)$$

we obtain the inclusion $a_{jk} \in \mathrm{C}^{2,\alpha}(\overline{\Omega})$.

Thus if we put $\mathcal{A}(x) = \mathcal{A}_{11}(x)$ in $\overline{\Omega}$, then $\mathcal{A} \in [\mathrm{C}^{2,\alpha}(\overline{\Omega})]^{m\times m}$ and $a_{jk} \in \mathrm{C}^{2,\alpha}(\overline{\Omega})$. □

In what follows we assume that elements of the matrices $\mathcal{A}$ and $((a_{jk}))$ belong to $\mathrm{C}^{2,\alpha}(\overline{\Omega})$.

Theorem 2.9 implies the following assertion.

COROLLARY 2.2. *If the maximum modulus principle*

$$\|\boldsymbol{u}\|_{[\mathrm{C}(\overline{\Omega})]^m} \le \|\boldsymbol{u}\big|_{\partial\Omega}\|_{[\mathrm{C}(\partial\Omega)]^m},$$

is valid for solutions $\boldsymbol{u} \in [\mathrm{C}^2(\Omega)]^m \cap [\mathrm{C}(\overline{\Omega})]^m$ of the system $\mathfrak{A}(D_x)\boldsymbol{u} = \boldsymbol{0}$ with constant coefficients, then

$$\mathcal{A}_{jk} = \mathcal{A}a_{jk}, \quad 1 \le j,k \le n,$$

where $\mathcal{A}$ and $((a_{jk}))$ are positive-definite $(m \times m)-$ and $(n \times n)-$matrices.

The next assertion shows that in the case of variable coefficients the validity of the equalities $\mathcal{A}_{jk}(x) = \mathcal{A}(x)a_{jk}(x), 1 \le j,k \le n$, everywhere in $\overline{\Omega}$ is not necessary for the inequality

$$\|\boldsymbol{u}\|_{[\mathrm{C}(\overline{\Omega})]^m} \le \|\boldsymbol{u}\big|_{\partial\Omega}\|_{[\mathrm{C}(\partial\Omega)]^m},$$

where $\boldsymbol{u}$ is the solution of the system $\mathfrak{A}(x, D_x)\boldsymbol{u} = \boldsymbol{0}$ from the class $[\mathrm{C}^2(\Omega)]^m \cap [\mathrm{C}(\overline{\Omega})]^m$. In other words, the classical maximum modulus principle may be valid for solutions of the system $\mathfrak{A}(x, D_x)\boldsymbol{u} = \boldsymbol{0}$ in Ω in the case when the equalities $\mathcal{A}_{jk}(x) = \mathcal{A}(x)a_{jk}(x)$ are violated inside of the domain.

THEOREM 2.10. *Let $\mathcal{A}_{ij}(x)$ be $(m \times m)$-matrices on $\overline{\mathbb{B}}$ with smooth elements and let the equalities*

$$\mathcal{A}_{ij}(x) = \delta_{ij} I$$

be valid in the spherical layer $\mathbb{B}\backslash\mathbb{B}_{1-\delta}$ for some $\delta > 0$, where I is the identity matrix of order m. Then any solution $\boldsymbol{u} \in [\mathrm{C}^2(\mathbb{B})]^m \cap [\mathrm{C}(\overline{\mathbb{B}})]^m$ of the strongly elliptic system

$$\sum_{i,j=1}^{n} \frac{\partial}{\partial x_i}\left(\mathcal{A}_{ij}(x)\frac{\partial \boldsymbol{u}}{\partial x_j}\right) - \lambda^2 \boldsymbol{u} = \boldsymbol{0}, \quad x \in \mathbb{B}, \tag{2.3.11}$$

with the boundary condition

$$\boldsymbol{u}\big|_{\partial\mathbb{B}} = \boldsymbol{\Phi}$$

satisfies the inequality

$$\|\boldsymbol{u}\|_{[\mathrm{C}(\mathbb{B})]^m} \leq \|\boldsymbol{\Phi}\|_{[\mathrm{C}(\partial\mathbb{B})]^m} \tag{2.3.12}$$

for sufficiently large $\lambda > 0$.

PROOF. *1. Auxiliary estimates.* Let U and V be open sets such that $\overline{U} \subset V \subset \mathbb{B}$ and let $\eta \in \mathrm{C}_0^\infty(V)$. Below we need two well-known local estimates

$$\sum_{k=0}^{l} \lambda^{l-k} \|\boldsymbol{u}\|_{[W_2^k(U)]^m} \leq c_0 \|\boldsymbol{u}\|_{[L_2(V)]^m}, \tag{2.3.13}$$

$$\int_U \eta^2 \left(|\nabla \boldsymbol{u}|^2 + \lambda^2 |\boldsymbol{u}|^2\right) dx \leq c_1 \int_V |\boldsymbol{u}|^2 |\nabla \eta|^2 dx, \tag{2.3.14}$$

where c_0 is a constant which depends on U and V but not on λ, c_1 does not depend on U, V and λ, and $\eta \in \mathrm{C}_0^\infty(V)$.

The estimate (2.3.13) can be obtained in the following way. We note that the vector-valued function $\boldsymbol{v}(x,z) = \exp(i\lambda z)\boldsymbol{u}(x)$, $z \in \mathbb{R}$, is a solution of the elliptic system

$$\sum_{i,j=1}^{n} \frac{\partial}{\partial x_i}\left(\mathcal{A}_{ij}(x)\frac{\partial \boldsymbol{v}}{\partial x_j}\right) + \frac{\partial^2 \boldsymbol{v}}{\partial z^2} = \boldsymbol{0}$$

in the cylinder $\mathbb{B} \times (-1,1)$. So the local estimate

$$\|\boldsymbol{v}\|_{[W_2^l(U\times(-1/2,1/2))]^m} \leq c_2 \|\boldsymbol{v}\|_{[L_2(V\times(-1,1))]^m}$$

is valid (see Agmon, Douglis and Nirenberg [**ADN2**]). Substituting here $\boldsymbol{v}$ expressed by $\boldsymbol{u}$, we arrive at (2.3.13).

In order to derive (2.3.14) it suffices to take the scalar product of (2.3.11) with $\eta^2 \boldsymbol{u}$, then integrate by parts and make use of the Cauchy inequality.

2. Proof of (2.3.12) *for large values of* λ. Consider the collection of balls

$$\mathbb{B}^{(k)} = \left\{ x \in \mathbb{R}^n : |x| < 1 - \frac{\delta}{2} + \frac{k\delta}{2\lambda} \right\}, \quad k = 1, 0, \ldots, [\lambda] - 1.$$

From (2.3.14) one can easily gets the inequality

$$\int_{\mathbb{B}^{(k)}} \left(|\nabla \boldsymbol{u}|^2 + \lambda^2 |\boldsymbol{u}|^2\right) dx \leq K\lambda^2 \int_{\mathbb{B}^{(k+1)}\backslash\mathbb{B}^{(k)}} |\boldsymbol{u}|^2 dx.$$

Here and in what follows in this proof K, K_1, K_2, K_3 are positive constants which depend on δ but not on k and λ.

Hence

$$\int_{\mathbb{B}^{(k)}} |\boldsymbol{u}|^2 dx \leq \frac{K}{K+1} \int_{\mathbb{B}^{(k+1)}} |\boldsymbol{u}|^2 dx$$

and so

$$\int_{\mathbb{B}_{1-\delta/2}} |\boldsymbol{u}|^2 dx \leq \left(\frac{K}{K+1}\right)^{[\lambda]} \int_{\mathbb{B}} |\boldsymbol{u}|^2 dx. \tag{2.3.15}$$

By the Sobolev imbedding theorem

$$\|\boldsymbol{u}\|_{[C(\overline{\mathbb{B}_{1-\delta}})]^m} \leq K_1 \|\boldsymbol{u}\|_{[W_2^l(\mathbb{B}_{1-\delta})]^m}$$

for $2l > n$, which together with (2.3.13) leads to

$$\|\boldsymbol{u}\|_{[C(\overline{\mathbb{B}_{1-\delta}})]^m} \leq K_2 \|\boldsymbol{u}\|_{[L_2(\mathbb{B}_{1-\delta/2})]^m} .$$

Estimating the right-hand side of the last inequality with the help of (2.3.15), we find

$$\|\boldsymbol{u}\|_{[C(\overline{\mathbb{B}_{1-\delta}})]^m} \leq \exp(-K_3\lambda) \|\boldsymbol{u}\|_{[C(\overline{\mathbb{B}})]^m}. \tag{2.3.16}$$

This and the maximum modulus principle for the operator $(\Delta - \lambda^2)I$ imply

$$\begin{aligned} \|\boldsymbol{u}\|_{[C(\overline{\mathbb{B}}\backslash\mathbb{B}_{1-\delta})]^m} &\leq \max\left\{\|\boldsymbol{u}\|_{[C(\partial\mathbb{B}_{1-\delta})]^m}, \|\boldsymbol{\Phi}\|_{[C(\partial\mathbb{B})]^m}\right\} \\ &\leq \max\left\{\exp(-K_3\lambda)\|\boldsymbol{u}\|_{[C(\overline{\mathbb{B}})]^m}, \|\boldsymbol{\Phi}\|_{[C(\partial\mathbb{B})]^m}\right\}. \end{aligned} \tag{2.3.17}$$

Combining (2.3.16) and (2.3.17), we arrive at the estimate

$$\begin{aligned} \|\boldsymbol{u}\|_{[C(\overline{\mathbb{B}})]^m} &= \max\left\{\|\boldsymbol{u}\|_{[C(\overline{\mathbb{B}_{1-\delta}})]^m}, \|\boldsymbol{u}\|_{[C(\overline{\mathbb{B}}\backslash\mathbb{B}_{1-\delta})]^m}\right\} \\ &\leq \max\left\{\exp(-K_3\lambda)\|\boldsymbol{u}\|_{[C(\overline{\mathbb{B}})]^m}, \|\boldsymbol{\Phi}\|_{[C(\partial\mathbb{B})]^m}\right\}, \end{aligned}$$

which completes the proof. □

2.3.1.3. *Necessary conditions for systems with principal part in the special form.* Now our aim is the proof of necessity in Theorems 1 and 2 stated in Introduction to the chapter.

Let $\mathfrak{A}^*(y, D_y)$ denote the operator conjugate of $\mathfrak{A}(y, D_y)$. We write the Green formula for the operators $\mathfrak{A}$ and $\mathfrak{A}^*$ in the form

$$\int_\Omega ((\mathfrak{A}\boldsymbol{u}, \boldsymbol{v}) - (\boldsymbol{u}, \mathfrak{A}^*\boldsymbol{v}))dy = \int_{\partial\Omega} ((Q\boldsymbol{u}, \boldsymbol{v}) - (\boldsymbol{u}, R\boldsymbol{v}))ds_y, \tag{2.3.18}$$

where Q and R are first order differential operators and $(,)$ is the inner product in $\mathbb{R}^m$.

By $G(x, y)$ we mean Green's matrix, i.e. the solution of the problem

$$\mathfrak{A}^*(y, D_y)G(x, y) = \delta(x - y)I \ \text{ for } \ (x, y) \in \Omega \times \Omega,$$

$$G(x, y) = 0 \ \text{ for } \ x \in \Omega, y \in \partial\Omega.$$

This and (2.3.18) imply

$$\boldsymbol{u}(x) = \int_\Omega G^*(x, y)\mathfrak{A}(y, D_y)\boldsymbol{u}(y)dy + \int_{\partial\Omega} [R(y, D_y)G(x, y)]^*\boldsymbol{u}(y)ds_y,$$

where $*$ means passage to the transposed matrix.

LEMMA 2.1. *Let the operator* $\mathfrak{A}(x, D_x)$ *be of the form*

$$\sum_{i,j=1}^{n} a_{ij}(x)\frac{\partial^2}{\partial x_i \partial x_j} - \sum_{j=1}^{n} \mathcal{A}_j(x)\frac{\partial}{\partial x_j} - \mathcal{A}_0(x),$$

where $((a_{ij}))$ *is a positive-definite matrix-valued function of order* n *in* $\overline{\Omega}$. *Let* $\mathcal{O}$ *be the origin,* $\mathcal{O} \in \Omega$. *If*

$$\boldsymbol{P}(x) = \boldsymbol{\alpha}_0 + \frac{1}{2}\sum_{i=1}^{n} \boldsymbol{\alpha}_i x_i + \sum_{i,j=1}^{n} \boldsymbol{\alpha}_{ij} x_i x_j,$$

where $\boldsymbol{\alpha}_{ij}, \boldsymbol{\alpha}_i, \boldsymbol{\alpha}_0$ *are vectors in* $\mathbb{R}^m, \boldsymbol{\alpha}_{ij} = \boldsymbol{\alpha}_{ji}, 1 \le i, j \le n$, *and if*

$$\mathfrak{A}(x, D_x)\boldsymbol{P}(x) = \boldsymbol{0} \quad \textit{for} \quad x = 0,$$

then there exists a vector-valued function $\boldsymbol{\Phi} \in [\mathrm{C}(\partial\Omega)]^m$ *such that the solution of the Dirichlet problem*

$$\mathfrak{A}(x, D_x)\boldsymbol{v} = \boldsymbol{0} \quad \textit{in} \quad \Omega, \quad \boldsymbol{v} = \boldsymbol{\Phi} \quad \textit{on} \ \partial\Omega$$

satisfies the relation

$$|\boldsymbol{v}(x) - \boldsymbol{P}(x)| = o(|x|^2) \quad \textit{as} \quad x \to 0.$$

PROOF. Let $\mathfrak{S}$ denote the subspace of $(\mathbb{R}^m)^N, N = (n+1)(n+2)/2$, spanned by vectors $\boldsymbol{\gamma} = (\boldsymbol{\gamma}_{ij}, \boldsymbol{\gamma}_i, \boldsymbol{\gamma}_0)$ such that $\boldsymbol{\gamma}_{ij} = \boldsymbol{\gamma}_{ji}$ and

$$\sum_{i,j=1}^{n} a_{ij}(0)\boldsymbol{\gamma}_{ij} + \sum_{i=1}^{n} \mathcal{A}_i(0)\boldsymbol{\gamma}_i + \mathcal{A}_0(0)\boldsymbol{\gamma}_0 = \boldsymbol{0}.$$

By $\mathfrak{K} = (\mathfrak{K}_{ij}, \mathfrak{K}_i, \mathfrak{K}_0)$ we denote the mapping

$$[\mathrm{C}(\partial\Omega)]^m \ni \boldsymbol{\Phi} \to \mathfrak{K}\boldsymbol{\Phi} \in \mathfrak{S}$$

defined by the equalities

$$(\mathfrak{K}\boldsymbol{\Phi})_{ij} = \int_{\partial\Omega} K_{ij}(0, y)\boldsymbol{\Phi}(y)ds_y,$$

$$(\mathfrak{K}\boldsymbol{\Phi})_{i} = \int_{\partial\Omega} K_{i}(0, y)\boldsymbol{\Phi}(y)ds_y,$$

$$(\mathfrak{K}\boldsymbol{\Phi})_{0} = \int_{\partial\Omega} K_{0}(0, y)\boldsymbol{\Phi}(y)ds_y,$$

where

$$K(x, y) = [R(y, D_y)G(x, y)]^*,$$

$$K_{ij}(x, y) = (\partial^2 K/\partial x_i \partial x_j)(x, y), \quad K_i(x, y) = (\partial K/\partial x_i)(x, y).$$

We show that $\mathfrak{K}$ maps onto $\mathfrak{S}$. Suppose the contrary assertion holds.

Let us introduce the vector $\boldsymbol{\beta} = (\boldsymbol{\beta}_{ij}, \boldsymbol{\beta}_{ij}, \boldsymbol{\beta}_0) \neq \boldsymbol{0}$ from $\mathfrak{S}$, orthogonal to the range $\mathfrak{R}(\mathfrak{K})$ of the operator $\mathfrak{K}$. Then for all $\boldsymbol{\Phi} \in [\mathrm{C}(\partial\Omega)]^m$ we have

$$\int_{\partial\Omega}\left(\boldsymbol{\Phi}(y), \sum_{i,j=1}^{n} K_{ij}^*(0, y)\boldsymbol{\beta}_{ij} + \sum_{i=1}^{n} K_i^*(0, y)\boldsymbol{\beta}_i + K^*(0, y)\boldsymbol{\beta}_0\right)ds_y = 0\,.$$

Consequently,

$$R(y, D_y)\left(\sum_{i,j=1}^{n} G_{ij}(0, y)\boldsymbol{\beta}_{ij} + \sum_{i=1}^{n} G_i(0, y)\boldsymbol{\beta}_i + G(0, y)\boldsymbol{\beta}_0\right) = \boldsymbol{0} \ \text{on} \ \partial\Omega,$$

where G_{ij} and G_i are the second and first order partial derivatives in x_i, x_j.

Since under the assumptions on the coefficients $a_{ij}, \mathcal{A}_i, \mathcal{A}_0$ of the operator $\mathfrak{A}^*(y, D_y)$ the uniqueness theorem for solution of the Cauchy problem is valid (see Hörmander [**Hör**], Ch. YIII, subsec. 8.9), then

$$\sum_{i,j=1}^{n} G_{ij}(0,y)\boldsymbol{\beta}_{ij} + \sum_{i=1}^{n} G_i(0,y)\boldsymbol{\beta}_i + G(0,y)\boldsymbol{\beta}_0 = \mathbf{0} \text{ on } \Omega\backslash\{0\}\,. \tag{2.3.19}$$

Without loss of generality one may assume that $\mathfrak{A}^*(0, D_y) = \Delta_y I$, where Δ_y is the Laplace operator.

Then

$$G(0,y) + \frac{|y|^{2-n}}{\omega_n(n-2)} I = O(|y|^{2-n+\varepsilon}) \text{ for } n > 2,$$

$$G(0,y) - 2\pi \log(c|y|)I = O(|y|^{\varepsilon}) \text{ for } n = 2, \tag{2.3.20}$$

$$G(0,y) - \big(|y|/2 + l(y)\big)I = O(|y|^{1+\varepsilon}) \text{ for } n = 1,$$

as $y \to 0$, where ω_n is the area of the unit sphere in $\mathbb{R}^n$, ε and c are positive constants and $l(y)$ is a linear function. The asymptotic representation may be differentiated twice.

So (2.3.19) can be rewritten in the form

$$\sum_{i,j=1}^{n} G_{ij}(0,y)\boldsymbol{\beta}_{ij} + \sum_{i=1}^{n} G_i(0,y)\boldsymbol{\beta}_i + G(0,y)\boldsymbol{\beta}_0 = \delta(y)\boldsymbol{V} \text{ on } \Omega, \tag{2.3.21}$$

where $\boldsymbol{V}$ is a constant vector. Since

$$\mathfrak{A}(x, D_x)G^*(x,y) = \delta(x-y)I, \quad (x,y) \in \Omega \times \Omega,$$

then for any constant vector $\boldsymbol{W}$

$$\left(\left(\sum_{i,j=1}^{n} a_{ij}(0)G^*_{ij}(0,y) + \sum_{i=1}^{n} \mathcal{A}_i(0)G^*_i(0,y) + \mathcal{A}_0(0)G^*(0,y)\right)\boldsymbol{W}, \boldsymbol{V}\right) = \delta(y)(\boldsymbol{W}, \boldsymbol{V}).$$

This implies that for all $y \in \Omega\backslash\{0\}$

$$\left(\sum_{i,j=1}^{n} G_{ij}(0,y)a_{ij}(0) + \sum_{i=1}^{n} G_i(0,y)\mathcal{A}^*_i(0) + G(0,y)\mathcal{A}^*_0(0)\right)\boldsymbol{V} = \delta(y)\boldsymbol{V}.$$

Comparing this equality with (2.3.21) we arrive at the identity

$$\begin{aligned}&\sum_{i,j=1}^{n} G_{ij}(0,y)(a_{ij}(0)\boldsymbol{V} - \boldsymbol{\beta}_{ij})\\ &+ \sum_{i=1}^{n} G_i(0,y)(\mathcal{A}^*_i(0)\boldsymbol{V} - \boldsymbol{\beta}_i) + G(0,y)(\mathcal{A}^*_0(0)\boldsymbol{V} - \boldsymbol{\beta}_0) = \mathbf{0} \text{ on } \Omega,\end{aligned} \tag{2.3.22}$$

which together with the asymptotic formula (2.3.20) yields

$$\sum_{i,j=1}^{n} \frac{\partial^2}{\partial y_i \partial_j} g(y)(a_{ij}(0)\boldsymbol{V} - \boldsymbol{\beta}_{ij}) = \mathbf{0} \text{ on } \mathbb{R}^n,$$

where g is the fundamental solution of the Laplace equation. Consequently, $\boldsymbol{\beta}_{ij} = a_{ij}(0)\boldsymbol{V}$. Further, from (2.3.20) and (2.3.22) we obtain the equality

$$\sum_{i=1}^{n} \frac{\partial}{\partial y_i} g(y)(\mathcal{A}_i^*(0)\boldsymbol{V} - \boldsymbol{\beta}_i) = \mathbf{0} \text{ on } \mathbb{R}^n,$$

from which it follows that $\boldsymbol{\beta}_i = \mathcal{A}_i^*(0)\boldsymbol{V}$. In a similar way we derive the equality $\boldsymbol{\beta}_0 = \mathcal{A}_0^*(0)\boldsymbol{V}$.

Substituting the expression for $\boldsymbol{\beta} \in \mathfrak{S}$ into

$$\sum_{i,j=1}^{n} a_{ij}(0)\boldsymbol{\beta}_{ij} + \sum_{i=1}^{n} \mathcal{A}_i(0)\boldsymbol{\beta}_i + \mathcal{A}_0(0)\boldsymbol{\beta}_0 = \mathbf{0},$$

we conclude that

$$\left(\sum_{i,j=1}^{n} a_{ij}^2(0)I + \sum_{i=1}^{n} \mathcal{A}_i(0)\mathcal{A}_i^*(0) + \mathcal{A}_0(0)\mathcal{A}_0^*(0) \right) \boldsymbol{v} = \mathbf{0}.$$

Hence

$$\sum_{i,j=1}^{n} |a_{ij}(0)\boldsymbol{v}|^2 + \sum_{i=1}^{n} |\mathcal{A}_i^*(0)\boldsymbol{v}|^2 + |\mathcal{A}_0^*(0)\boldsymbol{v}|^2 = \mathbf{0}$$

that is $\boldsymbol{\beta}_{ij} = \boldsymbol{\beta}_i = \boldsymbol{\beta}_0 = 0$. The contradiction we arrive at shows that $\mathfrak{R}(\mathfrak{K}) = \mathfrak{S}$.

Now we note that

$$\mathfrak{A}(x, D_x)\boldsymbol{P}(x) = \sum_{i,j=1}^{n} a_{ij}(0)\boldsymbol{\alpha}_{ij} + \sum_{i=1}^{n} \mathcal{A}_i(0)\boldsymbol{\alpha}_i + \mathcal{A}_0(0)\boldsymbol{\alpha}_0$$

for $x = 0$ and thus $(\boldsymbol{\alpha}_{ij}, \boldsymbol{\alpha}_i, \boldsymbol{\alpha}_0) \in \mathfrak{S}$. The proof of lemma is complete. □

THEOREM 2.11. *Let the system $\mathfrak{A}(x, D_x)\boldsymbol{u} = \mathbf{0}$ have the form*

$$\mathcal{A}(x) \sum_{j,k=1}^{n} a_{jk}(x) \frac{\partial^2 \boldsymbol{u}}{\partial x_j \partial x_k} - \sum_{j=1}^{n} \mathcal{A}_j(x) \frac{\partial \boldsymbol{u}}{\partial x_j} - \mathcal{A}_0(x)\boldsymbol{u} = \mathbf{0}, \tag{2.3.23}$$

where $\mathcal{A}$ and $((a_{jk}))$ are positive-definite matrices in $\overline{\Omega}$ of orders m and n, respectively, with elements from the space $\mathrm{C}^{2,\alpha}(\overline{\Omega})$.

If system (2.3.23) has no solutions $\boldsymbol{u} \in [\mathrm{C}^2(\Omega)]^m \cap [\mathrm{C}(\overline{\Omega})]^m$ with modulus attaining a strict local maximum at some point $x \in \Omega$, then:

(i) *for all $x \in \Omega$ and any $\boldsymbol{\xi}_j, \boldsymbol{\zeta} \in \mathbb{R}^m, j = 1, 2, \ldots, n$ with $(\boldsymbol{\xi}_j, \boldsymbol{\zeta}) = 0$ the inequality*

$$\sum_{j,k=1}^{n} a_{jk}(x)(\boldsymbol{\xi}_j, \boldsymbol{\zeta}) + \sum_{j=1}^{n} \left(\mathcal{A}^{-1}(x)\mathcal{A}_j(x)\boldsymbol{\xi}_j, \boldsymbol{\zeta}\right) + \left(\mathcal{A}^{-1}(x)\mathcal{A}_0(x)\boldsymbol{\zeta}, \boldsymbol{\zeta}\right) \geq 0$$

is valid.

PROOF. Let the origin be at a point $\mathcal{O} \in \Omega$. Take arbitrary $\boldsymbol{\zeta} \in \mathbb{R}^m \backslash \{\mathbf{0}\}, \boldsymbol{\xi}_j \in \mathbb{R}^m, \boldsymbol{\xi} = (\boldsymbol{\xi}_i, \ldots, \boldsymbol{\xi}_n)$, with $(\boldsymbol{\xi}_j, \boldsymbol{\zeta}) = 0, j = 1, 2, \ldots, n$, and let $\varepsilon = \text{const} > 0$.

We introduce the vector-valued function

$$\boldsymbol{P}(x, \boldsymbol{\xi}) = \begin{cases} \boldsymbol{\zeta} - \varepsilon\boldsymbol{\zeta}|x|^2 + \sum_{j=1}^{n} \boldsymbol{\xi}_j x_j + \boldsymbol{c}_1 \left| \sum_{j=1}^{n} \boldsymbol{\xi}_j x_j \right|^2 & \text{for } \boldsymbol{\xi} \neq \mathbf{0}, \\ \boldsymbol{\zeta} - \boldsymbol{c}_2|x|^2 & \text{for } \boldsymbol{\xi} = \mathbf{0}, \end{cases} \tag{2.3.24}$$

where the vectors $\boldsymbol{c}_1, \boldsymbol{c}_2 \in \mathbb{R}^m$ are defined by the equation

$$\mathfrak{A}(x, D_x)\boldsymbol{P}(x, \boldsymbol{\xi})|_{x=0} = \mathbf{0},$$

which yields

$$\boldsymbol{c}_1 = \frac{1}{2}\frac{\mathcal{A}^{-1}(0)\mathcal{A}_0(0)\boldsymbol{\zeta} + \mathcal{A}^{-1}(0)\sum_{j=1}^n \mathcal{A}_j(0)\boldsymbol{\xi}_j + 2\varepsilon\boldsymbol{\zeta}\sum_{j=1}^n a_{jj}(0)}{\sum_{j,k=1}^n a_{jk}(0)(\boldsymbol{\xi}_j, \boldsymbol{\xi}_k)}, \tag{2.3.25}$$

$$\boldsymbol{c}_2 = -\frac{1}{2}\frac{\mathcal{A}^{-1}(0)\mathcal{A}_0(0)\boldsymbol{\zeta}}{\sum_{j=1}^n a_{jj}(0)}. \tag{2.3.26}$$

From (2.3.24)-(2.3.26) we obtain

$$\begin{aligned} |\boldsymbol{P}(x, \boldsymbol{\xi})|^2 &= |\boldsymbol{\zeta}|^2(1 - 2\varepsilon|x|^2) + (1 + 2(\boldsymbol{\zeta}, \boldsymbol{c}_1))\left|\sum_{j=1}^n \boldsymbol{\xi}_j x_j\right|^2 + o(|x|^2) \\ &= (1 - 2\varepsilon|x|^2)|\boldsymbol{\zeta}|^2 + \left|\sum_{j=1}^n \boldsymbol{\xi}_j x_j\right|^2 \\ &\times\left\{\frac{2\varepsilon|\boldsymbol{\zeta}|^2\sum_{j=1}^n a_{jj}(0) + \sum_{i,j=1}^n a_{ij}(0)(\boldsymbol{\xi}_i, \boldsymbol{\xi}_j)}{\sum_{i,j=1}^n a_{ij}(0)(\boldsymbol{\xi}_i, \boldsymbol{\xi}_j)}\right. \\ &\left.+\frac{\left(\mathcal{A}^{-1}(0)\sum_{j=1}^n \mathcal{A}_j(0)\boldsymbol{\xi}_j, \boldsymbol{\zeta}\right) + (\mathcal{A}^{-1}(0)\mathcal{A}_0(0)\boldsymbol{\zeta}, \boldsymbol{\zeta})}{\sum_{i,j=1}^n a_{ij}(0)(\boldsymbol{\xi}_i, \boldsymbol{\xi}_j)}\right\} \\ &+o(|x|^2) \quad \text{as } x \to 0, \end{aligned} \tag{2.3.27}$$

where $\boldsymbol{\xi} \neq \mathbf{0}$ and

$$\begin{aligned} |\boldsymbol{P}(x, \mathbf{0})|^2 &= |\boldsymbol{\zeta}|^2 - 2(\boldsymbol{\zeta}, \boldsymbol{c}_2)|x|^2 + o(|x|^2) \\ &= |\boldsymbol{\zeta}|^2 + \frac{(\mathcal{A}^{-1}(0)\mathcal{A}_0(0)\boldsymbol{\zeta}, \boldsymbol{\zeta})}{\sum_{j=1}^n a_{jj}(0)}|x|^2 + o(|x|^2). \end{aligned} \tag{2.3.28}$$

Suppose for some $\boldsymbol{\zeta} \in \mathbb{R}^m\backslash\{\mathbf{0}\}, \boldsymbol{\xi}_j \in \mathbb{R}^m$ with $(\boldsymbol{\xi}_j, \boldsymbol{\zeta}) = 0, j = 1, \ldots, n$, we have

$$\sum_{j,k=1}^n a_{jk}(0)(\boldsymbol{\xi}_j, \boldsymbol{\xi}_k) + \sum_{j=1}^n \left(\mathcal{A}^{-1}(0)\mathcal{A}_j(0)\boldsymbol{\xi}_j, \boldsymbol{\zeta}\right) + \left(\mathcal{A}^{-1}(0)\mathcal{A}_0(0)\boldsymbol{\zeta}, \boldsymbol{\zeta}\right) < 0. \tag{2.3.29}$$

Then from (2.3.27) it follows for small enough ε that

$$\|\boldsymbol{P}(\cdot, \boldsymbol{\xi})\|_{[\mathrm{C}(\partial\mathbb{B}_r)]^m} \leq ((1 - 2\varepsilon r^2)|\boldsymbol{\zeta}|^2 + o(r^2))^{1/2} = (1 - \varepsilon r^2)|\boldsymbol{\zeta}| + o(r^2), \tag{2.3.30}$$

where $\boldsymbol{\xi} \neq \mathbf{0}$ and $r \to 0$. In a similar way from (2.3.28) we get

$$\|\boldsymbol{P}(\cdot, \mathbf{0})\|_{[\mathrm{C}(\partial\mathbb{B}_r)]^m} \leq ((1 - 2\eta r^2)|\boldsymbol{\zeta}|^2 + o(r^2))^{1/2} = (1 - \eta r^2)|\boldsymbol{\zeta}| + o(r^2), \tag{2.3.31}$$

where $r \to 0$ and

$$\eta = -\frac{1}{2}\frac{(\mathcal{A}^{-1}(0)\mathcal{A}_0(0)\boldsymbol{\zeta}, \boldsymbol{\zeta})}{|\boldsymbol{\zeta}|^2\sum_{j=1}^n a_{jj}(0)} > 0.$$

Thus from (2.3.30) and (2.3.31) it follows that

$$\|\boldsymbol{P}(\cdot, \boldsymbol{\xi})\|_{[\mathrm{C}(\partial\mathbb{B}_r)]^m} \leq (1 - \mu r^2)|\boldsymbol{\zeta}| + o(r^2) \tag{2.3.32}$$

as $r \to 0$, where $\mu = \varepsilon$ for $\boldsymbol{\xi} \neq \mathbf{0}$ and $\mu = \eta$ for $\boldsymbol{\xi} = \mathbf{0}$.

Lemma 2.1 can be applied to system (2.3.23) since $\mathcal{A}^{-1} \in [\mathrm{C}^{2,\alpha}(\overline{\Omega})]^{m\times m}$. Let $\boldsymbol{v}$ be the vector-valued function defined in Lemma 2.1. According to Lemma 2.1 and (2.3.32) we have

$$\|\boldsymbol{v}\|_{[\mathrm{C}(\partial\mathbb{B}_r)]^m} \leq \|\boldsymbol{v} - \boldsymbol{P}\|_{[\mathrm{C}(\partial\mathbb{B}_r)]^m} + \|\boldsymbol{P}\|_{[\mathrm{C}(\partial\mathbb{B}_r)]^m} = (1-\mu r^2)|\boldsymbol{\zeta}| + o(r^2)$$

as $r \to 0$. Taking into account that $|\boldsymbol{v}(0)| = |\boldsymbol{P}(0,\boldsymbol{\xi})| = |\boldsymbol{\zeta}|$ we get

$$|\boldsymbol{v}(0)| > \|\boldsymbol{v}|_{\partial\mathbb{B}_r}\|_{[\mathrm{C}(\partial\mathbb{B}_r)]^m}$$

for all $r \leq \rho$, where ρ is some sufficiently small positive constant.

Thus if (2.3.29) is valid then one can construct a solution of system (2.3.23) which has a strict local maximum in Ω. This contradicts the assumption of the theorem and proves the necessity of condition (i).

For $\boldsymbol{\zeta} = \mathbf{0}$ the inequality in (i) takes the form

$$\sum_{j,k=1}^{n} a_{jk}(x)(\boldsymbol{\xi}_j, \boldsymbol{\xi}_k) \geq 0,$$

and is valid by the positive-definiteness of the matrix $((a_{jk}(x)))$ in $\overline{\Omega}$. □

The next assertion immediately follows from Corollary 2.1 and Theorem 2.11.

THEOREM 2.12. *Let the classical maximum modulus principle*

$$\|\boldsymbol{u}\|_{[\mathrm{C}(\overline{\omega})]^m} \leq \|\boldsymbol{u}|_{\partial\omega}\|_{[\mathrm{C}(\partial\omega)]^m}$$

be valid for solutions of the system $\mathfrak{A}(x, D_x)\boldsymbol{u} = \mathbf{0}$ *in an arbitrary domain* $\omega \subset \Omega$ *with the boundary of the class* $\mathrm{C}^{2,\alpha}$. *Then:*

(i) *for all* $x \in \overline{\Omega}$ *the equality*

$$\mathcal{A}_{jk}(x) = \mathcal{A}(x) a_{jk}(x), \quad 1 \leq j,k \leq n,$$

is valid, where $\mathcal{A}$ *and* $((a_{jk}))$ *are positive-definite matrices in* $\overline{\Omega}$ *of orders* m *and* n, *respectively;*

(ii) *for all* $x \in \Omega$ *and any* $\boldsymbol{\xi}_j, \boldsymbol{\zeta} \in \mathbb{R}^m, j = 1, 2, \dots, n$ *with* $(\boldsymbol{\xi}_j, \boldsymbol{\zeta}) = 0$ *the inequality*

$$\sum_{j,k=1}^{n} a_{jk}(x)(\boldsymbol{\xi}_j, \boldsymbol{\xi}_k) + \sum_{j=1}^{n} \left(\mathcal{A}^{-1}(x)\mathcal{A}_j(x)\boldsymbol{\xi}_j, \boldsymbol{\zeta}\right) + \left(\mathcal{A}^{-1}(x)\mathcal{A}_0(x)\boldsymbol{\zeta}, \boldsymbol{\zeta}\right) \geq 0$$

is valid.

2.3.1.4. *A sufficient condition for systems with scalar principal part.*

THEOREM 2.13. *Let the system* $\mathfrak{A}(x, D_x)\boldsymbol{u} = \mathbf{0}$ *be of the form*

$$\sum_{j,k=1}^{n} a_{jk}(x)\frac{\partial^2 \boldsymbol{u}}{\partial x_j \partial x_k} - \sum_{j=1}^{n} \mathcal{A}_j(x)\frac{\partial \boldsymbol{u}}{\partial x_j} - \mathcal{A}_0(x)\boldsymbol{u} = \mathbf{0}, \tag{2.3.33}$$

where $((a_{jk}))$ *is a positive-definite matrix in* $\overline{\Omega}$ *of order* n *and let*

(i) *for all* $x \in \Omega$ *and any* $\boldsymbol{\xi}_j, \boldsymbol{\zeta} \in \mathbb{R}^m, j = 1, 2, \dots, n$ *with* $(\boldsymbol{\xi}_j, \boldsymbol{\zeta}) = 0$ *the inequality*

$$\sum_{j,k=1}^{n} a_{jk}(x)(\boldsymbol{\xi}_j, \boldsymbol{\xi}_k) + \sum_{j=1}^{n} \left(\mathcal{A}^{-1}(x)\mathcal{A}_j(x)\boldsymbol{\xi}_j, \boldsymbol{\zeta}\right) + \left(\mathcal{A}^{-1}(x)\mathcal{A}_0(x)\boldsymbol{\zeta}, \boldsymbol{\zeta}\right) \geq 0 \tag{2.3.34}$$

be valid. Then the classical maximum modulus principle

$$\|\boldsymbol{u}\|_{[\mathrm{C}(\overline{\Omega})]^m} \leq \|\boldsymbol{u}|_{\partial\Omega}\|_{[\mathrm{C}(\partial\Omega)]^m}$$

holds for solutions $\boldsymbol{u} \in [\mathrm{C}^2(\Omega)]^m \cap [\mathrm{C}(\overline{\Omega})]^m$ *of the system* $\mathfrak{A}(x, \partial x)\boldsymbol{u} = \mathbf{0}$ *and* $|\boldsymbol{u}|$ *may not attain a strict local maximum in* Ω.

PROOF. Consider the vector-valued function

$$\boldsymbol{v}(x) = \left(e^{\lambda\rho^2} - e^{\lambda|x-y|^2}\right)^{-1} \boldsymbol{u}(x), \tag{2.3.35}$$

where $\boldsymbol{u}$ is a solution of system (2.3.33) from the class $[\mathrm{C}^2(\Omega)]^m \cap [\mathrm{C}(\overline{\Omega})]^m$, ρ is the radius of a ball $\mathbb{B}_\rho(y)$ with center at $y \in \mathbb{R}^n \backslash \overline{\Omega}$ such that $\overline{\Omega} \subset \mathbb{B}_\rho(y)$ and $\lambda = \text{const} > 0$.

The straightforward calculation yields

$$\begin{aligned}
\mathfrak{A}(x, D_x)\boldsymbol{u} &= \left(e^{\lambda\rho^2} - e^{\lambda|x-y|^2}\right) \mathfrak{A}(x, D_x)\boldsymbol{v} - 2\lambda e^{\lambda|x-y|^2} \\
&\times \Bigg\{ \Bigg[I\sum_{j=1}^n a_{jj}(x) + 2\lambda I \sum_{j,k=1}^n a_{jk}(x)(x_j - y_j)(x_k - y_k) - \sum_{j=1}^n \mathcal{A}_j(x)(x_j - y_j) \Bigg] \boldsymbol{v} \\
&+ \sum_{j,k=1}^n a_{jk}(x) \Bigg[(x_j - y_j)\frac{\partial \boldsymbol{v}}{\partial x_k} + (x_k - y_k)\frac{\partial \boldsymbol{v}}{\partial x_j} \Bigg] \Bigg\},
\end{aligned} \tag{2.3.36}$$

where I is the identity matrix of order m.

Let

$$\mu = \min\left\{ \sum_{j,k=1}^n a_{jk}(x)\eta_j\eta_k : x \in \overline{\Omega}, \boldsymbol{\eta} = (\eta, \ldots, \eta) \in \mathbb{R}^n, |\boldsymbol{\eta}| = 1 \right\},$$

$$d_{max} = \max\left\{|x - y| : x \in \overline{\Omega}\right\}, \quad d_{min} = \min\left\{|x - y| : x \in \overline{\Omega}\right\},$$

and let λ_0 be a positive constant such that

$$n\mu + 2\lambda_0\mu d_{min}^2 \geq d_{max} \max\left\{ \sum_{j=1}^n \|\mathcal{A}_j(x)\|_s : \ x \in \overline{\Omega} \right\},$$

where $\|\cdot\|_s$ is the spectral norm of the matrix.

For all $x \in \Omega$ and $\boldsymbol{\sigma} \in \mathbb{R}^m \backslash \{\mathbf{0}\}$ we have

$$\begin{aligned}
&\left(\sum_{j=1}^n a_{jj}(x) + 2\lambda_0 \sum_{j,k=1}^n a_{jk}(x)(x_j - y_j)(x_k - y_k) \right) |\boldsymbol{\sigma}|^2 \\
&\qquad - \sum_{j=1}^n (\mathcal{A}_j(x)(x_j - y_j)\boldsymbol{\sigma}, \boldsymbol{\sigma}) > 0 \,.
\end{aligned} \tag{2.3.37}$$

Suppose $|\boldsymbol{v}|$ attains its global maximum at a point $x_0 \in \Omega$. One may assume, of course, that $|\boldsymbol{v}(x_0)| \neq 0$. Then

$$\begin{aligned}
&\frac{\partial}{\partial x_j}|\boldsymbol{v}|^2 \bigg|_{x=x_0} = 2\left(\frac{\partial \boldsymbol{v}}{\partial x_j}, \boldsymbol{v}\right)\bigg|_{x=x_0}, \\
&\sum_{j,k=1}^n a_{jk}(x)\frac{\partial^2 |\boldsymbol{v}|^2}{\partial x_j \partial x_k}\bigg|_{x=x_0} \leq 0.
\end{aligned} \tag{2.3.38}$$

Take the scalar product of (2.3.36) with $\boldsymbol{v}$ and use the identity

$$\frac{1}{2}\frac{\partial^2|\boldsymbol{v}|^2}{\partial x_j \partial x_k} = \left(\frac{\partial^2 \boldsymbol{v}}{\partial x_j \partial x_k}, \boldsymbol{v}\right) + \left(\frac{\partial \boldsymbol{v}}{\partial x_j}, \frac{\partial \boldsymbol{v}}{\partial x_k}\right).$$

As a result we obtain

$$\begin{aligned}
&\frac{1}{2}\sum_{j,k=1}^{n} a_{jk}(x)\frac{\partial^2|\boldsymbol{v}|^2}{\partial x_j \partial x_k}\bigg|_{x=x_0} \\
&= \left\{\sum_{j,k=1}^{n} a_{jk}(x)\left(\frac{\partial \boldsymbol{v}}{\partial x_j}, \frac{\partial \boldsymbol{v}}{\partial x_k}\right) + \sum_{j,k=1}^{n}\left(\mathcal{A}_j(x)\frac{\partial \boldsymbol{v}}{\partial x_j}, \boldsymbol{v}\right) + (\mathcal{A}_0(x)\boldsymbol{v}, \boldsymbol{v})\right\}\Bigg|_{x=x_0} \\
&+2\lambda e^{\lambda|x_0-y|^2}\left(e^{\lambda\rho^2} - e^{\lambda|x_0-y|^2}\right)^{-1}\left\{\left[\sum_{j=1}^{n} a_{jj}(x)\right.\right. \\
&\left.\left.+2\lambda\sum_{j,k=1}^{n} a_{jk}(x)(x_j-y_j)(x_k-y_k)\right]|\boldsymbol{v}|^2 - \sum_{j=1}^{n}(\mathcal{A}_j(x)(x_j-y_j)\boldsymbol{v}, \boldsymbol{v})\right\}\Bigg|_{x=x_0}.
\end{aligned}$$

This together with (2.3.34) and (2.3.37) yields the inequality

$$\sum_{j,k=1}^{n} a_{jk}(x)\frac{\partial^2|\boldsymbol{v}|^2}{\partial x_j \partial x_k}\bigg|_{x=x_0} > 0,$$

for $\lambda \geq \lambda_0$, which contradicts (2.3.38).

Thus if $\lambda \geq \lambda_0$ and if condition (i) is valid, then $|\boldsymbol{v}|$ may not attain its global maximum at some point $x \in \Omega$ and hence

$$\|\boldsymbol{v}\|_{[\mathrm{C}(\overline{\Omega})]^m} \leq \|\boldsymbol{v}|_{\partial\Omega}\|_{[\mathrm{C}(\partial\Omega)]^m}.$$

This and (2.3.35) imply

$$\max_{x\in\overline{\Omega}}\left(e^{\lambda\rho^2} - e^{\lambda|x-y|^2}\right)^{-1}|\boldsymbol{u}(x)| \leq \max_{x\in\partial\Omega}\left(e^{\lambda\rho^2} - e^{\lambda|x-y|^2}\right)^{-1}|\boldsymbol{u}(x)|,$$

where $\lambda \geq \lambda_0$. Consequently, for $\lambda \geq \lambda_0$ we have

$$\left(e^{\lambda\rho^2} - e^{\lambda d_{max}^2}\right)\|\boldsymbol{u}\|_{[\mathrm{C}(\overline{\Omega})]^m} \leq \left(e^{\lambda\rho^2} - e^{\lambda d_{min}^2}\right)\|\boldsymbol{u}|_{\partial\Omega}\|_{[\mathrm{C}(\partial\Omega)]^m},$$

that is

$$\|\boldsymbol{u}\|_{[\mathrm{C}(\overline{\Omega})]^m} \leq \frac{1-e^{-\lambda(\rho^2-d_{min}^2)}}{1-e^{-\lambda(\rho^2-d_{max}^2)}}\|\boldsymbol{u}|_{\partial\Omega}\|_{[\mathrm{C}(\partial\Omega)]^m}.$$

Using the arbitrariness of $\lambda \geq \lambda_0$ in the last inequality we find that the best constant in it is the unity.

Let $x \in \Omega$ and let $\overline{\mathbb{B}_r(x)} \subset \Omega$. Then the maximum modulus principle entails the inequality

$$\|\boldsymbol{u}\|_{[\mathrm{C}(\overline{\mathbb{B}_r(x)})]^m} \leq \|\boldsymbol{u}|_{\partial\mathbb{B}_r(x)}\|_{[\mathrm{C}(\partial\mathbb{B}_r(x))]^m},$$

where $\boldsymbol{u}$ is a solution of (2.3.33) from the class $[\mathrm{C}^2(\Omega)]^m \cap [\mathrm{C}(\overline{\Omega})]^m$. This implies that $|\boldsymbol{u}|$ may not attain its strict local maximum in Ω. $\square$

2.3.1.5. *Necessary and sufficient conditions for validity of the maximum modulus principle.* The next two theorems directly follow from Theorems 2.11, 2.12, 2.13.

THEOREM 2.14. *Let the system* $\mathfrak{A}(x, D_x)\boldsymbol{u} = \boldsymbol{0}$ *be of the form*

$$\mathcal{A}(x) \sum_{j,k=1}^{n} a_{jk}(x) \frac{\partial^2 \boldsymbol{u}}{\partial x_j \partial x_k} - \sum_{j=1}^{n} \mathcal{A}_j(x) \frac{\partial \boldsymbol{u}}{\partial x_j} - \mathcal{A}_0(x)\boldsymbol{u} = \boldsymbol{0},$$

where $\mathcal{A}$ *and* $((a_{jk}))$ *are positive-definite matrices in* $\overline{\Omega}$ *of orders* m *and* n, *respectively, with elements from the space* $\mathrm{C}^{2,\alpha}(\overline{\Omega})$.

The system $\mathfrak{A}(x, D_x)\boldsymbol{u} = \boldsymbol{0}$ *has no solutions* $\boldsymbol{u} \in [\mathrm{C}^2(\Omega)]^m \cap [\mathrm{C}(\overline{\Omega})]^m$ *with modulus attaining its strict local maximum at some point* $x \in \Omega$ *if and only if:*

(i) *for all* $x \in \Omega$ *and any* $\boldsymbol{\xi}_j$, $\boldsymbol{\zeta} \in \mathbb{R}^m$, $j = 1, \dots, n$, *with* $(\boldsymbol{\xi}_j, \boldsymbol{\zeta}) = 0$ *the inequality*

$$\sum_{j,k=1}^{n} a_{jk}(x)(\boldsymbol{\xi}_j, \boldsymbol{\xi}_k) + \sum_{j=1}^{n} (\mathcal{A}^{-1}(x)\mathcal{A}_j(x)\boldsymbol{\xi}_j, \boldsymbol{\zeta}) + (\mathcal{A}^{-1}(x)\mathcal{A}_0(x)\boldsymbol{\zeta}, \boldsymbol{\zeta}) \geq 0$$

is valid.

THEOREM 2.15. *The classical maximum modulus principle*

$$\|\boldsymbol{u}\|_{[\mathrm{C}(\overline{\omega})]^m} \leq \|\boldsymbol{u}|_{\partial\omega}\|_{[\mathrm{C}(\partial\omega)]^m},$$

holds for solutions of the system $\mathfrak{A}(x, D_x)\boldsymbol{u} = \boldsymbol{0}$ *in an arbitrary domain* $\omega \subset \Omega$ *with boundary from the class* $\mathrm{C}^{2,\alpha}$ *if and only if:*

(i) *for all* $x \in \overline{\Omega}$ *the equalities hold*

$$\mathcal{A}_{jk}(x) = \mathcal{A}(x) a_{jk}(x), \;\; 1 \leq j, \; k \leq n,$$

where $\mathcal{A}$ *and* $((a_{jk}))$ *are positive-definite matrices in* $\overline{\Omega}$ *of orders* m *and* n, *respectively;*

(ii) *for all* $x \in \Omega$ *and any* $\boldsymbol{\xi}_j$, $\boldsymbol{\zeta} \in \mathbb{R}^m$, $j = 1, \dots, n$, *with* $(\boldsymbol{\xi}_j, \boldsymbol{\zeta}) = 0$ *the inequality*

$$\sum_{j,k=1}^{n} a_{jk}(x)(\boldsymbol{\xi}_j, \boldsymbol{\xi}_k) + \sum_{j=1}^{n} (\mathcal{A}^{-1}(x)\mathcal{A}_j(x)\boldsymbol{\xi}_j, \boldsymbol{\zeta}) + (\mathcal{A}^{-1}(x)\mathcal{A}_0(x)\boldsymbol{\zeta}, \boldsymbol{\zeta}) \geq 0$$

is valid.

REMARK 2.4. Consider condition (ii) in Theorem 2.15 in the case $\mathcal{A}_0 = 0$. Clearly the inequality

$$\sum_{j,k=1}^{n} a_{jk}(x)(\boldsymbol{\xi}_j, \boldsymbol{\xi}_k) + \sum_{j=1}^{n} (\mathcal{A}^{-1}(x)\mathcal{A}_j(x)\boldsymbol{\xi}_j, \boldsymbol{\zeta}) \geq 0$$

is valid for all $\boldsymbol{\xi}_j$, $\boldsymbol{\zeta} \in \mathbb{R}^m$, $j = 1, \dots, n$, with $(\boldsymbol{\xi}_j, \boldsymbol{\zeta}) = 0$ if and only if

$$(\mathcal{A}^{-1}(x)\mathcal{A}_j(x)\boldsymbol{\xi}_j, \boldsymbol{\zeta}) = 0.$$

Let $\boldsymbol{e}_1, \dots, \boldsymbol{e}_m$ be an orthonormal basis in $\mathbb{R}^m$. Putting $\boldsymbol{\xi}_j = \boldsymbol{e}_i, \boldsymbol{\zeta} = \boldsymbol{e}_k, i \neq k, 1 \leq i, k \leq m$, at the last equality, we conclude that the matrix-valued functions $B_j = \mathcal{A}^{-1}\mathcal{A}_j$ are diagonal. Thus the condition (ii) implies

$$B_j^{(1,1)}(x)\xi_j^{(1)}\zeta^{(1)} + \cdots + B_j^{(m,m)}(x)\xi_j^{(m)}\zeta^{(m)} = 0,$$

$$\xi_j^{(1)}\zeta^{(1)} + \cdots + \xi_j^{(m)}\zeta^{(m)} = 0,$$

where $\xi_j^{(k)} = (\boldsymbol{\xi}_j, \boldsymbol{e}_k), \zeta^{(k)} = (\boldsymbol{\zeta}, \boldsymbol{e}_k), j = 1, \ldots, n$. Setting here $\xi_j^{(k)} = \zeta^{(k)} = 0$ for $i \neq m, k \neq i, k \neq m$, we find $B_j^{(i,i)}(x) = B_j^{(m,m)}(x)$ for all $i = 1, \ldots, m-1$. Consequently,

$$B_j(x) = \mathcal{A}^{-1}(x)\mathcal{A}_j(x) = a_j(x)I,$$

where

$$a_j(x) = B_j^{(1,1)}(x) = \cdots = B_j^{(m,m)}(x), \quad j = 1, \ldots, n.$$

Thus $\mathcal{A}_j(x) = \mathcal{A}(x)a_j(x)$.

The above remark leads directly to the following statements of Theorems 2.14 and 2.15 for $\mathcal{A}_0 = 0$.

COROLLARY 2.3. *Let the system* $\mathfrak{A}(x, D_x)\boldsymbol{u} = \boldsymbol{0}$ *be of the form*

$$\mathcal{A}(x) \sum_{j,k=1}^{n} a_{jk}(x) \frac{\partial^2 \boldsymbol{u}}{\partial x_j \partial x_k} - \sum_{j=1}^{n} \mathcal{A}_j(x) \frac{\partial \boldsymbol{u}}{\partial x_j} = \boldsymbol{0},$$

where $\mathcal{A}$ *and* $((a_{jk}))$ *are positive-definite real matrix-valued functions in* $\overline{\Omega}$ *of orders* m *and* n, *respectively, with elements from the space* $\mathrm{C}^{2,\alpha}(\overline{\Omega})$.

The system $\mathfrak{A}(x, D_x)\boldsymbol{u} = \boldsymbol{0}$ *has no solutions* $\boldsymbol{u} \in [\mathrm{C}^2(\Omega)]^m \cap [\mathrm{C}(\overline{\Omega})]^m$ *with modulus attaining its strict local maximum at some point* $x \in \Omega$ *if and only if*

$$\mathcal{A}_j(x) = \mathcal{A}(x)a_j(x), \quad 1 \leq j, \ k \leq n.$$

Here a_j *are scalar functions with the same smoothness as* $\mathcal{A}_j$.

COROLLARY 2.4. *The classical maximum modulus principle*

$$\|\boldsymbol{u}\|_{[\mathrm{C}(\overline{\omega})]^m} \leq \|\boldsymbol{u}|_{\partial\omega}\|_{[\mathrm{C}(\partial\omega)]^m}$$

holds for solutions of the system

$$\sum_{j,k=1}^{n} \mathcal{A}_{jk}(x) \frac{\partial^2 \boldsymbol{u}}{\partial x_j \partial x_k} - \sum_{j=1}^{n} \mathcal{A}_j(x) \frac{\partial \boldsymbol{u}}{\partial x_j} = \boldsymbol{0}$$

in an arbitrary domain $\omega \subset \Omega$ *with boundary from the class* $\mathrm{C}^{2,\alpha}$ *if and only if:*

$$\mathcal{A}_{jk}(x) = \mathcal{A}(x)a_{jk}(x), \quad \mathcal{A}_j(x) = \mathcal{A}(x)a_j(x), \quad 1 \leq j, \ k \leq n.$$

Here $\mathcal{A}$ *and* $((a_{jk}))$ *are positive-definite matrix-valued functions in* $\overline{\Omega}$ *of orders* m *and* n, *respectively, and* a_j *are scalar functions with the same smoothness as* $\mathcal{A}_j$.

REMARK 2.5. Minimizing the left-hand side of the inequality in condition (ii) of Theorem 2.15 over $\boldsymbol{\xi} = (\boldsymbol{\xi}_1, \ldots, \boldsymbol{\xi}_n), \boldsymbol{\xi}_j \in \mathbb{R}^m$, for a fixed $\boldsymbol{\zeta} \in \mathbb{R}^m$ with $(\boldsymbol{\xi}_j, \boldsymbol{\zeta}) = 0, j = 1, \ldots, n$, one can reformulate that condition.

This argument was used in C. Miranda [**Mi3**], where the maximum modulus principle was studied for elliptic systems with a scalar principal part.

One may assume that $\boldsymbol{\zeta} \in \mathbb{R}^m \backslash \{\boldsymbol{0}\}$ since the validity of the inequality

$$\sum_{j,k=1}^{n} a_{jk}(x)(\boldsymbol{\xi}_j, \boldsymbol{\xi}_k) \geq 0$$

for all $\boldsymbol{\xi}_j \in \mathbb{R}^m$ is guaranteed by the positive-definiteness of the matrix $((a_{jk}))$.

Let

$$\mathfrak{F}_{\boldsymbol{\zeta}}(\boldsymbol{\xi}_1, \ldots, \boldsymbol{\xi}_n) = \sum_{j,k=1}^{n} a_{jk}(x)(\boldsymbol{\xi}_j, \boldsymbol{\xi}_k) + \sum_{j=1}^{n}(M_j(x)\boldsymbol{\xi}_j, \boldsymbol{\zeta}) + (M_0(x)\boldsymbol{\zeta}, \boldsymbol{\zeta}),$$

where $M_j(x) = \mathcal{A}^{-1}\mathcal{A}_j(x), 0 \le j \le n$. Let $\xi_j^{(k)}$ and $\zeta^{(k)}$ denote the components of the vectors $\boldsymbol{\xi}_j$ and $\boldsymbol{\zeta}$, respectively. Then

$$\mathfrak{F}_{\zeta}(\boldsymbol{\xi}_1 \dots, \boldsymbol{\xi}_n) = \sum_{j,k=1}^{n} \sum_{i=1}^{m} a_{jk}(x)\xi_j^{(i)}\xi_k^{(i)} + \sum_{j=1}^{n} \sum_{i,k=1}^{m} M_j^{(i,k)}(x)\xi_j^{(k)}\zeta^{(i)} + \sum_{i,k=1}^{m} M_0^{(i,k)}(x)\zeta^{(k)}\zeta^{(i)}.$$

At a point of the constraint extremum of the function $\mathfrak{F}_{\zeta}(\boldsymbol{\xi}_1, \dots, \boldsymbol{\xi}_n)$ one has

$$\frac{\partial}{\partial \xi_k^{(i)}}\left(\mathfrak{F}_{\zeta}(\boldsymbol{\xi}_1 \dots, \boldsymbol{\xi}_n) - \sum_{j=1}^{n}\sum_{i=1}^{m} \lambda_j \xi_j^{(i)}\zeta^{(i)}\right) = 2\sum_{j=1}^{n} a_{jk}(x)\xi_j^{(i)}$$

$$+\sum_{j=1}^{m} M_k^{(j,i)}(x)\zeta^{(j)} - \lambda_k\zeta^{(i)} = 0, \tag{2.3.39}$$

where $k = 1, \dots, n,\ i = 1, \dots, m$ and the following constraint relations are valid

$$\sum_{l=1}^{m} \xi_j^{(l)}\zeta^{(l)} = 0, \quad j = 1, \dots, n.$$

Multiplying (2.3.39) by $\zeta^{(i)}$ and summing up over i from 1 to m we obtain

$$\lambda_k = |\boldsymbol{\zeta}|^{-2}(M_k^*(x)\boldsymbol{\zeta}, \boldsymbol{\zeta}),$$

where $*$ means passage to the transposed matrix. Consequently, condition (2.3.39), which defines $\boldsymbol{\xi}_j$, can be rewritten in the form

$$2\sum_{j=1}^{n} a_{jk}(x)\xi_j^{(i)} + \sum_{j=1}^{m} M_k^{(j,i)}(x)\zeta^{(j)} - |\boldsymbol{\zeta}|^{-2}(M_k^*(x)\boldsymbol{\zeta}, \boldsymbol{\zeta})\zeta^{(i)} = 0.$$

Taking into account the symmetry of the matrix $((a_{jk}(x)))$, from this we find

$$\boldsymbol{\xi}_j = \frac{1}{2}\sum_{k=1}^{n} b_{jk}(x)\left[|\boldsymbol{\zeta}|^{-2}(M_k^*(x)\boldsymbol{\zeta}, \boldsymbol{\zeta})\boldsymbol{\zeta} - M_k^*(x)\boldsymbol{\zeta}\right], \tag{2.3.40}$$

where $((b_{jk}(x))$ is the inverse matrix of $((a_{jk}(x))$.

The function $\mathfrak{F}_{\zeta}(\boldsymbol{\xi}_1 \dots, \boldsymbol{\xi}_n)$ attains its constraint minimum at the vectors (2.3.40) because of the positive-definiteness of the matrix $((a_{jk}(x))$. Calculating the values of $\mathfrak{F}_{\zeta}(\boldsymbol{\xi}_1 \dots, \boldsymbol{\xi}_n)$ at the vectors (2.3.40) we obtain

$$\min\left\{\mathfrak{F}_{\zeta}(\boldsymbol{\xi}_1, \dots, \boldsymbol{\xi}_n) : \boldsymbol{\xi}_1, \dots, \boldsymbol{\xi}_n \in \mathbb{R}^m,\ (\boldsymbol{\xi}_1, \boldsymbol{\zeta}) = 0, \dots, (\boldsymbol{\xi}_n, \boldsymbol{\zeta}) = 0\right\} = (M_0(x)\boldsymbol{\zeta}, \boldsymbol{\zeta})$$

$$+\frac{1}{4}\left\{|\boldsymbol{\zeta}|^{-2}\sum_{i,j=1}^{n} b_{ij}(x)(M_i(x)\boldsymbol{\zeta}, \boldsymbol{\zeta})(M_j(x)\boldsymbol{\zeta}, \boldsymbol{\zeta}) - \sum_{i,j=1}^{n} b_{ij}(x)(M_i^*(x)\boldsymbol{\zeta}, M_j^*(x)\boldsymbol{\zeta})\right\}.$$

Thus condition (i) in Theorem 2.14 and condition (ii) in Theorem 2.15 can be replaced by the following:

for all $x \in \Omega$ and for any $\boldsymbol{\zeta} \in \mathbb{R}^m \backslash \{\mathbf{0}\}$ the inequality holds:

$$|\boldsymbol{\zeta}|^{-2}\sum_{j,k=1}^{n} b_{jk}(x)(\mathcal{A}^{-1}(x)\mathcal{A}_j(x)\boldsymbol{\zeta}, \boldsymbol{\zeta})(\mathcal{A}^{-1}(x)\mathcal{A}_k(x)\boldsymbol{\zeta}, \boldsymbol{\zeta})$$

$$-\sum_{j,k=1}^{n} b_{jk}(x)(\mathcal{A}_j^*(x)(\mathcal{A}^*(x))^{-1}\boldsymbol{\zeta},\ \mathcal{A}_k^*(x)(\mathcal{A}^*(x))^{-1}\boldsymbol{\zeta}) + 4(\mathcal{A}^{-1}(x)\mathcal{A}_0(x)\boldsymbol{\zeta}, \boldsymbol{\zeta}) \ge 0.$$

2.3.2. The case of complex coefficients

In this section we extend basic results of Subsection 2.3.1 to system (2.1.2) with complex coefficients with solutions $\boldsymbol{u} = \boldsymbol{v} + \boldsymbol{w}$, where $\boldsymbol{v}$ and $\boldsymbol{w}$ are m-component vector-valued functions with real-valued components. Here, similarly to Subsection 2.3.1, we assume that Ω is a domain in $\mathbb{R}^n$ with compact closure $\overline{\Omega}$ and with boundary $\partial\Omega$ in the class $\mathrm{C}^{2,\alpha}, 0 < \alpha \leq 1$.

For the spaces of vector-valued and matrix-valued functions with complex components we retain the same notation as in the case of real components but use bold letters.

Let $\mathbb{C}^m$ be the complex linear space of dimension m with elements $\boldsymbol{a}+i\boldsymbol{b}$, where $\boldsymbol{a}, \boldsymbol{b} \in \mathbb{R}^m$ with the inner product $(\boldsymbol{c}, \boldsymbol{d}) = c_1\overline{d}_1 + \cdots + c_m\overline{d}_m, \boldsymbol{c} = (c_1, \ldots, c_m), \boldsymbol{d} = (d_1, \ldots, d_m)$. The length of the vector $\boldsymbol{d}$ in $\mathbb{C}^m$ is defined as $|\boldsymbol{d}| = (\boldsymbol{d}, \boldsymbol{d})^{1/2}$.

We introduce the operator

$$\mathfrak{C}(x, D_x) = \sum_{j,k=1}^{n} \mathcal{C}_{jk}(x)\frac{\partial^2}{\partial x_j \partial x_k} - \sum_{j=1}^{n} \mathcal{C}_j(x)\frac{\partial}{\partial x_j} - \mathcal{C}_0(x),$$

where $\mathcal{C}_{jk} = \mathcal{C}_{kj}, \mathcal{C}_j, \mathcal{C}_0$ are complex $(m \times m)$-matrix-valued functions in the spaces

$$[\mathbf{C}^{2,\alpha}(\overline{\Omega})]^{m\times m}, \qquad [\mathbf{C}^{1,\alpha}(\overline{\Omega})]^{m\times m}, \qquad [\mathbf{C}^{\alpha}(\overline{\Omega})]^{m\times m},$$

respectively. Suppose that the operator $\mathfrak{C}(x, D_x)$ is strongly elliptic in $\overline{\Omega}$, that is for all $x \in \overline{\Omega}$, $\boldsymbol{\zeta} = (\zeta_1, \ldots, \zeta_m) \in \mathbb{C}^m$, $\boldsymbol{\sigma} = (\sigma_1, \ldots, \sigma_n) \in \mathbb{R}^n$, with $\boldsymbol{\zeta},\ \boldsymbol{\sigma} \neq \boldsymbol{0}$, the inequality

$$\Re\left(\sum_{j,k=1}^{n} \mathcal{C}_{jk}(x)\sigma_j\sigma_k\boldsymbol{\zeta},\ \boldsymbol{\zeta} \right) > 0$$

holds.

Let $\mathcal{R}_{jk}, \mathcal{H}_{jk}, \mathcal{R}_j, \mathcal{H}_j, \mathcal{R}_0, \mathcal{H}_0$ be real $(m\times m)$-matrix-valued functions such that

$$\mathcal{C}_{jk} = \mathcal{R}_{jk} + i\mathcal{H}_{jk}, \quad \mathcal{C}_j = \mathcal{R}_j + i\mathcal{H}_j, \quad \mathcal{C}_0 = \mathcal{R}_0 + i\mathcal{H}_0.$$

We use the notation

$$\mathfrak{R}(x, D_x) = \sum_{j,k=1}^{n} \mathcal{R}_{jk}(x)\frac{\partial^2}{\partial x_j \partial x_k} - \sum_{j=1}^{n} \mathcal{R}_j(x)\frac{\partial}{\partial x_j} - \mathcal{R}_0(x),$$

$$\mathfrak{H}(x, D_x) = \sum_{j,k=1}^{n} \mathcal{H}_{jk}(x)\frac{\partial^2}{\partial x_j \partial x_k} - \sum_{j=1}^{n} \mathcal{H}_j(x)\frac{\partial}{\partial x_j} - \mathcal{H}_0(x).$$

Separating the real and imaginary parts of the system $\mathfrak{C}(x, \partial/\partial x)\boldsymbol{u} = \boldsymbol{0}$, where $\boldsymbol{u} = \boldsymbol{v} + i\boldsymbol{w}$, we get a system with real coefficients,

$$\mathfrak{R}(x, D_x)\boldsymbol{v} - \mathfrak{H}(x, D_x)\boldsymbol{w} = \boldsymbol{0}, \quad \mathfrak{H}(x, D_x)\boldsymbol{v} + \mathfrak{R}(x, D_x)\boldsymbol{w} = \boldsymbol{0},$$

which, like the original system, is strongly elliptic.

All the facts given below are resulted by application of the corresponding assertions on the maximum modulus for the systems with real coefficients to the systems obtained after the separation of real and imaginary parts. Thus all the assertions below are corollaries of the corresponding results in Subsection 2.3.1.

COROLLARY 2.5. *If the maximum modulus principle*

$$\|\boldsymbol{u}\|_{[\mathbf{C}(\overline{\Omega})]^m} \leq \|\boldsymbol{u}|_{\partial\Omega}\|_{[\mathbf{C}(\partial\Omega)]^m}$$

holds for solutions $\boldsymbol{u} \in [\mathbf{C}^2(\Omega)]^m \cap [\mathbf{C}(\overline{\Omega})]^m$ *of the system* $\mathfrak{C}(D_x)\boldsymbol{u} = \mathbf{0}$ *with constant coefficients, then*

$$\mathcal{C}_{jk} = \mathcal{C}a_{jk}, \quad 1 \leq j, k \leq n,$$

where $\mathcal{C}$ *is a constant complex* $(m \times m)$*-matrix such that* $\Re(\mathcal{C}\boldsymbol{\zeta}, \boldsymbol{\zeta}) > 0$ *for all* $\boldsymbol{\zeta} \in \mathbb{C}^m, \boldsymbol{\zeta} \neq \mathbf{0}$ *and* $((a_{jk}))$ *is a real positive-definite* $(n \times n)$*-matrix.*

THEOREM 2.16. *Let the system* $\mathfrak{C}(x, D_x)\boldsymbol{u} = \mathbf{0}$ *be of the form*

$$\mathcal{C}(x) \sum_{j,k=1}^{n} a_{jk}(x) \frac{\partial^2 \boldsymbol{u}}{\partial x_j \partial x_k} - \sum_{j=1}^{n} \mathcal{C}_j(x) \frac{\partial \boldsymbol{u}}{\partial x_j} - \mathcal{C}_0(x)\boldsymbol{u} = \mathbf{0},$$

where $\mathcal{C}$ *is a complex* $(m \times m)$*-matrix-valued function such that* $\Re(\mathcal{C}(x)\boldsymbol{\zeta}, \boldsymbol{\zeta}) > 0$ *for all* $\boldsymbol{\zeta} \in \mathbb{C}^m \backslash \{\mathbf{0}\}, x \in \overline{\Omega}$, $((a_{jk}))$ *is a real positive-definite matrix-valued function in* $\overline{\Omega}$ *of order* n *and let* $\mathcal{C} \in [\mathbf{C}^{2,\alpha}(\overline{\Omega})]^{m \times m}$, $((a_{jk})) \in [\mathrm{C}^{2,\alpha}(\overline{\Omega})]^{n \times n}$.

The system $\mathfrak{C}(x, D_x)\boldsymbol{u} = \mathbf{0}$ *has no solutions* $\boldsymbol{u} \in [\mathbf{C}^2(\Omega)]^m \cap [\mathbf{C}(\overline{\Omega})]^m$ *with modulus attaining its strict local maximum at some point* $x \in \Omega$ *if and only if*

(i) *for all* $x \in \Omega$ *and any* $\boldsymbol{\xi}_j, \boldsymbol{\zeta} \in \mathbb{C}^m, j = 1, \ldots, n,$ *such that* $\Re(\boldsymbol{\xi}_j, \boldsymbol{\zeta}) = 0$ *the inequality*

$$\Re\left\{\sum_{j,k=1}^{n} a_{jk}(x)(\boldsymbol{\xi}_j, \boldsymbol{\xi}_k) + \sum_{j=1}^{n}(\mathcal{C}^{-1}(x)\mathcal{C}_j(x)\boldsymbol{\xi}_j, \boldsymbol{\zeta}) + (\mathcal{C}^{-1}(x)\mathcal{C}_0(x)\boldsymbol{\zeta}, \boldsymbol{\zeta})\right\} \geq 0$$

is valid.

THEOREM 2.17. *The classical maximum modulus principle*

$$\|\boldsymbol{u}\|_{[\mathbf{C}(\overline{\omega})]^m} \leq \|\boldsymbol{u}|_{\partial\omega}\|_{[\mathbf{C}(\partial\omega)]^m}$$

is valid for solutions of the system $\mathfrak{C}(x, D_x)\boldsymbol{u} = \mathbf{0}$ *in an arbitrary domain* $\omega \subset \Omega$ *with boundary from the class* $\mathrm{C}^{2,\alpha}$ *if and only if:*

(i) *for all* $x \in \overline{\Omega}$ *the equalities*

$$\mathcal{C}_{jk}(x) = \mathcal{C}(x)a_{jk}(x), \quad 1 \leq j, k \leq n,$$

hold, where $\mathcal{C}$ *is a complex* $(m \times m)$*-matrix-valued function such that* $\Re(\mathcal{C}(x)\boldsymbol{\zeta}, \boldsymbol{\zeta}) > 0$ *for all* $x \in \overline{\Omega}, \boldsymbol{\zeta} \in \mathbb{C}^m \backslash \{\mathbf{0}\}$, $((a_{jk}))$ *is a real positive-definite* $(n \times n)$*-matrix-valued function in* $x \in \overline{\Omega}$ *of order* n*;*

(ii) *for all* $x \in \Omega$ *and any* $\boldsymbol{\xi}_j, \boldsymbol{\zeta} \in \mathbb{C}^m, j = 1, \ldots, n,$ *such that* $\Re(\boldsymbol{\xi}_j, \boldsymbol{\zeta}) = 0$ *the inequality*

$$\Re\left\{\sum_{j,k=1}^{n} a_{jk}(x)(\boldsymbol{\xi}_j, \boldsymbol{\xi}_k) + \sum_{j=1}^{n}(\mathcal{C}^{-1}(x)\mathcal{C}_j(x)\boldsymbol{\xi}_j, \boldsymbol{\zeta}) + (\mathcal{C}^{-1}(x)\mathcal{C}_0(x)\boldsymbol{\zeta}, \boldsymbol{\zeta})\right\} \geq 0$$

is valid.

COROLLARY 2.6. *Let the system* $\mathfrak{C}(x, D_x)\boldsymbol{u} = \mathbf{0}$ *be of the form*

$$\mathcal{C}(x) \sum_{j,k=1}^{n} a_{jk}(x) \frac{\partial^2 \boldsymbol{u}}{\partial x_j \partial x_k} - \sum_{j=1}^{n} \mathcal{C}_j(x) \frac{\partial \boldsymbol{u}}{\partial x_j} = \mathbf{0},$$

where $\mathcal{C}(x)$ *and* $((a_{jk}))$ *are the same matrix-valued functions as in Theorem* 2.16.

The system $\mathfrak{C}(x, D_x)\boldsymbol{u} = \mathbf{0}$ has no solutions $\boldsymbol{u} \in [\mathbf{C}^2(\Omega)]^m \cap [\mathbf{C}(\overline{\Omega})]^m$ with modulus attaining its strict local maximum at some point $x \in \Omega$ if and only if

$$\mathcal{C}_j(x) = \mathcal{C}(x)a_j(x), \quad 1 \le j,\ k \le n.$$

Here a_j are real scalar functions with the same smoothness as $\mathcal{C}_j$.

COROLLARY 2.7. *The classical maximum modulus principle*

$$\|\boldsymbol{u}\|_{[\mathbf{C}(\overline{\omega})]^m} \le \|\boldsymbol{u}|_{\partial\omega}\|_{[\mathbf{C}(\partial\omega)]^m}$$

holds for solutions of the system

$$\sum_{j,k=1}^{n} \mathcal{C}_{jk}(x)\frac{\partial^2 \boldsymbol{u}}{\partial x_j \partial x_k} - \sum_{j=1}^{n} \mathcal{C}_j(x)\frac{\partial \boldsymbol{u}}{\partial x_j} = \mathbf{0}$$

in an arbitrary domain $\omega \subset \Omega$ with boundary from the class $\mathrm{C}^{2,\alpha}$ if and only if

$$\mathcal{C}_{jk}(x) = \mathcal{C}(x)a_{jk}(x), \quad \mathcal{C}_j(x) = \mathcal{C}(x)a_j(x), \quad 1 \le j,\ k \le n.$$

Here $\mathcal{C}(x)$ and $((a_{jk}))$ are the matrix-valued functions defined in Theorem 2.16 *and a_j are real scalar functions with the same smoothness as $\mathcal{C}_j$.*

REMARK 2.6. It follows from Remark 2.5 that condition (i) in Theorem 2.16 and condition (ii) in Theorem 2.17 can be replaced by the following ones:
for all $x \in \Omega$ and for any $\boldsymbol{\zeta} \in \mathbb{C}^m \backslash \{\mathbf{0}\}$ the inequality

$$|\boldsymbol{\zeta}|^{-2} \sum_{j,k=1}^{n} b_{jk}(x)\Re(\mathcal{C}^{-1}(x)\mathcal{C}_j(x)\boldsymbol{\zeta}, \boldsymbol{\zeta})(\mathcal{C}^{-1}(x)\mathcal{C}_k(x)\boldsymbol{\zeta}, \boldsymbol{\zeta})$$
$$- \sum_{j,k=1}^{n} b_{jk}(x)(\mathcal{C}_j^*(x)(\mathcal{C}^*(x))^{-1}\boldsymbol{\zeta},\ \mathcal{C}_k^*(x)(\mathcal{C}^*(x))^{-1}\boldsymbol{\zeta}) + 4\Re(\mathcal{C}^{-1}(x)\mathcal{C}_0(x)\boldsymbol{\zeta}, \boldsymbol{\zeta}) \ge 0$$

is valid, where $((b_{jk}))$ is the inverse matrix of $((a_{jk}))$ and $\mathcal{C}_j^*(x)$ is the adjoint matrix of $\mathcal{C}_j(x)$.

We remark that the second sum is real by the symmetry of the matrix $((b_{jk}(x))$.

Consider the scalar uniformly elliptic equation with complex coefficients in Ω

$$c(x) \sum_{j,k=1}^{n} a_{jk}(x)\frac{\partial^2 u}{\partial x_j \partial x_k} - \sum_{j=1}^{n} c_j(x)\frac{\partial u}{\partial x_j} - c_0(x)u = 0, \tag{2.3.41}$$

where $\Re c(x) > 0$ for all $x \in \overline{\Omega}$ and $((a_{jk}))$ is a real positive-definite $(n \times n)$-matrix-valued function on $\overline{\Omega}$. By Theorem 2.16 and Remark 2.6 we arrive at the following assertion.

COROLLARY 2.8. *The equation* (2.3.41) *has no solutions $u \in \mathbf{C}^2(\Omega) \cap \mathbf{C}(\overline{\Omega})$ with modulus attaining its strict local maximum at some point $x \in \Omega$ if and only if for all $x \in \Omega$*

$$4\Re\left(\frac{c_0(x)}{c(x)}\right) \ge \sum_{j,k=1}^{n} b_{jk}(x)\Im\left(\frac{c_j(x)}{c(x)}\right)\Im\left(\frac{c_k(x)}{c(x)}\right), \tag{2.3.42}$$

where $((b_{jk}))$ is the $(n \times n)$-matrix inverse of $((a_{jk}))$.

We remark that (2.3.42) seems to be new even in the case $n = 1$, i.e. for ordinary differential equations with complex coefficients.

For a scalar uniformly elliptic equation with complex coefficients of the general form

$$\sum_{j,k=1}^{n} c_{jk}(x)\frac{\partial^2 u}{\partial x_j \partial x_k} - \sum_{j=1}^{n} c_j(x)\frac{\partial u}{\partial x_j} - c_0(x)u = 0, \tag{2.3.43}$$

Theorem 2.17 and Remark 2.6 imply

COROLLARY 2.9. *The classical maximum modulus principle*

$$\|\boldsymbol{u}\|_{\mathbf{C}(\overline{\omega})} \leq \|\boldsymbol{u}|_{\partial\omega}\|_{\mathbf{C}(\partial\omega)}$$

is valid for solutions of equation (2.3.43) *in an arbitrary domain* $\omega \subset \Omega$ *with boundary* $\partial\omega$ *of the class* $\mathrm{C}^{2,\alpha}$ *if and only if for all* $x \in \overline{\Omega}$

(i) $c_{jk}(x) = c(x)a_{jk}(x)$, $1 \leq j,\ k \leq n$, *where* $\Re c(x) > 0$ *for all* $x \in \overline{\Omega}$, *and* $((a_{jk}))$ *is a real positive-definite* $(n \times n)$*-matrix-valued function on* $\overline{\Omega}$;

(ii) *the inequality*

$$4\Re\left(\frac{c_0(x)}{c(x)}\right) \geq \sum_{j,k=1}^{n} b_{jk}(x)\Im\left(\frac{c_j(x)}{c(x)}\right)\Im\left(\frac{c_k(x)}{c(x)}\right)$$

holds, where $((b_{jk}))$ *is the* $(n \times n)$*-matrix inverse of* $((a_{jk}))$.

2.4. Comments to Chapter 2

The first articles aiming at the study and applications of the classical maximum modulus principle for solutions of elliptic second order systems concerned systems with scalar coefficients in the first and second derivatives of the unknown vector-function (see Bitsadze [**Bi1, Bi2**] , Pini [**Pini**] , Szeptycki [**Sze**]). These systems are weakly coupled: a system of partial differential equations is called weakly coupled if there are no derivatives in the coupling terms.

Sufficient conditions for validity of the maximum modulus principle, its modifications and generalizations for non-weakly coupled systems were given by Hile and Protter [**HP**], C. Miranda [**Mi3**], Rus [**Rus**], Stys [**Sty1**], Sabitov [**Sa1**], Wasowski [**Was**]. In particular, C. Miranda [**Mi3**] considered elliptic second order systems with a scalar principal part and with arbitrary coefficients in derivatives of order less than two. He found an algebraic inequality sufficient for the classical maximum modulus principle (condition (i) in Corollary 1 from introduction to the present chapter with the strict inequality sign). A survey of maximum principles for elliptic equations and systems with a scalar principal part is given by Protter [**Pr**].

An algebraic necessary and sufficient condition for validity of the maximum principle for the product $\alpha(x)|\boldsymbol{u}|$, where α is a certain function and $\boldsymbol{u}$ is a solution of the elliptic system with analytic coefficients, is due to Hong [**Hon**].

There is a number of results on the componentwise maximum principle for weakly coupled elliptic systems, in particular, on non-negativity of the components of a solution (cf. de Figueiredo and Mitidieri [**FM**], Lenhart and Schaefer [**LS**], López-Gómez and Molina-Meyer [**LoMo**], Mitidieri and Sweers [**MS**], Sirakov [**Si**] and bibliography there). Various maximum principles for weakly coupled systems are discussed in the book [**PW**] by Protter and Weinberger. Necessary and sufficient conditions for the component and for the so-called "stohastic" extremum principles for solutions of elliptic systems of the second order are given by Kamynin and Khimchenko in [**KK1**] and [**KK2**], respectively.

The maximum modulus principle makes it possible to obtain a pointwise estimate for solutions $u \in C(\overline{\Omega}) \cap C^2(\Omega)$ of elliptic equations with a bounded right-hand side in a bounded domain $\Omega \subset \mathbb{R}^n$ (see, e.g., Gilbarg and Trudinger [**GT**], Section 3.3, Theorem 3.7). The Alexandrov-Bakelman maximum principle (see Bakelman [**Bak1**]-[**Bak3**], Alexandrov [**Ale1, Ale2**], Nirenberg [**Nir2**], Gilbarg and Trudinger [**GT**], Section 9.1) is an improvement of the just mentioned a priori estimates for solutions of elliptic equations in non-divergence form with measurable bounded coefficients. There exist generalizations of the Alexandrov-Bakelman maximum principle for elliptic systems (see Sirakov [**Si**] and references therein).

Note that inequality (2.2.20) is a refinement of Polya's estimate in [**Po**]. This inequality shows that the classical maximum modulus principle does not hold for solutions of system (2.2.18), except for the case $\lambda + \mu = 0$.

CHAPTER 3

Sharp Constants in the Miranda-Agmon Inequalities for Solutions of Certain Systems of Mathematical Physics

3.1. Introduction

The classical maximum modulus principle for solutions of elliptic systems of the second order holds under rather heavy conditions on the coefficients (see Chapter 2 for a characterization of such systems). In particular, the Stokes and Lamé systems do not satisfy these conditions.

However, for a wide class of elliptic systems of the second order a more general Miranda-Agmon maximum principle holds. By this we mean the inequality

$$\sup_{\overline{\Omega}} |\mathbf{u}| \leq k(\Omega) \sup_{\partial\Omega} |\mathbf{u}|, \qquad \Omega \subset \mathbb{R}^n,$$

with a constant $k(\Omega)$ which is not necessarily equal to one.

In the present chapter we find best constants in the pointwise estimates of Miranda-Agmon maximum principle type for solutions of the Stokes, Lamé and viscoelasticity systems in a half-space, as well as for two components of the stress tensor of the planar deformed state in a half-plane.

It is worth noting that under certain conditions on the Dirichlet data at the boundary of the half-space the constant in the inequality of the Miranda-Agmon type maximum modulus principle may turn to be one. Hence one may say that the classical maximum modulus principle holds. We show, in particular, that if the velocity vector (the elastic displacement vector) is parallel to a constant vector at the boundary of a half-space and if the modulus of the boundary data does not exceed 1, then the velocity vector (the elastic displacement vector) is majorized by 1 at an arbitrary point of the half-space. An analogous classical maximum modulus principle is obtained for two components of the stress tensor of the planar deformed state.

Section 3.2 is devoted to linear strongly elliptic systems of the second order, without lower order terms, with constant coefficients and m unknown functions. We shall formulate one of the results which concerns systems with real coefficients.

Let $\mathbb{R}^n_+ = \{x = (x_1, \dots, x_n) : x_n > 0\}$. Consider the Dirichlet problem

$$\sum_{i,j=1}^{n} \mathcal{A}_{ij} \frac{\partial^2 \boldsymbol{u}}{\partial x_i \partial x_j} = \mathbf{0} \text{ in } \mathbb{R}^n_+, \quad \boldsymbol{u} = \psi_1 \boldsymbol{\xi}_1 + \cdots + \psi_\ell \boldsymbol{\xi}_\ell \text{ on } \partial\mathbb{R}^n_+$$

for strongly elliptic system, where $\mathcal{A}_{ij} = \mathcal{A}_{ji}$ are constant real $(m \times m)$-matrices, $\boldsymbol{u} = (u_1, \dots, u_m)$, $\boldsymbol{\xi}_1, \dots, \boldsymbol{\xi}_\ell$ is an orthogonal system of normed m-component vectors, $\psi_1, \dots, \psi_\ell$ are continuous and bounded functions on $\partial\mathbb{R}^n_+$, $1 \leq \ell \leq m$. The solution $\boldsymbol{u}_{\boldsymbol{\xi}}$ of this problem, where $\boldsymbol{\xi} = (\boldsymbol{\xi}_1, \dots, \boldsymbol{\xi}_\ell)$, can be expressed (see

Subsection 2.2.1) as the vector-valued double layer potential

$$\boldsymbol{u}_{\boldsymbol{\xi}}(x)=\int_{\partial\mathbb{R}^n_+}\mathcal{M}_{\boldsymbol{\xi}}(\boldsymbol{e}_{xy})\boldsymbol{\psi}(y')\omega(x,dy'),$$

where $y=(y',0), y'=(y_1,\dots,y_{n-1})$, $\mathcal{M}_{\boldsymbol{\xi}}$ is a continuous $(m\times\ell)$-matrix-valued function with bounded components on the hemisphere $\mathbb{S}^{n-1}_-=\{x\in\mathbb{R}^n:\ |x|=1,\ x_n<0\}$, $\boldsymbol{e}_{xy}=(y-x)|y-x|^{-1}$, $\boldsymbol{\psi}=(\psi_1,\dots,\psi_\ell)$, and

$$\omega(x,\mathcal{B})=\int_{\mathcal{B}}\frac{x_n}{|y-x|^n}\,dy'$$

is the solid angle at which a Borel set $\mathcal{B}\subset\partial\mathbb{R}^n_+$ is seen from the point $x\in\mathbb{R}^n_+$. We note that $\omega(x,\mathcal{B})\geq 0$, and that $\omega(x,\cdot)$ is a finite regular Borel measure on $\partial\mathbb{R}^n_+$ for any fixed $x\in\mathbb{R}^n_+$. We prove the following assertion.

THEOREM 1. *Let x be an arbitrary fixed point in $\mathbb{R}^n_+$ and let $\boldsymbol{z}\in\mathbb{R}^m, |\boldsymbol{z}|=1$. The sharp constant $K_\ell(\boldsymbol{\xi},\boldsymbol{z})$ in the inequality*

$$|(\boldsymbol{z},u_{\boldsymbol{\xi}}(x))|\leq K_\ell(\boldsymbol{\xi},\boldsymbol{z})\sup\{|\boldsymbol{u}_{\boldsymbol{\xi}}(x',0)|:x'\in\partial\mathbb{R}^n_+\},$$

is given by

$$K_\ell(\boldsymbol{\xi},\boldsymbol{z})=\int_{\mathbb{S}^{n-1}_-}|\mathcal{M}^*_{\boldsymbol{\xi}}(\boldsymbol{e}_\sigma)\boldsymbol{z}|d\sigma,$$

where $\boldsymbol{e}_\sigma$ is the n-dimensional unit vector joining the origin to a point $\sigma\in\mathbb{S}^{n-1}_-$, and $*$ denotes the transposition of a matrix.

The sharp constant $K_\ell(\boldsymbol{\xi})$ in the inequality

$$|\boldsymbol{u}_{\boldsymbol{\xi}}(x)|\leq K_\ell(\boldsymbol{\xi})\sup\{|\boldsymbol{u}_{\boldsymbol{\xi}}(x',0)|:x'\in\partial\mathbb{R}^n_+\},$$

is given by

$$K_\ell(\boldsymbol{\xi})=\sup_{|\boldsymbol{z}|=1}\int_{\mathbb{S}^{n-1}_-}|\mathcal{M}^*_{\boldsymbol{\xi}}(\boldsymbol{e}_\sigma)\boldsymbol{z}|d\sigma.$$

In Section 3.3 we consider the Lamé system

$$\mu\Delta\boldsymbol{u}+(\lambda+\mu)\mathrm{grad}\ \mathrm{div}\ \boldsymbol{u}=\boldsymbol{0}, \tag{3.1.1}$$

and the Stokes system

$$\nu\Delta\boldsymbol{u}-\mathrm{grad}\,p=\boldsymbol{0},\quad \mathrm{div}\ \boldsymbol{u}=0, \tag{3.1.2}$$

in $\mathbb{R}^n_+$, $n\geq 2$, with the boundary condition

$$\boldsymbol{u}\big|_{x_n=0}=\psi_1\boldsymbol{\xi}_1+\dots+\psi_\ell\boldsymbol{\xi}_\ell. \tag{3.1.3}$$

Here λ and μ are the Lamé constants, ν is the kinematic coefficient of viscosity, $\boldsymbol{u}=(u_1,\dots,u_n)$ is either the displacement vector of an elastic medium or the velocity vector of a fluid, p is the pressure in the fluid, $\boldsymbol{\xi}_1,\dots,\boldsymbol{\xi}_\ell$ is an orthogonal system of normed n-component vectors, $\psi_1,\dots,\psi_\ell$ are continuous and bounded functions on $\partial\mathbb{R}^n_+$, $1\leq\ell\leq n$.

The next assertion, proved by Kresin and Maz'ya in [**KM9**], concerns the solution $\boldsymbol{u}_{\boldsymbol{\xi}}$ of the problems (3.1.1), (3.1.3) and (3.1.2), (3.1.3). Whenever the constants do not depend on $\boldsymbol{\xi}$ we shall write K_ℓ instead of $K_\ell(\boldsymbol{\xi})$.

THEOREM 2. *Let x be an arbitrary fixed point in $\mathbb{R}^n_+$ and let $\boldsymbol{z}\in\mathbb{R}^n, |\boldsymbol{z}|=1$. Let, further, $\boldsymbol{u}_{\boldsymbol{\xi}}$ be a solution of problem* (3.1.1), (3.1.3) *or the vector component of a solution of problem* (3.1.2), (3.1.3).

The best constant $K_\ell(\boldsymbol{\xi}, \boldsymbol{z})$ *in the inequality*

$$|(\boldsymbol{z}, u_{\boldsymbol{\xi}}(x))| \le K_\ell(\boldsymbol{\xi}, \boldsymbol{z}) \sup\{|\boldsymbol{u}_{\boldsymbol{\xi}}(x', 0)| : x' \in \partial\mathbb{R}^n_+\} \tag{3.1.4}$$

is given by

$$K_\ell(\boldsymbol{\xi}, \boldsymbol{z}) = \frac{1}{\omega_n} \int_{\mathbb{S}^{n-1}} \left\{ \sum_{i=1}^{\ell} \Big[(1-\kappa)(\boldsymbol{\xi}_i, \boldsymbol{z}) + m\kappa(\boldsymbol{e}_\sigma, \boldsymbol{z})(\boldsymbol{e}_\sigma, \boldsymbol{\xi}_i) \Big]^2 \right\}^{1/2} d\sigma,$$

where $\kappa = 1$ *for the Stokes system,* $\kappa = (\lambda + \mu)(\lambda + 3\mu)^{-1}$ *for the Lamé system,* ω_n *being the area of* $\mathbb{S}^{n-1} = \big\{x \in \mathbb{R}^n : |x| = 1\big\}$, *and* $\boldsymbol{e}_\sigma$ *is the* n*-dimensional unit vector joining the origin to a point* $\sigma \in \mathbb{S}^{n-1}$.

The best constant $K_\ell(\boldsymbol{\xi})$ *in the inequality*

$$|\boldsymbol{u}_{\boldsymbol{\xi}}(x)| \le K_\ell(\boldsymbol{\xi}) \sup\{|\boldsymbol{u}_{\boldsymbol{\xi}}(x', 0)| : x' \in \partial\mathbb{R}^n_+\}$$

is given by

$$K_\ell(\boldsymbol{\xi}) = \frac{1}{\omega_n} \sup_{|\boldsymbol{z}|=1} \int_{\mathbb{S}^{n-1}} \left\{ \sum_{i=1}^{\ell} \Big[(1-\kappa)(\boldsymbol{\xi}_i, \boldsymbol{z}) + m\kappa(\boldsymbol{e}_\sigma, \boldsymbol{z})(\boldsymbol{e}_\sigma, \boldsymbol{\xi}_i) \Big]^2 \right\}^{1/2} d\sigma.$$

Here $K_1(\xi) = 1$ *and* $K_\ell(\boldsymbol{\xi}) > 1$ *for all* $2 \le \ell \le n$ *and* $0 < \kappa \le 1$.

This theorem implies the next two assertions.

COROLLARY 1. *Let* x *be an arbitrary point in* $\mathbb{R}^n_+$, *and let* $\boldsymbol{u} = \boldsymbol{u}_{\boldsymbol{\xi}}$ *be a solution of problem* (3.1.1), (3.1.3) *or the vector component of a solution of problem* (3.1.2), (3.1.3) *with* $\ell = n$. *The best constant* K_n *in the inequality*

$$|\boldsymbol{u}(x)| \le K_n \sup\{|\boldsymbol{u}(x', 0)| : x' \in \partial\mathbb{R}^n_+\}$$

is

$$K_n = \frac{2\Gamma(n/2)}{\sqrt{\pi}\,\Gamma((n-1)/2)} \int_0^{\pi/2} \Big[(1-\kappa)^2 + n\kappa(n\kappa - 2\kappa + 2)\cos^2\theta \Big]^{1/2} \sin^{n-2}\theta d\theta,$$

and, in particular,

$$K_2 = \frac{2}{\pi}(1+\kappa)\mathbf{E}\left(\frac{2\sqrt{\kappa}}{1+\kappa}\right) = 1 + \frac{1}{2^2}\kappa^2 + \frac{1}{2^2 4^2}\kappa^4 + \dots + \left[\frac{(2k-3)!!}{2^k k!}\right]^2 \kappa^{2k} + \dots,$$

$$K_3 = \frac{1}{2}\left[1 + 2\kappa + \frac{(1-\kappa)^2}{\sqrt{3\kappa(\kappa+2)}} \log \frac{1 + 2\kappa + \sqrt{3\kappa(\kappa+2)}}{1-\kappa}\right],$$

where

$$\mathbf{E}(k) = \int_0^{\pi/2} \sqrt{1 - k^2 \sin^2\varphi}\, d\varphi$$

is the complete elliptic integral of the second kind.

In the case $\kappa = 1$, *i.e., for the* n*-dimensional Stokes system,*

$$K_n = \frac{2}{\sqrt{\pi}} \frac{\Gamma(\frac{n}{2}+1)}{\Gamma(\frac{n+1}{2})}$$

and, in particular, $K_2 = 4/\pi$ *and* $K_3 = 3/2$.

COROLLARY 2. *Let* $\ell = n-1$ *and let* $(\boldsymbol{z}, \boldsymbol{\xi}_1) = \dots = (\boldsymbol{z}, \boldsymbol{\xi}_{n-1}) = 0, |\boldsymbol{z}| = 1$. *Then the sharp constant* $K_{n-1}(\boldsymbol{\xi}, \boldsymbol{z}) = \mathcal{K}_{n-1}$ *in* (3.1.4) *takes the form*

$$\mathcal{K}_{n-1} = \frac{2\kappa\Gamma(\frac{n}{2})}{\sqrt{\pi}\Gamma(\frac{n-1}{2})}.$$

In particular, $\mathcal{K}_1 = 2/\pi, \mathcal{K}_2 = 1$ *for the two-dimensional and three-dimensional Stokes systems, respectively.*

In Section 3.4, we consider the system of equations in $\mathbb{R}^2_+$ for the stresses in a planar deformed state

$$\partial\sigma_{11}/\partial x_1 + \partial\sigma_{12}/\partial x_2 = 0,$$
$$\partial\sigma_{12}/\partial x_1 + \partial\sigma_{22}/\partial x_2 = 0,$$
$$\Delta(\sigma_{11} + \sigma_{22}) = 0,$$

where σ_{11}, σ_{12} and σ_{22} are the components of the stress tensor in the half-plane $\mathbb{R}^2_+$.

Using the relation between solutions of the two-dimensional Stokes system and the elements of the stress tensor of the planar deformed state, we arrive at the following result

THEOREM 3. *Let* $(\sigma_{12}(x_1, 0), \sigma_{22}(x_1, 0)) = \psi(x_1)\boldsymbol{\xi}$, *where* $\psi \in \mathrm{C}(\partial\mathbb{R}^2_+)$ *and* $\boldsymbol{\xi}$ *is a two-dimensional unit vector. Then the inequality*

$$\|(\sigma_{12}^2 + \sigma_{22}^2)^{1/2}\|_{\mathrm{C}(\overline{\mathbb{R}^2_+})} \leq \|(\sigma_{12}^2 + \sigma_{22}^2)^{1/2}\|_{\mathrm{C}(\partial\mathbb{R}^2_+)}$$

is valid.

In the general case, the best constant in the inequality

$$\|(\sigma_{12}^2 + \sigma_{22}^2)^{1/2}\|_{\mathrm{C}(\overline{\mathbb{R}^2_+})} \leq \mathcal{K}\|(\sigma_{12}^2 + \sigma_{22}^2)^{1/2}\|_{\mathrm{C}(\partial\mathbb{R}^2_+)}$$

is equal to $4/\pi$.

In Section 3.5 we consider the first boundary value problem for equations of the quasistatic linear viscoelasticity for the medium described by the three parametric model in the half-space $\mathbb{R}^3_+$. We present the best value $\mathcal{K}(t)$ of the coefficient in the inequality

$$|\boldsymbol{u}(x,t)| \leq \mathcal{K}(t)\sup\{|\boldsymbol{u}(y,\tau)| : y \in \partial\mathbb{R}^3_+,\ 0 \leq \tau \leq t\},$$

where $\boldsymbol{u} = (u_1, u_2, u_3)$ is the vector of viscoelastic displacements and $x \in \mathbb{R}^3_+$. In particular, we give formulas for $\mathcal{K}(t)$ in the case of the Maxwell medium and in the case of the medium with purely elastic behaviour under volume compression (the last assumption is often used in applications).

3.2. Best constants in the Miranda-Agmon inequalities for solutions of strongly elliptic systems in a half-space

3.2.1. Systems with constant real coefficients

We introduce notation used henceforth. Let $x \in \mathbb{R}^n_+ = \{x = (x', x_n) : x' = (x_1, \dots, x_{n-1}) \in \mathbb{R}^{n-1}, x_n > 0\}$ and $\mathbb{S}^{n-1} = \{x \in \mathbb{R}^n : |x| = 1\}$. By $[\mathrm{C}^2(\mathbb{R}^n_+)]^m$ we denote the space of m-component vector-valued functions with continuous derivatives up to the second order in $\mathbb{R}^n_+$. Further, $[\mathrm{C_b}(\overline{\mathbb{R}^n_+})]^m$ and $[\mathrm{C_b}(\partial\mathbb{R}^n_+)]^m$ will stand for the spaces of continuous and bounded m-component vector-valued functions on $\overline{\mathbb{R}^n_+}$ and $\partial\mathbb{R}^n_+$, respectively. In the case $m = 1$ the corresponding index will be omitted. The notation $Q = [Q_1, \dots, Q_\ell]$ stands for a matrix Q with columns $Q_1, \dots, Q_\ell$.

We introduce the strongly elliptic operator

$$\mathfrak{A}_0(D_x) = \sum_{i,j=1}^n \mathcal{A}_{ij}\frac{\partial^2}{\partial x_i \partial x_j}, \tag{3.2.1}$$

where $D_x = (\partial/\partial x_1, \dots, \partial/\partial x_n)$ and $\mathcal{A}_{ij} = \mathcal{A}_{ji}$ are constant real $(m\times m)$-matrices. The strong ellipticity of $\mathfrak{A}_0(D_x)$ means that the inequality

$$\Big(\sum_{i,j=1}^{n} \mathcal{A}_{ij}\sigma_i\sigma_j \boldsymbol{\zeta}, \boldsymbol{\zeta} \Big) > 0$$

is valid for all $\boldsymbol{\zeta} = (\zeta_1, \dots, \zeta_m) \in \mathbb{R}^m \backslash \{\mathbf{0}\}$ and $\boldsymbol{\sigma} = (\sigma_1, \dots, \sigma_n) \in \mathbb{R}^n \backslash \{\mathbf{0}\}$.

Henceforth, we use the notation $\boldsymbol{e}_{xy} = (y-x)|y-x|^{-1}$. We write solution (2.2.4) of the Dirichlet problem (2.2.2) for the system $\mathfrak{A}_0(D_x)\boldsymbol{u} = \mathbf{0}$ in $\mathbb{R}^n_+$, which was considered in Subsection 2.2.1, as

$$\boldsymbol{u}(x) = \int_{\partial\mathbb{R}^n_+} \mathcal{M}(\boldsymbol{e}_{xy}) \boldsymbol{f}(y')\, \omega(x, dy'). \tag{3.2.2}$$

Let $\boldsymbol{\xi} = (\boldsymbol{\xi}_1, \dots, \boldsymbol{\xi}_\ell)$, where $\boldsymbol{\xi}_1, \dots, \boldsymbol{\xi}_\ell$ is an orthogonal system of fixed normed vectors in $\mathbb{R}^m$, $1 \le \ell \le m$. Let $\boldsymbol{\psi} = (\psi_1, \dots, \psi_\ell) \in [\mathrm{C_b}(\partial\mathbb{R}^n_+)]^\ell$.

We replace $\boldsymbol{f}(y')$ in (3.2.2) by

$$\boldsymbol{\Psi}_{\boldsymbol{\xi}}(y') = \sum_{i=1}^{\ell} \psi_i(y')\boldsymbol{\xi}_i \tag{3.2.3}$$

and use the notation

$$\boldsymbol{u}_{\boldsymbol{\xi}}(x) = \int_{\partial\mathbb{R}^n_+} \mathcal{M}(\boldsymbol{e}_{xy}) \boldsymbol{\Psi}_{\boldsymbol{\xi}}(y') \omega(x, dy'). \tag{3.2.4}$$

Introducing the $(m \times \ell)$-matrix-valued function

$$\mathcal{M}_{\boldsymbol{\xi}}(\boldsymbol{e}_{xy}) = \Big[\mathcal{M}(\boldsymbol{e}_{xy})\boldsymbol{\xi}_1, \dots, \mathcal{M}(\boldsymbol{e}_{xy})\boldsymbol{\xi}_\ell\Big] \tag{3.2.5}$$

and taking into account (3.2.3), we rewrite (3.2.4) as

$$\boldsymbol{u}_{\boldsymbol{\xi}}(x) = \int_{\partial\mathbb{R}^n_+} \mathcal{M}_{\boldsymbol{\xi}}(\boldsymbol{e}_{xy}) \boldsymbol{\psi}(y') \omega(x, dy'). \tag{3.2.6}$$

Since $\boldsymbol{\Psi}_{\boldsymbol{\xi}}$ denotes the boundary data of the Dirichlet problem (2.2.2), we have

$$\boldsymbol{u}_{\boldsymbol{\xi}}(x', 0) = \boldsymbol{\Psi}_{\boldsymbol{\xi}}(x').$$

This together with (3.2.3) implies

$$|\boldsymbol{u}_{\boldsymbol{\xi}}(x', 0)| = |\boldsymbol{\psi}(x')|. \tag{3.2.7}$$

The operator in the right-hand side of (3.2.6)

$$T_{\boldsymbol{\xi},x}(\boldsymbol{\psi}) = \int_{\partial\mathbb{R}^n_+} \mathcal{M}_{\boldsymbol{\xi}}(\boldsymbol{e}_{xy}) \boldsymbol{\psi}(y') \omega(x, dy'), \tag{3.2.8}$$

where $\boldsymbol{\xi}$ and x are some parameters, maps $[\mathrm{C_b}(\partial\mathbb{R}^n_+)]^\ell$ into $\mathbb{R}^m$ and is of type (1.2.29). Let $\boldsymbol{z}$ be a fixed unit vector in $\mathbb{R}^m$. Similarly to the general case, we introduce the functional $\mathcal{T}_{\boldsymbol{\xi},x;\boldsymbol{z}}$ on $[\mathrm{C_b}(\partial\mathbb{R}^n_+)]^\ell$ by the equality

$$\mathcal{T}_{\boldsymbol{\xi},x;\boldsymbol{z}}(\boldsymbol{\psi}) = (\boldsymbol{z}, T_{\boldsymbol{\xi},x}(\boldsymbol{\psi})). \tag{3.2.9}$$

THEOREM 3.1. *Let x be an arbitrary fixed point in $\mathbb{R}^n_+$ and let $\boldsymbol{z} \in \mathbb{R}^m, |\boldsymbol{z}| = 1$. The sharp constant $K_{\boldsymbol{\xi},\boldsymbol{z}}$ in the inequality*

$$|(\boldsymbol{z}, \boldsymbol{u}_{\boldsymbol{\xi}}(x))| \leq K_\ell(\boldsymbol{\xi}, \boldsymbol{z}) \sup\{|\boldsymbol{u}_{\boldsymbol{\xi}}(x', 0)| : x' \in \partial\mathbb{R}^n_+\}, \tag{3.2.10}$$

is given by

$$K_\ell(\boldsymbol{\xi}, \boldsymbol{z}) = \int_{\mathbb{S}^{n-1}_-} |\mathcal{M}^*_{\boldsymbol{\xi}}(\boldsymbol{e}_\sigma)\boldsymbol{z}| d\sigma, \tag{3.2.11}$$

where $\boldsymbol{e}_\sigma$ is the n-dimensional unit vector joining the origin with a point $\sigma \in \mathbb{S}^{n-1}_-$.

The sharp constant K_ξ in the inequality

$$|\boldsymbol{u}_{\boldsymbol{\xi}}(x)| \leq K_\ell(\boldsymbol{\xi}) \sup\{|\boldsymbol{u}_{\boldsymbol{\xi}}(x', 0)| : x' \in \partial\mathbb{R}^n_+\}, \tag{3.2.12}$$

is given by

$$K_\ell(\boldsymbol{\xi}) = \sup_{|\boldsymbol{z}|=1} \int_{\mathbb{S}^{n-1}_-} |\mathcal{M}^*_{\boldsymbol{\xi}}(\boldsymbol{e}_\sigma)\boldsymbol{z}| d\sigma. \tag{3.2.13}$$

PROOF. By (3.2.6)-(3.2.9),

$$\begin{aligned} |(\boldsymbol{z}, \boldsymbol{u}_\xi(x))| &= |(\boldsymbol{z}, T_{\boldsymbol{\xi},x}(\psi))| = |\mathcal{T}_{\boldsymbol{\xi},x;\boldsymbol{z}}(\boldsymbol{\psi})| \\ &\leq \|\mathcal{T}_{\boldsymbol{\xi},x;z}\| \, \|\boldsymbol{\psi}\| = \|\mathcal{T}_{\boldsymbol{\xi},x;\boldsymbol{z}}\| \, \|\boldsymbol{u}_{\boldsymbol{\xi}}\big|_{x_n=0}\|. \end{aligned} \tag{3.2.14}$$

Applying Corollary 1.2 combined with Remark 1.2 to the functional $\mathcal{T}_{\boldsymbol{\xi},x;\boldsymbol{z}}$, we find

$$\|\mathcal{T}_{\boldsymbol{\xi},x;\boldsymbol{z}}\| = \int_{\partial\mathbb{R}^n_+} |\mathcal{M}^*_{\boldsymbol{\xi}}(\boldsymbol{e}_{xy})z|\omega(x, dy') = \int_{\mathbb{S}^{n-1}_-(x)} |\mathcal{M}^*_{\boldsymbol{\xi}}(\boldsymbol{e}_{xy})z| d\sigma(y),$$

where $\mathbb{S}^{n-1}_-(x)$ is the lower half of the unit sphere centered at a point $x \in \mathbb{R}^n$. Since the last integral does not depend on x, it follows from (3.2.14) with $K_{\boldsymbol{\xi},\boldsymbol{z}} := \|\mathcal{T}_{\boldsymbol{\xi},x;\boldsymbol{z}}\|$ that (3.2.10) is valid with the sharp constant (3.2.11).

Similarly, using the inequality

$$|\boldsymbol{u}_{\boldsymbol{\xi}}(x)| = |T_{\boldsymbol{\xi},x}(\boldsymbol{\psi})| \leq \|T_{\boldsymbol{\xi},x}\| \, \|\boldsymbol{\psi}\| = \|T_{\boldsymbol{\xi},x}\| \, \|\boldsymbol{u}_{\boldsymbol{\xi}}\big|_{x_n=0}\|$$

and applying Corollary 1.2 to the operator (3.2.8), we arrive at (3.2.12) with the sharp constant (3.2.13). □

In particular, for $\boldsymbol{\xi}_1 = \boldsymbol{e}_1, \ldots, \boldsymbol{\xi}_\ell = \boldsymbol{e}_\ell$ and $\ell = m$, equality (3.2.3) should be replaced by

$$\boldsymbol{f}(y') = \sum_{i=1}^m f_i(y')\boldsymbol{e}_i,$$

where $\boldsymbol{e}_i$ is the unit vector of the i-th coordinate axis. Therefore by (3.2.5) we have $\mathcal{M}_{\boldsymbol{e}} = \mathcal{M}$ with $\boldsymbol{e} = (\boldsymbol{e}_1, \ldots, \boldsymbol{e}_m)$. Since $\boldsymbol{u}_{\boldsymbol{e}} = \boldsymbol{u}$, by Theorem 3.1 we obtain

COROLLARY 3.1. *Let x be an arbitrary fixed point in $\mathbb{R}^n_+$ and let $\boldsymbol{z} \in \mathbb{R}^m, |\boldsymbol{z}| = 1$. The sharp constant $K_m(\boldsymbol{e}, \boldsymbol{z})$ in the inequality*

$$|(\boldsymbol{z}, \boldsymbol{u}(x))| \leq K_m(\boldsymbol{e}, \boldsymbol{z}) \sup\{|\boldsymbol{u}(x', 0)| : x' \in \partial\mathbb{R}^n_+\}, \tag{3.2.15}$$

is given by

$$K_m(\boldsymbol{e}, \boldsymbol{z}) = \int_{\mathbb{S}^{n-1}_-} |\mathcal{M}^*(\boldsymbol{e}_\sigma)\boldsymbol{z}| d\sigma, \tag{3.2.16}$$

where $\boldsymbol{e}_\sigma$ is the n-dimensional unit vector joining the origin with a point $\sigma \in \mathbb{S}^{n-1}_-$.

The sharp constant $K_m(\boldsymbol{e})$ *in the inequality*

$$|\boldsymbol{u}(x)| \leq K_m(\boldsymbol{e}) \sup\{|\boldsymbol{u}(x',0)| : x' \in \partial\mathbb{R}^n_+\}, \tag{3.2.17}$$

is given by

$$K_m(\boldsymbol{e}) = \sup_{|\boldsymbol{z}|=1} \int_{\mathbb{S}^{n-1}_-} |\mathcal{M}^*(\boldsymbol{e}_\sigma)\boldsymbol{z}| d\sigma. \tag{3.2.18}$$

3.2.2. Systems with constant complex coefficients

We introduce the strongly elliptic operator

$$\mathfrak{C}_0(D_x) = \sum_{i,j=1}^{n} \mathcal{C}_{ij} \frac{\partial^2}{\partial x_i \partial x_j},$$

where $D_x = (\partial/\partial x_1, \ldots, \partial/\partial x_n)$ and $\mathcal{C}_{ij} = \mathcal{C}_{ji}$ are constant complex $(m \times m)$-matrices. The strong ellipticity of $\mathfrak{C}_0(D_x)$ means that the inequality

$$\Re \left(\sum_{i,j=1}^{n} \mathcal{C}_{ij}\sigma_i\sigma_j\boldsymbol{\zeta}, \boldsymbol{\zeta} \right) > 0$$

holds for all $\boldsymbol{\zeta} = (\zeta_1, \ldots, \zeta_m) \in \mathbb{C}^m \backslash \{\boldsymbol{0}\}$, $\boldsymbol{\sigma} = (\sigma_1, \ldots, \sigma_n) \in \mathbb{R}^n \backslash \{\boldsymbol{0}\}$.

We understand the notations $[\mathbf{C}^2(\mathbb{R}^n_+)]^m$, $[\mathbf{C_b}(\overline{\mathbb{R}^n_+})]^m$ and $[\mathbf{C_b}(\partial\mathbb{R}^n_+)]^m$ for spaces of vector-valued functions with complex components in the same sense as in the previous subsection.

There exists a solution of the Dirichlet problem (see Subsections 2.2.1, 2.2.2)

$$\mathfrak{C}_0(D_x)\boldsymbol{w} = \boldsymbol{0} \text{ in } \mathbb{R}^n_+, \quad \boldsymbol{w} = \boldsymbol{g} \text{ on } \partial\mathbb{R}^n_+, \tag{3.2.19}$$

where $\boldsymbol{g} \in [\mathbf{C_b}(\partial\mathbb{R}^n_+)]^m$, which is bounded on $\overline{\mathbb{R}^n_+}$ and continuous up to $\partial\mathbb{R}^n_+$. This solution can be represented in the form

$$\boldsymbol{w}(x) = \int_{\partial\mathbb{R}^n_+} \mathcal{G}(\boldsymbol{e}_{xy})\boldsymbol{g}(y')\, \omega(x, dy'). \tag{3.2.20}$$

Here $\mathcal{G}$ is a continuous $(m \times m)$-matrix-valued function with bounded complex components given on the hemisphere $\mathbb{S}^{n-1}_-$. Similarly to the case of real coefficients, the solution is unique in the class $[\mathbf{C}^2(\mathbb{R}^n_+)]^m \cap [\mathbf{C_b}(\overline{\mathbb{R}^n_+})]^m$.

Let $\boldsymbol{\varphi} = (\varphi_1, \ldots, \varphi_\ell) \in [\mathbf{C_b}(\partial\mathbb{R}^n_+)]^\ell$ and let $\boldsymbol{\eta} = (\boldsymbol{\eta}_1, \ldots, \boldsymbol{\eta}_\ell)$, where $\boldsymbol{\eta}_1, \ldots, \boldsymbol{\eta}_\ell$ is a fixed orthogonal system of normed vectors in $\mathbb{C}^m, 1 \leq \ell \leq m$. The solution $\boldsymbol{w}_{\boldsymbol{\eta}}$ of the Dirichlet problem (3.2.19) with the boundary data

$$\boldsymbol{\Phi}_{\boldsymbol{\eta}}(y') = \sum_{i=1}^{\ell} \varphi_i(y')\boldsymbol{\eta}_i$$

can be written as

$$\boldsymbol{w}_{\boldsymbol{\eta}}(x) = \int_{\partial\mathbb{R}^n_+} \mathcal{G}(\boldsymbol{e}_{xy})\boldsymbol{\Phi}_{\boldsymbol{\eta}}(y')\omega(x, dy') \tag{3.2.21}$$

in view of (3.2.20). Introducing the $(m \times \ell)$-matrix-valued function

$$\mathcal{G}_{\boldsymbol{\eta}}(\boldsymbol{e}_{xy}) = \Big[\mathcal{G}(\boldsymbol{e}_{xy})\boldsymbol{\eta}_1, \ldots, \mathcal{G}(\boldsymbol{e}_{xy})\boldsymbol{\eta}_\ell\Big],$$

we rewrite (3.2.21) as

$$\boldsymbol{w}_{\boldsymbol{\eta}}(x) = T_{\boldsymbol{\eta},x}(\boldsymbol{\varphi}) = \int_{\partial\mathbb{R}^n_+} \mathcal{G}_{\boldsymbol{\eta}}(\boldsymbol{e}_{xy})\boldsymbol{\varphi}(y')\omega(x, dy'), \tag{3.2.22}$$

with $\boldsymbol{\eta}$ and x as parameters.

The operator $T_{\boldsymbol{\eta},x}(\boldsymbol{\varphi})$ maps $[\mathbf{C_b}(\partial\mathbb{R}^n_+)]^\ell$ into $\mathbb{C}^m$ and is of type (1.2.29). For any fixed $\boldsymbol{z}\in\mathbb{C}^m$, we define the functional $\mathcal{T}_{\boldsymbol{\eta},x;\boldsymbol{z}}(\boldsymbol{\varphi})$on $[\mathbf{C_b}(\partial\mathbb{R}^n_+)]^\ell$ by the equality $\mathcal{T}_{\boldsymbol{\eta},x;\boldsymbol{z}}(\boldsymbol{\varphi})=(\boldsymbol{z},T_{\boldsymbol{\eta},x}(\boldsymbol{\varphi}))$. Similarly to the previous subsection, one can establish an analogue of Theorem 3.1 for $\boldsymbol{w_\eta}$.

THEOREM 3.2. *Let x be an arbitrary fixed point in $\mathbb{R}^n_+$ and let $\boldsymbol{z}\in\mathbb{C}^m$, $|\boldsymbol{z}|=1$. The sharp constant $\boldsymbol{K}_{\boldsymbol{\eta},\boldsymbol{z}}$ in the inequality*

$$|(\boldsymbol{z},\boldsymbol{w_\eta}(x))|\leq \boldsymbol{K}_\ell(\boldsymbol{\eta},\boldsymbol{z})\sup\{|\boldsymbol{w_\eta}(x',0)|: x'\in\partial\mathbb{R}^n_+\},$$

is given by

$$\boldsymbol{K}_\ell(\boldsymbol{\eta},\boldsymbol{z})=\int_{\mathbb{S}^{n-1}_-}|\mathcal{G}^*_{\boldsymbol{\eta}}(\boldsymbol{e}_\sigma)\boldsymbol{z}|d\sigma,$$

where $\mathcal{G}^*$ stands for the adjoint matrix of $\mathcal{G}$.

The sharp constant $\boldsymbol{K}_\eta$ in the inequality

$$|\boldsymbol{w_\eta}(x)|\leq \boldsymbol{K}_\ell(\boldsymbol{\eta})\sup\{|\boldsymbol{w_\eta}(x',0)|: x'\in\partial\mathbb{R}^n_+\},$$

is given by

$$\boldsymbol{K}_\ell(\boldsymbol{\eta})=\sup_{|\boldsymbol{z}|=1}\int_{\mathbb{S}^{n-1}_-}|\mathcal{G}^*_{\boldsymbol{\eta}}(\boldsymbol{e}_\sigma)\boldsymbol{z}|d\sigma.$$

Similarly to Subsection 3.2.1 we obtain

COROLLARY 3.2. *Let x be an arbitrary fixed point in $\mathbb{R}^n_+$ and let $\boldsymbol{z}\in\mathbb{C}^m$, $|\boldsymbol{z}|=1$. The sharp constant $\boldsymbol{K}_m(\boldsymbol{e},\boldsymbol{z})$ in the inequality*

$$|(\boldsymbol{z},\boldsymbol{w}(x))|\leq \boldsymbol{K}_m(\boldsymbol{e},\boldsymbol{z})\sup\{|\boldsymbol{w}(x',0)|: x'\in\partial\mathbb{R}^n_+\},$$

is given by

$$\mathcal{K}_m(\boldsymbol{e},\boldsymbol{z})=\int_{\mathbb{S}^{n-1}_-}|\mathcal{G}^*(\boldsymbol{e}_\sigma)\boldsymbol{z}|d\sigma,$$

where $\mathcal{G}^*$ stands for the adjoint matrix of $\mathcal{G}$.

The sharp constant $\boldsymbol{K}_m(\boldsymbol{e})$ in the inequality

$$|\boldsymbol{w}(x)|\leq \boldsymbol{K}_m(\boldsymbol{e})\sup\{|\boldsymbol{w}(x',0)|: x'\in\partial\mathbb{R}^n_+\},$$

is given by

$$\boldsymbol{K}_m(\boldsymbol{e})=\sup_{|\boldsymbol{z}|=1}\int_{\mathbb{S}^{n-1}_-}|\mathcal{G}^*(\boldsymbol{e}_\sigma)\boldsymbol{z}|d\sigma.$$

Now, we consider elliptic systems with complex coefficients and with real boundary data. We write vector-valued function $\boldsymbol{w}_\eta$ in (3.2.21) as $\boldsymbol{w_\eta}(x)=\boldsymbol{u_\eta}(x)+i\boldsymbol{v_\eta}(x)$, where $\boldsymbol{u_\eta}$ and $\boldsymbol{v_\eta}$ are real m component vector-valued functions. Suppose that $\boldsymbol{\varphi}=(\varphi_1,\dots,\varphi_\ell)\in[\mathrm{C_b}(\partial\mathbb{R}^n_+)]^\ell$ and that $\boldsymbol{\eta}_1,\dots,\boldsymbol{\eta}_\ell$ is an orthogonal system of normed vectors in $\mathbb{R}^m$. Then $\boldsymbol{v_\eta}(x',0)=\mathbf{0}$ and

$$\boldsymbol{w_\eta}(x',0)=\boldsymbol{u_\eta}(x',0)=\sum_{i=1}^{\ell}\varphi_i(x')\boldsymbol{\eta}_i.$$

Let $T^{\langle R\rangle}_{\boldsymbol{\eta},x}$ denote the restriction of the operator $T_{\boldsymbol{\eta},x}$ defined by (3.2.22) on $[\mathrm{C_b}(\partial\mathbb{R}^n_+)]^\ell$. Then $\boldsymbol{w_\eta}(x)=T^{\langle R\rangle}_{\boldsymbol{\eta},x}(\boldsymbol{\varphi})$. For any fixed unit vector $\boldsymbol{z}\in\mathbb{C}^m$, we introduce the functional $\mathcal{T}^{\langle\Re\rangle}_{\boldsymbol{\eta},x;\boldsymbol{z}}$ on $[\mathrm{C_b}(\partial\mathbb{R}^n_+)]^\ell$ by the equality $\mathcal{T}^{\langle\Re\rangle}_{\boldsymbol{\eta},x;\boldsymbol{z}}(\boldsymbol{\varphi})=\Re(\boldsymbol{z},T^{\langle R\rangle}_{\boldsymbol{\eta},x}(\boldsymbol{\varphi}))$. Therefore, $\Re(\boldsymbol{z},\boldsymbol{w_\eta}(x))=\mathcal{T}^{\langle\Re\rangle}_{\boldsymbol{\eta},x;\boldsymbol{z}}(\boldsymbol{\varphi})$.

The next assertion contains sharp estimates of $|\Re(\boldsymbol{z}, \boldsymbol{w}_{\boldsymbol{\eta}}(x))|$ and $|\boldsymbol{w}_{\boldsymbol{\eta}}(x)|$ under the above assumptions on $\boldsymbol{w}_{\boldsymbol{\eta}}(x', 0)$.

THEOREM 3.3. *Let x be arbitrary fixed point in $\mathbb{R}^n_+$ and let $\boldsymbol{z} \in \mathbb{C}^m, |z| = 1$. Further, let $\boldsymbol{\varphi} = (\varphi_1, \ldots, \varphi_\ell) \in [\mathrm{C_b}(\partial\mathbb{R}^n_+)]^\ell$ and let $\boldsymbol{\eta}_1, \ldots, \boldsymbol{\eta}_\ell$ be an orthogonal system of normed vectors in $\mathbb{R}^m$. Then the sharp constant $\boldsymbol{K}^{\langle R\rangle}_\ell(\boldsymbol{\eta}, \boldsymbol{z})$ in the inequality*

$$|\Re(\boldsymbol{z}, \boldsymbol{w}_{\boldsymbol{\eta}}(x))| \le \boldsymbol{K}^{\langle R\rangle}_\ell(\boldsymbol{\eta}, \boldsymbol{z}) \sup\{|\boldsymbol{w}_{\boldsymbol{\eta}}(x', 0)| : x' \in \partial\mathbb{R}^n_+\}, \tag{3.2.23}$$

is given by

$$\boldsymbol{K}^{\langle R\rangle}_\ell(\boldsymbol{\eta}, \boldsymbol{z}) = \int_{\mathbb{S}^{n-1}_-} \left|\Re\left\{\mathcal{G}^*_{\boldsymbol{\eta}}(\boldsymbol{e}_\sigma)\boldsymbol{z}\right\}\right| d\sigma, \tag{3.2.24}$$

where $\mathcal{G}^*$ stands for the adjoint matrix of $\mathcal{G}$.

The sharp constant $\boldsymbol{K}^{\langle R\rangle}_\ell(\boldsymbol{\eta})$ in the inequality

$$|\boldsymbol{w}_{\boldsymbol{\eta}}(x)| \le \boldsymbol{K}^{\langle R\rangle}_\ell(\boldsymbol{\eta}) \sup\{|\boldsymbol{w}_{\boldsymbol{\eta}}(x', 0)| : x' \in \partial\mathbb{R}^n_+\} \tag{3.2.25}$$

is given by

$$\boldsymbol{K}^{\langle R\rangle}_\ell(\boldsymbol{\eta}) = \sup_{|\boldsymbol{z}|=1} \int_{\mathbb{S}^{n-1}_-} \left|\Re\left\{\mathcal{G}^*_\eta(\boldsymbol{e}_\sigma)\boldsymbol{z}\right\}\right| d\sigma. \tag{3.2.26}$$

PROOF. We have

$$|\Re(\boldsymbol{z}, \boldsymbol{w}_{\boldsymbol{\eta}}(x))| = |\mathcal{T}^{\langle\Re\rangle}_{\boldsymbol{\eta},x;\boldsymbol{z}}(\boldsymbol{\varphi})| \le \|\mathcal{T}^{\langle\Re\rangle}_{\boldsymbol{\eta},x;\boldsymbol{z}}\|\ \|\boldsymbol{\varphi}\| = \|\mathcal{T}^{\langle\Re\rangle}_{\boldsymbol{\eta},x;\boldsymbol{z}}\|\ \|\boldsymbol{w}_{\boldsymbol{\eta}}\big|_{x_n=0}\|. \tag{3.2.27}$$

Applying Corollary 1.3 combined with Remark 1.4 to the functional $\mathcal{T}^{\langle\Re\rangle}_{\boldsymbol{\eta},x;\boldsymbol{z}}$, we obtain

$$\|\mathcal{T}^{\langle\Re\rangle}_{\boldsymbol{\eta},x;\boldsymbol{z}}\| = \int_{\partial\mathbb{R}^n_+} |\Re\{\mathcal{G}^*_{\boldsymbol{\eta}}(\boldsymbol{e}_{xy})\boldsymbol{z}\}|\omega(x, dy') = \int_{\mathbb{S}^{n-1}_-(x)} |\Re\{\mathcal{G}^*_{\boldsymbol{\eta}}(\boldsymbol{e}_{xy})\boldsymbol{z}\}| d\sigma(y).$$

Since the integral on the right-hand side of the last equality does not depend on x, then (3.2.27) with $\|\mathcal{T}^{\langle\Re\rangle}_{\boldsymbol{\eta},x;\boldsymbol{z}}\| = \boldsymbol{K}^{\langle R\rangle}_\ell(\boldsymbol{\eta}, \boldsymbol{z})$ implies (3.2.23) with the sharp constant (3.2.24).

Similarly, using

$$|\boldsymbol{w}_{\boldsymbol{\eta}}(x)| = |T^{\langle R\rangle}_{\boldsymbol{\eta},x}(\boldsymbol{\varphi})| \le \|T^{\langle R\rangle}_{\boldsymbol{\eta},x}\|\ \|\boldsymbol{\varphi}\| = \|T^{\langle R\rangle}_{\boldsymbol{\eta},x}\|\ \|\boldsymbol{w}_{\boldsymbol{\eta}}\big|_{x_n=0}\|,$$

and applying Corollary 1.3 combined with Remark 1.4 to the operator $T^{\langle R\rangle}_{\boldsymbol{\eta},x}$, we arrive at (3.2.25) with the sharp constant (3.2.26). □

REMARK 3.1. Putting $\boldsymbol{z} = \boldsymbol{\zeta} \in \mathbb{R}^m$ in (3.2.23) and (3.2.24), we obtain

$$|(\boldsymbol{\zeta}, \Re\boldsymbol{w}_{\boldsymbol{\eta}}(x))| \le \boldsymbol{K}^{\langle R\rangle}_\ell(\boldsymbol{\eta}, \boldsymbol{z}) \sup\{|\boldsymbol{w}_{\boldsymbol{\eta}}(x', 0)| : x' \in \partial\mathbb{R}^n_+\} \tag{3.2.28}$$

with the sharp constant

$$\boldsymbol{K}^{\langle R\rangle}_\ell(\boldsymbol{\eta}, \boldsymbol{z}) = \int_{\mathbb{S}^{n-1}_-} \left|\Re\left\{\mathcal{G}^*_{\boldsymbol{\eta}}(\boldsymbol{e}_\sigma)\right\}\boldsymbol{\zeta}\right| d\sigma. \tag{3.2.29}$$

Similarly, putting $\boldsymbol{z} = i\boldsymbol{\zeta}, \boldsymbol{\zeta} \in \mathbb{R}^m$ in (3.2.23) and (3.2.24), we arrive at

$$|(\boldsymbol{\zeta}, \Im\boldsymbol{w}_{\boldsymbol{\eta}}(x))| \le \boldsymbol{K}^{\langle I\rangle}_\ell(\boldsymbol{\eta}, \boldsymbol{\zeta}) \sup\{|\boldsymbol{w}_{\boldsymbol{\eta}}(x', 0)| : x' \in \partial\mathbb{R}^n_+\}, \tag{3.2.30}$$

with the sharp constant

$$\boldsymbol{K}_\ell^{\langle I\rangle}(\boldsymbol{\eta},\boldsymbol{\zeta}) = \int_{\mathbb{S}_-^{n-1}} \left|\Im\left\{\mathcal{G}_{\boldsymbol{\eta}}^*(\boldsymbol{e}_\sigma)\right\}\boldsymbol{\zeta}\right| d\sigma. \tag{3.2.31}$$

Note that inequalities (3.2.28), (3.2.30) together with (3.2.29), (3.2.31) imply the estimates

$$|\Re\boldsymbol{w}_{\boldsymbol{\eta}}(x)| \le \boldsymbol{K}_\ell^{\langle R\rangle}(\boldsymbol{\eta})\sup\{|\boldsymbol{w}_{\boldsymbol{\eta}}(x',0)| : x'\in\partial\mathbb{R}_+^n\},$$

$$|\Im\boldsymbol{w}_{\boldsymbol{\eta}}(x)| \le \boldsymbol{K}_\ell^{\langle I\rangle}(\boldsymbol{\eta})\sup\{|\boldsymbol{w}_{\boldsymbol{\eta}}(x',0)| : x'\in\partial\mathbb{R}_+^n\},$$

with the sharp constants

$$\boldsymbol{K}_\ell^{\langle R\rangle}(\boldsymbol{\eta}) = \sup_{|\boldsymbol{\zeta}|=1}\int_{\mathbb{S}_-^{n-1}} \left|\Re\left\{\mathcal{G}_{\boldsymbol{\eta}}^*(\boldsymbol{e}_\sigma)\right\}\boldsymbol{\zeta}\right| d\sigma, \quad \boldsymbol{K}_\ell^{\langle I\rangle}(\boldsymbol{\eta}) = \sup_{|\boldsymbol{\zeta}|=1}\int_{\mathbb{S}_-^{n-1}} \left|\Im\left\{\mathcal{G}_{\boldsymbol{\eta}}^*(\boldsymbol{e}_\sigma)\right\}\boldsymbol{\zeta}\right| d\sigma,$$

where $\boldsymbol{\zeta}\in\mathbb{R}^m$.

3.3. The Lamé and Stokes systems in a half-space

In the half-space $\mathbb{R}_+^n$, $n\ge 2$, consider the Lamé system

$$\mu\Delta\boldsymbol{u} + (\lambda+\mu)\text{grad div }\boldsymbol{u} = \boldsymbol{0}, \tag{3.3.1}$$

and the Stokes system

$$\nu\Delta\boldsymbol{u} - \text{grad }p = \boldsymbol{0}, \quad \text{div }\boldsymbol{u} = 0, \tag{3.3.2}$$

with the boundary condition

$$\boldsymbol{u}\big|_{x_n=0} = \boldsymbol{f}, \tag{3.3.3}$$

where λ and μ are the Lamé constants, ν is the kinematic coefficient of viscosity, $\boldsymbol{f}\in[\mathrm{C_b}(\partial\mathbb{R}_+^n)]^n$, $\boldsymbol{u}=(u_1,\dots,u_n)$ is the displacement vector of an elastic medium or the velocity vector of a fluid, and p is the pressure in the fluid.

Solution $\boldsymbol{u}\in[\mathrm{C}^2(\mathbb{R}_+^n)]^n\cap[\mathrm{C_b}(\overline{\mathbb{R}_+^n})]^n$ of problem (3.3.1), (3.3.3) and vector component $\boldsymbol{u}\in[\mathrm{C}^2(\mathbb{R}_+^n)]^n\cap[\mathrm{C_b}(\overline{\mathbb{R}_+^n})]^n$ of solution $\{\boldsymbol{u},p\}$ of problem (3.3.2), (3.3.3) admit the representation

$$\boldsymbol{u}(x) = \int_{\partial\mathbb{R}_+^n}\mathcal{H}\left(\frac{y-x}{|y-x|}\right)\frac{x_n}{|y-x|^n}\boldsymbol{f}(y')dy', \tag{3.3.4}$$

where $x\in\mathbb{R}_+^n$, $y=(y',0)$, $y'=(y_1,\dots,y_{n-1})$. Here $\mathcal{H}$ is the $(n\times n)$-matrix-valued function on $\mathbb{S}^{n-1}=\{x\in\mathbb{R}^n : |x|=1\}$ with the elements

$$\frac{2}{\omega_n}\left((1-\kappa)\delta_{ij} + n\kappa\frac{(y_i-x_i)(y_j-x_j)}{|y-x|^2}\right), \tag{3.3.5}$$

$\kappa=1$ for the Stokes system, $\kappa=(\lambda+\mu)(\lambda+3\mu)^{-1}$ for the Lamé system, and ω_n being the area of $\mathbb{S}^{n-1}$.

REMARK 3.2. We assume that $\mu>0, \lambda+\mu>0$ that is $\kappa\in(0,1]$ in (3.3.5). Note that in the elasticity theory it is assumed that $\mu>0, 3\lambda+2\mu>0$ (see Kupradze, Gegelia, Basheleishvili and Burchuladze [**KGBB**]).

Representation (3.3.4) is well known in the two- and three-dimensional cases (see Kupradze, Gegelia, Basheleishvili and Burchuladze [**KGBB**], Ladyzhenskaya [**Lad**], Parton and Perlin [**PP1**]). For an arbitrary $n\ge 2$ formula (3.3.4) results by solving the boundary integral equation for the density of the vector-valued double layer potential (7.4.65) with $\Omega = \mathbb{R}^n_+$ in the same way as the Poisson formula in $\mathbb{R}^n_+$ is derived in the theory of harmonic potentials.

The expression (3.3.4) of the displacement vector satisfying the Dirichlet problem (3.3.1), (3.3.3) and the velocity vector satisfying the Dirichlet problem (3.3.2), (3.3.3) is a particular case of (2.2.3).

Similarly to Subsection 3.2.1, we fix an orthogonal system of the normed vectors $\boldsymbol{\xi}_1, \dots, \boldsymbol{\xi}_\ell \in \mathbb{R}^n, 1 \le \ell \le n$ and use the notation $\boldsymbol{\xi} = (\boldsymbol{\xi}_1, \dots, \boldsymbol{\xi}_\ell)$. We introduce a vector-valued function $\boldsymbol{\psi} = (\psi_1, \dots, \psi_\ell) \in [\mathrm{C_b}(\partial\mathbb{R}^n_+)]^\ell$ and put

$$\boldsymbol{\Phi}_\xi(y') = \sum_{i=1}^{\ell} \psi_i(y')\boldsymbol{\xi}_i. \tag{3.3.6}$$

We construct a solution $\boldsymbol{u}_{\boldsymbol{\xi}}$ of the Dirichlet problem (3.3.1), (3.3.3) and the vector component $\boldsymbol{u}_{\boldsymbol{\xi}}$ of the solution of the Dirichlet problem (3.3.2), (3.3.3) with the boundary data $\boldsymbol{\Phi}_{\boldsymbol{\xi}}$, i.e.

$$\boldsymbol{u}_{\boldsymbol{\xi}}(x) = \int_{\partial\mathbb{R}^n_+} \mathcal{H}\left(\boldsymbol{e}_{xy}\right) \boldsymbol{\Phi}_{\boldsymbol{\xi}}(y')\omega(x, dy'). \tag{3.3.7}$$

Here

$$\omega(x, \mathcal{B}) = \int_{\mathcal{B}} \frac{x_n}{|y-x|^n} dy'$$

is the solid angle at which a Borel set $\mathcal{B} \subset \partial\mathbb{R}^n_+$ is seen from the point $x \in \mathbb{R}^n_+$ and $\boldsymbol{e}_{xy} = (y-x)|y-x|^{-1}$. We put

$$\mathcal{H}_{\boldsymbol{\xi}}(\boldsymbol{e}_{xy}) = \Big[\mathcal{H}(\boldsymbol{e}_{xy})\boldsymbol{\xi}_1, \dots, \mathcal{H}(\boldsymbol{e}_{xy})\boldsymbol{\xi}_\ell\Big] \tag{3.3.8}$$

and write (3.3.7) in the form

$$\boldsymbol{u}_{\boldsymbol{\xi}}(x) = \int_{\partial\mathbb{R}^n_+} \mathcal{H}_{\boldsymbol{\xi}}\left(\boldsymbol{e}_{xy}\right) \boldsymbol{\psi}(y')\omega(x, dy'), \tag{3.3.9}$$

where

$$\boldsymbol{u}_{\boldsymbol{\xi}}(x', 0) = \sum_{i=1}^{\ell} \psi_i(x')\boldsymbol{\xi}_i. \tag{3.3.10}$$

Hence, in view of the orthogonality of $\boldsymbol{\xi}_1, \dots, \boldsymbol{\xi}_\ell$,

$$|\boldsymbol{u}_{\boldsymbol{\xi}}(x', 0)| = |\boldsymbol{\psi}(x')|. \tag{3.3.11}$$

THEOREM 3.4. *Let x be an arbitrary point in $\mathbb{R}^n_+$ and let $\boldsymbol{z} \in \mathbb{R}^n, |\boldsymbol{z}| = 1$. The best constant $K_\ell(\boldsymbol{\xi}, \boldsymbol{z})$ in the inequality*

$$|(\boldsymbol{z}, \boldsymbol{u}_{\boldsymbol{\xi}}(x))| \le K_\ell(\boldsymbol{\xi}, \boldsymbol{z}) \sup\{|\boldsymbol{u}_{\boldsymbol{\xi}}(x', 0)| : x' \in \partial\mathbb{R}^n_+\} \tag{3.3.12}$$

is given by

$$K_\ell(\boldsymbol{\xi}, \boldsymbol{z}) = \frac{1}{\omega_n} \int_{\mathbb{S}^{n-1}} \left\{ \sum_{i=1}^{\ell} \Big[(1-\kappa)(\boldsymbol{\xi}_i, \boldsymbol{z}) + n\kappa(\boldsymbol{e}_\sigma, \boldsymbol{z})(\boldsymbol{e}_\sigma, \boldsymbol{\xi}_i)\Big]^2 \right\}^{1/2} d\sigma, \tag{3.3.13}$$

where $\boldsymbol{e}_\sigma$ is the n-dimensional unit vector joining the origin with a point $\sigma \in \mathbb{S}^{n-1}$.

The best constant $K_\ell(\boldsymbol{\xi})$ in the inequality

$$|\boldsymbol{u}_{\boldsymbol{\xi}}(x)| \le K_\ell(\boldsymbol{\xi}) \sup\{|\boldsymbol{u}_{\boldsymbol{\xi}}(x', 0)| : x' \in \partial\mathbb{R}^n_+\} \tag{3.3.14}$$

is given by

$$K_\ell(\boldsymbol{\xi}) = \frac{1}{\omega_n} \sup_{|\boldsymbol{z}|=1} \int_{\mathbb{S}^{n-1}} \left\{ \sum_{i=1}^{\ell} \Big[(1-\kappa)(\boldsymbol{\xi}_i, \boldsymbol{z}) + n\kappa(\boldsymbol{e}_\sigma, \boldsymbol{z})(\boldsymbol{e}_\sigma, \boldsymbol{\xi}_i) \Big]^2 \right\}^{1/2} d\sigma. \tag{3.3.15}$$

Here $K_1(\boldsymbol{\xi}) = 1$ *and* $K_\ell(\boldsymbol{\xi}) > 1$ *for all* $2 \le \ell \le n$ *and* $0 < \kappa \le 1$.

PROOF. By Theorem 3.1, the sharp constant $K_n(\boldsymbol{\xi}, \boldsymbol{z})$ in (3.3.12) is given by

$$K_\ell(\boldsymbol{\xi}, \boldsymbol{z}) = \int_{\mathbb{S}_-^{n-1}} |\mathcal{H}_{\boldsymbol{\xi}}^*(\boldsymbol{e}_\sigma)\boldsymbol{z}| d\sigma, \tag{3.3.16}$$

where the matrix-valued function $\mathcal{H}_{\boldsymbol{\xi}}$ is defined by (3.3.8) and the elements of the $(n \times n)$-matrix $\mathcal{H}$ generating $\mathcal{H}_{\boldsymbol{\xi}}$ have the form (3.3.5).

In view of (3.3.5),

$$\mathcal{H}(\boldsymbol{e}_{xy})\boldsymbol{\xi}_i = \frac{2}{\omega_n}\Big[(1-\kappa)\boldsymbol{\xi}_i + n\kappa \boldsymbol{e}_{xy}(\boldsymbol{e}_{xy}, \boldsymbol{\xi}_i) \Big]. \tag{3.3.17}$$

By this equality and by (3.3.8), (3.3.16)

$$K_\ell(\boldsymbol{\xi}, \boldsymbol{z}) = \frac{2}{\omega_n} \int_{\mathbb{S}_-^{n-1}} \left\{ \sum_{i=1}^{\ell} \Big[(1-\kappa)(\boldsymbol{\xi}_i, \boldsymbol{z}) + n\kappa(\boldsymbol{e}_\sigma, \boldsymbol{z})(\boldsymbol{e}_\sigma, \boldsymbol{\xi}_i) \Big]^2 \right\}^{1/2} d\sigma.$$

Since the integrand is even with respect to $\boldsymbol{e}_\sigma$, it follows that

$$K_\ell(\boldsymbol{\xi}, \boldsymbol{z}) = \frac{1}{\omega_n} \int_{\mathbb{S}^{n-1}} \left\{ \sum_{i=1}^{\ell} \Big[(1-\kappa)(\boldsymbol{\xi}_i, \boldsymbol{z}) + n\kappa(\boldsymbol{e}_\sigma, \boldsymbol{z})(\boldsymbol{e}_\sigma, \boldsymbol{\xi}_i) \Big]^2 \right\}^{1/2} d\sigma,$$

which proves that the sharp constant in (3.3.12) is given by (3.3.13).

By Theorem 3.1, the sharp constant in (3.3.14) is defined by

$$K_\ell(\boldsymbol{\xi}) = \sup_{|\boldsymbol{z}|=1} \int_{\mathbb{S}_-^{n-1}} |\mathcal{H}_{\boldsymbol{\xi}}^*(\boldsymbol{e}_\sigma)\boldsymbol{z}| d\sigma\ , \tag{3.3.18}$$

which together with (3.3.8) and (3.3.17) implies

$$K_\ell(\boldsymbol{\xi}) = \frac{2}{\omega_n} \sup_{|\boldsymbol{z}|=1} \int_{\mathbb{S}_-^{n-1}} \left\{ \sum_{i=1}^{\ell} \Big[(1-\kappa)(\boldsymbol{\xi}_i, \boldsymbol{z}) + n\kappa(\boldsymbol{e}_\sigma, \boldsymbol{z})(\boldsymbol{e}_\sigma, \boldsymbol{\xi}_i) \Big]^2 \right\}^{1/2} d\sigma.$$

Since the integrand is even with respect to $\boldsymbol{e}_\sigma$, we arrive at (3.3.15).

Our next goal is to prove the assertion of the present theorem concerning the values of $K_\ell(\boldsymbol{\xi})$ for different $\ell, 1 \le \ell \le n$. Putting $\ell = 1$ in (3.3.15), we find

$$K_1(\boldsymbol{\xi}) = \frac{1}{\omega_n} \sup_{|\boldsymbol{z}|=1} \int_{\mathbb{S}_-^{n-1}} |(1-\kappa)(\boldsymbol{\xi}, \boldsymbol{z}) + n\kappa(\boldsymbol{e}_\sigma, \boldsymbol{z})(\boldsymbol{e}_\sigma, \boldsymbol{\xi})|\, d\sigma. \tag{3.3.19}$$

This implies

$$K_1(\boldsymbol{\xi}) \le \frac{1-\kappa}{\omega_n} \sup_{|\boldsymbol{z}|=1} \int_{\mathbb{S}^{n-1}} |(\boldsymbol{\xi}, \boldsymbol{z})| d\sigma + \frac{n\kappa}{\omega_n} \sup_{|\boldsymbol{z}|=1} \int_{\mathbb{S}^{n-1}} |(\boldsymbol{e}_\sigma, \boldsymbol{z})|\, |(\boldsymbol{e}_\sigma, \boldsymbol{\xi})| d\sigma.$$

By the Cauchy inequality applied to the last integral,

$$K_1(\boldsymbol{\xi}) \le 1 - \kappa + \frac{n\kappa}{\omega_n} \sup_{|\boldsymbol{z}|=1} \left\{ \int_{\mathbb{S}^{n-1}} (\boldsymbol{e}_\sigma, \boldsymbol{z})^2 d\sigma \int_{\mathbb{S}^{n-1}} (\boldsymbol{e}_\sigma, \boldsymbol{\xi})^2 d\sigma \right\}^{1/2}. \tag{3.3.20}$$

Let θ_1 be the angle between $\boldsymbol{e}_\sigma$ and $\boldsymbol{z}$. We have

$$\int_{\mathbb{S}^{n-1}} (\boldsymbol{e}_\sigma, \boldsymbol{z})^2 d\sigma$$

$$= \int_0^\pi \cos^2\theta_1 \sin^{n-2}\theta_1 d\theta_1 \int_0^\pi \sin^{n-3}\theta_2 d\theta_2 \ldots \int_0^\pi \sin\theta_{n-2} d\theta_{n-2} \int_0^{2\pi} d\theta_{n-1}$$

$$= 2\pi \int_0^\pi \cos^2\theta_1 \sin^{n-2}\theta_1 d\theta_1 \prod_{k=1}^{n-3} \int_0^\pi \sin^k\theta d\theta$$

$$= 2\pi \frac{\sqrt{\pi}\,\Gamma\left(\frac{n-1}{2}\right)}{n\Gamma\left(\frac{n}{2}\right)} \prod_{k=1}^{n-3} \frac{\sqrt{\pi}\,\Gamma\left(\frac{k+1}{2}\right)}{\Gamma\left(\frac{k+2}{2}\right)} = \frac{2\pi^{3/2}\Gamma\left(\frac{n-1}{2}\right)}{n\Gamma\left(\frac{n}{2}\right)} \frac{\pi^{(n-3)/2}}{\Gamma\left(\frac{n-1}{2}\right)} = \frac{2\pi^{n/2}}{n\Gamma\left(\frac{n}{2}\right)}.$$

Thus,

$$\int_{\mathbb{S}^{n-1}} (\boldsymbol{e}_\sigma, \boldsymbol{z})^2 d\sigma = \frac{\omega_n}{n}\,. \tag{3.3.21}$$

Using (3.3.21) in (3.3.20), we find $K_1(\boldsymbol{\xi}) \leq 1$. Hence $K_1(\boldsymbol{\xi}) = 1$.

Next, we show that $K_\ell(\boldsymbol{\xi}) > 1$ for all $\ell = 2, \ldots, n$ if $0 < \kappa \leq 1$. We put $\boldsymbol{z} = \boldsymbol{\xi}_1$ in (3.3.15). In view of (3.3.21), it follows from (3.3.15) that

$$\begin{aligned} K_\ell(\boldsymbol{\xi}) &\geq \frac{1}{\omega_n} \int_{\mathbb{S}^{n-1}} \left\{ [(1-\kappa) + n\kappa(\boldsymbol{e}_\sigma, \boldsymbol{\xi}_1)^2]^2 + n^2\kappa^2 \sum_{i=2}^{\ell} (\boldsymbol{e}_\sigma, \boldsymbol{\xi}_1)^2 (\boldsymbol{e}_\sigma, \boldsymbol{\xi}_i)^2 \right\}^{1/2} d\sigma \\ &> \frac{1}{\omega_n} \int_{\mathbb{S}^{n-1}} [(1-\kappa) + n\kappa(\boldsymbol{e}_\sigma, \boldsymbol{\xi}_1)^2] d\sigma = \frac{1}{\omega_n}\left[(1-\kappa)\omega_n + n\kappa\frac{\omega_n}{n}\right] = 1. \end{aligned}$$

Therefore, $K_\ell(\boldsymbol{\xi}) > 1$ for all $\ell = 2, \ldots, n$, and $0 < \kappa \leq 1$. □

Corollary 3.3. *Let $\ell = n-1$ and $(\boldsymbol{z}, \boldsymbol{\xi}_1) = \cdots = (\boldsymbol{z}, \boldsymbol{\xi}_{n-1}) = 0, |\boldsymbol{z}| = 1$. The sharp constant $K_{n-1}(\boldsymbol{\xi}, \boldsymbol{z}) = \mathcal{K}_{n-1}$ in* (3.3.12) *is given by*

$$\mathcal{K}_{n-1} = \frac{2\kappa\,\Gamma(\frac{n}{2})}{\sqrt{\pi}\;\Gamma(\frac{n-1}{2})}. \tag{3.3.22}$$

In particular, $\mathcal{K}_1 = 2/\pi$ and $\mathcal{K}_2 = 1$ for the two-dimensional and three-dimensional Stokes systems.

Proof. Putting $\ell = n-1$ in (3.3.13) and using the equalities $(\boldsymbol{z}, \boldsymbol{\xi}_1) = \cdots = (\boldsymbol{z}, \boldsymbol{\xi}_{n-1}) = 0$, we find

$$\begin{aligned} K_{n-1}(\boldsymbol{\xi}, \boldsymbol{z}) &= \frac{1}{\omega_n} \int_{\mathbb{S}^{n-1}} \left\{ n^2\kappa^2 (\boldsymbol{e}_\sigma, \boldsymbol{z})^2 \sum_{i=1}^{n-1} (\boldsymbol{e}_\sigma, \boldsymbol{\xi}_i)^2 \right\}^{1/2} d\sigma \\ &= \frac{n\kappa}{\omega_n} \int_{\mathbb{S}^{n-1}} |(\boldsymbol{e}_\sigma, \boldsymbol{z})| \left[1 - (\boldsymbol{e}_\sigma, \boldsymbol{z})^2\right]^{1/2} d\sigma \\ &= \frac{n\kappa}{\omega_n} \int_0^\pi |\cos\theta_1| \sin^{n-1}\theta_1 d\theta_1 \int_0^\pi \sin^{n-3}\theta_2 d\theta_2 \ldots \int_0^\pi \sin\theta_{n-2} d\theta_{n-2} \int_0^{2\pi} d\theta_{n-1} \\ &= \frac{2\pi n\kappa}{\omega_n} \cdot \frac{2}{n} \prod_{k=1}^{n-3} \int_0^\pi \sin^k\theta d\theta = \frac{4\pi\kappa}{\omega_n} \prod_{k=1}^{n-3} \frac{\sqrt{\pi}\,\Gamma(\frac{k+1}{2})}{\Gamma(\frac{k+2}{2})} = \frac{4\pi\kappa}{\omega_n} \cdot \frac{\pi^{(n-3)/2}}{\Gamma(\frac{n-1}{2})}, \end{aligned}$$

which implies (3.3.22). □

For $\ell = n$, (3.3.15) becomes

$$K_n(\boldsymbol{\xi}) = \frac{1}{\omega_n} \sup_{|\boldsymbol{z}|=1} \int_{\mathbb{S}^{n-1}} \left[(1-\kappa)^2 + n\kappa(n\kappa - 2\kappa + 2)(\boldsymbol{e}_\sigma, \boldsymbol{z})^2\right]^{1/2} d\sigma. \tag{3.3.23}$$

Since the right-hand side in (3.3.23) does not depend on $\boldsymbol{\xi}$, we omit the notation of $\boldsymbol{\xi}$ in its left-hand side. By Theorem 3.4, $K_n > 1$ for $0 < \kappa \le 1$.

COROLLARY 3.4. *Let x be an arbitrary point in $\mathbb{R}^n_+$. The best constant K_n in the inequality*

$$|\boldsymbol{u}(x)| \le K_n \sup\{|\boldsymbol{u}(x', 0)| : x' \in \partial\mathbb{R}^n_+\} \tag{3.3.24}$$

has the form

$$K_n = \frac{2\Gamma\left(\frac{n}{2}\right)}{\sqrt{\pi}\,\Gamma\left(\frac{n-1}{2}\right)} \int_0^{\pi/2} \left[(1-\kappa)^2 + n\kappa(n\kappa - 2\kappa + 2)\cos^2\theta\right]^{1/2} \sin^{n-2}\theta d\theta, \tag{3.3.25}$$

and, in particular,

$$K_2 = \frac{2}{\pi}(1+\kappa)\,\mathbf{E}\left(\frac{2\sqrt{\kappa}}{1+\kappa}\right) = 1 + \frac{1}{2^2}\kappa^2 + \frac{1}{2^2 4^2}\kappa^4 + \cdots + \left[\frac{(2k-3)!!}{2^k k!}\right]^2 \kappa^{2k} + \ldots,$$

$$K_3 = \frac{1}{2}\left[1 + 2\kappa + \frac{(1-\kappa)^2}{\sqrt{3\kappa(\kappa+2)}} \log\frac{1 + 2\kappa + \sqrt{3\kappa(\kappa+2)}}{1-\kappa}\right],$$

where $\mathbf{E}(k)$ is the complete elliptic integral of the second kind.

In the case $\kappa = 1$, i.e., for an n-dimensional Stokes system,

$$K_n = \frac{2}{\sqrt{\pi}} \frac{\Gamma(\frac{n}{2}+1)}{\Gamma(\frac{n+1}{2})},$$

and, in particular, $K_2 = 4/\pi$ and $K_3 = 3/2$.

PROOF. By (3.3.23) we have

$$K_n = \frac{1}{\omega_n} \sup_{|\boldsymbol{z}|=1} \int_{\mathbb{S}^{n-1}} \left[(1-\kappa)^2 + n\kappa(n\kappa - 2\kappa + 2)(\boldsymbol{e}_\sigma, \boldsymbol{z})^2\right]^{1/2} d\sigma.$$

Hence,

$$K_n = \frac{1}{\omega_n} \sup_{|\boldsymbol{z}|=1} \int_{\mathbb{S}^{n-1}} \left[(1-\kappa)^2 + n\kappa(n\kappa - 2\kappa + 2)\cos^2\theta(\sigma)\right]^{1/2} d\sigma, \tag{3.3.26}$$

where $\theta(\sigma)$ is the angle between the vectors $\boldsymbol{e}_\sigma$ and $\boldsymbol{z}$. Since the integral does not depend on $\boldsymbol{z}$,

$$\begin{aligned}
K_n &= \frac{1}{\omega_n}\int_0^\pi \dots \int_0^\pi \int_0^{2\pi} \Big[(1-\kappa)^2 + n\kappa(n\kappa - 2\kappa + 2)\cos^2\theta\Big]^{1/2} \sin^{n-2}\theta \\
&\qquad\qquad \times \sin^{n-3}\theta_1 \dots \sin\theta_{n-3} d\theta d\theta_1 \dots d\theta_{n-2} \\
&= \frac{4\pi}{\omega_n}\Big(\prod_{k=1}^{n-3}\int_0^{\pi/2} \sin^k\varphi d\varphi\Big) \\
&\quad \times \int_0^{\pi/2} \Big[(1-\kappa)^2 + n\kappa(n\kappa - 2\kappa + 2)\cos^2\theta\Big]^{1/2} \sin^{n-2}\theta d\theta \\
&= \frac{2}{\sqrt{\pi}}\frac{\Gamma(n/2)}{\Gamma((n-1)/2)}\int_0^{\pi/2} \Big[(1-\kappa)^2 + n\kappa(n\kappa - 2\kappa + 2)\cos^2\theta\Big]^{1/2} \sin^{n-2}\theta d\theta.
\end{aligned}$$

□

3.4. Planar deformed state

We assume here that the displacement vector depends (in certain Cartesian coordinates) on two variables only, say x_1 and x_2. Supposing $u_3 = 0$, that is, there is no displacement along the x_3-axis, we have the so-called planar deformation.

Let σ_{11}, σ_{12} and σ_{22} be the components of stress tensor in the half-plane $\mathbb{R}^2_+$. Consider the system of equations in $\mathbb{R}^2_+$ for the stresses in the planar deformed state (see, for example, Muskhelishvili [**Mu**]):

$$\partial\sigma_{11}/\partial x_1 + \partial\sigma_{12}/\partial x_2 = 0, \tag{3.4.1}$$

$$\partial\sigma_{12}/\partial x_1 + \partial\sigma_{22}/\partial x_2 = 0, \tag{3.4.2}$$

$$\Delta(\sigma_{11} + \sigma_{22}) = 0, \tag{3.4.3}$$

with the boundary conditions

$$\sigma_{12}(x_1, 0) = g_1(x_1), \quad \sigma_{22}(x_1, 0) = g_2(x_1), \tag{3.4.4}$$

where g_1 and g_2 are continuous and bounded functions on $\partial\mathbb{R}^2_+$.

In the following assertion (see Kresin and Maz'ya [**KM9**], Maz'ya and Kresin [**MK**]) we prove, in particular, that the classical maximum principle for the function $(\sigma_{12}^2 + \sigma_{22}^2)^{1/2}$, which holds when its boundary values have a special form.

THEOREM 3.5. *Let $(\sigma_{12}(x_1, 0), \sigma_{22}(x_1, 0)) = \psi(x_1)\boldsymbol{\xi}$, where $\psi \in C(\partial\mathbb{R}^2_+)$ and $\boldsymbol{\xi}$ is a two-dimensional unit vector. Then the inequality*

$$\|(\sigma_{12}^2 + \sigma_{22}^2)^{1/2}\|_{C_b(\overline{\mathbb{R}^2_+})} \le \|(\sigma_{12}^2 + \sigma_{22}^2)^{1/2}\|_{C_b(\partial\mathbb{R}^2_+)} \tag{3.4.5}$$

holds.

In the general case, for boundary data (3.4.4) the best constant in the inequality

$$\|(\sigma_{12}^2 + \sigma_{22}^2)^{1/2}\|_{C_b(\overline{\mathbb{R}^2_+})} \le \mathcal{K}\|(\sigma_{12}^2 + \sigma_{22}^2)^{1/2}\|_{C_b(\partial\mathbb{R}^2_+)} \tag{3.4.6}$$

is equal to $4/\pi$.

PROOF. Using (3.4.1), (3.4.2), we get

$$\frac{\partial}{\partial x_1}\left(\frac{\partial\sigma_{11}}{\partial x_1}+\frac{\partial\sigma_{12}}{\partial x_2}\right)-\frac{\partial}{\partial x_2}\left(\frac{\partial\sigma_{12}}{\partial x_1}+\frac{\partial\sigma_{22}}{\partial x_2}\right)=0,$$

i.e.,

$$\frac{\partial^2\sigma_{11}}{\partial x_1^2}=\frac{\partial^2\sigma_{22}}{\partial x_2^2}.$$

Substituting the last equality into (3.4.3), we obtain

$$\Delta\sigma_{22}+\frac{\partial}{\partial x_2}\left(\frac{\partial\sigma_{11}}{\partial x_2}+\frac{\partial\sigma_{22}}{\partial x_2}\right)=0. \tag{3.4.7}$$

By (3.4.1) and (3.4.2),

$$\frac{\partial}{\partial x_2}\left(\frac{\partial\sigma_{11}}{\partial x_1}+\frac{\partial\sigma_{12}}{\partial x_2}\right)+\frac{\partial}{\partial x_1}\left(\frac{\partial\sigma_{12}}{\partial x_1}+\frac{\partial\sigma_{22}}{\partial x_2}\right)=0, \tag{3.4.8}$$

i.e.

$$\Delta\sigma_{12}+\frac{\partial}{\partial x_1}\left(\frac{\partial\sigma_{11}}{\partial x_2}+\frac{\partial\sigma_{22}}{\partial x_2}\right)=0. \tag{3.4.9}$$

Let $\sigma_{12}=v_1$, $\sigma_{22}=v_2$, $\boldsymbol{v}=(v_1,v_2)$ and $p=-(\partial\sigma_{11}/\partial x_2+\partial\sigma_{22}/\partial x_2)$. Then equalities (3.4.2), (3.4.7), and (3.4.9) can be written as the Stokes system

$$\Delta\boldsymbol{v}-\text{grad}\,p=0,\quad \text{div}\,\boldsymbol{v}=0. \tag{3.4.10}$$

Hence, for any functions σ_{11}, σ_{12}, and σ_{22} satisfying the boundary problem (3.4.1)-(3.4.4) there exists the vector component $(v_1,v_2)=(\sigma_{12},\sigma_{22})$ of the solution of (3.4.10) subject to the boundary conditions

$$v_1(x_1,0)=g_1(x_1),\quad v_2(x_1,0)=g_2(x_1). \tag{3.4.11}$$

If

$$\begin{aligned} V &= \{\boldsymbol{v}\in[\mathrm{C}^2(\mathbb{R}^2_+)]^2\cap[\mathrm{C}(\overline{\mathbb{R}^2_+})]^2 : \Delta\boldsymbol{v}-\text{grad}\,p=0,\quad \text{div}\,\boldsymbol{v}=0\,\},\\ S &= \{(\sigma_{12},\sigma_{22})\in[\mathrm{C}^2(\mathbb{R}^2_+)]^2\cap[\mathrm{C}(\overline{\mathbb{R}^2_+})]^2 : \partial\sigma_{11}/\partial x_1+\partial\sigma_{12}/\partial x_2=0,\\ &\qquad \partial\sigma_{12}/\partial x_1+\partial\sigma_{22}/\partial x_2=0,\ \Delta(\sigma_{11}+\sigma_{22})=0\,\}, \end{aligned}$$

then the last assertion can be written as $S\subset V$.

By Theorem 3.4, if $\boldsymbol{v}(x_1,0)=\psi(x_1)\boldsymbol{\xi}$, then, for any $x\in\mathbb{R}^2_+$, there holds

$$|\boldsymbol{v}(x)|\le\|\boldsymbol{v}|_{x_2=0}\|.$$

Putting here $\boldsymbol{v}=(\sigma_{12},\sigma_{22})$, we arrive at (3.4.5), where the constant 1 in front of the norm on the right-hand side is the best possible.

Assume now that a solution of problem (3.4.10), (3.4.11) is known. Let $v_1=\sigma_{12}$, $v_2=\sigma_{22}$ and $\boldsymbol{v}=(\sigma_{12},\sigma_{22})$. Then the equality div $\boldsymbol{v}=0$ becomes (3.4.2), and (3.4.10) can be rewritten in the form

$$\frac{\partial\sigma_{12}}{\partial x_1}+\frac{\partial\sigma_{22}}{\partial x_2}=0,\quad \Delta\sigma_{12}-\frac{\partial p}{\partial x_1}=0,\quad \Delta\sigma_{22}-\frac{\partial p}{\partial x_2}=0. \tag{3.4.12}$$

Let us substitute

$$\frac{\partial^2\sigma_{12}}{\partial x_1^2}=-\frac{\partial^2\sigma_{22}}{\partial x_1\partial x_2}$$

into the second equation in (3.4.12). As a result,

$$\frac{\partial^2\sigma_{12}}{\partial x_2^2} - \frac{\partial^2\sigma_{22}}{\partial x_1\partial x_2} - \frac{\partial p}{\partial x_1} = 0,$$

i.e.

$$\frac{\partial}{\partial x_1}\left(\frac{\partial\sigma_{22}}{\partial x_2} + p\right) = \frac{\partial}{\partial x_2}\left(\frac{\partial\sigma_{12}}{\partial x_2}\right).$$

Therefore, there exists a function σ_{11} such that

$$\frac{\partial\sigma_{11}}{\partial x_2} = -\left(\frac{\partial\sigma_{22}}{\partial x_2} + p\right), \quad \frac{\partial\sigma_{11}}{\partial x_1} = -\frac{\partial\sigma_{12}}{\partial x_2}. \tag{3.4.13}$$

The system (3.4.1)-(3.4.3) is obtained from (3.4.12) and (3.4.13), and so $V \subset S$, and $S = V$ because $S \subset V$. An immediate consequence of this is that the exact constant $\mathcal{K}$ in (3.4.6) must coincide with the constant $K_2 = 4/\pi$ for the Stokes system (Corollary 3.4). □

3.5. The system of quasistatic viscoelasticity

The equilibrium equations for the viscoelastic medium described by the three parametric model under quasistatic assumptions has the form (see Germain [**Ge**], Rabotnov [**Ra**])

$$\int_0^t \left\{M(t-\tau)\Delta\frac{\partial \boldsymbol{u}(x,\tau)}{\partial\tau} + \left(C(t-\tau) + \frac{M(t-\tau)}{3}\right)\text{grad div}\frac{\partial \boldsymbol{u}(x,\tau)}{\partial\tau}\right\}d\tau \tag{3.5.1}$$
$$+\left\{M(t)\Delta\boldsymbol{u}(x,\ +0) + \left(C(t) + \frac{M(t)}{3}\right)\text{grad div}\ \boldsymbol{u}(x,\ +0)\right\} = \boldsymbol{0},$$

where $M(\tau) = \mu(1-\alpha+\alpha\exp(-\rho\tau))$, $C(\tau) = \kappa(1-\beta+\beta\exp(-\theta\tau))$ are relaxation functions under displacement and isotropic compression, respectively, $\mu, \kappa, \alpha, \beta, \rho, \theta$ are parameters of the model ($\mu, \kappa, \rho, \theta > 0$, $0 \le \alpha, \beta \le 1$), $\boldsymbol{u}(x,\ +0)$ is the limit of $\boldsymbol{u}(x,t)$ as $t \to +0$. We assume that $\alpha + \beta > 0$.

The Maxwell medium ($\alpha = \beta = 1$) and the medium with purely elastic behaviour under volume compression ($\beta = 0$) (see, e.g., Germain [**Ge**], Rabotnov [**Ra**]) are particular cases of the three parametric model.

We supply (3.5.1) with the boundary condition

$$\boldsymbol{u}\Big|_{x_3=0,t>0} = \boldsymbol{\varphi}, \tag{3.5.2}$$

where $\boldsymbol{\varphi} = (\varphi_1, \varphi_2, \varphi_3)$ is a continuous and bounded vector-valued function on $\partial\mathbb{R}^3_+ \times \overline{\mathbb{R}^1_+}$.

A bounded in $\mathbb{R}^3_+ \times \mathbb{R}^1_+$ solution of the problem (3.5.1), (3.5.2) has the form

$$\boldsymbol{u}(x,t) = \frac{1}{2\pi}\int_{y_3=0} \mathfrak{M}_\gamma(\boldsymbol{e}_{xy})\boldsymbol{\varphi}(y',t)\frac{x_3}{|y-x|^3}dy'$$
$$+\frac{1}{2\pi}\int_0^t\int_{y_3=0} \mathcal{H}(t-\tau)\mathfrak{N}(\boldsymbol{e}_{xy})\boldsymbol{\varphi}(y',\tau)\frac{x_3}{|y-x|^3}dy'd\tau,$$

where $\boldsymbol{e}_{xy} = (y-x)|y-x|^{-1}$, $y' = (y_1, y_2)$, $\mathfrak{M}_\gamma$ and $\mathfrak{N}$ are (3×3)-matrices with elements

$$(1-\gamma)\delta_{ij} + 3\gamma(\boldsymbol{e}, \boldsymbol{e}_i)(\boldsymbol{e}, \boldsymbol{e}_j) \quad\text{and}\quad -\delta_{ij} + 3(\boldsymbol{e}, \boldsymbol{e}_i)(\boldsymbol{e}, \boldsymbol{e}_j),$$

respectively,

$$\gamma = (\mu + 3\kappa)(7\mu + 3\kappa)^{-1}, \quad \mathcal{H}(\tau) = \eta_1\exp(-\xi_1\tau) + \eta_2\exp(-\xi_2\tau)$$

with

$$\eta_1 = \frac{c(a+\sqrt{a^2-b})-d}{2\sqrt{a^2-b}}, \ \eta_2 = -\frac{c(a-\sqrt{a^2-b})-d}{2\sqrt{a^2-b}},$$

$$\xi_1 = a+\sqrt{a^2-b}, \ \xi_2 = a-\sqrt{a^2-b}.$$

Here

$$a = \frac{7\mu(\theta+\rho-\alpha\rho)+3\kappa(\theta+\rho-\alpha\theta)}{2(7\mu+3\kappa)}, \quad b = \frac{\rho\theta(7\mu(1-\alpha)+3\kappa(1-\beta))}{7\mu+3\kappa},$$

$$c = \frac{18\kappa\mu(\alpha\rho-\beta\theta)}{(7\mu+3\kappa)^2}, \quad d = \frac{18\kappa\mu\rho\theta(\alpha-\beta)}{(7\mu+3\kappa)^2}.$$

THEOREM 3.6. *The best constant in*

$$|\boldsymbol{u}(x,t)| \le \mathcal{K}(t)\sup\{|\boldsymbol{u}(y,\tau)| : y\in\partial\mathbb{R}^3_+, \ 0\le\tau\le t\}, \tag{3.5.3}$$

for the three parametric model of the viscoelastic medium is

$$\mathcal{K}(t) = \left(1+\frac{\sqrt{3}}{6}\log(2+\sqrt{3})\right)\mathcal{F}(t)$$

$$+\frac{1}{2}\left(1+2\gamma+\frac{(1-\gamma)^2}{\sqrt{3\gamma(\gamma+2)}}\log\frac{1+2\gamma+\sqrt{3\gamma(\gamma+2)}}{1-\gamma}\right),$$

where

$$\mathcal{F}(t) = \int_0^t |\mathcal{H}(\tau)|d\tau = \int_0^t |\eta_1 \mathit{exp}(-\xi_1\tau)+\eta_2\mathit{exp}(-\xi_2\tau)|d\tau.$$

In particular,

$$\mathcal{F}(t) = \eta(1-\exp(-\xi t))$$

with

$$\eta = \frac{18\kappa\mu|\rho-\theta|}{(7\mu\theta+3\kappa\rho)(7\mu+3\kappa)}, \quad \xi = \frac{7\mu\theta+3\kappa\rho}{7\mu+3\kappa}$$

for the Maxwell medium $(\alpha=\beta=1)$ *and*

$$\mathcal{F}(t) = \lambda(1-\mathit{exp}(-\chi t))$$

with

$$\lambda = \frac{18\alpha\kappa\mu}{(7\mu+3\kappa)(7\mu(1-\alpha)+3\kappa)}, \quad \chi = \frac{\rho(7\mu(1-\alpha)+3\kappa)}{7\mu+3\kappa}$$

for the medium with purely elastic behaviour under volume compression $(\beta=0)$.

PROOF. Let (x,t) be a fixed point in $\mathbb{R}^3_+\times\mathbb{R}^1_+$. We find the norm $|||\boldsymbol{u}(x,t)|||$ of the mapping $[\mathrm{C_b}(\partial\mathbb{R}^3_+\times\overline{\mathbb{R}^1_+})]^3 \ni \boldsymbol{\varphi}\to\boldsymbol{u}(x,t)\in\mathbb{R}^3$. Using the properties of the inner product in $\mathbb{R}^n$ and the fact that the supremum operations commute, we find that

$$\begin{aligned}
2\pi|||\boldsymbol{u}(x,t)||| &= \sup_{|\boldsymbol{\varphi}|\le 1}\left|\int_{y_3=0}\mathfrak{M}_\gamma(\boldsymbol{e}_{xy})\boldsymbol{\varphi}(y',t)\frac{x_3}{|y-x|^3}dy'\right.\\
&+\left.\int_0^t\int_{y_3=0}\mathcal{H}(t-\tau)\mathfrak{N}(\boldsymbol{e}_{xy})\boldsymbol{\varphi}(y',\tau)\frac{x_3}{|y-x|^3}dy'd\tau\right|\\
&= \sup_{|\boldsymbol{\varphi}|\le 1}\sup_{|\boldsymbol{z}|=1}\left(\boldsymbol{z},\int_{y_3=0}\mathfrak{M}_\gamma(\boldsymbol{e}_{xy})\boldsymbol{\varphi}(y',t)\frac{x_3}{|y-x|^3}dy'\right.\\
&+\left.\int_0^t\int_{y_3=0}\mathcal{H}(t-\tau)\mathfrak{N}(\boldsymbol{e}_{xy})\boldsymbol{\varphi}(y',\tau)\frac{x_3}{|y-x|^3}dy'd\tau\right)\\
&= \sup_{|\boldsymbol{\varphi}|\le 1}\sup_{|z|=1}\left[\int_{y_3=0}(\boldsymbol{z},\,\mathfrak{M}_\gamma(\boldsymbol{e}_{xy})\boldsymbol{\varphi}(y',t))\,\frac{x_3}{|y-x|^3}dy'\right.\\
&+\left.\int_0^t\int_{y_3=0}\mathcal{H}(t-\tau)\,(\boldsymbol{z},\,\mathfrak{N}(\boldsymbol{e}_{xy})\boldsymbol{\varphi}(y',\tau))\,\frac{x_3}{|y-x|^3}dy'd\tau\right]\\
&= \sup_{|\boldsymbol{z}|=1}\sup_{|\boldsymbol{\varphi}|\le 1}\left[\int_{y_3=0}(\mathfrak{M}^*_\gamma(\boldsymbol{e}_{xy})\boldsymbol{z},\,\boldsymbol{\varphi}(y',t))\,\frac{x_3}{|y-x|^3}dy'\right.\\
&+\left.\int_0^t\int_{y_3=0}\mathcal{H}(t-\tau)\,(\mathfrak{N}^*(\boldsymbol{e}_{xy})z,\,\boldsymbol{\varphi}(y',\tau))\,\frac{x_3}{|y-x|^3}dy'd\tau\right], \qquad (3.5.4)
\end{aligned}$$

where $\mathfrak{M}^*_\gamma$ and $\mathfrak{N}^*$ are the transposed matrices of $\mathfrak{M}_\gamma$ and $\mathfrak{N}$, respectively. Therefore,

$$\begin{aligned}
2\pi|||\boldsymbol{u}(x,t)||| &\le \sup_{|\boldsymbol{z}|=1}\left[\int_{y_3=0}|\mathfrak{M}^*_\gamma(\boldsymbol{e}_{xy})\boldsymbol{z}|\,\frac{x_3}{|y-x|^3}dy'\right.\\
&+\left.\int_0^t\int_{y_3=0}|\mathcal{H}(t-\tau)|\,|\mathfrak{N}^*(\boldsymbol{e}_{xy})\boldsymbol{z}|\,\frac{x_3}{|y-x|^3}dy'd\tau\right]. \qquad (3.5.5)
\end{aligned}$$

On the other hand, for any $\varepsilon, 0<\varepsilon<t$, from (3.5.4) it follows the inequality

$$\begin{aligned}
2\pi|||\boldsymbol{u}(x,t)||| &\ge \sup_{|\boldsymbol{z}|=1}\left|\int_{y_3=0}(\mathfrak{M}^*_\gamma(\boldsymbol{e}_{xy})\boldsymbol{z},\,\boldsymbol{\varphi}(y',t))\,\frac{x_3}{|y-x|^3}dy'\right.\\
&+\left.\int_0^{t-\varepsilon}\int_{y_3=0}\mathcal{H}(t-\tau)\,(\mathfrak{N}^*(\boldsymbol{e}_{xy})\boldsymbol{z},\,\varphi(y',\tau))\,\frac{x_3}{|y-x|^3}dy'd\tau\right|-c(\varepsilon),
\end{aligned}$$

where $|\boldsymbol{\varphi}|\le 1$ and

$$c(\varepsilon)=\left|\int_{t-\varepsilon}^t\int_{y_3=0}\mathcal{H}(t-\tau)\,(\mathfrak{N}^*(\boldsymbol{e}_{xy})\boldsymbol{z},\,\boldsymbol{\varphi}(y',\tau))\,\frac{x_3}{|y-x|^3}dy'd\tau\right|\le 4\pi(|\eta_1|+|\eta|_2)\varepsilon.$$

The last estimate is obtained by using the equality

$$|\mathfrak{N}^*(\boldsymbol{e}_{xy})\boldsymbol{z}|=[1+3(\boldsymbol{e}_{xy},z)^2]^{1/2}, \qquad (3.5.6)$$

which follows from the definition of the matrix $\mathfrak{N}$.

Therefore,

$$\begin{aligned}
2\pi|||\boldsymbol{u}(x,t)||| &\ge \liminf_{\varepsilon\to 0}\sup_{|\boldsymbol{z}|=1}\left|\int_{y_3=0}(\mathfrak{M}^*_\gamma(\boldsymbol{e}_{xy})\boldsymbol{z},\,\boldsymbol{\varphi}(y',t))\,\frac{x_3}{|y-x|^3}dy'\right.\\
&+\left.\int_0^{t-\varepsilon}\int_{y_3=0}\mathcal{H}(t-\tau)\,(\mathfrak{N}^*(\boldsymbol{e}_{xy})\boldsymbol{z},\,\boldsymbol{\varphi}(y',\tau))\,\frac{x_3}{|y-x|^3}dy'd\tau\right|. \qquad (3.5.7)
\end{aligned}$$

Using the definition of the matrix $\mathfrak{M}_\gamma$ and setting $|\boldsymbol{z}| = 1$, we obtain

$$|\mathfrak{M}^*_\gamma(\boldsymbol{e}_{xy})\boldsymbol{z}| = [(1-\gamma)^2 + 3\gamma(\gamma+2)(\boldsymbol{e}_{xy}, \boldsymbol{z})^2]^{1/2}. \tag{3.5.8}$$

Since $\mu > 0, \kappa > 0$, then $\gamma \in (0,1)$. Consequently, we can suppose that

$$|\mathfrak{M}^*_\gamma(\boldsymbol{e}_{xy})\boldsymbol{z}| > 0.$$

By $\alpha_\varepsilon(t)$ we denote a function from the class $C^\infty(\overline{\mathbb{R}}^1_+)$ such that $0 \le \alpha_\varepsilon(t) \le 1$ and $\alpha_\varepsilon(t) = 1$ for $t \le \varepsilon/2$, $\alpha_\varepsilon(t) = 0$ for $t \ge \varepsilon$. Let $\psi(\tau)$ denote the function continuous on $[0,t]$ with $|\psi(\tau)| \le 1$.

We define the vector-valued function

$$[1-\alpha_\varepsilon(t-\tau)]\frac{\mathfrak{N}^*(\boldsymbol{e}_{xy})\boldsymbol{z}}{|\mathfrak{N}^*(\boldsymbol{e}_{xy})\boldsymbol{z}|}\psi(\tau) + \alpha_\varepsilon(t-\tau)\frac{\mathfrak{M}^*_\gamma(\boldsymbol{e}_{xy})\boldsymbol{z}}{|\mathfrak{M}^*_\gamma(\boldsymbol{e}_{xy})\boldsymbol{z}|}$$

and substitute it in (3.5.7) instead of $\varphi(y',\tau)$. As a result we obtain

$$\begin{aligned} 2\pi|||\boldsymbol{u}(x,t)||| \ \ge\ & \liminf_{\varepsilon\to 0} \sup_{|\boldsymbol{z}|=1} \bigg| \int_{y_3=0} |\mathfrak{M}^*_\gamma(\boldsymbol{e}_{xy})\boldsymbol{z}| \frac{x_3}{|y-x|^3} dy' \\ & + \int_0^{t-\varepsilon}\int_{y_3=0} \mathcal{H}(t-\tau)|\mathfrak{N}^*(\boldsymbol{e}_{xy})\boldsymbol{z}| \frac{x_3}{|y-x|^3} dy' d\tau \bigg|. \end{aligned}$$

From this, by the arbitrariness of ψ, $|\psi| \le 1$ and by (3.5.5), we get

$$\begin{aligned} |||\boldsymbol{u}(x,t)||| \ =\ & \frac{1}{2\pi} \sup_{|\boldsymbol{z}|=1} \bigg[\int_{y_3=0} |\mathfrak{M}^*_\gamma(\boldsymbol{e}_{xy})\boldsymbol{z}| \frac{x_3}{|y-x|^3} dy' \\ & + \int_0^{t}\int_{y_3=0} |\mathcal{H}(t-\tau)|\, |\mathfrak{N}^*(\boldsymbol{e}_{xy})\boldsymbol{z}| \frac{x_3}{|y-x|^3} dy' d\tau \bigg]. \end{aligned} \tag{3.5.9}$$

We denote by $\mathbb{S}^2_-(x)$ the lower half of the unit sphere $\mathbb{S}^2(x)$ centered at a point $x \in \mathbb{R}^3_+$. Let σ be the intersection of $\mathbb{S}^2(x)$ with the vector $y - x$. Substituting (3.5.6) and (3.5.8) into (3.5.9), we arrive at

$$\begin{aligned} |||\boldsymbol{u}(x,t)||| \ =\ & \frac{1}{2\pi} \sup_{|\boldsymbol{z}|=1} \bigg[\int_{\mathbb{S}^2_-(x)} [(1-\gamma)^2 + 3\gamma(\gamma+2)(\boldsymbol{e}_{x\sigma}, \boldsymbol{z})^2]^{1/2} d\sigma \\ & + \int_0^{t}\int_{\mathbb{S}^2_-(x)} |\mathcal{H}(t-\tau)|\, [1+3(\boldsymbol{e}_{x\sigma}, \boldsymbol{z})^2]^{1/2} d\tau d\sigma \bigg]. \end{aligned}$$

Since the integrands are even with respect to $\boldsymbol{e}_{x\sigma}$, it follows that

$$\begin{aligned} |||\boldsymbol{u}(x,t)||| \ =\ & \frac{1}{4\pi} \sup_{|\boldsymbol{z}|=1} \bigg[\int_{\mathbb{S}^2(x)} [(1-\gamma)^2 + 3\gamma(\gamma+2)(\boldsymbol{e}_{x\sigma}, \boldsymbol{z})^2]^{1/2} d\sigma \\ & + \int_0^{t} |\mathcal{H}(t-\tau)| d\tau \int_{\mathbb{S}^2(x)} [1+3(\boldsymbol{e}_{x\sigma}, \boldsymbol{z})^2]^{1/2} d\sigma \bigg] \\ =\ & \frac{1}{4\pi} \sup_{|\boldsymbol{z}|=1} \bigg[\int_{\mathbb{S}^2(x)} [(1-\gamma)^2 + 3\gamma(\gamma+2)\cos^2\theta(\sigma)]^{1/2} d\sigma \\ & + \int_0^{t} |\mathcal{H}(t-\tau)| d\tau \int_{\mathbb{S}^2(x)} [1+3\cos^2\theta(\sigma)]^{1/2} d\sigma \bigg], \end{aligned}$$

where $\theta(\sigma)$ is the angle between the vectors $\boldsymbol{e}_{x\sigma}$ and $\boldsymbol{z}$. Therefore,

$$\begin{aligned}\mathcal{K}(t) &= \frac{1}{4\pi}\sup_{|\boldsymbol{z}|=1}\left[\int_{\mathbb{S}^2(x)}[(1-\gamma)^2+3\gamma(\gamma+2)\cos^2\theta(\sigma)]^{1/2}d\sigma\right.\\ &+ \left.\int_0^t|\mathcal{H}(t-\tau)|d\tau\int_{\mathbb{S}^2(x)}[1+3\cos^2\theta(\sigma)]^{1/2}d\sigma\right].\end{aligned}$$

The last integral does not depend on $\boldsymbol{z}$. Hence

$$\begin{aligned}\mathcal{K}(t) &= \frac{1}{4\pi}\left[\int_0^\pi\int_0^{2\pi}[(1-\gamma)^2+3\gamma(\gamma+2)\cos^2\theta]^{1/2}\sin\theta\; d\theta d\omega\right.\\ &+ \left.\mathcal{F}(t)\int_0^\pi\int_0^{2\pi}[1+3\cos^2\omega]^{1/2}\sin\theta\; d\theta d\omega\right]\\ &= \int_0^{\pi/2}[(1-\gamma)^2+3\gamma(\gamma+2)\cos^2\theta]^{1/2}\sin\theta\; d\theta\\ &+ \mathcal{F}(t)\int_0^{\pi/2}[1+3\cos^2\theta]^{1/2}\sin\theta\; d\theta.\end{aligned}$$

□

REMARK 3.3. A similar result holds for the plane quasistatic viscoelasticity problem in $\mathbb{R}^2_+$. In this case the best value of $\mathcal{K}(t)$ in the inequality analogous to (3.5.3) is

$$\begin{aligned}\mathcal{K}(t) &= \mathcal{F}(t)+\frac{2}{\gamma}(1+\gamma)\mathbf{E}\left(\frac{2\sqrt{\gamma}}{1+\gamma}\right)\\ &= \mathcal{F}(t)+1+\frac{1}{2^2}\gamma^2+\frac{1}{2^24^2}\gamma^4+\cdots+\left[\frac{(2m-3)!!}{2^m m!}\right]^2\gamma^{2m}+\ldots,\end{aligned}$$

where γ and $\mathcal{F}(t)$ are the same as in the preceding theorem, and $\mathbf{E}(k)$ is the complete elliptic integral of the second kind.

REMARK 3.4. The value $\mathcal{K}(t)$ in the two above formulas is greater than one (see Theorem 3.4 and Corollary 3.4) which implies that the classical maximum modulus principle is not valid for the viscoelastic medium described by the three parametric model.

3.6. Comments to Chapter 3

Polya's paper [**Po**] contains an example (cf. Remark 2.2) showing that the classical maximum modulus principle fails for the Lamé system. Moreover, Polya's example demonstrates that the best factor in the inequality between the absolute value of the elastic displacement inside the three-dimensional ball and its maximum value on the boundary of that ball exceeds 1. A similar inequality with coefficient depending on the domain holds for domains with smooth boundary, and this inequality for solutions of the Lamé system is called Fichera's maximum principle (see Fichera [**Fi**]). Fichera's maximum principle is a particular case of the Miranda-Agmon maximum principle for elliptic systems of the second and higher order (see Agmon, Douglis and Nirenberg [**ADN1, ADN2**], Agmon [**Ag2**], C. Miranda [**Mi2**], Schulze [**Sch**], Zhou [**Zh1**], Cannarsa [**Can**], Albinus [**Alb1**]).

There are works on the Miranda-Agmon maximum principle for elliptic systems in domains with singularities at the boundary (Maz'ya and Plamenevskiĭ [**MPl**], Albinus [**Alb2**], Maz'ya and Rossmann [**MR2**]-[**MR4**] and the bibliography there).

Estimates for the maximum modulus of solutions to the Navier-Stokes system where obtained by Solonnikov [**So5**] for smooth domains, by Maz'ya and Rossmann [**MR1**] for polyhedral domains, by Russo [**Russ1**] for Lipschitz domains. Miranda-Agmon maximum principle as well as existence and uniqueness of solutions to Stokes system and elastostatics were treated by Maremonti and Russo [**MaRu1, MaRu2**], and Tartaglione [**Ta**] (cf. also bibliography there). A survey of maximum principles for the elasticity theory is given by Wheeler [**Wh**].

CHAPTER 4

Sharp Pointwise Estimates for Solutions of Elliptic Systems with Boundary Data from L^p

4.1. Introduction

In this chapter we consider solutions $\boldsymbol{u} = (u_1, \dots, u_m)$ of the strongly elliptic system

$$\sum_{i,j=1}^{n} \mathcal{A}_{ij} \frac{\partial^2 \boldsymbol{u}}{\partial x_i \partial x_j} = \boldsymbol{0} \tag{4.1.1}$$

with constant real $(m \times m)$-matrix-valued coefficients $\mathcal{A}_{ij} = \mathcal{A}_{ji}$ in the half-space $\mathbb{R}^n_+ = \{x = (x_1, \dots, x_n : x_n > 0\}$ as well as in a domain $\Omega \subset \mathbb{R}^n$ with smooth boundary $\partial\Omega$ and compact closure $\overline{\Omega}$.

We find a representation for the sharp constant $\mathcal{C}_p$ in the inequality

$$|\boldsymbol{u}(x)| \leq \mathcal{C}_p \, x_n^{(1-n)/p} \, \|\boldsymbol{u}|_{x_n=0}\|_p, \tag{4.1.2}$$

where x is an arbitrary point in $\mathbb{R}^n_+$, $\boldsymbol{u}$ is the solution of system (4.1.1) with boundary data from $[L^p(\partial\mathbb{R}^n_+)]^m$, $|\cdot|$ is the length of a vector in an m-dimensional Euclidean space, and $\|\cdot\|_p$ is the L^p-norm of the modulus of an m-component vector-valued function, $1 \leq p \leq \infty$.

The value $\mathcal{C}_p$ is connected with the asymptotic behaviour of solutions to system (4.1.1) near the boundary $\partial\Omega$. In particular, we show that

$$\lim_{x \to \mathcal{O}_x} \left| x - \mathcal{O}_x \right|^{(n-1)/p} \sup \left\{ |\boldsymbol{u}(x)| : \|\boldsymbol{u}|_{\partial\Omega}\|_p \leq 1 \right\} = \mathcal{C}_p(\mathcal{O}_x), \tag{4.1.3}$$

where $\mathcal{O}_x$ is a point at $\partial\Omega$ nearest to $x \in \Omega$, $\boldsymbol{u}$ is a solution of the Dirichlet problem in Ω for the strongly elliptic system (4.1.1) with boundary data from $[L^p(\partial\Omega)]^m$, and $\mathcal{C}_p(\mathcal{O}_x)$ is the sharp constant in inequality (4.1.2) for the tangent space $\mathbb{R}^n_+(\mathcal{O}_x)$ to $\partial\Omega$ at $\mathcal{O}_x$.

The derivation of explicit formulas for the sharp constant in inequality (4.1.2) for the displacement vector and the velocity vector of a fluid is reduced to an optimization problem on the unit sphere $\mathbb{S}^{n-1}$. This problem can be explicitly solved for any $p \in [1, \infty]$ in the case of the Stokes system. We obtain also estimates with sharp constants for the modulus of solutions to the Stokes and Lamé systems in the center of a ball by the spherical means of order p of the modulus of boundary values on the sphere.

In Section 4.2, applying Proposition 1.2 from Chapter 1, we obtain a representation for the constant $\mathcal{C}_p$ in inequality (4.1.2) for solutions of the Dirichlet problem for system (4.1.1) in the half-space $\mathbb{R}^n_+$. A representation of the sharp constant in (4.1.2) for system (4.1.1) in the case $p = \infty$ was derived earlier in Section 2.2.

It is shown, that for any solution $\boldsymbol{u}$ of the Dirichlet problem in Ω for (4.1.1) with boundary data from $[L^p(\partial\Omega)]^m$ and all $x \in \Omega$ the relation

$$\sup\left\{|\boldsymbol{u}(x)| : \|\boldsymbol{u}|_{\partial\Omega}\|_p \leq 1\right\} = \mathcal{C}_p(\mathcal{O}_x)|x - \mathcal{O}_x|^{-(n-1)/p} + O\left(|x - \mathcal{O}_x|^{\varepsilon-(n-1)/p}\right) \tag{4.1.4}$$

holds for some $\varepsilon > 0$. Here $\mathcal{O}_x$ is a point on $\partial\Omega$ nearest to $x \in \Omega$, and $\mathcal{C}_p(\mathcal{O}_x)$ is the best constant in (4.1.2) for the half-space $\mathbb{R}^n_+(\mathcal{O}_x)$. Equality (4.1.3) is an immediate consequence of (4.1.4).

In Section 4.3 we consider the Stokes system

$$\nu\Delta\boldsymbol{u} - \text{grad}\, p = \boldsymbol{0}, \quad \text{div}\, \boldsymbol{u} = 0,$$

in the half-space $\mathbb{R}^n_+$, $n \geq 2$, with the boundary condition

$$\boldsymbol{u}\Big|_{x_n=0} = \boldsymbol{f},$$

where ν is the kinematic coefficient of viscosity, $\boldsymbol{u} = (u_1, \ldots, u_n)$ is the velocity vector of a fluid, p is the pressure in the fluid, and $\boldsymbol{f} \in [L^p(\partial\mathbb{R}^n_+)]^n$. Despite the fact that the Stokes system is not strongly elliptic, the representation for the velocity vector $\boldsymbol{u}$ in the half-space is of the same nature as in the case of strongly elliptic systems. Hence, the result obtained in Section 4.2 applies to the Stokes system. It is shown, that for any $x \in \mathbb{R}^n_+$ the sharp coefficient $\mathcal{C}_p$ in the inequality (4.1.2) for the velocity vector $\boldsymbol{u}$ defined by a solution $(\boldsymbol{u}, p)$ of the Stokes system is given by

$$\mathcal{C}_1 = \frac{2\Gamma\left(\frac{n+2}{2}\right)}{\pi^{n/2}}, \qquad \mathcal{C}_\infty = \frac{2\Gamma\left(\frac{n+2}{2}\right)}{\sqrt{\pi}\,\Gamma\left(\frac{n+1}{2}\right)},$$

and

$$\mathcal{C}_p = \frac{2\Gamma\left(\frac{n+2}{2}\right)}{\pi^{(n+p-1)/(2p)}} \left\{ \frac{\Gamma\left(\frac{2p+n-1}{2p-2}\right)}{\Gamma\left(\frac{n+1}{2p-2}p\right)} \right\}^{(p-1)/p}$$

for $1 < p < \infty$. In particular, $\mathcal{C}_\infty = 4/\pi$ for $n = 2$ and $\mathcal{C}_\infty = 3/2$ for $n = 3$.

In Section 4.4 we find a representation of the sharp constant in the inequality

$$|\boldsymbol{u}(0)| \leq \mathring{B}_p\, M_p(\boldsymbol{u}; \partial\mathbb{B}_r),$$

where $\boldsymbol{u}$ is the velocity vector in a three-dimensional ball, and $M_p(\boldsymbol{u}; \partial\mathbb{B}_r)$ is the spherical mean of order p for the modulus of $\boldsymbol{u}$. In particular, it is shown that

$$\mathring{B}_1 = 6, \qquad \mathring{B}_2 = \frac{3}{2}\sqrt{6}.$$

Section 4.5 concerns the Lamé system

$$\mu\Delta\boldsymbol{u} + (\lambda + \mu)\text{grad div}\, \boldsymbol{u} = \boldsymbol{0}$$

in the half-space $\mathbb{R}^n_+$ with the boundary condition

$$\boldsymbol{u}\Big|_{x_n=0} = \boldsymbol{f},$$

where λ and μ are the Lamé constants, $\boldsymbol{u} = (u_1, \ldots, u_n)$ is the displacement vector of an elastic medium, and $\boldsymbol{f} \in [L^p(\partial\mathbb{R}^n_+)]^n$. We find a representation for the sharp constant $\mathcal{C}_p$ in (4.1.2). In particular, we show that

$$\mathcal{C}_{1,\varkappa} = \frac{[1+\varkappa(n-1)]\Gamma(n/2)}{\pi^{n/2}},$$

$$\mathcal{C}_{2,\varkappa} = \left\{ \frac{\Gamma\left(\frac{n}{2}\right)}{2^n \pi^{n/2}} \left[1 + (n-1)\varkappa^2 + \left(1+(n-1)\varkappa\right)^2 \right] \right\}^{1/2},$$

where $\varkappa = (\lambda+\mu)(\lambda+3\mu)^{-1}$.

The concluding Section 4.6 is devoted specifically to the sharp constant in

$$|\boldsymbol{u}(0)| \le B_{p,\varkappa}\, M_p(\boldsymbol{u}; \partial\mathbb{B}_r)$$

where $\boldsymbol{u}$ is a solution of the three-dimensional Lamé system in a ball. We show, in particular, that

$$B_{1,\varkappa} = \frac{3(1+3\varkappa)}{3-\varkappa}, \qquad B_{2,\varkappa} = \frac{\sqrt{3}}{3-\varkappa}\left(17\varkappa^2 - 2\varkappa + 3\right)^{1/2}.$$

4.2. Best constants in pointwise estimates for solutions of strongly elliptic systems with boundary data from L^p

4.2.1. Elliptic systems in a half-space

We remind some notations used in previous chapters. Let $x \in \mathbb{R}^n_+ = \{x = (x', x_n) : x' = (x_1, \dots, x_{n-1}) \in \mathbb{R}^{n-1}, x_n > 0\}$ and $\mathbb{S}^{n-1} = \{x \in \mathbb{R}^n : |x| = 1\}$. By $[\mathrm{C}^2(\mathbb{R}^n_+)]^m$ we denote the space of m-component vector-valued functions with continuous derivatives up to the second order in $\mathbb{R}^n_+$. Further, $[\mathrm{C_b}(\overline{\mathbb{R}^n_+})]^m$ and $[\mathrm{C_b}(\partial\mathbb{R}^n_+)]^m$ will stand for the spaces of continuous and bounded m-component vector-valued functions on $\overline{\mathbb{R}^n_+}$ and $\partial\mathbb{R}^n_+$, respectively.

By $\|\cdot\|_p$ we denote the norm in the space $[L^p(\partial\mathbb{R}^n_+)]^m$, that is

$$\|\boldsymbol{f}\|_p = \left\{ \int_{\partial\mathbb{R}^n_+} |\boldsymbol{f}(x')|^p dx' \right\}^{1/p},$$

if $1 \le p < \infty$, and

$$\|\boldsymbol{f}\|_\infty = \text{ess sup}\{|\boldsymbol{f}(x')| : x' \in \partial\mathbb{R}^n_+\}.$$

We introduce the strongly elliptic operator

$$\mathfrak{A}_0(D_x) = \sum_{i,j=1}^n \mathcal{A}_{ij} \frac{\partial^2}{\partial x_i \partial x_j}, \tag{4.2.1}$$

where $D_x = (\partial/\partial x_1, \dots, \partial/\partial x_n)$ and $\mathcal{A}_{ij} = \mathcal{A}_{ji}$ are constant real $(m \times m)$-matrices. As before, the strong ellipticity of $\mathfrak{A}_0(D_x)$ means that the inequality

$$\left(\sum_{i,j=1}^n \mathcal{A}_{ij}\sigma_i\sigma_j \boldsymbol{\zeta}, \boldsymbol{\zeta} \right) > 0$$

is valid for all $\boldsymbol{\zeta} = (\zeta_1, \dots, \zeta_m) \in \mathbb{R}^m \backslash \{\mathbf{0}\}$ and $\boldsymbol{\sigma} = (\sigma_1, \dots, \sigma_n) \in \mathbb{R}^n \backslash \{\mathbf{0}\}$.

Similarly to Subsection 2.2.1, we start by considering a bounded solution of the problem

$$\mathfrak{A}_0(D_x)\boldsymbol{u} = \mathbf{0} \text{ in } \mathbb{R}^n_+, \quad \boldsymbol{u} = \boldsymbol{f} \text{ on } \partial\mathbb{R}^n_+, \tag{4.2.2}$$

where $\boldsymbol{f} \in [\mathrm{C_b}(\partial\mathbb{R}^n_+)]^m$, which is continuous up to $\partial\mathbb{R}^n_+$ and can be represented as

$$\boldsymbol{u}(x) = \int_{\partial\mathbb{R}^n_+} \mathcal{M}\left(\frac{y-x}{|y-x|}\right)\frac{x_n}{|y-x|^n}\boldsymbol{f}(y')dy'. \tag{4.2.3}$$

Here $y=(y',0), y'=(y_1,\ldots,y_{n-1})$, and $\mathcal{M}$ is an $(m\times m)$-matrix-valued function with continuous components on the closure of the hemisphere $\mathbb{S}^{n-1}_- = \big\{x \in \mathbb{R}^n : |x|=1,\ x_n<0\big\}$.

The uniqueness of a solution to the Dirichlet problem (4.2.2) in the class $[\mathrm{C}^2(\mathbb{R}^n_+)]^m$ with boundary data from $[L^p(\partial\mathbb{R}^n_+)]^m$ can be derived by means of a standard argument from (4.2.3) and from local estimates for derivatives of solutions to elliptic systems (see Agmon, Douglis and Nirenberg [**ADN2**], Solonnikov [**So1**]). In particular, this relates the Lamé system. Moreover, this is also true for the Stokes system, although it is not of the form (4.1.1).

PROPOSITION 4.1. *Let x be an arbitrary point in $\mathbb{R}^n_+$ and let $\boldsymbol{z}\in\mathbb{R}^m$. The sharp coefficient $\mathcal{K}_p(x)$ in the inequality*

$$|\boldsymbol{u}(x)| \le \mathcal{K}_p(x)\|\boldsymbol{u}|_{x_n=0}\|_p \tag{4.2.4}$$

is given by

$$\mathcal{K}_p(x) = \mathcal{C}_p\, x_n^{(1-n)/p}, \tag{4.2.5}$$

where

$$\mathcal{C}_1 = \sup_{|\boldsymbol{z}|=1}\ \sup_{\sigma\in\mathbb{S}^{n-1}_-} |\mathcal{M}^*(\boldsymbol{e}_\sigma)\boldsymbol{z}|(\boldsymbol{e}_\sigma,-\boldsymbol{e}_n)^n, \tag{4.2.6}$$

$$\mathcal{C}_\infty = \sup_{|\boldsymbol{z}|=1}\int_{\mathbb{S}^{n-1}_-}|\mathcal{M}^*(\boldsymbol{e}_\sigma)\boldsymbol{z}|d\sigma, \tag{4.2.7}$$

and

$$\mathcal{C}_p = \sup_{|\boldsymbol{z}|=1}\left\{\int_{\mathbb{S}^{n-1}_-}|\mathcal{M}^*(\boldsymbol{e}_\sigma)\boldsymbol{z}|^{p/(p-1)}(\boldsymbol{e}_\sigma,-\boldsymbol{e}_n)^{n/(p-1)}d\sigma\right\}^{(p-1)/p} \tag{4.2.8}$$

for $1<p<\infty$. Here $\boldsymbol{e}_\sigma$ is the n-dimensional unit vector joining the origin with a point $\sigma\in\mathbb{S}^{n-1}_-$, and $*$ denotes the transposition of a matrix.

PROOF. By Proposition 1.2 and (4.2.3), the sharp constant $\mathcal{K}_p(x)$ in (4.2.4) is given by

$$\mathcal{K}_p(x) = \sup_{|\boldsymbol{z}|=1}\left\{\int_{\partial\mathbb{R}^n_+}\left|\mathcal{M}^*\left(\frac{y-x}{|y-x|}\right)\boldsymbol{z}\right|^q\frac{x_n^q}{|y-x|^{nq}}dy'\right\}^{1/q},$$

where $1<p\le\infty$. Putting $\rho=|y'-x'|$, we write the last equality as follows

$$\mathcal{K}_p(x) = \sup_{|\boldsymbol{z}|=1}\left\{\int_{\mathbb{S}^{n-2}(x')}d\sigma\int_0^\infty\left|\mathcal{M}^*\left(\frac{\rho\boldsymbol{e}'_\sigma - x_n\boldsymbol{e}_n}{\big[\,\rho^2+x_n^2\big]^{1/2}}\right)\boldsymbol{z}\right|^q\frac{x_n^q}{\big[\,\rho^2+x_n^2\big]^{nq/2}}\,\rho^{n-2}d\rho\right\}^{1/q},$$

where

$$\boldsymbol{e}'_\sigma = \left(\frac{y'-x'}{|y'-x'|}\,,\,0\right),$$

and $\mathbb{S}^{n-2}(x')$ is the $(n-2)$-dimensional unit sphere centered at the point x'.

Now we make the change of variable $\rho = x_n \tan\varphi$ in the last representation for $\mathcal{K}_p(x)$ with φ standing for the angle between the vectors $\rho \boldsymbol{e}'_\sigma - x_n \boldsymbol{e}_n$ and $-\boldsymbol{e}_n$, and obtain

$$\mathcal{K}_p(x) = x_n^{(1-n)(q-1)/q}$$

$$\times \sup_{|\boldsymbol{z}|=1} \left\{ \int_{\mathbb{S}^{n-2}(x')} d\sigma \int_0^{\pi/2} \left| \mathcal{M}^* \left(\boldsymbol{e}'_\sigma \sin\varphi - \boldsymbol{e}_n \cos\varphi \right) \boldsymbol{z} \right|^q \cos^{n(q-1)}\varphi \, \sin^{n-2}\varphi \, d\varphi \right\}^{1/q} .$$

Using the independence of the integral on x' and the notation

$$\boldsymbol{e}_\sigma = \boldsymbol{e}'_\sigma \sin\varphi - \boldsymbol{e}_n \cos\varphi,$$

we arrive at (4.2.5) with $\mathcal{C}_p$ defined by (4.2.8). In particular, (4.2.8) becomes (4.2.7) for $p = \infty$.

Next note that by Proposition 1.2 and (4.2.3), the sharp constant in (4.2.4) with $p = 1$ can be written as

$$\mathcal{K}_1(x) = \sup_{|\boldsymbol{z}|=1} \sup_{y \in \partial\mathbb{R}^n_+} \left| \mathcal{M}^* \left(\frac{y - x}{|y - x|} \right) \boldsymbol{z} \right| \frac{x_n}{|y - x|^n}.$$

Setting here $|y' - x'| = x_n \tan\varphi$, $0 \le \varphi < \pi/2$, we find

$$\mathcal{K}_1(x) = x_n^{1-n} \sup_{|\boldsymbol{z}|=1} \sup_{\mathbb{S}^{n-1}_-} |\mathcal{M}^*(\boldsymbol{e}_\sigma)\boldsymbol{z}| \, (\boldsymbol{e}_\sigma, -\boldsymbol{e}_n)^n,$$

which implies (4.2.5) with $p = 1$ and with $\mathcal{C}_1$ defined by (4.2.6). □

4.2.2. Asymptotic formula involving the Poisson matrix in a domain

In what follows, by smoothness we mean the membership in C^∞. Suppose Ω is a domain in $\mathbb{R}^n$ with smooth boundary $\partial\Omega$ and compact closure $\overline{\Omega}$. By $\boldsymbol{\nu}(y)$ we denote the unit interior normal to $\partial\Omega$ at a point $y \in \partial\Omega$. Let $\mathbb{R}^n_+(y) = \{x \in \mathbb{R}^n : (x, \boldsymbol{\nu}(y)) > 0\}$, $\mathbb{R}^n_-(y) = \{x \in \mathbb{R}^n : (x, \boldsymbol{\nu}(y)) < 0\}$ and $\mathbb{R}^{n-1}(y) = \partial\mathbb{R}^n_+(y)$.

We consider the Dirichlet problem

$$\mathfrak{A}_0(D_x)\boldsymbol{u_g} = \boldsymbol{0} \ \text{ in } \ \Omega, \quad \boldsymbol{u_g}\big|_{\partial\Omega} = \boldsymbol{g} \tag{4.2.9}$$

for the strongly elliptic operator $\mathfrak{A}_0(D_x)$, defined by (4.2.1), with $\boldsymbol{g} \in [L^p(\partial\Omega)]^m$.

Let x be a fixed point in Ω, and let

$$\mathcal{K}_p(x, \Omega) = \sup \frac{|\boldsymbol{u_g}(x)|}{\|\boldsymbol{g}\|_{[L^p(\partial\Omega)]^m}}, \tag{4.2.10}$$

where the supremum is taken over all vector-valued functions $\boldsymbol{g}$ in the class $[L^p(\partial\Omega)]^m$.

THEOREM 4.1. *For all* $x \in \Omega$

$$\mathcal{K}_p(x, \Omega) = \mathcal{C}_p(\mathcal{O}_x)|x - \mathcal{O}_x|^{-(n-1)/p} + O\Big(|x - \mathcal{O}_x|^{\varepsilon-(n-1)/p}\Big)$$

with some $\varepsilon > 0$. *Here* $\mathcal{O}_x$ *is a point on* $\partial\Omega$ *nearest to* $x \in \Omega$, *and* $\mathcal{C}_p(\mathcal{O}_x)$ *is the best constant in* (4.2.5) *for the half-space* $\mathbb{R}^n_+(\mathcal{O}_x)$.

Before giving a proof of this theorem, we formulate its obvious corollary.

COROLLARY 4.1. *The equality*

$$\lim_{x\to\mathcal{O}_x} \left|x-\mathcal{O}_x\right|^{(n-1)/p} \mathcal{K}_p(x,\Omega) = \mathcal{C}_p\big(\mathcal{O}_x\big)$$

holds.

As a direct consequence of the last equality we have

$$\sup_{x\in\Omega}\left|x-\mathcal{O}_x\right|^{(n-1)/p} \mathcal{K}_p(x,\Omega) \geq \sup_{x\in\Omega}\mathcal{C}_p\big(\mathcal{O}_x\big). \tag{4.2.11}$$

In particular,

$$\sup_{x\in\Omega} \mathcal{K}_\infty(x,\Omega) \geq \sup_{x\in\Omega}\mathcal{C}_\infty\big(\mathcal{O}_x\big). \tag{4.2.12}$$

Proof of the Theorem. The Poisson matrix of problem (4.2.9) with singularity at the point $y\in\partial\Omega$ will be denoted by $P_\Omega(x,y)$. In other words, P_Ω satisfies the problem

$$\mathfrak{A}_0(D_x)P_\Omega(x,y) = \mathbf{0} \ \text{ for } \ x\in\Omega, \qquad P_\Omega(x,y) = \delta(y-x)I \ \text{ for } \ x\in\partial\Omega,$$

where δ is the Dirac function and I is the $(m\times m)$-identity matrix. We put $P_\Omega(x,y)=0$ for $x\in\mathbb{R}^n\backslash\overline{\Omega}$.

The notation $\Pi(x,y)$ will be used for the Poisson matrix with singularity at y of the Dirichlet problem for the operator $\mathfrak{A}_0(D_x)$ in the half-space $\mathbb{R}^n_+(y)$. The matrix-valued function $x\to\Pi(x,y)$ is extended by zero to the half-space $\mathbb{R}^n_-(y)$. Clearly,

$$\Pi(x,y) = \frac{(y-x,\boldsymbol{\nu}(y))}{|y-x|^n}\,\mathcal{F}\left(\frac{y-x}{|y-x|};\ \boldsymbol{\nu}(y)\right), \tag{4.2.13}$$

where

$$\left\|\mathcal{F}\left(\frac{y-x}{|y-x|};\ \boldsymbol{\nu}(y)\right) - \mathcal{F}\left(\frac{y-x}{|y-x|};\ \boldsymbol{\nu}(\mathcal{O}_x)\right)\right\| \leq c\,|y-\mathcal{O}_x|, \tag{4.2.14}$$

and $\|\cdot\|$ is the matrix norm induced by the Euclidean norm in $\mathbb{R}^m$. It is well-known (see Krasovskiĭ [**Kras1, Kras2**], Solonnikov [**So2, So3**]) that for $x\in\Omega$ and $y\in\partial\Omega$

$$\|P_\Omega(x,y)-\Pi(x,y)\| \leq c(\varepsilon_o)\,|x-y|^{2-n-\varepsilon_o}, \tag{4.2.15}$$

where ε_o is an arbitrary positive number. Therefore, for $x\in\Omega$ and any $\boldsymbol{z}\in\mathbb{S}^{m-1}$,

$$\begin{aligned} \|P^*_\Omega(x,\cdot)\boldsymbol{z}-\Pi^*(x,\cdot)\boldsymbol{z}\|_{[L^q(\partial\Omega)]^m} &\leq c(\varepsilon_o)\,\|\,|x-\cdot\,|^{2-n-\varepsilon_o}\|_{L^q(\partial\Omega)} \\ &\leq c_1(\varepsilon)|x-\mathcal{O}_x|^{\varepsilon-(n-1)/p} \end{aligned} \tag{4.2.16}$$

with some $\varepsilon>0$. Using (4.2.13) and (4.2.14), we arrive at the estimate

$$\begin{aligned} &\left|\left\|\frac{(\cdot-x,\boldsymbol{\nu}(\cdot))}{|\cdot-x|^n}\,\mathcal{F}^*\left(\frac{\cdot-x}{|\cdot-x|};\ \boldsymbol{\nu}(\cdot)\right)\boldsymbol{z}\right\|_{[L^q(\partial\Omega)]^m}\right. \\ &\qquad\left. - \left\|\frac{(\cdot-x,\boldsymbol{\nu}(\mathcal{O}_x))}{|\cdot-x|^n}\,\mathcal{F}^*\left(\frac{\cdot-x}{|\cdot-x|};\ \boldsymbol{\nu}(\mathcal{O}_x)\right)\boldsymbol{z}\right\|_{[L^q(\partial\Omega)]^m}\right| \\ &\qquad\leq c_2(\varepsilon)|x-\mathcal{O}_x|^{\varepsilon-(n-1)/p}. \end{aligned} \tag{4.2.17}$$

Since $\partial\mathbb{R}^n_+(\mathcal{O}_x)$ is tangent to $\partial\Omega$ at the point $\mathcal{O}_x$, one can see that

$$\Bigg| \left\| \frac{(\cdot - x, \boldsymbol{\nu}(\mathcal{O}_x))}{|\cdot - x|^n} \mathcal{F}^*\left(\frac{\cdot - x}{|\cdot - x|}; \boldsymbol{\nu}(\mathcal{O}_x)\right) \boldsymbol{z} \right\|_{[L^q(\partial\Omega)]^m}$$

$$- \left\| \frac{(\cdot - x, \boldsymbol{\nu}(\mathcal{O}_x))}{|\cdot - x|^n} \mathcal{F}^*\left(\frac{\cdot - x}{|\cdot - x|}; \boldsymbol{\nu}(\mathcal{O}_x)\right) \boldsymbol{z} \right\|_{[L^q(\mathbb{R}^{n-1}(\mathcal{O}_x))]^m} \Bigg|$$

$$\leq c_3(\varepsilon)|x - \mathcal{O}_x|^{\varepsilon-(n-1)/p}. \tag{4.2.18}$$

Using (4.2.13) and combining (4.2.16), (4.2.17) and (4.2.18), we obtain for any $\boldsymbol{z} \in \mathbb{S}^{m-1}$

$$\Bigg| \|P^*_\Omega(x, \cdot)\boldsymbol{z}\|_{[L^q(\partial\Omega)]^m} - \left\| \frac{(\cdot - x, \boldsymbol{\nu}(\mathcal{O}_x))}{|\cdot - x|^n} \mathcal{F}^*\left(\frac{\cdot - x}{|\cdot - x|}; \boldsymbol{\nu}(\mathcal{O}_x)\right) \boldsymbol{z} \right\|_{[L^q(\mathbb{R}^{n-1}(\mathcal{O}_x))]^m} \Bigg|$$

$$\leq c_4(\varepsilon)|x - \mathcal{O}_x|^{\varepsilon-(n-1)/p}. \tag{4.2.19}$$

The second norm in (4.2.19) is equal to

$$|x - \mathcal{O}_x|^{-(n-1)/p} \left\{ \int_{\mathbb{R}^{n-1}(\mathcal{O}_x)} \left| \mathcal{F}^*\left(\frac{(\boldsymbol{\sigma}', -1)}{(1 + |\boldsymbol{\sigma}'|^2)^{1/2}}; \boldsymbol{\nu}(\mathcal{O}_x)\right) \boldsymbol{z} \right|^q \frac{d\sigma'}{(1 + |\boldsymbol{\sigma}'|^2)^{nq/2}} \right\}^{1/q},$$

which, after taking the supremum over $\boldsymbol{z} \in \mathbb{S}^{m-1}$, becomes

$$\mathcal{C}_p(\mathcal{O}_x)|x - \mathcal{O}_x|^{-(n-1)/p} = \mathcal{K}_p(\mathcal{O}_x)$$

by Proposition 4.1. Combining this with (4.2.19), we complete the proof of Theorem 4.1. □

4.3. The Stokes system in a half-space

Consider the Stokes system

$$\nu\Delta\boldsymbol{u} - \text{grad}\, p = \boldsymbol{0}, \quad \text{div}\, \boldsymbol{u} = 0 \quad \text{in } \mathbb{R}^n_+, \quad n \geq 2, \tag{4.3.1}$$

with the boundary condition

$$\boldsymbol{u}\big|_{x_n=0} = \boldsymbol{f}, \tag{4.3.2}$$

where ν is the kinematic coefficient of viscosity, $\boldsymbol{u} = (u_1, \dots, u_n)$ is the velocity vector of a fluid, p is the pressure in the fluid, and $\boldsymbol{f} = (f_1, \dots, f_n)$ is a continuous and bounded vector-valued function on $\partial\mathbb{R}^n_+$.

The vector component $\boldsymbol{u} \in [\mathrm{C}^2(\mathbb{R}^n_+)]^n$ of the solution $\{\boldsymbol{u}, p\}$ to the Dirichlet problem for the Stokes system in the half-space $\mathbb{R}^n_+$ which is bounded and continuous up to $\partial\mathbb{R}^n_+$ admits the representation (see Remark 3.2)

$$\boldsymbol{u}(x) = \int_{\partial\mathbb{R}^n_+} S\left(\frac{y - x}{|y - x|}\right) \frac{x_n}{|y - x|^n} \boldsymbol{f}(y')dy', \tag{4.3.3}$$

where $x \in \mathbb{R}^n_+$, $y = (y', 0)$, $y' = (y_1, \dots, y_{n-1})$, and $S(\boldsymbol{e}_\sigma)$ is the $(n \times n)$-matrix-valued function on the unit sphere $\mathbb{S}^{n-1}$ of $\mathbb{R}^n$ with the elements

$$\frac{2n}{\omega_n}(\boldsymbol{e}_\sigma, \boldsymbol{e}_i)(\boldsymbol{e}_\sigma, \boldsymbol{e}_j), \tag{4.3.4}$$

and $\omega_n = 2\pi^{n/2}/\Gamma(n/2)$ is the area of $\mathbb{S}^{n-1}$.

THEOREM 4.2. *Let x be an arbitrary point in $\mathbb{R}^n_+$. The sharp coefficient $\mathcal{K}_p(x)$ in the inequality*

$$|\boldsymbol{u}(x)| \le \mathcal{K}_p(x) \|\boldsymbol{u}|_{x_n=0}\|_p \tag{4.3.5}$$

for the velocity vector $\boldsymbol{u}$ defined by a solution $(\boldsymbol{u}, p)$ of the Stokes system is given by

$$\mathcal{K}_p(x) = \mathcal{C}_p \, x_n^{(1-n)/p}, \tag{4.3.6}$$

where

$$\mathcal{C}_1 = \frac{2\Gamma\left(\frac{n+2}{2}\right)}{\pi^{n/2}}, \qquad \mathcal{C}_\infty = \frac{2\Gamma\left(\frac{n+2}{2}\right)}{\sqrt{\pi}\,\Gamma\left(\frac{n+1}{2}\right)}, \tag{4.3.7}$$

and

$$\mathcal{C}_p = \frac{2\Gamma\left(\frac{n+2}{2}\right)}{\pi^{(n+p-1)/(2p)}} \left\{ \frac{\Gamma\left(\frac{2p+n-1}{2p-2}\right)}{\Gamma\left(\frac{n+1}{2p-2}p\right)} \right\}^{(p-1)/p} \tag{4.3.8}$$

for $1 < p < \infty$. In particular,

$$\mathcal{C}_2 = \left\{ \frac{(n+1)\Gamma\left(\frac{n+2}{2}\right)}{2^{n-1}\pi^{n/2}} \right\}^{1/2}. \tag{4.3.9}$$

PROOF. Since the vector component $\boldsymbol{u}$ of the solution $\{\boldsymbol{u}, p\}$ to the Dirichlet problem (4.3.1), (4.3.2) is given by (4.3.3), it obeys Proposition 4.1. By (4.3.4) we have

$$|S^*(\boldsymbol{e}_\sigma)\boldsymbol{z}| = \frac{2n}{\omega_n}\big|(\boldsymbol{e}_\sigma, \boldsymbol{z})\big|, \tag{4.3.10}$$

which together with (4.2.8) implies

$$\mathcal{C}_p = \frac{2n}{\omega_n} \sup_{|\boldsymbol{z}|=1} \left\{ \int_{\mathbb{S}^{n-1}_-} \big|(\boldsymbol{e}_\sigma, \boldsymbol{z})\big|^{p/(p-1)} (\boldsymbol{e}_\sigma, -\boldsymbol{e}_n)^{n/(p-1)} d\sigma \right\}^{(p-1)/p},$$

where $1 < p \le \infty$. Noting that the function $|(\boldsymbol{e}_\sigma, -\boldsymbol{e}_n)|$ is even on the sphere $\mathbb{S}^{n-1}$, we can write $\mathcal{C}_p$ as

$$\mathcal{C}_p = \frac{2^{1/p}\, n}{\omega_n} \sup_{|\boldsymbol{z}|=1} \left\{ \int_{\mathbb{S}^{n-1}} \big|(\boldsymbol{e}_\sigma, \boldsymbol{z})\big|^{p/(p-1)} \big|(\boldsymbol{e}_\sigma, \boldsymbol{e}_n)\big|^{n/(p-1)} d\sigma \right\}^{(p-1)/p}. \tag{4.3.11}$$

This immediately implies the lower estimate

$$\mathcal{C}_p \ge \frac{2^{1/p}\, n}{\omega_n} \left\{ \int_{\mathbb{S}^{n-1}} \big|(\boldsymbol{e}_\sigma, \boldsymbol{e}_n)\big|^{(p+n)/(p-1)} d\sigma \right\}^{(p-1)/p}. \tag{4.3.12}$$

Next we derive the upper estimate for $\mathcal{C}_p$. Setting

$$l = \frac{p+n}{n}, \qquad s = \frac{p+n}{p},$$

and noting that $1/s + 1/l = 1$, we have by Hölder's inequality

$$\int_{\mathbb{S}^{n-1}} \big|(\boldsymbol{e}_\sigma, \boldsymbol{z})\big|^{p/(p-1)} \big|(\boldsymbol{e}_\sigma, \boldsymbol{e}_n)\big|^{n/(p-1)} d\sigma$$

$$\leq \left\{ \int_{\mathbb{S}^{n-1}} \big|(\boldsymbol{e}_\sigma, \boldsymbol{z})\big|^{ps/(p-1)} d\sigma \right\}^{1/s} \left\{ \int_{\mathbb{S}^{n-1}} \big|(\boldsymbol{e}_\sigma, \boldsymbol{e}_n)\big|^{nl/(p-1)} d\sigma \right\}^{1/l}.$$

Taking into account that $ps = nl = p + n$ and that the first integral on the right-hand side is independent on $\boldsymbol{z}$, we find

$$\int_{\mathbb{S}^{n-1}} \big|(\boldsymbol{e}_\sigma, \boldsymbol{z})\big|^{p/(p-1)} \big|(\boldsymbol{e}_\sigma, \boldsymbol{e}_n)\big|^{n/(p-1)} d\sigma \leq \int_{\mathbb{S}^{n-1}} \big|(\boldsymbol{e}_\sigma, \boldsymbol{e}_n)\big|^{(p+n)/(p-1)} d\sigma,$$

which together with (4.3.11) leads to

$$\mathcal{C}_p \leq \frac{2^{1/p}\, n}{\omega_n} \left\{ \int_{\mathbb{S}^{n-1}} \big|(\boldsymbol{e}_\sigma, \boldsymbol{e}_n)\big|^{(p+n)/(p-1)} d\sigma \right\}^{(p-1)/p}.$$

Combining this with (4.3.12), we arrive at the equality

$$\mathcal{C}_p = \frac{2^{1/p}\, n}{\omega_n} \left\{ \int_{\mathbb{S}^{n-1}} \big|(\boldsymbol{e}_\sigma, \boldsymbol{e}_n)\big|^{(p+n)/(p-1)} d\sigma \right\}^{(p-1)/p}, \tag{4.3.13}$$

where $1 < p < \infty$. The formula (4.3.8) follows from (4.3.13) and

$$\int_{\mathbb{S}^{n-1}} \big|(\boldsymbol{e}_\sigma, \boldsymbol{e}_n)\big|^{(p+n)/(p-1)} d\sigma = 2\omega_{n-1} \int_0^{\pi/2} \cos^{\frac{p+n}{p-1}} \vartheta \sin^{n-2} \vartheta d\vartheta$$

$$= \frac{2\pi^{(n-1)/2}}{\Gamma\left(\frac{n-1}{2}\right)}\, B\left(\frac{2p+n-1}{2p-2}, \frac{n-1}{2} \right) = 2\pi^{(n-1)/2} \frac{\Gamma\left(\frac{2p+n-1}{2p-2}\right)}{\Gamma\left(\frac{n+1}{2p-2}p\right)},$$

where $B(u, v)$ is the Beta-function. Passing to the limit in (4.3.8) as $p \to \infty$, we arrive at the second equality in (4.3.7).

Unifying (4.2.6) and (4.3.10), we obtain

$$\begin{aligned} \mathcal{C}_1 &= \sup_{|\boldsymbol{z}|=1} \sup_{\sigma \in \mathbb{S}^{n-1}_-} |S^*(\boldsymbol{e}_\sigma)\boldsymbol{z}| (\boldsymbol{e}_\sigma, -\boldsymbol{e}_n)^n = \sup_{\sigma \in \mathbb{S}^{n-1}_-} \sup_{|\boldsymbol{z}|=1} \frac{2n}{\omega_n} \big|(\boldsymbol{e}_\sigma, \boldsymbol{z})\big| (\boldsymbol{e}_\sigma, -\boldsymbol{e}_n)^n \\ &= \frac{2n}{\omega_n} = \frac{n\Gamma(n/2)}{\pi^{n/2}}, \end{aligned}$$

which proves the first equality in (4.3.7). □

4.4. The Stokes system in a ball

A representation of the vector component $\boldsymbol{u} \in [\mathrm{C}^2(\mathbb{B}_r)]^3 \cap [\mathrm{C}(\overline{\mathbb{B}}_r)]^3$ of the solution $\{\boldsymbol{u}, p\}$ of the first boundary value problem for the Stokes system

$$\nu \Delta \boldsymbol{u} - \operatorname{grad} p = \boldsymbol{0}, \quad \operatorname{div} \boldsymbol{u} = 0$$

in a ball $\mathbb{B}_r = \{x \in \mathbb{R}^3 : |x| < r\}$ is given in the paper [**CG**] by Chichinadze and Gegelia. In particular, the value of the velocity $\boldsymbol{u}$ at the center of the ball $\mathbb{B}_r$ satisfies

$$\boldsymbol{u}(0) = \frac{3}{8\pi r^2} \int_{\partial \mathbb{B}_r} T(\sigma) \boldsymbol{u}(\sigma) d\sigma, \tag{4.4.1}$$

where T is (3×3)-matrix-valued function on $\partial\mathbb{B}_r$ with the elements

$$-\delta_{ij} + 5\,\frac{\sigma_i \sigma_j}{|\sigma|^2}, \tag{4.4.2}$$

and $\boldsymbol{u}$ obeys the orthogonality condition:

$$\int_{\partial\mathbb{B}_r} \big(\boldsymbol{u}(\sigma), \boldsymbol{\sigma}\big) d\sigma = 0. \tag{4.4.3}$$

Let us denote the set of 3-component vector-valued functions on $\partial\mathbb{B}_r$, for which (4.4.3) holds, by $\boldsymbol{N}_r$.

The estimate of $|\boldsymbol{u}(0)|$ given below contains the spherical mean of order p for the modulus of $\boldsymbol{u}$

$$M_p(\boldsymbol{u}; \partial\mathbb{B}_r) = \left\{ \frac{1}{4\pi r^2} \int_{\partial\mathbb{B}_r} |\boldsymbol{u}(\sigma)|^p d\sigma \right\}^{1/p}.$$

PROPOSITION 4.2. *The sharp constant* $\mathring{B}_p$ *in the inequality*

$$|\boldsymbol{u}(0)| \leq \mathring{B}_p\, M_p(\boldsymbol{u}; \partial\mathbb{B}_r) \tag{4.4.4}$$

is given by

$$\mathring{B}_1 = 6, \tag{4.4.5}$$

and

$$\mathring{B}_p = \frac{3}{2} \left\{ \int_0^1 \left(1 + 15u^2\right)^{\frac{p}{2(p-1)}} du \right\}^{\frac{p-1}{p}} \tag{4.4.6}$$

for $1 < p \leq \infty$. *In particular,*

$$\mathring{B}_2 = \frac{3}{2}\sqrt{6}, \qquad \mathring{B}_\infty = 3 + \frac{\sqrt{15}}{20} \log\left(4 + \sqrt{15}\,\right).$$

PROOF. By (4.4.1) and the arguments in the proof of Proposition 1.2, the sharp constant in (4.4.4) can be written as

$$\mathring{B}_p = \frac{3\left(4\pi r^2\right)^{1/p}}{8\pi r^2} \sup_{|\boldsymbol{z}|=1} \sup \left\{ \int_{\partial\mathbb{B}_r} \Big(T^*(\sigma)\boldsymbol{z}, \boldsymbol{u}(\sigma)\Big) d\sigma : \|\boldsymbol{u}\|_{[L_p(\partial\mathbb{B}_r)]^3} \leq 1,\ \boldsymbol{u} \in \boldsymbol{N}_r \right\}.$$

We introduce one more constant

$$B_p = \frac{3\left(4\pi r^2\right)^{1/p}}{8\pi r^2} \sup_{|\boldsymbol{z}|=1} \sup \left\{ \int_{\partial\mathbb{B}_r} \Big(T^*(\sigma)\boldsymbol{z}, \boldsymbol{u}(\sigma)\Big) d\sigma : \|\boldsymbol{u}\|_{[L_p(\partial\mathbb{B}_r)]^3} \leq 1 \right\}, \tag{4.4.7}$$

whose definition contains no condition $\boldsymbol{u} \in \boldsymbol{N}_r$.

Using Remark 1.5, we write the function $\boldsymbol{v}_{\boldsymbol{z}}$ providing the supremum in $\boldsymbol{v}$ in (4.4.7) as:

$$\boldsymbol{u}_{\boldsymbol{z}}(\sigma) = \begin{cases} \dfrac{T^*(\sigma)\boldsymbol{z}|T^*(\sigma)\boldsymbol{z}|^{q-2}}{\|T^*(\sigma)\boldsymbol{z}\|_q^{q/p}} & \text{for} \quad |T^*(\sigma)\boldsymbol{z}| \neq 0, \\ \boldsymbol{0} & \text{for} \quad |T^*(\sigma)\boldsymbol{z}| = 0. \end{cases} \tag{4.4.8}$$

By (4.4.2),

$$T^*(\sigma)\boldsymbol{z} = -\boldsymbol{z} + 5\,\frac{\boldsymbol{\sigma}(\boldsymbol{\sigma}, \boldsymbol{z})}{|\boldsymbol{\sigma}|^2}.$$

This and (4.4.8) imply $\boldsymbol{u_z}(-\sigma) = \boldsymbol{u_z}(\sigma)$. Hence $\boldsymbol{u_z} \in \boldsymbol{N}_r$, and therefore $\mathring{B}_p = B_p$. Thus, by (4.4.7) and Proposition 1.2,

$$\mathring{B}_p = \frac{3\big(4\pi r^2\big)^{1/p}}{8\pi r^2} \sup_{|\boldsymbol{z}|=1} \|T^* \boldsymbol{z}\|_{[L_q(\partial\mathbb{B}_r)]^3}. \tag{4.4.9}$$

Hence, taking into account that

$$|T^*(\sigma)\boldsymbol{z}| = \left[1 + 15\frac{(\boldsymbol{\sigma}, \boldsymbol{z})^2}{|\boldsymbol{\sigma}|^2}\right]^{1/2} \tag{4.4.10}$$

for any $|\boldsymbol{z}| = 1$ and $p \in (1, \infty]$, we obtain

$$\begin{aligned}
\mathring{B}_p &= \frac{3\big(4\pi r^2\big)^{1/p}}{8\pi r^2} \sup_{|\boldsymbol{z}|=1} \left\{\int_{\partial\mathbb{B}_r} \left[1 + 15\frac{(\boldsymbol{\sigma}, \boldsymbol{z})^2}{|\boldsymbol{\sigma}|^2}\right]^{q/2} d\sigma\right\}^{1/q} \\
&= \frac{3\big(4\pi r^2\big)^{1/p} r^{2/q}}{8\pi r^2} \sup_{|\boldsymbol{z}|=1} \left\{\int_{\mathbb{S}^2} \left[1 + 15(\boldsymbol{e}_\sigma, \boldsymbol{z})^2\right]^{q/2} d\sigma\right\}^{1/q} \\
&= \frac{3}{2(4\pi)^{(p-1)/p}} \left\{4\pi \int_0^{\pi/2} \left(1 + 15\cos^2\vartheta\right)^{\frac{p}{2(p-1)}} \sin\vartheta\, d\vartheta\right\}^{\frac{p-1}{p}},
\end{aligned}$$

which implies (4.4.6).

To get (4.4.5) we combine (4.4.9) and (4.4.10)

$$\mathring{B}_1 = \frac{3}{2} \sup_{|\boldsymbol{z}|=1} \sup_{\sigma\in\mathbb{S}^2} \left[1 + 15(\boldsymbol{e}_\sigma, \boldsymbol{z})^2\right]^{1/2} = 6. \tag{4.4.11}$$

□

4.5. The Lamé system in a half-space

Consider the Lamé system

$$\mu\Delta\boldsymbol{u} + (\lambda + \mu)\text{grad div } \boldsymbol{u} = \boldsymbol{0} \text{ in } \mathbb{R}^n_+, \quad n \geq 2, \tag{4.5.1}$$

with the boundary condition

$$\boldsymbol{u}\big|_{x_n=0} = \boldsymbol{f}, \tag{4.5.2}$$

where λ and μ are the Lamé constants, $\boldsymbol{u} = (u_1, \ldots, u_n)$ is the displacement vector of an elastic medium, $\boldsymbol{f} = (f_1, \ldots, f_n)$ is a continuous and bounded vector-valued function on $\partial\mathbb{R}^n_+$.

The solution $\boldsymbol{u}$ of the Dirichlet problem for the Lamé system in the half-space $\mathbb{R}^n_+$ which is bounded and continuous up to $\partial\mathbb{R}^n_+$ admits the representation (see Remark 3.2)

$$\boldsymbol{u}(x) = \int_{\partial\mathbb{R}^n_+} \mathcal{H}\left(\frac{y - x}{|y - x|}\right) \frac{x_n}{|y - x|^n}\, \boldsymbol{f}(y')dy', \tag{4.5.3}$$

Here $\mathcal{H}(\boldsymbol{e}_\sigma)$ is the $(n \times n)$-matrix-valued function on the sphere $\mathbb{S}^{n-1}$ of $\mathbb{R}^n$ with the elements

$$\frac{2}{\omega_n}\left[(1 - \varkappa)\delta_{ij} + n\varkappa(\boldsymbol{e}_\sigma, \boldsymbol{e}_i)(\boldsymbol{e}_\sigma, \boldsymbol{e}_j)\right], \tag{4.5.4}$$

where $\varkappa = (\lambda + \mu)(\lambda + 3\mu)^{-1}$, and ω_n being the area of $\mathbb{S}^{n-1}$.

THEOREM 4.3. *Let x be an arbitrary point in $\mathbb{R}^n_+$ and let $\boldsymbol{z} \in \mathbb{R}^n$. The sharp coefficient $\mathcal{K}_{p,\varkappa}(x)$ in the inequality*

$$|\boldsymbol{u}(x)| \leq \mathcal{K}_{p,\varkappa}(x)\|\boldsymbol{u}|_{x_n=0}\|_p \tag{4.5.5}$$

for a solution $\boldsymbol{u}$ of the Lamé system is given by

$$\mathcal{K}_{p,\varkappa}(x) = \mathcal{C}_{p,\varkappa}\, x_n^{(1-n)/p}, \tag{4.5.6}$$

where

$$\mathcal{C}_{1,\varkappa} = \frac{[1+\varkappa(n-1)]\Gamma\,(n/2)}{\pi^{n/2}}, \tag{4.5.7}$$

$$\mathcal{C}_{\infty,\varkappa} = \frac{2\Gamma\left(\frac{n}{2}\right)}{\sqrt{\pi}\,\Gamma\left(\frac{n-1}{2}\right)} \int_0^{\pi/2} \left[(1-\varkappa)^2 + n\varkappa(n\varkappa-2\varkappa+2)\cos^2\theta\right]^{1/2} \sin^{n-2}\theta d\theta, \tag{4.5.8}$$

and

$$\mathcal{C}_{p,\varkappa} = \frac{2^{1/p}}{\omega_n} \sup_{|\boldsymbol{z}|=1} \left\{ \int_{\mathbb{S}^{n-1}} F^{p/(2(p-1))}\left((\boldsymbol{e}_\sigma, \boldsymbol{z})\right) |(\boldsymbol{e}_\sigma, \boldsymbol{e}_n)|^{n/(p-1)} d\sigma \right\}^{(p-1)/p} \tag{4.5.9}$$

for $1 < p < \infty$. Here

$$F(t) = (1-\varkappa)^2 + n\varkappa(n\varkappa - 2\varkappa + 2)t^2. \tag{4.5.10}$$

In particular,

$$\mathcal{C}_{\frac{2k}{2k-1},\varkappa} = \frac{2^{(2k-1)/(2k)}}{\omega_n} \left\{ \int_{\mathbb{S}^{n-1}} F^k\left((\boldsymbol{e}_\sigma, \boldsymbol{e}_n)\right) |(\boldsymbol{e}_\sigma, \boldsymbol{e}_n)|^{(2k-1)n} d\sigma \right\}^{1/(2k)}, \tag{4.5.11}$$

where k is a positive integer.

As a particular case of (4.5.11) *one has*

$$\mathcal{C}_{2,\varkappa} = \left\{ \frac{\Gamma\left(\frac{n}{2}\right)}{2^n \pi^{n/2}} \left[1 + (n-1)\varkappa^2 + \left(1 + (n-1)\varkappa\right)^2\right] \right\}^{1/2}. \tag{4.5.12}$$

PROOF. Since the solution of the Dirichlet problem (4.5.1), (4.5.2) is given by (4.5.3), it follows that it obeys Proposition 4.1. Taking into account that the elements of the matrix $\mathcal{H}(\boldsymbol{e}_\sigma)$ are given by (4.5.4), we find

$$|\mathcal{H}^*(\boldsymbol{e}_\sigma)\boldsymbol{z}| = \frac{2}{\omega_n}\left[(1-\varkappa)^2 + n\varkappa(n\varkappa - 2\varkappa + 2)(\boldsymbol{e}_\sigma, \boldsymbol{z})^2\right]^{1/2}, \tag{4.5.13}$$

which together with (4.2.8) leads to

$$\mathcal{C}_{p,\varkappa} = \frac{2}{\omega_n} \sup_{|\boldsymbol{z}|=1} \left\{ \int_{\mathbb{S}^{n-1}_-} \left[F\left((\boldsymbol{e}_\sigma, \boldsymbol{z})\right)\right]^{\frac{p}{2(p-1)}} (\boldsymbol{e}_\sigma, -\boldsymbol{e}_n)^{n/(p-1)} d\sigma \right\}^{(p-1)/p}$$

for $1 < p \leq \infty$. The function $|(\boldsymbol{e}_\sigma, -\boldsymbol{e}_n)|$ is even on the sphere $\mathbb{S}^{n-1}$, therefore the last equality can be written as (4.5.9).

Passing to the limit in (4.5.9) as $p \to \infty$, we find

$$\mathcal{C}_{\infty,\varkappa} = \frac{1}{\omega_n} \sup_{|\boldsymbol{z}|=1} \int_{\mathbb{S}^{n-1}} \left[(1-\varkappa)^2 + n\varkappa(n\varkappa - 2\varkappa + 2)(\boldsymbol{e}_\sigma, \boldsymbol{z})^2\right]^{1/2} d\sigma,$$

which implies, by the independence of the last integral on $\boldsymbol{z}$,

$$\mathcal{C}_{\infty,\varkappa} = \frac{2\omega_{n-1}}{\omega_n}\int_0^{\pi/2}\Big[(1-\varkappa)^2 + n\varkappa(n\varkappa - 2\varkappa + 2)\cos^2\vartheta\Big]^{1/2}\sin^{n-2} d\vartheta.$$

Thus (4.5.8) follows.

By (4.2.6) and (4.5.13) we have

$$\begin{aligned}\mathcal{C}_{1,\varkappa} &= \sup_{|\boldsymbol{z}|=1}\ \sup_{\sigma\in\mathbb{S}^{n-1}_-} |\mathcal{H}^*(\boldsymbol{e}_\sigma)\boldsymbol{z}|(\boldsymbol{e}_\sigma, -\boldsymbol{e}_n)^n \\ &= \sup_{\sigma\in\mathbb{S}^{n-1}_-}\ \sup_{|\boldsymbol{z}|=1} \frac{2}{\omega_n}\Big[(1-\varkappa)^2 + n\varkappa(n\varkappa - 2\varkappa + 2)(\boldsymbol{e}_\sigma, \boldsymbol{z})^2\Big]^{1/2}(\boldsymbol{e}_\sigma, -\boldsymbol{e}_n)^n \\ &= \frac{2}{\omega_n}\Big[(1-\varkappa)^2 + n\varkappa(n\varkappa - 2\varkappa + 2)\Big]^{1/2} = \frac{[1+\varkappa(n-1)]\Gamma(n/2)}{\pi^{n/2}},\end{aligned}$$

which results in (4.5.7).

Consider a particular case of (4.5.9) for $p = 2k(2k-1)^{-1}$. The lower estimate

$$\mathcal{C}_{\frac{2k}{2k-1},\varkappa} \geq \frac{2^{\frac{2k-1}{2k}}}{\omega_n}\left\{\int_{\mathbb{S}^{n-1}}\Big[F\big((\boldsymbol{e}_\sigma, \boldsymbol{e}_n)\big)\Big]^k |(\boldsymbol{e}_\sigma, \boldsymbol{e}_n)|^{(2k-1)n} d\sigma\right\}^{1/(2k)} \tag{4.5.14}$$

is a direct corollary of (4.5.9) for $p = 2k(2k-1)^{-1}$. Now we derive an upper estimate for the constant $\mathcal{C}_{\frac{2k}{2k-1},\varkappa}$. By (4.5.9) and (4.5.10)

$$\mathcal{C}_{\frac{2k}{2k-1},\varkappa} \leq \frac{2^{\frac{2k-1}{2k}}}{\omega_n}\left\{\sum_{j=0}^{k}\mathcal{T}_{jk}(\varkappa)\int_{\mathbb{S}^{n-1}}(\boldsymbol{e}_\sigma, \boldsymbol{z})^{2j}|(\boldsymbol{e}_\sigma, \boldsymbol{e}_n)|^{(2k-1)n} d\sigma\right\}^{1/(2k)}, \tag{4.5.15}$$

where

$$\mathcal{T}_{jk}(\varkappa) = \binom{k}{j}(1-\varkappa)^{2(k-j)}[n\varkappa(n\varkappa - 2\varkappa + 2)]^j. \tag{4.5.16}$$

Adopting the notation

$$P_{jk} = \frac{2j + (2k-1)n}{2j}, \qquad Q_{jk} = \frac{2j + (2k-1)n}{(2k-1)n},$$

where $j = 1, 2, \dots, k$, we see that

$$\frac{1}{P_{jk}} + \frac{1}{Q_{jk}} = 1, \tag{4.5.17}$$

and

$$2j\ P_{jk} = (2k-1)n\ Q_{jk} = 2j + (2k-1)n. \tag{4.5.18}$$

Taking into account (4.5.17), we obtain by Hölder's inequality

$$\begin{aligned}&\int_{\mathbb{S}^{n-1}}(\boldsymbol{e}_\sigma, \boldsymbol{z})^{2j}\big|(\boldsymbol{e}_\sigma, \boldsymbol{e}_n)\big|^{(2k-1)n} d\sigma \\ &\qquad \leq \left\{\int_{\mathbb{S}^{n-1}}\big|(\boldsymbol{e}_\sigma, \boldsymbol{z})\big|^{2jP_{jk}} d\sigma\right\}^{1/P_{jk}}\left\{\int_{\mathbb{S}^{n-1}}\big|(\boldsymbol{e}_\sigma, \boldsymbol{e}_n)\big|^{(2k-1)nQ_{jk}} d\sigma\right\}^{1/Q_{jk}}.\end{aligned}$$

By (4.5.17), (4.5.18) as well as by independence of the first integral on the right-hand side on $\boldsymbol{z}$, we find

$$\int_{\mathbb{S}^{n-1}} (\boldsymbol{e}_\sigma, \boldsymbol{z})^{2j} \big|(\boldsymbol{e}_\sigma, \boldsymbol{e}_n)\big|^{(2k-1)n} d\sigma \le \int_{\mathbb{S}^{n-1}} \big|(\boldsymbol{e}_\sigma, \boldsymbol{e}_n)\big|^{2j+(2k-1)n} d\sigma,$$

which together with (4.5.15) leads to

$$\mathcal{C}_{\frac{2k}{2k-1},\varkappa} \le \frac{2^{\frac{2k-1}{2k}}}{\omega_n} \left\{ \sum_{j=0}^{k} \mathcal{T}_{jk}(\varkappa) \int_{\mathbb{S}^{n-1}} \big|(\boldsymbol{e}_\sigma, \boldsymbol{e}_n)\big|^{2j+(2k-1)n} d\sigma \right\}^{1/(2k)}.$$

Combined with (4.5.16), this estimate can be written as

$$\mathcal{C}_{\frac{2k}{2k-1},\varkappa} \le \frac{2^{\frac{2k-1}{2k}}}{\omega_n} \left\{ \int_{\mathbb{S}^{n-1}} \Big[F\big((\boldsymbol{e}_\sigma, \boldsymbol{e}_n)\big)\Big]^k |(\boldsymbol{e}_\sigma, \boldsymbol{e}_n)|^{(2k-1)n} d\sigma \right\}^{1/(2k)},$$

which by (4.5.14) results in (4.5.11). □

REMARK 4.1. In the case $p = \infty$ we have the sharp constant in the inequality

$$|\boldsymbol{u}(x)| \le \mathcal{C}_{\infty,\varkappa} \sup\{|\boldsymbol{u}(x')| : x' \in \partial\mathbb{R}^n_+\}$$

(see Agmon, Douglis and Nirenberg [**ADN1**], Agmon [**Ag2**], Fichera [**Fi**], C. Miranda [**Mi2**]).

For instance, (4.5.8) implies the formulas, obtained in [**MK**]:

$$\mathcal{C}_{\infty,\varkappa} = \frac{2}{\pi}(1+\varkappa)\mathbf{E}\Big(\frac{2\sqrt{\varkappa}}{1+\varkappa}\Big)$$

for $n = 2$, and

$$\mathcal{C}_{\infty,\varkappa} = \frac{1}{2}\left(1 + 2\varkappa + \frac{(1-\varkappa)^2}{\sqrt{3\varkappa(\varkappa+2)}} \log \frac{1 + 2\varkappa + \sqrt{3\varkappa(\varkappa+2)}}{1-\varkappa}\right)$$

for $n = 3$, where $\mathbf{E}(k)$ is the complete elliptic integral of the second kind.

REMARK 4.2. The constant $\mathcal{C}_{p,\varkappa}$ in the previous assertion is defined by solving an optimization problem on the unit sphere $\mathbb{S}^{n-1}$.

In particular, it is shown in the proof of Theorem 4.3 that for $p = 2k(2k-1)^{-1}$, where k is a positive integer number, the supremum of the integral in the representation for $\mathcal{C}_{p,\varkappa}$ is attained on the vectors $\boldsymbol{z} = \boldsymbol{e}_n$ and $\boldsymbol{z} = -\boldsymbol{e}_n$. Note also that (4.5.11) can be written in the form

$$\begin{aligned}\mathcal{C}_{\frac{2k}{2k-1},\varkappa} &= \frac{2^{\frac{2k-1}{2k}}}{\omega_n} \left\{ 2\omega_{n-1} \int_0^{\pi/2} F^k(\cos\vartheta) \cos^{(2k-1)n}\vartheta \, \sin^{n-2}\vartheta \, d\vartheta \right\}^{1/(2k)} \\ &= \frac{\Gamma\left(\frac{n}{2}\right)}{\pi^{n/2}} \left\{ \frac{2\pi^{(n-1)/2}}{\Gamma\left(\frac{n-1}{2}\right)} \int_0^{\pi/2} F^k(\cos\vartheta) \cos^{(2k-1)n}\vartheta \, \sin^{n-2}\vartheta \, d\vartheta \right\}^{1/(2k)}.\end{aligned}$$

4.6. The Lamé system in a ball

A representation for the solution of the Dirichlet problem for the Lamé system in the ball, which is analogous to the Poisson formula, was obtained in the paper [**Na**] by Natroshvili (cf. also Kupradze, Gegelia, Basheleishvili and Burchuladze [**KGBB**], Ch.14, Chichinadze and Gegelia [**CG**], Ch. 2). As its direct corollary one obtains the following formula for the elastic displacement in the center of the ball $\mathbb{B}_r = \{x \in \mathbb{R}^3 : |x| < r\}$

$$\boldsymbol{u}(0) = \frac{3}{4\pi(3-\varkappa)r^2} \int_{\partial\mathbb{B}_r} \mathcal{N}(\sigma)\boldsymbol{u}(\sigma)d\sigma, \tag{4.6.1}$$

where $\mathcal{N}$ is the (3×3)-matrix-valued function on $\partial\mathbb{B}_r$ with the elements

$$(1-2\varkappa)\delta_{ij} + 5\varkappa\frac{\sigma_i\sigma_j}{|\sigma|^2}, \tag{4.6.2}$$

and

$$\varkappa = (\lambda+\mu)(\lambda+3\mu)^{-1}, \text{ and } \boldsymbol{u} \in [\mathrm{C}(\partial\mathbb{B}_\mathrm{r})]^3.$$

Proposition 4.3. *The sharp constant $\mathcal{B}_{p,\varkappa}$ in the inequality*

$$|\boldsymbol{u}(0)| \le B_{p,\varkappa}\, M_p(\boldsymbol{u};\partial\mathbb{B}_r)$$

is given by

$$B_{1,\varkappa} = \frac{3(1+3\varkappa)}{3-\varkappa}, \tag{4.6.3}$$

$$B_{\infty,\varkappa} = \frac{3}{2(3-\varkappa)}\left[1+3\varkappa+\frac{(1-2\varkappa)^2}{\sqrt{5\varkappa(\varkappa+2)}}\log\frac{1+3\varkappa+\sqrt{5\varkappa(\varkappa+2)}}{|1-2\varkappa|}\right],$$

and

$$B_{p,\varkappa} = \frac{3}{3-\varkappa}\left\{\int_0^1 \left[(1-2\varkappa)^2+5\varkappa(\varkappa+2)u^2\right]^{\frac{p}{2(p-1)}} du\right\}^{\frac{p-1}{p}} \tag{4.6.4}$$

for $1<p<\infty$.

In particular,

$$B_{2,\varkappa} = \frac{\sqrt{3}}{3-\varkappa}\left(17\varkappa^2-2\varkappa+3\right)^{1/2}.$$

Proof. According to (4.6.1) and Proposition 1.2, the sharp constant in

$$|\boldsymbol{u}(0)| \le B_{p,\varkappa}\, M_p(\boldsymbol{u};\partial\mathbb{B}_r)$$

can be written as

$$B_{p,\varkappa} = \frac{3\left(4\pi r^2\right)^{1/p}}{4\pi(3-\varkappa)r^2} \sup_{|\boldsymbol{z}|=1} \|\mathcal{N}^*\boldsymbol{z}\|_{[L_q(\partial\mathbb{B}_r)]^3}. \tag{4.6.5}$$

Hence, taking into account that

$$|\mathcal{N}^*(\sigma)\boldsymbol{z}| = \left[(1-2\varkappa)^2+5\varkappa(\varkappa+2)\frac{(\boldsymbol{\sigma},\boldsymbol{z})^2}{|\boldsymbol{\sigma}|^2}\right]^{1/2}, \tag{4.6.6}$$

for $|\boldsymbol{z}|=1$ and $p\in(1,\infty]$ we find

$$\begin{aligned} B_{p,\varkappa} &= \frac{3\big(4\pi r^2\big)^{1/p}}{4\pi(3-\varkappa)r^2}\sup_{|\boldsymbol{z}|=1}\left\{\int_{\partial\mathbb{B}_r}\left[(1-2\varkappa)^2+5\varkappa(\varkappa+2)\frac{(\boldsymbol{\sigma},\boldsymbol{z})^2}{|\boldsymbol{\sigma}|^2}\right]^{q/2}d\sigma\right\}^{1/q} \\ &= \frac{3\big(4\pi r^2\big)^{1/p}r^{2/q}}{4\pi(3-\varkappa)r^2}\sup_{|\boldsymbol{z}|=1}\left\{\int_{\mathbb{S}^2}\left[(1-2\varkappa)^2+5\varkappa(\varkappa+2)(\boldsymbol{e}_\sigma,\boldsymbol{z})^2\right]^{q/2}d\sigma\right\}^{1/q}, \end{aligned}$$

i.e.,

$$B_{p,\varkappa}=\frac{3}{\big(4\pi\big)^{(p-1)/p}(3-\varkappa)}\left\{4\pi\int_0^{\pi/2}\left[(1-2\varkappa)^2+5\varkappa(\varkappa+2)\cos^2\vartheta\right]^{\frac{p}{2(p-1)}}\sin\vartheta d\vartheta\right\}^{\frac{p-1}{p}},$$

which implies (4.6.4).

By (4.6.5) and (4.6.6),

$$B_{1,\varkappa}=\frac{3}{3-\varkappa}\sup_{|\boldsymbol{z}|=1}\left[(1-2\varkappa)^2+5\varkappa(\varkappa+2)(\boldsymbol{e}_\sigma,\boldsymbol{z})^2\right]^{1/2}=\frac{3(1+3\varkappa)}{3-\varkappa},$$

which proves (4.6.3). □

4.7. Comments to Chapter 4

Elliptic equations with boundary data from L^p were studied, in particular, by Strichartz [**Str**].

A representation of the vector component $\boldsymbol{u}\in[\mathrm{C}^2(\mathbb{B})]^n\cap[\mathrm{C}(\overline{\mathbb{B}})]^n$ of the solution $\{\boldsymbol{u},p\}$ to the Dirichlet problem for the system

$$\nu\Delta\boldsymbol{u}-\operatorname{grad}p=\boldsymbol{0},\quad \operatorname{div}\boldsymbol{u}=\text{const} \tag{4.7.1}$$

in the unit ball $\mathbb{B}=\{x\in\mathbb{R}^n:|x|<1\}$ was obtained by Kratz [**Krat1, Krat3**]. The vector $\boldsymbol{u}$ subject to (4.7.1) is sometimes called the Stokes function. Kratz [**Krat1**]-[**Krat4**]) is interested in a constant in the Miranda-Agmon maximum principle. He finds the best constant for the unit disk [**Krat2**]. The sharp constant $c_n(0)$ in the inequality

$$|\boldsymbol{u}(0)|\le c_n(0)\max_{x\in\partial\mathbb{B}}|\boldsymbol{u}(x)|$$

was obtained in [**Krat4**].

We note, that the sharp constants in (4.1.2) for the Lamé and Stokes systems in the case $p=\infty$ are given in Section 3.3 (see Corollary (3.4)).

CHAPTER 5

Sharp Constant in the Miranda-Agmon Type Inequality for Derivatives of Solutions to Higher Order Elliptic Equations

5.1. Introduction

The present chapter is devoted to sharp Miranda-Agmon type estimate for solutions of higher order elliptic equations with constant coefficients. By the Miranda-Agmon maximum principle for a homogeneous elliptic equation of the order 2ℓ one means either the estimate

$$\max_{\overline{\Omega}} \mid \nabla_{\ell-1} u \mid \leq c(\Omega) \max_{\partial\Omega} \mid \nabla_{\ell-1} u \mid, \qquad \Omega \subset \mathbb{R}^n,$$

or a similar estimate with the right-hand side containing an additional weak norm of the solution.

In Section 5.2 we consider the elliptic operator

$$P(D_x) = \sum_{|\beta| \leq 2\ell} a_{\beta} D_x^{\beta}$$

with constant complex coefficients, where $D_x^{\beta} = \partial^{|\beta|}/\partial x_1^{\beta_1} \dots \partial x_n^{\beta_n}$, and $\beta = (\beta_1, \dots, \beta_n)$ is a multi-index of order $|\beta| = \beta_1 + \dots + \beta_n$. By $P_0(\xi)$ we denote the principal homogeneous part of the polynomial $P(\xi)$. For $n = 2$ we assume also that all ξ_2-roots of the polynomial $P_0(\xi)$ are distributed equally in the upper and lower parts of the plane for all $\xi_1 \in \mathbb{R}\backslash\{0\}$. Everywhere in this chapter, by smoothness we mean the membership in C^{∞}.

Let $\mathbb{R}^n_+(\boldsymbol{\nu}) = \{x \in \mathbb{R}^n : (x, \boldsymbol{\nu}) > 0\}$, where $\boldsymbol{\nu}$ is a unit vector and let $K(\boldsymbol{\nu})$ be the best constant in the Miranda-Agmon inequality

$$\sup_{\mathbb{R}^n_+(\boldsymbol{\nu})} |\nabla_{\ell-1} u| \leq \mathcal{K} \sup_{\partial\mathbb{R}^n_+(\boldsymbol{\nu})} |\nabla_{\ell-1} u|. \tag{5.1.1}$$

Here

$$|\nabla_{\ell-1} u| = \left(\sum_{|\beta| = \ell-1} \frac{(\ell-1)!}{\beta!} |D_x^{\beta} u|^2 \right)^{1/2},$$

and u is an arbitrary solution of the equation $P_0(D_x)u = 0$, smooth in $\overline{\mathbb{R}^n_+(\boldsymbol{\nu})}$ and such that $u(x) = O(|x|^{\ell-1})$ for large $|x|$. Further, let Ω be a domain in $\mathbb{R}^n$ with smooth boundary $\partial\Omega$ and compact closure $\overline{\Omega}$. For solutions u of the equation $P(D_x)u = 0$, smooth in $\overline{\Omega}$, we obtain a weak form of the Miranda-Agmon inequality

$$\max_{\overline{\Omega}} |\nabla_{\ell-1} u| \leq \left(\sup_{\boldsymbol{\nu} \in \mathbb{S}^{n-1}} K(\boldsymbol{\nu}) + \varepsilon \right) \max_{\partial\Omega} |\nabla_{\ell-1} u| + c(\varepsilon) \|u\|_{L^1(\Omega)}, \tag{5.1.2}$$

where ε is any positive number and $c(\varepsilon)$ is a positive constant independent of u.

As an illustration, in Section 5.3 we find the best constant in (5.1.2) for the biharmonic equation. To be specific, we obtain the sharp constant (not depending on $\boldsymbol{\nu}$) in the C. Miranda inequality

$$\sup_{\overline{\mathbb{R}^n_+}} |\nabla u| \le K \sup_{\partial\mathbb{R}^n_+} |\nabla u|, \tag{5.1.3}$$

where $\mathbb{R}^n_+ = \{x = (x_1, \dots, x_n) : x_n > 0\}$, u is a solution of the biharmonic equation in $\mathbb{R}^n_+$ from $\mathrm{C}^\infty(\overline{\mathbb{R}^n_+})$ with $\nabla u(x_1, \dots, x_{n-1}, 0)$ being a bounded vector-valued function on $\partial\mathbb{R}^n_+$ and $u(x) = O(|x|)$ for large $|x|$. We show that

$$K = \frac{2\Gamma(\frac{n}{2})}{\sqrt{\pi}\,\Gamma(\frac{n-1}{2})} \int_0^{\pi/2} \left[4 + n(n-4)\cos^2\vartheta\right]^{1/2} \sin^{n-2}\vartheta \, d\vartheta.$$

In particular, $K = 4/\pi$ for $n = 2, K = 1/2 + 2\pi\sqrt{3}/9$ for $n = 3$ and $K = 2$ for $n = 4$.

Besides, in the two-dimensional case we prove the classical maximum principle for $|\nabla u|$, which holds if $\nabla u(x_1, 0)$ is parallel to a constant vector in $\mathbb{R}^2$. Namely, we show that a biharmonic function in $\mathbb{R}^2_+$ such that

$$\nabla u(x_1, 0) = \psi(x_1)\boldsymbol{\xi},$$

where $\psi \in \mathrm{C}(\partial\mathbb{R}^2_+)$ and $\boldsymbol{\xi}$ is a unit two-dimensional vector, obeys the inequality

$$\|\nabla\, u\|_{[\mathrm{C}_b(\overline{\mathbb{R}^2_+})]^2} \le \|\nabla\, u\|_{[\mathrm{C}_b(\partial\mathbb{R}^2_+)]^2}.$$

5.2. Weak form of the Miranda-Agmon inequality with the sharp constant

Suppose Ω is a domain in $\mathbb{R}^n$ with smooth boundary $\partial\Omega$ and a compact closure $\overline{\Omega}$. We consider the elliptic operator

$$P(D_x) = \sum_{|\beta| \le 2\ell} a_\beta D_x^\beta$$

with constant complex coefficients, where $D_x^\beta = \partial^{|\beta|}/\partial x_1^{\beta_1} \dots \partial x_n^{\beta_n}$, and $\beta = (\beta_1, \dots, \beta_n)$ is a multi-index of order $|\beta| = \beta_1 + \dots + \beta_n$. By $P_0(\xi)$ we denote the principal homogeneous part of the polynomial $P(\xi)$. For $n = 2$ we assume additionally that all ξ_2-roots of the polynomial $P_0(\xi)$ are distributed equally in the upper and lower parts of the plane for all $\xi_1 \in \mathbb{R}\backslash\{0\}$. We use the notation introduced in Section 5.1.

In this section we deal with the best constant in a weak form of the Miranda-Agmon inequality

$$\max_{\overline{\Omega}} |\nabla_{\ell-1} u| \le K \max_{\partial\Omega} |\nabla_{\ell-1} u| + C\|u\|_{L^1(\Omega)}. \tag{5.2.1}$$

We prove the equality

$$\inf K = \sup_{\boldsymbol{\nu} \in \mathbb{S}^{n-1}} K(\boldsymbol{\nu}),$$

where the infimum on the left-hand side is taken over all numbers K for which there exists such constant C that for all smooth solutions of the equation $P(D_x)u = 0$ in Ω the estimate (5.2.1) is valid.

LEMMA 5.1. *Suppose $G(x,y)$ is Green's function of the Dirichlet problem for the operator $P_0(D_x)$ in $\mathbb{R}^n_+ = \{x = (x', x_n) : x' = (x_1, \dots, x_{n-1}) \in \mathbb{R}^{n-1},\ x_n > 0\}$. Then for any multi-indices $\alpha, \beta, |\alpha| = |\beta| = \ell - 1$,*

$$\sum_{i=1}^{n-1} \left| \frac{\partial}{\partial y_i} D_x^\alpha D_y^\beta G(x,y) \right| \leq \begin{cases} |x-y|^{1-n} & \text{if } |x-y| < 2\min\{x_n, y_n\}, \\ \dfrac{c x_n y_n}{|x-y|^{n+1}} & \text{if } |x-y| \geq 2\min\{x_n, y_n\}, \end{cases}$$

where $x, y \in \mathbb{R}^n_+$.

The proof follows from the estimate

$$\left| D_x^\gamma D_y^\delta G(x,y) \right| \leq c|x-y|^{-1-n},\ |\gamma| = \ell,\ |\delta| = \ell + 1$$

(see Agmon, Douglis and Nirenberg [**ADN1**]) and from the homogeneous Dirichlet conditions for $G(x,y)$ with respect to x and y.

The following assertion gives the best constant in a weak form of the Miranda-Agmon inequality.

THEOREM 5.1. *For any solution of the equation $P(D_x)u = 0$, smooth on $\overline{\Omega}$, the inequality*

$$\max_{\overline{\Omega}} |\nabla_{\ell-1} u| \leq \left(\sup_{\boldsymbol{\nu} \in \mathbb{S}^{n-1}} K(\boldsymbol{\nu}) + \varepsilon \right) \max_{\partial\Omega} |\nabla_{\ell-1} u| + c(\varepsilon) \|u\|_{L^1(\Omega)}$$

is valid, where ε is any positive number and $c(\varepsilon)$ is a positive constant independent of u.

PROOF. Suppose the maximum of $|\nabla_{\ell-1} u|$ on $\overline{\Omega}$ is attained at a point $a \in \overline{\Omega}$. Let $\mathbb{B}_r(a)$ denote the open ball with center a and sufficiently small radius r depending on ε. We introduce smooth functions φ and ψ with supports in $\mathbb{B}_r(a)$ such that $\varphi\psi = \varphi$, $\varphi(a) = 1$. It suffices to consider the case when the set $\mathbb{B}_r(a) \bigcap \partial\Omega$ is nonempty. We place the origin $\mathcal{O}$ in $\mathbb{B}_r(a) \bigcap \partial\Omega$ and we direct the $\mathcal{O}x_n$ axis along the inner normal $\boldsymbol{\nu}$ to $\partial\Omega$. Suppose the surface $\partial\Omega$ is defined near $\mathcal{O}$ by the equation $x_n = f(x')$. We assume the radius of the ball $\mathbb{B}_r(a)$ to be so small that $|\nabla f(x')| < \varepsilon$ on $\mathbb{B}_r(a)$.

We introduce the mapping

$$\varkappa : \mathbb{B}_r(a) \bigcap \overline{\Omega} \ni x \to y \in \overline{\mathbb{R}}^n_+,$$

where $y = (y', y_n)$, $y' = x'$, $y_n = x_n - f(x')$. We will denote the functions $u \circ \varkappa^{-1}$, $\varphi \circ \varkappa^{-1}$, $\psi \circ \varkappa^{-1}$ by u, φ, ψ as before. In $\mathbb{R}^n_+$ the function u satisfies the equation

$$P\left(D_{y'} - \nabla f(y') \frac{\partial}{\partial y_n}, \frac{\partial}{\partial y_n} \right) u = 0. \tag{5.2.2}$$

We multiply (5.2.2) by φ and rewrite the equality obtained as $P_0(D_y)(\varphi u) = F_1 + F_2$, where

$$F_1 = \sum_{i,j=1}^{n-1} \sum_{|\beta|=\ell-1} \frac{\partial}{\partial y_i} D_y^\alpha \left(\frac{\partial f}{\partial y_j} Q_{i,j,\alpha}(y', D_y)(\varphi u) \right), \tag{5.2.3}$$

$F_2 = R(y, D_y)(\psi u)$, $Q_{i,j,\alpha}$ and R are differential operators with smooth coefficients, ord $Q_{i,j,\alpha} = \ell$, ord $R = 2\ell - 1$. Clearly, $\varphi u = v + w$, where v is a solution of $P_0(D_y)v = 0$ in $\mathbb{R}^n_+$,

$$v = \varphi u, \quad \frac{\partial v}{\partial y_n} = \frac{\partial(\varphi u)}{\partial y_n}, \dots, \frac{\partial^{\ell-1} v}{\partial y_n^{\ell-1}} = \frac{\partial^{\ell-1}(\varphi u)}{\partial y_n^{\ell-1}} \quad \text{on } \partial\mathbb{R}^n_+,$$

and w is a solution of $P_0(D_y)w = F_1 + F_2$ in $\mathbb{R}^n_+$,

$$w = 0, \quad \frac{\partial w}{\partial y_n} = 0, \dots, \frac{\partial^{\ell-1} w}{\partial y_n^{\ell-1}} = 0 \text{ on } \partial\mathbb{R}^n_+,$$

such that $v(y) = O(|y|^{\ell-1})$, $w(y) = O(|y|^{\ell-1})$ for large $|y|$.

By definition of the constant $K(\boldsymbol{\nu})$ the following inequality is true:

$$\sup_{\mathbb{R}^n_+} |\nabla_{\ell-1} v| \le K(\boldsymbol{\nu}) \sup_{\partial\mathbb{R}^n_+} |\nabla_{\ell-1}(\varphi u)|. \tag{5.2.4}$$

We obtain an estimate for $|\nabla_{\ell-1} w|$. Suppose, as in Lemma 5.1, that $G(x, y)$ is Green's function for the Dirichlet problem for $P_0(D_x)$ in $\mathbb{R}^n_+$ (for convenience here $x = (x', x_n)$ denotes a point of the half-space $\mathbb{R}^n_+$, in contrast to the beginning of the proof of the theorem). For all $y \in \mathbb{R}^n_+$ and any multi-index β of order $\ell - 1$ the following equality is true:

$$(D_y^\beta w)(y) = \int_{\mathbb{R}^n_+} D_y^\beta G(y, x)\big(F_1(x) + F_2(x)\big)dx. \tag{5.2.5}$$

Hence from (5.2.3) we find

$$\int_{\mathbb{R}^n_+} D_y^\beta G(y, x) F_1(x)dx = -\sum_{i,j=1}^{n-1} \sum_{|\alpha|=\ell-1} \int_{\mathbb{R}^n_+} D_y^\beta \frac{\partial}{\partial x_i} D_x^\alpha G(y, x) \frac{\partial}{\partial x_j} Q_{i,j,\alpha}(x', D_x)(\varphi u)dx.$$

We represent the integral

$$I(y) = \int_{\mathbb{R}^n_+} \left| D_y^\beta \frac{\partial}{\partial x_j} D_x^\alpha G(y, x) \right| x_n^{-1} dx,$$

as the sum of integrals $I_1(y) + I_2(y)$, the first of which is over the set $S = \{x : |x - y| < 2\min\{x_n, y_n\}\}$. Since in this set either $y_n \le x_n < 3y_n$, or $x_n \le y_n < 3x_n$, it follows by Lemma 5.1 that

$$I_1(y) \le c y_n^{-1} \int_{|y-x|<2y_n} \frac{dx}{|x - y|^{n-1}} \le c.$$

Again applying Lemma 5.1, we obtain

$$\begin{aligned} I_2(y) &\le c\, y_n \int_{\mathbb{R}^n_+ \setminus S} \frac{dx}{|x - y|^{n+1}} \\ &\le c\, y_n \left(\int_{\mathbb{R}^n_+ \setminus S,\, 3x_n < y_n} \frac{dx}{(y_n + |x' - y'|)^{n+1}} + \int_{\mathbb{R}^n_+ \setminus S,\, 3x_n > y_n} \frac{dx}{|x - y|^{n+1}} \right) \\ &\le c\, y_n^2 \int_{\mathbb{R}^{n-1}} \frac{dx'}{(y_n + |x' - y'|)^{n+1}} + c y_n \int_{y_n/3}^\infty \frac{dx_n}{x_n^2} \le c. \end{aligned}$$

Consequently the integral $I(y)$ is uniformly bounded. Now it is clear that

$$\left|\int_{\mathbb{R}^n_+} D_y^\beta G(y,x)F_1(x)dx\right| \le c\,\varepsilon \max_{\mathbb{R}^n_+} y_n \sum_{k=0}^{\ell} |\nabla_k(\varphi u)(y)|. \tag{5.2.6}$$

Equation (5.2.2) and the known local estimate for solutions of elliptic equations (see Agmon, Douglis and Nirenberg [**ADN1**], Theorem 15.3) imply the estimate

$$\max_{\mathbb{R}^n_+} y_n |\nabla_\ell(\varphi u)(y)| \le c \max_{\mathbb{R}^n_+} \sum_{k=0}^{\ell-1} |\nabla_k(\psi u)(y)|,$$

which in combination with (5.2.6) gives the inequality

$$\left|\int_{\mathbb{R}^n_+} D_y^\beta G(y,x)F_1(x)dx\right| \le c\,\varepsilon \max_{\mathbb{R}^n_+} \sum_{k=0}^{\ell-1} |\nabla_k(\psi u)(y)|. \tag{5.2.7}$$

We proceed to an estimate of the integral

$$\int_{\mathbb{R}^n_+} D_y^\beta G(y,x)F_2(x)dx, \qquad |\beta| = \ell-1.$$

Its absolute value does not exceed

$$\begin{aligned} c \sum_{|\alpha|=\ell} \int_{\mathbb{R}^n_+} |D_y^\beta D_x^\alpha G(y,x)| \sum_{k=0}^{\ell-1} |\nabla_k(\psi u)(x)|dx \\ \le \; c \int_{\mathbb{R}^n_+} |y-x|^{1-n} \sum_{k=0}^{\ell-1} |\nabla_k(\psi u)(x)|dx \le c\|\psi u\|_{W_p^{\ell-1}(\mathbb{R}^n_+)}, \end{aligned}$$

where $p > n$. Combining this with (5.2.7), we deduce the following inequality from (5.2.5):

$$\sup_{\overline{\mathbb{R}^n_+}} |\nabla_{\ell-1} w| \le c\,\varepsilon \max_{\mathbb{R}^n_+} \sum_{k=0}^{\ell-1} |\nabla_k(\psi u)| + c(\varepsilon)\|\psi u\|_{W_p^{\ell-1}(\mathbb{R}^n_+)}.$$

Hence (5.2.4) implies

$$\begin{aligned} \max_{\overline{\mathbb{R}^n_+}} |\nabla_{\ell-1}(\varphi u)| \;&\le\; K(\boldsymbol{\nu}) \max_{\mathbb{R}^{n-1}} |\nabla_{\ell-1}(\varphi u)| \\ &+\; c\,\varepsilon \max_{\mathbb{R}^n_+} \sum_{k=0}^{\ell-1} |\nabla_k(\psi u)| + c(\varepsilon)\|\psi u\|_{W_p^{\ell-1}(\mathbb{R}^n_+)}. \end{aligned}$$

Returning to the variable $x \in \overline{\Omega}$ and applying Sobolev's theorem on the imbedding of $W_p^{\ell-1}(\Omega)$ in $C^{\ell-2}(\overline{\Omega})$ for $\ell \ge 2$, $p > n$, we find

$$\max_{\overline{\Omega}} |\nabla_{\ell-1} u| = |\nabla_{\ell-1}(\varphi u)(a)| \le \big(K(\boldsymbol{\nu}) + c\varepsilon\big) \max_{\partial\Omega} |\nabla_{\ell-1} u| + c(\varepsilon)\|u\|_{W_p^{\ell-1}(\Omega)}.$$

Hence by increasing the coefficient for ε we obtain that in the second term on the right-hand side it is possible to replace Ω by a domain Ω_ε, $\overline{\Omega}_\varepsilon \subset \Omega$. Using the local estimate (see Agmon, Douglis and Nirenberg [**ADN1**], Theorem 15.1″)

$$\|u\|_{W_p^{\ell-1}(\Omega_\varepsilon)} \le c(\varepsilon)\|u\|_{L^1(\Omega)},$$

we complete the proof of the theorem. □

5.3. Sharp constants for biharmonic functions

5.3.1. Classical maximum principle for the gradient of biharmonic function in the half-plane

First, we shall prove the classical maximum principle for $|\nabla u|$, which holds if $\nabla u(x_1, 0)$ is parallel to a constant vector in $\mathbb{R}^2$. Namely, the following consequence of Theorem 3.4 holds.

COROLLARY 5.1. *Let u be a biharmonic function in $\mathbb{R}^2_+$ such that*

$$\nabla u(x_1, 0) = \psi(x_1)\boldsymbol{\xi},$$

where $\psi \in \mathrm{C}(\partial\mathbb{R}^2_+)$ and $\boldsymbol{\xi}$ is a unit two-dimensional vector. Then the inequality

$$\|\nabla\, u\|_{[\mathrm{C_b}(\overline{\mathbb{R}^2_+})]^2} \leq \|\nabla\, u\|_{[\mathrm{C_b}(\partial\mathbb{R}^2_+)]^2} \tag{5.3.1}$$

holds.

PROOF. Let u be a solution of the equation $\Delta^2 u = 0$ in $\mathbb{R}^2_+$ such that $\nabla u = \boldsymbol{f}$ on $\partial\mathbb{R}^2_+$, where $\boldsymbol{f} = (f_1, f_2) \in [\mathrm{C}(\partial\mathbb{R}^2_+)]^2$.

Putting $v_1 = -\partial u/\partial x_2$, $v_2 = \partial u/\partial x_1$ and $\boldsymbol{v} = (v_1, v_2)$ we have

$$\mathrm{div}\ \boldsymbol{v} = 0.$$

We write $\Delta^2 u = 0$ in the form

$$\frac{\partial}{\partial x_1}\Delta\frac{\partial u}{\partial x_1} + \frac{\partial}{\partial x_2}\Delta\frac{\partial u}{\partial x_2} = \frac{\partial}{\partial x_1}\Delta v_2 - \frac{\partial}{\partial x_2}\Delta v_1 = 0.$$

This implies the existence of a function p such that

$$\Delta v_1 - \frac{\partial p}{\partial x_1} = 0, \quad \Delta v_2 - \frac{\partial p}{\partial x_2} = 0.$$

Here, the boundary conditions $v_1 = -f_2$ and $v_2 = f_1$ are satisfied on $\partial\mathbb{R}^2_+$.

Hence, for any biharmonic functions u in $\mathbb{R}^2_+$ such that ∇u on $\partial\mathbb{R}^2_+$ is a function $\boldsymbol{f} \in [\mathrm{C}(\partial\mathbb{R}^2_+)]^2$, there exists the solution $\boldsymbol{v} = (-\partial u/\partial x_2,\ \partial u/\partial x_1)$ of the boundary value problem

$$\Delta\boldsymbol{v} - \mathrm{grad}\ p = \boldsymbol{0}, \ \ \mathrm{div}\ \boldsymbol{v} = 0 \ \text{ in } \mathbb{R}^2_+, \quad \boldsymbol{v} = (-f_2, f_1) \text{ on } \partial\mathbb{R}^2_+.$$

Since $\nabla u(x_1, 0) = (f_1(x_1), f_2(x_1)) = \psi(x_1)\boldsymbol{\xi}$, it follows that

$$\boldsymbol{v}(x_1, 0) = (-f_2(x_1), f_1(x_1)) = \psi(x_1)\boldsymbol{\xi}',$$

where $\boldsymbol{\xi}'$ is a constant two-dimensional unit vector.

By Theorem 3.4, if $\boldsymbol{v}(x_1, 0) = \psi(x_1)\boldsymbol{\xi}'$, then, for any $x \in \mathbb{R}^2_+$,

$$|\boldsymbol{v}(x)| \leq \|\boldsymbol{v}|_{x_2=0}\|.$$

Putting here $\boldsymbol{v} = (-\partial u/\partial x_2,\ \partial u/\partial x_1)$, we arrive at (5.3.1) with the constant 1 in front of the norm of gradient on the right-hand side. □

5.3.2. Representation for the gradient of biharmonic function in the half-space

Now we consider biharmonic functions in the n-dimensional half-space. In the following lemma we obtain a representation for the gradient of a biharmonic function in the half-space $\mathbb{R}^n_+$.

Let $M(\boldsymbol{e})$ denotes an $(n\times n)$-matrix-valued function on the $(n-1)$-dimensional Euclidean sphere $\mathbb{S}^{n-1}$ with the elements of the form

$$2\omega_n^{-1}[2\delta_{ij} - n(\boldsymbol{e}, \boldsymbol{e}_i)(\boldsymbol{e}, \boldsymbol{e}_j)],$$

where ω_n is the area of the sphere $\mathbb{S}^{n-1}$, $\boldsymbol{e}_k$ is the unit vector of the k-th coordinate axis. Further let $\boldsymbol{e}_{xy} = (y-x)|y-x|^{-1}$ and $x' = (x_1, \dots, x_{n-1})$.

We will say that the function $u \in \mathrm{C}^\infty(\overline{\mathbb{R}^n_+})$ belongs to the class $\mathfrak{A}$, if $\nabla u(x', 0)$ is a bounded vector-valued function on $\mathbb{R}^{n-1}$ and $u(x) = O(|x|)$ for large $|x|$.

LEMMA 5.2. *Suppose u is a solution of the biharmonic equation in $\mathbb{R}^n_+$ in the class $\mathfrak{A}$. Then*

$$u(x) = \frac{2}{\omega_n}\left\{ n\int_{\partial\mathbb{R}^n_+} \frac{x_n^3}{|y-x|^{n+2}}\, u(y', 0)\, dy' + \int_{\partial\mathbb{R}^n_+} \frac{x_n^2}{|y-x|^n} \frac{\partial u}{\partial y_n}(y', 0)\, dy' \right\}$$

and

$$\nabla u(x) = \int_{\partial\mathbb{R}^n_+} M(\boldsymbol{e}_{xy})\, \nabla u(y', 0)\, \frac{x_n}{|y-x|^n}\, dy',$$

where $x \in \mathbb{R}^n_+$ and $y = (y', 0)$.

PROOF. We first consider the Dirichlet problem

$$\Delta^2 u = 0 \ \text{ in } \ \mathbb{R}^n_+, \ \ u|_{x_n=0} = f_1(x'), \ \ \left.\frac{\partial u}{\partial x_n}\right|_{x_n=0} = f_2(x'),$$

where $f_1, f_2 \in \mathrm{C}_0^\infty(\mathbb{R}^{n-1})$, i.e, are infinitely differentiable functions with compact supports. Applying the Fourier transform in x' to the biharmonic equation and boundary conditions, we obtain a boundary value problem for the ordinary differential equation

$$\frac{d^4F[u]}{dx_n^4} - 2|\xi'|^2\frac{d^2F[u]}{dx_n^2} + |\xi'|^4\, F[u] = 0 \tag{5.3.2}$$

$$F[u]\big|_{x_n=0} = F[f_1], \quad \left.\frac{dF[u]}{dx_n}\right|_{x_n=0} = F[f_2]\,.$$

The general solution of this equation is given by

$$F[u](\xi', x_n) = \big(A_1(\xi') + A_2(\xi')x_n\big)e^{-|\xi'|x_n} + \big(B_1(\xi') + B_2(\xi')x_n\big)e^{|\xi'|x_n}. \tag{5.3.3}$$

To obtain a solution of the original Dirichlet boundary value problem it is necessary to set $B_1(\xi') = B_2(\xi') = 0$. From (5.3.2) and (5.3.3) it follows that

$$F[u](\xi', x_n) = f_1(\xi')(1 + x_n|\xi'|)e^{-|\xi'|x_n} + f_2(\xi')x_n e^{-|\xi'|x_n}. \tag{5.3.4}$$

We find the inverse Fourier transform of this function. We pass to spherical coordinates, directing the polar axis along the vector x:

$$\begin{aligned} F^{-1}\left(e^{-|\xi'|x_n}\right) &= \frac{1}{(2\pi)^{n-1}} \int_{\mathbb{R}^{n-1}} e^{-|\xi'|x_n} e^{i(x',\,\xi')}\, d\xi' \\ &= \frac{\omega_{n-2}}{(2\pi)^{n-1}} \int_0^\infty \left[\int_0^\pi e^{ir\rho\cos\theta} \sin^{n-3}\theta\, d\theta\right] e^{-rx_n} r^{n-2}\, dr, \end{aligned}$$

where $\rho = |x'|$, $r = |\xi'|$, θ is the angle between the vectors x' and ξ'. Let $J_\nu(x)$ be the Bessel function of the first kind. Applying known formulas for the integrals (see, e.g., Gradshtein and Ryzhik [**GR**], **8.411.7** and **6.623.2**),

$$J_\nu(z) = \frac{\left(\frac{z}{2}\right)^\nu}{\Gamma\left(\nu+\frac{1}{2}\right)\Gamma\left(\frac{1}{2}\right)} \int_0^\pi e^{\pm iz\cos\varphi} \sin^{2\nu}\varphi\, d\varphi, \qquad \Re\left(\nu+\frac{1}{2}\right) > 0 \tag{5.3.5}$$

and

$$\int_0^\infty e^{-\alpha x} J_\nu(\beta x) x^{\nu+1} dx = \frac{2\alpha(2\beta)^\nu \Gamma\left(\nu+\frac{3}{2}\right)}{\sqrt{\pi}(\alpha^2+\beta^2)^{(2\nu+3)/2}}, \quad \Re\,\nu > -1, \tag{5.3.6}$$

we find

$$\begin{aligned} F^{-1}\left(e^{-|\xi'|x_n}\right) &= \frac{\rho^{(3-n)/2}}{(2\pi)^{(n-1)/2}} \int_0^\infty r^{(n-1)/2}\, e^{-rx_n} J_{\frac{n-3}{2}}(\rho r)\, dr \\ &= \frac{\Gamma(n/2)}{\pi^{n/2}} \frac{x_n}{|x|^n} = \frac{2}{\omega_n} \frac{x_n}{|x|^n}. \end{aligned} \tag{5.3.7}$$

Hence

$$F^{-1}\left(|\xi'|e^{-|\xi'|x_n}\right) = -\frac{\partial}{\partial x_n} F^{-1}\left(e^{-|\xi'|x_n}\right) = -\frac{2}{\omega_n}\left(1 - n\frac{x_n^2}{|x|^2}\right)\frac{1}{|x|^n}. \tag{5.3.8}$$

By (5.3.4)-(5.3.8) a solution of the Dirichlet problem for $\Delta^2 u = 0$ in $\mathbb{R}^n_+$ with boundary functions in $\mathrm{C}_0^\infty(\mathbb{R}^{n-1})$ has the form

$$u(x) = \frac{2}{\omega_n}\left\{ n \int_{\partial\mathbb{R}^n_+} \frac{x_n^3}{|y-x|^{n+2}} f_1(y')\, dy' + \int_{\partial\mathbb{R}^n_+} \frac{x_n^2}{|y-x|^n} f_2(y')\, dy' \right\}. \tag{5.3.9}$$

In Agmon, Douglis and Nirenberg [**ADN1**], Ch. 1, is shown that the last equality determines a unique solution of the Dirichlet problem $\Delta^2 u = 0$ in $\mathbb{R}^n_+$ in the class $\mathrm{C}^\infty(\overline{\mathbb{R}^n_+})$, $u(x) = O(|x|)$ for large $|x|$ in the case when both $f_1, f_2 \in \mathrm{C}^\infty(\mathbb{R}^{n-1})$ are bounded functions.

Suppose $u \in \mathfrak{A}$ and $\Delta^2 u = 0$ in $\mathbb{R}^n_+$. Then (5.3.9) holds, where

$$f_1(x') = u(x', 0), \quad f_2(x') = \frac{\partial u}{\partial x_n}(x', 0).$$

By differentiating (5.3.9) we obtain

$$\begin{aligned} \frac{\partial u}{\partial x_i} = \frac{2}{\omega_n}\Bigg\{ & n \int_{\partial\mathbb{R}^n_+} \left(\frac{3x_n^2\delta_{in}}{|y-x|^{n+2}} + (n+2)\frac{x_n^3(y_i-x_i)}{|y-x|^{n+4}}\right) f_1(y')\, dy' \\ & + \int_{\partial\mathbb{R}^n_+} \left(\frac{2x_n^2\delta_{in}}{|y-x|^n} + n\frac{x_n^2(y_i-x_i)}{|y-x|^{n+2}}\right) f_2(y')\, dy' \Bigg\}, \quad i = 1, \ldots, n. \end{aligned} \tag{5.3.10}$$

We note that for any points $x \in \mathbb{R}^n_+$ and $y \in \partial\mathbb{R}^n_+$ the following equality holds

$$\frac{2nx_n^2}{\omega_n|y-x|^{n+2}}\left(3\delta_{in}+(n+2)\frac{x_n(y_i-x_i)}{|y-x|^2}\right)$$

$$=-\frac{2}{\omega_n}\sum_{j=1}^{n-1}\frac{\partial}{\partial y_j}\left(\left[2\delta_{ij}-\frac{n(y_i-x_i)(y_j-x_j)}{|y-x|^2}\right]\frac{x_n}{|y-x|^n}\right)$$

$$=-\sum_{j=1}^{n-1}\frac{\partial}{\partial y_j}\left(M_{ij}(\boldsymbol{e}_{xy})\frac{x_n}{|y-x|^n}\right),\qquad i=1,\dots,n.$$

The first of the integrals in (5.3.10) containing f_1 converges for any function f_1,

$$|f_1(y')|\le c(1+|y'|),$$

and for all points $x \in \mathbb{R}^n_+$. Therefore by the last equality, integrating by parts we obtain

$$\frac{2n}{\omega_n}\int_{\partial\mathbb{R}^n_+}\left[\frac{3x_n^2\delta_{in}}{|y-x|^{n+2}}+(n+2)\frac{x_n^3(y_i-x_i)}{|y-x|^{n+4}}\right]f_1(y')\,dy'$$

$$=\frac{2n}{\omega_n}\lim_{r\to\infty}\int_{|y'-x'|\le r}\left[\frac{3x_n^2\delta_{in}}{|y-x|^{n+2}}+(n+2)\frac{x_n^3(y_i-x_i)}{|y-x|^{n+4}}\right]f_1(y')\,dy'$$

$$=\lim_{r\to\infty}\left\{\int_{|y'-x'|\le r}\sum_{j=1}^{n-1}M_{ij}(\boldsymbol{e}_{xy})\frac{x_n}{|y-x|^n}\frac{\partial f_1}{\partial y_j}\,dy'\right.$$

$$\left.-\int_{|y'-x'|=r}\sum_{j=1}^{n-1}M_{ij}(\boldsymbol{e}_{xy})\frac{x_n}{|y-x|^n}(\boldsymbol{\nu}(y'),\boldsymbol{e}_j)f_1(y')\,ds(y')\right\},$$

where $\boldsymbol{\nu}(y')$ is the exterior normal at the point y' to the $(n-2)$-dimensional sphere in $\mathbb{R}^{n-1}$ with center at the point x' and radius r, $ds(y')$ is the area element of this sphere. The first integral in the last equality has a limit as $r\to\infty$ by virtue of boundedness of $\nabla' f_1 = \{\partial f_1/\partial x_1,\dots,\partial f_1/\partial x_{n-1}\}$, and the integral over the $(n-2)$-dimensional sphere is $O(1/r)$ as $r\to\infty$ for any function f_1 admitting the estimate $|f_1(y')|\le c(1+|y'|)$. Therefore

$$\frac{2n}{\omega_n}\int_{\partial\mathbb{R}^n_+}\left[\frac{3x_n^2\delta_{in}}{|y-x|^{n+2}}+(n+2)\frac{x_n^3(y_i-x_i)}{|y-x|^{n+4}}\right]f_1(y')\,dy'$$

$$=\int_{\partial\mathbb{R}^n_+}\sum_{j=1}^{n-1}M_{ij}(\boldsymbol{e}_{xy})\frac{x_n}{|y-x|^n}\frac{\partial f_1}{\partial y_j}\,dy',$$

which together with (5.3.10) leads to the representation for the gradient of a biharmonic function indicated in the statement of the lemma. □

5.3.3. Sharp constant in the C. Miranda inequality

In the next theorem we find the exact constant K in the C. Miranda inequality

$$\sup_{\overline{\mathbb{R}^n_+}}|\nabla u|\le \mathcal{K}\sup_{\partial\mathbb{R}^n_+}|\nabla u|, \tag{5.3.11}$$

where u is a solution of the biharmonic equation in $\mathbb{R}^n_+$ from $C^\infty(\overline{\mathbb{R}^n_+})$ with $\nabla u(x',0)$ being a bounded vector-valued function on $\partial\mathbb{R}^n_+$ and $u(x) = O(|x|)$ for large $|x|$.

THEOREM 5.2. *The exact constant K is given by*

$$K = \frac{2\Gamma(\frac{n}{2})}{\sqrt{\pi}\,\Gamma(\frac{n-1}{2})} \int_0^{\pi/2} \big[4 + n(n-4)\cos^2\vartheta\big]^{1/2} \sin^{n-2}\vartheta d\vartheta.$$

In particular, $K = 4/\pi$ for $n = 2, K = 1/2 + 2\pi\sqrt{3}/9$ for $n = 3$ and $K = 2$ for $n = 4$.

PROOF. Clearly,

$$K = \sup\{\|\nabla u(x)\|_{[C_b(\overline{\mathbb{R}^n_+})]^n} : u \in \mathfrak{A},\ \|\nabla u|_{x_n=0}\|_{[C_b(\partial\mathbb{R}^n_+)]^n} \le 1,\ \Delta^2 u = 0\}.$$

Using the permutability of the operation of taking the supremum and by the continuity of $\nabla u(x)$ in $\mathbb{R}^n_+$, we obtain

$$\begin{aligned} K &= \sup\{\sup_{\overline{\mathbb{R}^n_+}} |\nabla u(x)| : u \in \mathfrak{A},\ \|\nabla u|_{x_n=0}\|_{[C_b(\partial\mathbb{R}^n_+)]^n} \le 1,\ \Delta^2 u = 0\} \\ &= \sup_{\mathbb{R}^n_+} \sup\{|\nabla u(x)| : u \in \mathfrak{A},\ \|\nabla u|_{x_n=0}\|_{[C_b(\partial\mathbb{R}^n_+)]^n} \le 1,\ \Delta^2 u = 0\}. \end{aligned} \tag{5.3.12}$$

By properties of the inner product in $\mathbb{R}^n$, for any fixed point $x \in \mathbb{R}^n_+$ we have

$$\begin{aligned} |\nabla u(x)| &= \left| \int_{\partial\mathbb{R}^n_+} M(\boldsymbol{e}_{xy})\, \nabla u(y',0)\, \frac{x_n}{|y-x|^n}\, dy' \right| \\ &= \sup_{|\boldsymbol{z}|=1} \left(\boldsymbol{z},\ \int_{\partial\mathbb{R}^n_+} M(\boldsymbol{e}_{xy})\, \nabla u(y',0)\, \frac{x_n}{|y-x|^n}\, dy' \right) \\ &= \sup_{|\boldsymbol{z}|=1} \int_{\partial\mathbb{R}^n_+} (\boldsymbol{z},\ M(\boldsymbol{e}_{xy})\, \nabla u(y',0))\, \frac{x_n}{|y-x|^n}\, dy' \\ &= \sup_{|\boldsymbol{z}|=1} \int_{\partial\mathbb{R}^n_+} (M^*(\boldsymbol{e}_{xy})\boldsymbol{z},\ \nabla u(y',0))\, \frac{x_n}{|y-x|^n}\, dy', \end{aligned} \tag{5.3.13}$$

where M^* is the transpose of the matrix M. Hence (5.3.12) implies the upper estimate of K

$$K \le \sup_{\mathbb{R}^n_+} \sup_{|\boldsymbol{z}|=1} \int_{\partial\mathbb{R}^n_+} |M^*(\boldsymbol{e}_{xy})\boldsymbol{z}| \frac{x_n}{|y-x|^n}\, dy'. \tag{5.3.14}$$

Let $\mathbb{S}^{n-1}_-$ denote the lower half of the unit sphere $\mathbb{S}^{n-1}$. By the definition of the matrix-valued function $M(\boldsymbol{e})$ we have

$$|M^*(\boldsymbol{e})\boldsymbol{z}| = \frac{2}{\omega_n}[4 + n(n-4)(\boldsymbol{e},\boldsymbol{z})^2]^{1/2}, \tag{5.3.15}$$

where $|\boldsymbol{z}| = 1$. Hence (5.3.14) implies the estimate

$$K \le \frac{2}{\omega_n} \sup_{\mathbb{R}^n_+} \sup_{|\boldsymbol{z}|=1} \int_{\mathbb{S}^{n-1}_-} [4 + n(n-4)(\boldsymbol{e}_\sigma,\boldsymbol{z})^2]^{1/2}\, d\sigma.$$

The integrand in the integral over the half-sphere is even with respect to $\boldsymbol{e}_\sigma$. Therefore,

$$K \le \frac{1}{\omega_n} \sup_{\mathbb{R}^n_+} \sup_{|\boldsymbol{z}|=1} \int_{\mathbb{S}^{n-1}} [4 + n(n-4)(\boldsymbol{e}_\sigma,\boldsymbol{z})^2]^{1/2}\, d\sigma.$$

Since the integral in the last inequality does not depend on $x \in \mathbb{R}^n_+$ and $\boldsymbol{z}$, $|\boldsymbol{z}| = 1$, then

$$K \leq \frac{1}{\omega_n} \int_{\mathbb{S}^{n-1}} [4 + n(n-4)(\boldsymbol{e}_\sigma, \boldsymbol{z})^2]^{1/2}\, d\sigma, \tag{5.3.16}$$

where $\boldsymbol{z}$ is an arbitrary fixed n-dimensional unit vector.

We now obtain a lower estimate for K. To do this, we consider the vector-valued function

$$\boldsymbol{w}(y') = g_\varepsilon(|y' - x'|)\, \frac{M^*(\boldsymbol{e}_{xy})\boldsymbol{e}_n}{|M^*(\boldsymbol{e}_{xy})\boldsymbol{e}_n|}\Bigg|_{y_n=0}, \tag{5.3.17}$$

where $\boldsymbol{e}_n = (0, 0, \ldots, 1)$, x is a fixed point in $\mathbb{R}^n_+$, $g_\varepsilon \in \mathrm{C}^\infty(\mathbb{R}^1_+)$, $0 \leq g_\varepsilon(t) \leq 1$, $g_\varepsilon(t) = 1$ for $t \geq \varepsilon$ and $g_\varepsilon(t) = 0$ for $t < \varepsilon/2$. The function g_ε is introduced in order to single out the case $n = 2$, since $M^*(\boldsymbol{e}_{xy})\boldsymbol{e}_n = 0$ for $n = 2$ and $(\boldsymbol{e}_{xy}, \boldsymbol{e}_n) = \pm 1$.

From the definition of $M(\boldsymbol{e})$ we have the equalities

$$\begin{aligned} w_j(y') &= g_\varepsilon(|y' - x'|)\Bigg\{ 2\delta_{jn} + n\frac{x_n((1-\delta_{jn})y_j - x_j)}{|y' - x'|^2 + x_n^2} \\ &\times \left[4 + n(n-4)\frac{x_n^2}{|y' - x'|^2 + x_n^2}\right]^{-1/2}\Bigg\}, \quad j = 1, 2, \ldots, n. \end{aligned}$$

It is directly verified that $\partial w_j/\partial y_i = \partial w_i/\partial y_j$ for $i, j = 1, 2, \ldots, n-1$. Therefore the vector-valued function $\boldsymbol{w}' = \{w_1, \ldots, w_{n-1}\}$ determines a potential field on $\mathbb{R}^{n-1}$, i.e., there exists a scalar function ψ on $\mathbb{R}^{n-1}$ such that $\boldsymbol{w}' = \nabla'\psi$.

The gradient of a solution of the boundary value problem

$$\Delta^2 v = 0 \ \text{ in } \ \mathbb{R}^n_+, \ \ v(x', 0) = \psi(x'), \ \ \frac{\partial v}{\partial x_n}(x', 0) = w_n(x'),$$

where $v \in \mathfrak{A}$, is by Lemma 5.2 represented in the form

$$\nabla v(x) = \int_{\partial\mathbb{R}^n_+} M(\boldsymbol{e}_{xy})\, \boldsymbol{w}(y')\, \frac{x_n}{|y - x|^n}\, dy'.$$

From (5.3.12) and (5.3.13) we obtain

$$K \geq |\nabla v(x)| \geq (\nabla v(x), \boldsymbol{e}_n) = \int_{\partial\mathbb{R}^n_+} (M^*(\boldsymbol{e}_{xy})\boldsymbol{e}_n,\ \boldsymbol{w}(y'))\, \frac{x_n}{|y - x|^n}\, dy'.$$

Combining this with (5.3.17), we obtain

$$\begin{aligned} K &\geq \int_{\partial\mathbb{R}^n_+} g_\varepsilon(|y' - x'|)|M^*(\boldsymbol{e}_{xy})\boldsymbol{e}_n|\, \frac{x_n}{|y - x|^n}\, dy' \\ &= \int_{|y' - x'| \geq \varepsilon} |M^*(\boldsymbol{e}_{xy})\boldsymbol{e}_n|\, \frac{x_n}{|y - x|^n}\, dy \\ &+ \int_{|y' - x'| < \varepsilon} g_\varepsilon(|y' - x'|)|M^*(\boldsymbol{e}_{xy})\boldsymbol{e}_n|\, \frac{x_n}{|y - x|^n}\, dy'. \end{aligned}$$

Estimating the second integral on the right-hand side of the last equality

$$\int_{|y'-x'|<\varepsilon} g_\varepsilon(|y'-x'|)|M^*(\boldsymbol{e}_{xy})\boldsymbol{e}_n| \frac{x_n}{|y-x|^n}\, dy'$$

$$= \frac{2}{\omega_n}\int_{|y'-x'|<\varepsilon} [4+n(n-4)(\boldsymbol{e}_{xy},\boldsymbol{e}_n)^2]^{1/2} \frac{x_n}{|y-x|^n}\, dy' \leq c\left(\frac{\varepsilon}{x_n}\right)^{n-1}$$

and keeping in mind that ε is arbitrary, we arrive at the estimate

$$K \geq \int_{\partial\mathbb{R}^n_+} |M^*(\boldsymbol{e}_{xy})\boldsymbol{e}_n| \frac{x_n}{|y-x|^n}\, dy'.$$

Just as in the derivation of (5.3.16), we obtain

$$K \geq \frac{1}{\omega_n}\int_{\mathbb{S}^{n-1}} [4+n(n-4)(\boldsymbol{e}_\sigma,\boldsymbol{z})^2]^{1/2}\, d\sigma.$$

Since it is possible to set $\boldsymbol{z}=\boldsymbol{e}_n$ in (5.3.16),

$$K = \frac{1}{\omega_n}\int_{\mathbb{S}^{n-1}} [4+n(n-4)(\boldsymbol{e}_\sigma,\boldsymbol{z})^2]^{1/2}\, d\sigma.$$

Let ϑ denote the angle between $\boldsymbol{e}_\sigma$ and $\boldsymbol{e}_n$. We pass to spherical coordinates and arrive at

$$\begin{aligned} K &= \frac{\Gamma(n/2)}{2\pi^{n/2}}\int_0^\pi \cdots \int_0^\pi \int_0^{2\pi} \left[4+n(n-4)\cos^2\vartheta\right]^{1/2} \sin^{n-2}\vartheta \\ &\quad \times \sin^{n-3}\vartheta_1 \ldots \sin\vartheta_{n-3}\, d\vartheta\, d\vartheta_1 \ldots\, d\vartheta_{n-2} \\ &= 4\pi\frac{\Gamma(n/2)}{2\pi^{n/2}}\left(\prod_{l=1}^{n-3} 2\int_0^{\pi/2} \sin^l\omega\, d\omega\right)\int_0^{\pi/2} \left[4+n(n-4)\cos^2\vartheta\right]^{1/2}\sin^{n-2}\vartheta \\ &= \frac{2\Gamma(n/2)}{\sqrt{\pi}\,\Gamma(\frac{n-1}{2})}\int_0^{\pi/2} \left[4+n(n-4)\cos^2\vartheta\right]^{1/2}\sin^{n-2}\vartheta. \end{aligned}$$

□

5.4. Comments to Chapter 5

An estimate similar to (5.1.1) for the biharmonic equation with two variables was proved by C. Miranda [**Mi1**]. For strongly elliptic equations in the two-dimensional case, a result of type (5.1.1) was obtained by C. Miranda in [**Mi2**] with the help of Agmon's result in [**Ag1**]. In the general case, (5.1.1) was established by Agmon, Douglis and Nirenberg [**ADN1**], Ch. 1.

CHAPTER 6

Sharp Pointwise Estimates for Directional Derivatives and Khavinson's Type Extremal Problems for Harmonic Functions

6.1. Introduction

Sharp pointwise estimates for the gradient of harmonic functions are of use in problems relating electrostatics as well as hydrodynamics of ideal fluid, elasticity and hydrodynamics of the viscous incompressible fluid (see, e.g. books by Protter and Weinberger [**PW**], Landau and Lifshitz [**LL1, LL2**]).

A certain estimate of such a type appeared in a D. Khavinson's problem for harmonic functions which will be describe now. In [**Kh**], Khavinson found the sharp constant $K(x)$ in the inequality for the radial derivative of a harmonic function u in the ball $B = \{x \in \mathbb{R}^3 : |x| < 1\}$:

$$\left|\frac{\partial u(x)}{\partial |x|}\right| \leq K(x) \sup_{|y|<1} |u(y)|, \qquad x \in B,$$

and, in a conversation, made a conjecture that the same constant $K(x)$ should appear in the stronger sharp inequality

$$|\nabla u(x)| \leq \mathcal{K}(x) \sup_{|y|<1} |u(y)|,$$

i.e. $\mathcal{K}(x) = K(x)$.

Now, consider the sharp inequality

$$\left|\frac{\partial u(x)}{\partial \boldsymbol{\ell}}\right| \leq K(x; \boldsymbol{\ell}) \sup_{|y|<1} |u(y)|,$$

where u is a harmonic function in B, $x \in B$, and $\boldsymbol{\ell}$ is an arbitrary unit vector in $\mathbb{R}^3$. By Khavinson's conjecture, given any $x \in B$, the maximum value of $K(x; \boldsymbol{\ell})$ with respect to $\boldsymbol{\ell}$ is attained at a radial direction. By Khavinson's extremal problem for harmonic functions we mean the problem of finding the direction of $\boldsymbol{\ell}$ as well as the maximal value of $K(x; \boldsymbol{\ell})$. Note that at present no solution of this particular problem is known.

In this chapter we obtain sharp pointwise estimates for directional derivatives of harmonic functions in multidimensional domains. We find solutions to various Khavinson's type extremal problems for harmonic functions in the unit ball and in a half-space which, for any domain in $\mathbb{R}^n$, can be described as follows.

Let $\mathcal{H}$ be a class of harmonic functions in a domain $\mathcal{D} \subset \mathbb{R}^n$, let $x \in \mathcal{D}$ be a fixed point, and let Φ be a positively homogeneous and nonnegative functional defined on functions in $\mathcal{H}$ and such that

$$\big|\big(\nabla u(x),\ \boldsymbol{\ell}\big)\big| \leq C(x)\Phi(u)$$

for all $|\boldsymbol{\ell}| = 1$ and $u \in \mathcal{H}$. One is looking for directions $\boldsymbol{\ell}$ for which the sharp coefficient $\mathcal{K}_\Phi(x;\boldsymbol{\ell})$ in

$$\big|\big(\nabla u(x),\ \boldsymbol{\ell}\big)\big| \le \mathcal{K}_\Phi(x;\boldsymbol{\ell})\Phi(u), \qquad u \in \mathcal{H},$$

attains its largest value, and maximal value of $\mathcal{K}_\Phi(x;\boldsymbol{\ell})$.

It is worthily of noting that an example of Khavinson's type extremal problem for the disk (see Comments to the chapter) shows that with a certain choice of the functional Φ (for example, as the sup-norm of the Neumann boundary data), the direction at which $\mathcal{K}_\Phi(x;\boldsymbol{\ell})$ attains its largest value is not radial.

Also note that sharp estimates for harmonic functions considered in the present chapter are multidimensional generalizations of known inequalities from analytic function theory, stated in Comments to this chapter.

In Section 6.2 we give such multidimensional generalizations for the gradients of bounded or semibounded harmonic functions. In particular, we present pointwise estimates with sharp coefficients for the gradient of bounded or semibounded harmonic functions in a multidimensional ball. It is shown, for instance, that the inequality

$$|\nabla u(x)| \le \frac{n + (n-2)|x|}{1 - |x|^2}\, \mathcal{Q}_{\mathbb{B}}(u;x) \tag{6.1.1}$$

holds with a sharp coefficient in front of $\mathcal{Q}_{\mathbb{B}}(u;x)$ for any harmonic function u on the ball $\mathbb{B}$. Here $\mathcal{Q}_{\mathbb{B}}(u;x)$ is either of the following expressions:

(i) $\sup\limits_{y\in\mathbb{B}} u(y) - u(x)$,

(ii) $u(x) - \inf\limits_{y\in\mathbb{B}} u(y)$,

(iii) $\sup\limits_{y\in\mathbb{B}} |u(y)| - |u(x)|$.

As a corollary, we obtain similar sharp estimates in a half-space $\mathbb{H} \subset \mathbb{R}^n$, as well as improved estimates (in comparison with the known ones) for the gradient of harmonic functions in an arbitrary domain $G \subset \mathbb{R}^n$. It is shown that Khavinson's hypothesis on the role of the radial direction in the ball and the normal direction in the half-space holds true in estimates for directional derivatives of bounded or semibounded harmonic functions.

In Section 6.3 we find a representation for the sharp coefficient $\mathcal{C}_p(x)$ in the inequality

$$|\nabla u(x)| \le \mathcal{C}_p(x)\|u\|_p\,, \tag{6.1.2}$$

where u is a harmonic function in the half-space $\mathbb{R}^n_+ = \big\{x = (x', x_n) : x' \in \mathbb{R}^{n-1}, x_n > 0\big\}$, represented by the Poisson integral with boundary values in $L^p(\mathbb{R}^{n-1})$, $\|\cdot\|_p$ is the norm in $L^p(\mathbb{R}^{n-1})$, $1 \le p \le \infty$, $x \in \mathbb{R}^n_+$. It is shown that

$$\mathcal{C}_p(x) = C_p\, x_n^{(1-n-p)/p} \tag{6.1.3}$$

and explicit formulas for C_p in (6.1.3) for $p = 1, 2, \infty$ are given.

Note that a direct consequence of (6.1.2) is the following sharp limit relation for the gradient of a harmonic function in the n-dimensional domain Ω with smooth boundary:

$$\lim_{x\to\mathcal{O}_x} \big|x - \mathcal{O}_x\big|^{(n+p-1)/p} \sup\big\{|\nabla u(x)| : \|u|_{\partial\Omega}\|_p \le 1\big\} = C_p,$$

where $\mathcal{O}_x$ is a point on $\partial\Omega$ nearest to $x \in \Omega$ (compare with Corollary 4.1, where a relation of the same nature for the values of solutions to elliptic systems was obtained).

We pass to a description of Section 6.3.

In Subsection 6.3.1 we characterize C_p in terms of an extremal problem on the unit sphere in $\mathbb{R}^n$.

In Subsection 6.3.2 we reduce this problem to that of finding the supremum of a certain double integral, depending on a scalar parameter and show that

$$C_1 = \frac{2(n-1)}{\omega_n} = \frac{(n-1)\Gamma\left(n/2\right)}{\pi^{n/2}}, \qquad C_2 = \sqrt{\frac{n(n-1)}{2^n\omega_n}} = \sqrt{\frac{(n-1)\Gamma\left(\frac{n+2}{2}\right)}{2^n\pi^{n/2}}},$$

where ω_n is the area of the unit sphere in $\mathbb{R}^n$.

In Subsection 6.3.4 we treat the more difficult case $p = \infty$. We anticipate the proof of the main result by deriving in Subsection 6.3.3 an algebraic inequality to be used for finding an explicit formula for C_∞. Solving the variational problem stated in Subsection 6.3.2, we find

$$C_\infty = \frac{4(n-1)^{(n-1)/2}\,\omega_{n-1}}{n^{n/2}\,\omega_n} = \frac{4(n-1)^{(n-1)/2}\Gamma\left(\frac{n}{2}\right)}{\sqrt{\pi}n^{n/2}\Gamma\left(\frac{n-1}{2}\right)}.$$

In particular,

$$C_\infty = \frac{4}{3\sqrt{3}}, \qquad C_\infty = \frac{3\sqrt{3}}{2\pi}.$$

for $n = 3$ and $n = 4$, respectively.

As a trivial corollary of the inequality

$$|\nabla u(x)| \le \frac{4(n-1)^{(n-1)/2}\,\omega_{n-1}}{n^{n/2}\,\omega_n x_n} \sup_{y\in\mathbb{R}^n_+} |u(y)|, \tag{6.1.4}$$

which is equivalent to (6.1.2) with $p = \infty$, we find

$$|\nabla u(x)| \le \frac{2(n-1)^{(n-1)/2}\,\omega_{n-1}}{n^{n/2}\,\omega_n x_n} \operatorname{osc}_{\mathbb{R}^n_+}(u), \tag{6.1.5}$$

where $\operatorname{osc}_{\mathbb{R}^n_+}(u)$ is the oscillation of u on $\mathbb{R}^n_+$.

When dealing in Subsection 6.3.4 with the Khavinson's type problem for the multidimensional half-space, we show that in fact, the sharp constants in pointwise estimates for the absolute value of the normal derivative and of the modulus of the gradient of a harmonic function coincide. We also show in Subsection 6.3.2 that similar assertions hold for $p = 1$ and $p = 2$.

In Section 6.4 we find a representation for the sharp coefficient $\mathcal{K}_p(x; \boldsymbol{\ell})$ in the inequality

$$\big|\left(\nabla u(x),\, \boldsymbol{\ell}\right)\big| \le \mathcal{K}_p(x; \boldsymbol{\ell})\|u\|_p, \tag{6.1.6}$$

where x is an arbitrary point in the multidimensional unit ball $\mathbb{B} = \{x \in \mathbb{R}^n : r = |x| < 1\}$, $\boldsymbol{\ell}$ is an arbitrary unit vector in $\mathbb{R}^n$, u is a harmonic function in $\mathbb{B}$, represented by the Poisson integral with boundary values in $L^p(\partial\mathbb{B})$, $\|\cdot\|_p$ is the norm in $L^p(\partial\mathbb{B})$, $1 \le p \le \infty$. The representation in question is specified for the normal and tangential directions with respect to the sphere $|x| = r < 1$ at an arbitrary point. As a consequence, we obtain explicit formulas for the sharp coefficient $\mathcal{K}_p(x)$ in the inequality

$$|\nabla u(x)| \leq \mathcal{K}_p(x)\|u\|_p \tag{6.1.7}$$

for the cases $p = 1$ and $p = 2$. We show that for $p = 1, 2$ and for any n the maximum value of $\mathcal{K}_p(x; \boldsymbol{\ell})$ with respect to $\boldsymbol{\ell}$ is attained at the radial direction. In other words, we show that the sharp constant $\mathcal{K}_p(x)$ in (6.1.7) coincides with the sharp constant $\mathcal{K}_p(x; \boldsymbol{\ell})$ in (6.1.6) for $p = 1, 2$, where $\boldsymbol{\ell}$ is the radial direction. This fact proves the Khavinson's type hypothesis concerning the role of the radial direction in (6.1.6) for $p = 1$ and $p = 2$.

Obviously, the norm $\|u\|_p$ in (6.1.6) and (6.1.7) can be replaced by the best approximation of the boundary value by a constant in the norm of $L^p(\mathbb{S}^{n-1})$:

$$\Lambda_p(u) = \min_{c \in \mathbb{R}} \|u - c\|_p \, .$$

In particular, the norms $\|u\|_2$ and $\|u\|_\infty$ can be replaced by

$$\Lambda_2(u) = \|u - u(0)\|_2 \quad \text{and} \quad \Lambda_\infty(u) = \frac{1}{2} \operatorname{osc}_{\partial\mathbb{B}}(u),$$

respectively.

Next we describe the content of Section 6.4.

In Subsection 6.4.1 we find a representation for the best constant in (6.1.6) as an integral over the sphere $\mathbb{S}^{n-1}$. As a corollary, we characterize the sharp constant $\mathcal{K}_p(x)$ in inequality (6.1.7) in terms of an extremal problem on the unit sphere in $\mathbb{R}^n$.

In Subsection 6.4.2 we obtain the sharp inequality

$$|\nabla u(0)| \leq \mathcal{K}_p(0)\|u\|_p, \tag{6.1.8}$$

where $\mathcal{K}_1(0) = n/\omega_n$, and

$$\mathcal{K}_p(0) = \frac{n}{\omega_n} \left\{ \frac{2\pi^{(n-1)/2}\Gamma\left(\frac{q+1}{2}\right)}{\Gamma\left(\frac{n+q}{2}\right)} \right\}^{1/q}, \tag{6.1.9}$$

where $1 < p \leq \infty$, $1/q + 1/p = 1$. We also consider the case $p = 1$ and show that

$$\mathcal{K}_1(x) = \mathcal{K}_1(x;\ \boldsymbol{\nu}_x) = \frac{n + (n-2)r}{\omega_n(1-r)^n},$$

where $\boldsymbol{\nu}_x$ is the normal to the sphere $|x| = r < 1$ at a point x.

Subsection 6.4.3 is devoted to a simplification of the representation of $\mathcal{K}_p(x; \boldsymbol{\ell})$, reducing the integral over $\mathbb{S}^{n-1}$ to a double integral depending on the tangential and normal projections of a unit vector $\boldsymbol{\ell}$. As a corollary, we reduce the extremal problem for $\mathcal{K}_p(x; \ell)$ on the unit sphere in $\mathbb{R}^n$ to that of finding the supremum of a certain double integral, depending on a scalar parameter. We also obtain a representation of $\mathcal{K}_p(x; \boldsymbol{\nu}_x)$ as a definite integral as well as a formula for the best constant in the estimate of the modulus of derivative in a tangential direction $\boldsymbol{\tau}_x$:

$$\mathcal{K}_p(x; \boldsymbol{\tau}_x) = \mathcal{K}_p(0)(1 - r^2) \left\{ F\left(\frac{n(q-1) + q + 2}{2}, \frac{(n+2)q}{2}; \frac{n+q}{2}; r^2 \right) \right\}^{1/q},$$

where $F(a, b; c; x)$ is the hypergeometric Gauss function, and $\mathcal{K}_p(0)$ is defined by formula (6.1.9).

In Subsection 6.4.4 we show that

$$\mathcal{K}_2(x) = \mathcal{K}_2(x;\ \boldsymbol{\nu}_x) = \left(\frac{n + (n^2+n-4)r^2 + (n-2)^2 r^4}{\omega_n (1-r^2)^{n+1}}\right)^{1/2}$$

and

$$\mathcal{K}_2(x; \boldsymbol{\tau}_x) = \left\{\frac{n(1+r^2)}{\omega_n(1-r^2)^{n+1}}\right\}^{1/2}.$$

Moreover, $\mathcal{K}_2(x;\boldsymbol{\tau}_x) \le \mathcal{K}_2(x;\ \boldsymbol{\ell}) \le \mathcal{K}_2(x;\ \boldsymbol{\nu}_x)$.

In Subsection 6.4.5 we derive explicit formulas for $\mathcal{K}_\infty(x;\ \boldsymbol{\nu}_x)$ and $\mathcal{K}_\infty(x;\boldsymbol{\tau}_x)$. In particular, we find

$$\mathcal{K}_\infty(x;\boldsymbol{\nu}_x) = \frac{1}{r^2}\left(\frac{(1+r^2/3)^{3/2}}{1-r^2} - 1\right),$$

$$\mathcal{K}_\infty(x;\boldsymbol{\tau}_x) = \frac{2}{\pi(1-r^2)}\left[2\mathbf{K}(r) - (1+r^2)\mathbf{D}(r)\right]$$

for $n = 3$, and

$$\mathcal{K}_\infty(x;\boldsymbol{\nu}_x) = \frac{1}{\pi r^3}\left(\frac{r(2+r^2)(4-r^2)^{1/2}}{(1-r^2)} - 2\arccos\frac{(2-r^2)^2-2}{2}\right),$$

$$\mathcal{K}_\infty(x;\boldsymbol{\tau}_x) = \frac{8(1-r^2)}{\pi r^2}\left[\frac{1+r^2}{(1-r^2)^2} - \frac{1}{2r}\log\frac{1+r}{1-r}\right]$$

for $n = 4$. Here

$$\mathbf{K}(k) = \int_0^{\pi/2} \frac{d\varphi}{\sqrt{1-k^2\sin^2\varphi}}$$

is the complete elliptic integral of the first kind, and

$$\mathbf{D}(k) = \int_0^{\pi/2} \frac{\sin^2\varphi}{\sqrt{1-k^2\sin^2\varphi}}\, d\varphi$$

is the complete elliptic integral (see Prudnikov, Brychkov and Marichev [**PBM1**]).

In Section 6.5 we consider the inequality with the sharp coefficient

$$|\nabla u(x)| \le \mathcal{C}_{n,p}(x) \left|\left|\frac{\partial u}{\partial \boldsymbol{\nu}}\right|\right|_p, \tag{6.1.10}$$

where $x \in \mathbb{R}^n_+ = \{x = (x_1, \dots, x_n) : x_n > 0\}$ and u is a solution of the Neumann problem for the Laplace equation in $\mathbb{R}^n_+$ with boundary data in $L^p(\mathbb{R}^{n-1}), 1 \le p < \infty$. We show that

$$\mathcal{C}_{n,p}(x) = \frac{C_{n,p}}{x_n^{(n-1)/p}},$$

and find a representation for $C_{n,p}$ in terms of an extremal problem on the unit sphere in $\mathbb{R}^n$. By solving that problem for $p \in [1, n]$, we find

$$C_{n,p} = \begin{cases} \dfrac{1}{\omega_n} & \text{for} \quad p = 1, \\[2ex] \dfrac{2^{1/p}}{\omega_n}\left\{\dfrac{2\pi^{(n-1)/2}\Gamma\left(\frac{n+p-1}{2p-2}\right)}{\Gamma\left(\frac{np}{2p-2}\right)}\right\}^{1-\frac{1}{p}} & \text{for} \quad 1 < p \le n. \end{cases}$$

Note that the maximal value of the sharp coefficient $\mathcal{C}_{n,p}(x;\boldsymbol{\ell})$ in

$$\left|\frac{\partial u(x)}{\partial \boldsymbol{\ell}}\right| \leq \mathcal{C}_{n,p}(x;\boldsymbol{\ell}) \left|\left|\frac{\partial u}{\partial \boldsymbol{\nu}}\right|\right|_p$$

for a fixed x is attained at the direction normal to $\partial\mathbb{R}^n_+$ if $1 \leq p < n$. It turns out that $\mathcal{C}_{n,n}(x;\boldsymbol{\ell})$ is independent of $\boldsymbol{\ell}$.

6.2. Khavinson's type extremal problem for bounded or semibounded harmonic functions in a ball and a half-space

6.2.1. Sharp parametric estimates for the gradient and the radial derivative of bounded or semibounded harmonic functions in the unit ball

By $\boldsymbol{\nu}_x$ we denote the normal to the sphere $|x| = r < 1$ at a point x. The sharp inequalities for $|\nabla u(x)|$ and $\big|\big(\nabla u(x), \boldsymbol{\nu}_x\big)\big|$ in Proposition below are parameter dependent. The role of a parameter is played by a number $\lambda \in [0,\ 1]$.

Introduction of the parameter enables one to unify several variants of the right-hand side.

We introduce some notation used henceforth. Let $|\cdot|$ be the length of a vector in the n-dimensional Euclidean space $\mathbb{R}^n$. Let, further, $\mathbb{B}_\rho = \big\{x \in \mathbb{R}^n : |x| < \rho\big\}$, and $\mathbb{S}_\rho = \big\{x \in \mathbb{R}^n : |x| = \rho\big\}$. We put $\mathbb{B} = \mathbb{B}_1$.

The main assertion of the section is

PROPOSITION 6.1. *Let x be an arbitrary point in $\mathbb{B}$, and let $0 \leq \lambda \leq 1$. Then, for any harmonic function u on $\mathbb{B}$, the inequalities*

$$|\nabla u(x)| \leq \mathcal{C}(\lambda, x)\ \mathcal{Q}_{\mathbb{B}}(u; \lambda x)\ ,$$

(6.2.1)

$$|(\nabla u(x), \boldsymbol{\nu}_x)| \leq \mathcal{C}(\lambda, x)\ \mathcal{Q}_{\mathbb{B}}(u; \lambda x)$$

hold with the same sharp coefficient

$$\mathcal{C}(\lambda, x) = \frac{\big(n + (n-2)|x|\big)(1 - \lambda|x|)^{n-1}}{(1-|x|)^n(1+\lambda|x|)}$$

in front of $\mathcal{Q}_{\mathbb{B}}(u; \lambda x)$. Here $\mathcal{Q}_{\mathbb{B}}(u; x)$ is either of the following expressions:

(i) $\sup\limits_{y\in\mathbb{B}} u(y) - u(x)$,

(ii) $u(x) - \inf\limits_{y\in\mathbb{B}} u(y)$,

(iii) $\sup\limits_{y\in\mathbb{B}} |u(y)| - |u(x)|$.

In particular, the sharp inequalities

$$|\nabla u(x)| \leq \frac{n + (n-2)|x|}{1-|x|^2}\ \mathcal{Q}_{\mathbb{B}}(u; x)\ ,$$

(6.2.2)

$$\big|\big(\nabla u(x), \boldsymbol{\nu}_x\big)\big| \leq \frac{n + (n-2)|x|}{1-|x|^2}\ \mathcal{Q}_{\mathbb{B}}(u; x)$$

and

$$|\nabla u(x)| \leq \frac{n+(n-2)|x|}{(1-|x|)^n}\, \mathcal{Q}_{\mathbb{B}}(u;0)\,,$$

(6.2.3)

$$\big|(\nabla u(x), \boldsymbol{\nu}_x)\big| \leq \frac{n+(n-2)|x|}{(1-|x|)^n}\, \mathcal{Q}_{\mathbb{B}}(u;0)$$

hold.

PROOF. *1. Proof of inequalities* (6.2.1). Let u be a harmonic function on $\mathbb{B}$. Let, further, x be an arbitrary point in $\mathbb{B}$, and $\rho \in (|x|, 1)$. Using the Poisson formula

$$u(x) = \int_{\mathbb{S}_\rho} \mathcal{P}_\rho(x,y)u(y)d\sigma_y, \qquad \mathcal{P}_\rho(x,y) = \frac{1}{\omega_n \rho}\frac{\rho^2-|x|^2}{|y-x|^n},$$

where ω_n is the area of $\mathbb{S}_1$, we find

$$\nabla u(x) = \frac{1}{\omega_n\rho}\int_{\mathbb{S}_\rho} \frac{n\left(\rho^2-|x|^2\right)(y-x)-2|y-x|^2x}{|y-x|^{n+2}}\,u(y)d\sigma_y.$$

This equality can be written as

$$\nabla u(x) = \int_{\mathbb{S}_\rho} \frac{\left(n\left(\rho^2-|x|^2\right)(y-x)-2|y-x|^2x\right)|y-\lambda x|^n}{|y-x|^{n+2}(\rho^2-\lambda^2|x|^2)}\,\mathcal{P}_\rho(\lambda x,y)u(y)d\sigma_y,$$

which implies

$$(\nabla u(x), \boldsymbol{\ell}) = \int_{\mathbb{S}_\rho} \frac{\left(n\left(\rho^2-|x|^2\right)(y-x)-2|y-x|^2x, \boldsymbol{\ell}\right)|y-\lambda x|^n}{|y-x|^{n+2}(\rho^2-\lambda^2|x|^2)}\mathcal{P}_\rho(\lambda x,y)u(y)d\sigma_y$$

with an arbitrary $\boldsymbol{\ell} \in \mathbb{R}^n$. Hence,

$$|\nabla u(x)| \leq \mathcal{M}_\rho(\lambda, x)\|\mathcal{P}_\rho(\lambda x,\cdot)u\|_{L^1(\mathbb{S}_\rho)}, \tag{6.2.4}$$

where

$$\mathcal{M}_\rho(\lambda,x) = \max_{|\boldsymbol{\ell}|=1}\max_{y\in\mathbb{S}_\rho} \frac{\left|\left(n\left(\rho^2-|x|^2\right)(y-x)-2|y-x|^2x,\ \boldsymbol{\ell}\right)\right|\,|y-\lambda x|^n}{(\rho^2-\lambda^2|x|^2)|y-x|^{n+2}}$$

$$= \frac{1}{\rho^2-\lambda^2|x|^2}\max_{y\in\mathbb{S}_\rho}\frac{\left|n\left(\rho^2-|x|^2\right)(y-x)-2|y-x|^2x\right|\,|y-\lambda x|^n}{|y-x|^{n+2}}. \tag{6.2.5}$$

It follows by (6.2.5) that

$$\mathcal{M}_\rho(\lambda,x) \leq \frac{1}{\rho^2-\lambda^2|x|^2}\max_{y\in\mathbb{S}_\rho}\mathcal{N}_\rho(y;\lambda,x)\max_{y\in\mathbb{S}_\rho}\left(\frac{|y-\lambda x|}{|y-x|}\right)^n, \tag{6.2.6}$$

with

$$\mathcal{N}_\rho(y;\lambda,x) = \frac{\left|n\left(\rho^2-|x|^2\right)(y-x)-2|y-x|^2x\right|}{|y-x|^2}. \tag{6.2.7}$$

We find the first maximum in (6.2.6). Taking into account that the numerator on the right-hand side of (6.2.7) can be written in the form

$$|y-x|\left[n^2\left(\rho^2-|x|^2\right)^2-4n\left(\rho^2-|x|^2\right)(y-x,x)+4|y-x|^2|x|^2\right]^{1/2},$$

and using the identity

$$2(y-x,x) = \rho^2-|y-x|^2-|x|^2,$$

we obtain

$$\mathcal{N}_\rho(y;\lambda,x)=\frac{\left[n(n-2)\left(\rho^2-|x|^2\right)^2+2|y-x|^2\left(n(\rho^2-|x|^2)+2|x|^2\right)\right]^{1/2}}{|y-x|}.$$

Therefore,

$$\begin{aligned}\max_{y\in\mathbb{S}_\rho}\mathcal{N}_\rho(y;\lambda,x) &= \left[n(n-2)\left(\rho+|x|\right)^2+2\left(n(\rho^2-|x|^2)+2|x|^2\right)\right]^{1/2}\\ &= n\rho+(n-2)|x|. \qquad (6.2.8)\end{aligned}$$

Now, consider the second maximum in (6.2.6). Since for any $y\in\mathbb{S}_\rho$

$$\frac{|y-\lambda x|}{|y-x|}\le\frac{|y-x|+(1-\lambda)|x|}{|y-x|}=1+\frac{(1-\lambda)|x|}{|y-x|}\le 1+\frac{(1-\lambda)|x|}{\rho-|x|}=\frac{\rho-\lambda|x|}{\rho-|x|},$$

it follows that

$$\max_{y\in\mathbb{S}_\rho}\frac{|y-\lambda x|}{|y-x|}=\frac{\rho-\lambda|x|}{\rho-|x|}.$$

This equality in combination with (6.2.8) enables one to write (6.2.6) in the form

$$\mathcal{M}_\rho(\lambda,x)\le\frac{\big(n\rho+(n-2)|x|\big)(\rho-\lambda|x|)^{n-1}}{(\rho+\lambda|x|)(\rho-|x|)^n},$$

which, together with (6.2.4), leads to the inequality

$$|\nabla u(x)|\le\frac{\big(n\rho+(n-2)|x|\big)(\rho-\lambda|x|)^{n-1}}{(\rho+\lambda|x|)(\rho-|x|)^n}\|\mathcal{P}_\rho(\lambda x,\cdot)u\|_{L^1(\mathbb{S}_\rho)}. \qquad (6.2.9)$$

We introduce the notation

$$\mathcal{S}_\rho(u)=\sup\{u(y):y\in\mathbb{B}_\rho\},$$

and change u for $\mathcal{S}_\rho(u)-u$ in (6.2.9). Then

$$|\nabla u(x)|\le\frac{\big(n\rho+(n-2)|x|\big)(\rho-\lambda|x|)^{n-1}}{(\rho+\lambda|x|)(\rho-|x|)^n}\left\|\mathcal{P}_\rho(\lambda x,\cdot)\left(\mathcal{S}_\rho(u)-u\right)\right\|_{L^1(\mathbb{S}_\rho)},$$

and we arrive at the estimate

$$|\nabla u(x)|\le\frac{\big(n\rho+(n-2)|x|\big)(\rho-\lambda|x|)^{n-1}}{(\rho+\lambda|x|)(\rho-|x|)^n}\left\{\mathcal{S}_\rho(u)-u(\lambda x)\right\}. \qquad (6.2.10)$$

Similarly, using the notation

$$\mathcal{I}_\rho(u)=\inf\{u(y):y\in\mathbb{B}_\rho\},$$

and replacing u by $u-\mathcal{I}_\rho(u)$ in (6.2.9), we obtain

$$|\nabla u(x)|\le\frac{\big(n\rho+(n-2)|x|\big)(\rho-\lambda|x|)^{n-1}}{(\rho+\lambda|x|)(\rho-|x|)^n}\left\{u(\lambda x)-\mathcal{I}_\rho(u)\right\}. \qquad (6.2.11)$$

Further, by (6.2.10) we have the inequality:

$$|\nabla u(x)|\le\frac{\big(n\rho+(n-2)|x|\big)(\rho-\lambda|x|)^{n-1}}{(\rho+\lambda|x|)(\rho-|x|)^n}\sup_{y\in\mathbb{B}_\rho}\left\{\,|u(y)|-u(\lambda x)\right\},$$

which, together with the change of u for $-u$, leads to

$$|\nabla u(x)|\le\frac{\big(n\rho+(n-2)|x|\big)(\rho-\lambda|x|)^{n-1}}{(\rho+\lambda|x|)(\rho-|x|)^n}\sup_{y\in\mathbb{B}_\rho}\left\{\,|u(y)|-|u(\lambda x)|\right\}. \qquad (6.2.12)$$

Passing to the limit as $\rho \uparrow 1$ in (6.2.10)-(6.2.12), we arrive at (6.2.1). Thus, the sharp coefficient in

$$|\nabla u(x)| \leq \mathcal{C}(\lambda, x)\ \mathcal{Q}_{\mathbb{B}}(u; \lambda x) \tag{6.2.13}$$

and hence in

$$|(\nabla u(x), \boldsymbol{\nu}_x)| \leq \mathcal{C}(\lambda, x)\ \mathcal{Q}_{\mathbb{B}}(u; \lambda x) \tag{6.2.14}$$

satisfies the inequality

$$\mathcal{C}(\lambda, x) \leq \frac{\Big(n + (n-2)|x|\Big)(1-\lambda|x|)^{n-1}}{(1-|x|)^n(1+\lambda|x|)}. \tag{6.2.15}$$

2. Sharpness of inequalities (6.2.1). Obviously, it suffices to show the sharpness only for the second inequality in (6.2.1). We introduce the function family

$$\mathcal{F}_{\varrho,y}(x) = \omega_n \varrho \mathcal{P}_\varrho(x, y) = \frac{\varrho^2 - |x|^2}{|y-x|^n}, \tag{6.2.16}$$

where $|x| < 1$, and a point y, situated on the sphere $|y| = \varrho > 1$, plays the role of parameter.

We fix a point x^* on the sphere $|x| = r$. Let y^* be the intersection of the sphere $|y| = \varrho$ and the ray passing through the center of the ball $\mathbb{B}$ and x^*.

Using the equality

$$\nabla \mathcal{F}_{\varrho,y^*}(x) = \frac{n(\varrho^2 - |x|^2)(y^* - x) - 2|y^* - x|^2 x}{|y^* - x|^{n+2}}, \tag{6.2.17}$$

we obtain

$$(\nabla \mathcal{F}_{\varrho,y^*}(x^*), \boldsymbol{\nu}_{x^*}) = \left(\nabla \mathcal{F}_{\varrho,y^*}(x^*), \frac{x^*}{|x^*|}\right) = \frac{n(\varrho^2 - |x^*|^2) - 2|y^* - x^*||x^*|}{|y^* - x^*|^{n+1}}. \tag{6.2.18}$$

Further, by harmonicity of $\mathcal{F}_{\varrho,y^*}(x)$ in $\overline{\mathbb{B}}$ and by the maximum principle, we find

$$\inf_{x \in \mathbb{B}} \mathcal{F}_{\varrho,y^*}(x) = \frac{\varrho - 1}{(\varrho+1)^{n-1}},$$

which implies

$$\mathcal{F}_{\varrho,y^*}(\lambda x^*) - \inf_{x \in \mathbb{B}} \mathcal{F}_{\varrho,y^*}(x) = \frac{\varrho^2 - \lambda^2|x^*|^2}{|y^* - \lambda x^*|^n} - \frac{\varrho - 1}{(\varrho+1)^{n-1}}. \tag{6.2.19}$$

By (6.2.18) and (6.2.19), we have

$$\Big|(\nabla \mathcal{F}_{\varrho,y^*}(x^*), \boldsymbol{\nu}_{x^*})\Big| \left(\mathcal{F}_{\varrho,y^*}(\lambda x^*) - \inf_{x \in \mathbb{B}} \mathcal{F}_{\varrho,y^*}(x)\right)^{-1} = H_{\varrho,y^*}(x^*), \tag{6.2.20}$$

where

$$H_{\varrho,y}(x) = \frac{\big[n(\varrho^2 - |x|^2) - 2|y-x||x|\big]|y - \lambda x|^n(\varrho+1)^{n-1}}{|y-x|^{n+1}\big[(\varrho^2 - \lambda^2|x|^2)(\varrho+1)^{n-1} - |y - \lambda x|^n(\varrho - 1)\big]}. \tag{6.2.21}$$

Since the function $\mathcal{G}_{\rho,y}(x) = -\mathcal{F}_{\rho,y}(x)$ satisfies

$$\sup_{x \in \mathbb{B}} \mathcal{G}_{\varrho,y^*}(x) - \mathcal{G}_{\varrho,y^*}(\lambda x^*) = \mathcal{F}_{\varrho,y^*}(\lambda x^*) - \inf_{x \in \mathbb{B}} \mathcal{F}_{\varrho,y^*}(x),$$

we deduce from (6.2.20) that

$$\Big|(\nabla \mathcal{G}_{\varrho,y^*}(x^*), \boldsymbol{\nu}_{x^*})\Big| \left(\sup_{x \in \mathbb{B}} \mathcal{G}_{\varrho,y^*}(x) - \mathcal{G}_{\varrho,y^*}(\lambda x^*)\right)^{-1} = H_{\varrho,y^*}(x^*). \tag{6.2.22}$$

Let us consider the family of functions

$$\mathcal{R}_{\varrho,y^*}(x) = \mathcal{F}_{\varrho,y^*}(x) - \frac{(\varrho+1)^n + (\varrho-1)^n}{2(\varrho^2-1)^{n-1}} \,,$$

which satisfies

$$\sup_{x\in\mathbb{B}} \mathcal{R}_{\varrho,y^*}(x) = -\inf_{x\in\mathbb{B}} \mathcal{R}_{\varrho,y^*}(x) = \frac{(\varrho+1)^n - (\varrho-1)^n}{2(\varrho^2-1)^{n-1}} \,,$$

and

$$\begin{aligned}\sup_{x\in\mathbb{B}} |\mathcal{R}_{\varrho,y^*}(x)| - |\mathcal{R}_{\varrho,y^*}(\lambda x^*)| &\le \sup_{x\in\mathbb{B}} |\mathcal{R}_{\varrho,y^*}(x)| + \mathcal{R}_{\varrho,y^*}(\lambda x^*) \\ &= \frac{\varrho^2 - \lambda^2|x^*|^2}{|y^*-\lambda x^*|^n} - \frac{\varrho-1}{(\varrho+1)^{n-1}} \,. \end{aligned} \tag{6.2.23}$$

Since $\mathcal{R}_{\rho,y}(x)$ and $\mathcal{F}_{\rho,y}(x)$ differ only by a constant term, by (6.2.18) and (6.2.23) we have

$$\Big|\big(\nabla\mathcal{R}_{\varrho,y^*}(x^*), \boldsymbol{\nu}_{x^*}\big)\Big| \left(\sup_{x\in\mathbb{B}} |\mathcal{R}_{\varrho,y^*}(x)| - |\mathcal{R}_{\varrho,y^*}(\lambda x^*)|\right)^{-1} \ge H_{\varrho,y^*}(x^*), \tag{6.2.24}$$

with $H_{\varrho,y}(x)$ defined by (6.2.21).

We pass to the limit in (6.2.20), (6.2.22) and (6.2.24) as $\varrho \downarrow 1$. That results in the sharp estimate for the best possible coefficient in (6.2.14) and hence in (6.2.13)

$$\mathcal{C}(\lambda, x^*) \ge \frac{\Big(n + (n-2)|x^*|\Big)(1-\lambda|x^*|)^{n-1}}{(1-|x^*|)^n(1+\lambda|x^*|)} \,.$$

This, being combined with (6.2.15), proves the sharpness of the coefficient in (6.2.1) by arbitrariness of x^*. □

REMARK 6.1. The variant (i) in Proposition 6.1 of the first inequality in (6.2.2)

$$|\nabla u(x)| \le \frac{n+(n-2)|x|}{1-|x|^2} \sup_{|y|<1} \{u(y) - u(x)\},$$

is an extension of the Lindelöf estimate (6.6.12) to the n-dimensional case.

REMARK 6.2. Similarly, the variant (i) in Proposition 6.1 of the first estimate in (6.2.3)

$$|\nabla u(x)| \le \frac{n+(n-2)|x|}{(1-|x|)^n} \sup_{|y|<1} \{u(y) - u(0)\},$$

is a multidimensional generalization of the Hadamard-Borel-Carathéodory type inequality (6.6.5) for the derivative of an analytic function.

REMARK 6.3. Now, suppose that a harmonic function u is positive on $\mathbb{B}$. Then, by the variant (ii) in Proposition 6.1 of the first inequality in (6.2.2) and (6.2.3) we have

$$|\nabla u(x)| \le \frac{n+(n-2)|x|}{1-|x|^2}\, u(x), \qquad |\nabla u(x)| \le \frac{n+(n-2)|x|}{(1-|x|)^n}\, u(0),$$

respectively. The first of these estimates is an extension to the multidimensional case of the Ruscheweyh inequality (6.6.6) written for the gradient of a harmonic function. Note that the estimate $|\nabla u(0)| \le n\, u(0)$ follows directly from the classical Harnack inequality in the unit ball.

REMARK 6.4. The variant (iii) in Proposition 6.1 of the first estimate in (6.2.3)

$$|\nabla u(x)| \leq \frac{n+(n-2)|x|}{(1-|x|)^n} \sup_{|y|<1} \{ |u(y)| - |u(0)| \},$$

is a multidimensional generalization of one of Landau's type inequalities for the first derivative of an analytic function (see Kresin and Maz'ya [**KM11**], p. 70, pp.81-85).

6.2.2. Corollaries of Proposition 6.1

By dilation, we obtain the following result, equivalent to Proposition 6.1 and involving the ball $\mathbb{B}_R$ with an arbitrary R.

PROPOSITION 6.2. *Let x be a point in $\mathbb{B}_R$, and let $0 \leq \lambda \leq 1$. Then, for any harmonic function on $\mathbb{B}_R$, the inequalities*

$$\begin{aligned} &|\nabla u(x)| \leq \mathcal{C}_R(\lambda, x)\ \mathcal{Q}_{\mathbb{B}_R}(u; \lambda x)\ , \\ &|(\nabla u(x), \boldsymbol{\nu}_x)| \leq \mathcal{C}_R(\lambda, x)\ \mathcal{Q}_{\mathbb{B}_R}(u; \lambda x) \end{aligned} \tag{6.2.25}$$

hold with the same sharp coefficient

$$\mathcal{C}_R(\lambda, x) = \frac{\Big(nR+(n-2)|x|\Big)(R-\lambda|x|)^{n-1}}{(R-|x|)^n(R+\lambda|x|)}$$

in front of $\mathcal{Q}_{\mathbb{B}_R}(u; \lambda x)$. Here $\mathcal{Q}_{\mathbb{B}_R}(u; x)$ is either of the following expressions:

(i) $\sup_{y \in \mathbb{B}_R} u(y) - u(x)$,

(ii) $u(x) - \inf_{y \in \mathbb{B}_R} u(y)$,

(iii) $\sup_{y \in \mathbb{B}_R} |u(y)| - |u(x)|$.

In particular, the sharp inequalities

$$\begin{aligned} &|\nabla u(x)| \leq \frac{nR+(n-2)|x|}{R^2-|x|^2}\ \mathcal{Q}_{\mathbb{B}_R}(u; x), \\ &\big|(\nabla u(x), \boldsymbol{\nu}_x)\big| \leq \frac{nR+(n-2)|x|}{R^2-|x|^2}\ \mathcal{Q}_{\mathbb{B}_R}(u; x) \end{aligned} \tag{6.2.26}$$

and

$$\begin{aligned} &|\nabla u(x)| \leq \frac{nR+(n-2)|x|}{R^{2-n}(R-|x|)^n}\ \mathcal{Q}_{\mathbb{B}_R}(u; 0), \\ &\big|(\nabla u(x), \boldsymbol{\nu}_x)\big| \leq \frac{nR+(n-2)|x|}{R^{2-n}(R-|x|)^n}\ \mathcal{Q}_{\mathbb{B}_R}(u; 0) \end{aligned} \tag{6.2.27}$$

hold.

Now, we formulate two corollaries of Proposition 6.2. The first of them is an analogue of inequalities (6.2.26) for an open half-space $\mathbb{H}$ in $\mathbb{R}^n$.

We note first that with the notation $d_x = R - |x|$, one can write (6.2.26) in the form

$$|\nabla u(x)| \le \frac{nR + (n-2)(R-d_x)}{d_x(2R-d_x)} \mathcal{Q}_{\mathbb{B}_R}(u;x)\,,$$

(6.2.28)

$$\big|(\nabla u(x), \boldsymbol{\nu}_x)\big| \le \frac{nR + (n-2)(R-d_x)}{d_x(2R-d_x)} \mathcal{Q}_{\mathbb{B}_R}(u;x)\,.$$

Let $\boldsymbol{\nu}$ stand for the unit exterior normal to the half-space $\mathbb{H} \subset \mathbb{R}^n$. Passing to the limit as $R \to \infty$ in (6.2.28) so that $d_x = \text{const}$, we arrive at the following assertion.

COROLLARY 6.1. *Let x be an arbitrary point in an open half-space $\mathbb{H} \subset \mathbb{R}^n$, and let $d_x = \text{dist}(x, \partial\mathbb{H})$. Then, for any harmonic function on $\mathbb{H}$, the inequalities*

$$|\nabla u(x)| \le \frac{n-1}{d_x} \mathcal{Q}_{\mathcal{H}}(u;x),$$

$$|(\nabla u(x), \boldsymbol{\nu})| \le \frac{n-1}{d_x} \mathcal{Q}_{\mathcal{H}}(u;x)$$

hold with the same sharp coefficient in front of $\mathcal{Q}_{\mathcal{H}}(u;x)$. Here $\mathcal{Q}_{\mathcal{H}}(u;x)$ is either of the following expressions:

(i) $\sup\limits_{y\in\mathbb{H}} u(y) - u(x)$,

(ii) $u(x) - \inf\limits_{y\in\mathbb{H}} u(y)$,

(iii) $\sup\limits_{y\in\mathbb{H}} |u(y)| - |u(x)|$.

The second corollary of Proposition 6.2 contains an improvement of the estimate

$$|\nabla u(x)| \le \frac{n}{d_x} \sup_{y\in\mathcal{D}} \{u(y) - u(x)\} \tag{6.2.29}$$

and similar estimates (see, e.g., Gilbarg and Trudinger [**GT**]) for the gradient of a harmonic function u on a domain $\mathcal{D} \subset \mathbb{R}^n$. Here $x \in \mathcal{D}$, and $d_x = \text{dist}(x, \partial\mathcal{D})$.

The improvement just mentioned of the estimates for the gradient of a harmonic function in $\mathcal{D}$ stems directly from (6.2.28). It is formulated in the next assertion.

COROLLARY 6.2. *Let x be a fixed point in a domain $\mathcal{D} \subset \mathbb{R}^n$, and let ξ_x be a point on $\partial\mathcal{D}$ for which $|\xi_x - x| = d_x$. Further, let R be the radius of the largest ball lying entirely in $\mathcal{D}$ with center on the straight line L passing through x and ξ_x. Then, for any harmonic function u on $\mathcal{D}$, the inequality*

$$|\nabla u(x)| \le \frac{nR + (n-2)(R-d_x)}{d_x(2R-d_x)} \mathcal{Q}_{\mathcal{D}}(u;x) \tag{6.2.30}$$

holds, where $\mathcal{Q}_{\mathcal{D}}(u;x)$ is either of the expressions:

(i) $\sup\limits_{y\in\mathcal{D}} u(y) - u(x)$,

(ii) $u(x) - \inf\limits_{y\in\mathcal{D}} u(y)$,

(iii) $\sup\limits_{y\in\mathcal{D}} |u(y)| - |u(x)|$.

We note that the factor

$$\frac{nR + (n-2)(R - d_x)}{2R - d_x}$$

before $1/d_x$ in (6.2.30) tends to $n-1$, as $x \to \xi_x$, $x \in L$, whereas the similar coefficient in (6.2.29) is equal to the constant n.

6.3. Sharp estimates for directional derivatives and Khavinson's type extremal problem in a half-space with boundary data from L^p

6.3.1. Auxiliary assertion

We introduce some notation used henceforth. Let $\mathbb{R}^n_+ = \{x = (x', x_n) : x' = (x_1, \dots, x_{n-1}) \in \mathbb{R}^{n-1}, x_n > 0\}$, $\mathbb{S}^{n-1} = \{x \in \mathbb{R}^n : |x| = 1\}$, $\mathbb{S}^{n-1}_+ = \{x \in \mathbb{R}^n : |x| = 1,\ x_n > 0\}$ and $\mathbb{S}^{n-1}_- = \{x \in \mathbb{R}^n : |x| = 1,\ x_n < 0\}$. Let $\boldsymbol{e}_\sigma$ stand for the n-dimensional unit vector joining the origin with a point σ on the sphere $\mathbb{S}^{n-1}$.

By $\|\cdot\|_p$ we denote the norm in the space $L^p(\mathbb{R}^{n-1})$, that is

$$\|f\|_p = \left\{ \int_{\mathbb{R}^{n-1}} |f(x')|^p \, dx' \right\}^{1/p},$$

if $1 \le p < \infty$, and

$$\|f\|_\infty = \text{ess sup}\{|f(x')| : x' \in \mathbb{R}^{n-1}\}.$$

Let $\boldsymbol{\ell}$ denote a unit vector in $\mathbb{R}^n$ and let $\boldsymbol{\nu}$ be the unit vector of the interior normal to the hyperplane $\partial\mathbb{R}^n_+$. Similarly, by $\boldsymbol{\tau}$ we mean a unit vector in $\partial\mathbb{R}^n_+$ such that $\boldsymbol{\ell} = \ell_\tau \boldsymbol{\tau} + \ell_\nu \boldsymbol{\nu}$. In what follows we assume that the Cartesian coordinate system is chosen in such a way that $\boldsymbol{e}_n = \boldsymbol{\nu}$.

Next, by $h^p(\mathbb{R}^n_+)$ we denote the Hardy space of harmonic functions on $\mathbb{R}^n_+$, which can be represented as the Poisson integral

$$u(x) = \frac{2}{\omega_n} \int_{\mathbb{R}^{n-1}} \frac{x_n}{|y - x|^n} u(y') dy' \tag{6.3.1}$$

with boundary values in $L^p(\mathbb{R}^{n-1})$, $1 \le p \le \infty$, where $y = (y', 0)$, $y' \in \mathbb{R}^{n-1}$.

Now, we find a representation for the best coefficient $\mathcal{C}_p(x; \boldsymbol{\ell})$ in the inequality for the absolute value of the derivative of $u \in h^p(\mathbb{R}^n_+)$ in an arbitrary direction $\boldsymbol{\ell} \in \mathbb{S}^{n-1}$. In particular, we obtain a formula for the constant in a similar inequality for the modulus of the gradient.

LEMMA 6.1. *Let $u \in h^p(\mathbb{R}^n_+)$, and let x be an arbitrary point in $\mathbb{R}^n_+$. The sharp coefficient $\mathcal{C}_p(x; \boldsymbol{\ell})$ in the inequality*

$$|\,(\nabla u(x), \boldsymbol{\ell})\,| \le \mathcal{C}_p(x; \boldsymbol{\ell}) \|u\|_p$$

is given by

$$\mathcal{C}_p(x; \boldsymbol{\ell}) = C_p(\boldsymbol{\ell}) x_n^{(1-n-p)/p}, \tag{6.3.2}$$

where

$$C_1(\boldsymbol{\ell}) = \frac{2}{\omega_n} \sup_{\sigma \in \mathbb{S}^{n-1}_+} \left|\left(\boldsymbol{e}_n - n(\boldsymbol{e}_\sigma, \boldsymbol{e}_n)\boldsymbol{e}_\sigma,\ \boldsymbol{\ell}\right)\right| \left(\boldsymbol{e}_\sigma, \boldsymbol{e}_n\right)^n, \tag{6.3.3}$$

$$C_p(\boldsymbol{\ell}) = \frac{2}{\omega_n} \left\{ \int_{\mathbb{S}^{n-1}_+} \left|\left(\boldsymbol{e}_n - n(\boldsymbol{e}_\sigma, \boldsymbol{e}_n)\boldsymbol{e}_\sigma, \boldsymbol{\ell}\right)\right|^{p/(p-1)} \left(\boldsymbol{e}_\sigma, \boldsymbol{e}_n\right)^{n/(p-1)} d\sigma \right\}^{(p-1)/p} \tag{6.3.4}$$

for $1 < p < \infty$, *and*

$$C_\infty(\boldsymbol{\ell}) = \frac{2}{\omega_n} \int_{\mathbb{S}^{n-1}_+} \big|(\boldsymbol{e}_n - n(\boldsymbol{e}_\sigma, \boldsymbol{e}_n)\boldsymbol{e}_\sigma,\ \boldsymbol{\ell})\big|\, d\sigma. \tag{6.3.5}$$

In particular, the sharp coefficient $\mathcal{C}_p(x)$ *in the inequality*

$$|\nabla u(x)| \le \mathcal{C}_p(x)\|u\|_p \tag{6.3.6}$$

is given by

$$\mathcal{C}_p(x) = C_p\, x_n^{(1-n-p)/p}, \tag{6.3.7}$$

where

$$C_p = \sup_{|\boldsymbol{\ell}|=1} C_p(\boldsymbol{\ell}). \tag{6.3.8}$$

PROOF. Let $x = (x', x_n)$ be a fixed point in $\mathbb{R}^n_+$. The representation (6.3.1) implies

$$\frac{\partial u}{\partial x_i} = \frac{2}{\omega_n} \int_{\mathbb{R}^{n-1}} \left[\frac{\delta_{ni}}{|y-x|^n} + \frac{n x_n (y_i - x_i)}{|y-x|^{n+2}} \right] u(y')dy',$$

that is

$$\begin{aligned} \nabla u(x) &= \frac{2}{\omega_n} \int_{\mathbb{R}^{n-1}} \left[\frac{\boldsymbol{e}_n}{|y-x|^n} + \frac{n x_n (y-x)}{|y-x|^{n+2}} \right] u(y')dy' \\ &= \frac{2}{\omega_n} \int_{\mathbb{R}^{n-1}} \frac{\boldsymbol{e}_n - n(\boldsymbol{e}_{xy}, \boldsymbol{e}_n)\boldsymbol{e}_{xy}}{|y-x|^n}\, u(y')dy', \end{aligned}$$

where $\boldsymbol{e}_{xy} = (y-x)|y-x|^{-1}$. For any $\boldsymbol{\ell} \in \mathbb{S}^{n-1}$,

$$(\nabla u(x), \boldsymbol{\ell}) = \frac{2}{\omega_n} \int_{\mathbb{R}^{n-1}} \frac{(\boldsymbol{e}_n - n(\boldsymbol{e}_{xy}, \boldsymbol{e}_n)\boldsymbol{e}_{xy},\ \boldsymbol{\ell})}{|y-x|^n}\, u(y')dy'. \tag{6.3.9}$$

Hence,

$$\mathcal{C}_1(x; \boldsymbol{\ell}) = \frac{2}{\omega_n} \sup_{y \in \mathbb{R}^{n-1}} \frac{|(\boldsymbol{e}_n - n(\boldsymbol{e}_{xy}, \boldsymbol{e}_n)\boldsymbol{e}_{xy},\ \boldsymbol{\ell})|}{|y-x|^n} \tag{6.3.10}$$

and

$$\mathcal{C}_p(x; \boldsymbol{\ell}) = \frac{2}{\omega_n} \left\{ \int_{\mathbb{R}^{n-1}} \frac{|(\boldsymbol{e}_n - n(\boldsymbol{e}_{xy}, \boldsymbol{e}_n)\boldsymbol{e}_{xy}, \boldsymbol{\ell})|^q}{|y-x|^{nq}}\, dy' \right\}^{1/q} \tag{6.3.11}$$

for $1 < p \le \infty$, where $p^{-1} + q^{-1} = 1$.

Taking into account the equality

$$\frac{x_n}{|y-x|} = (\boldsymbol{e}_{xy}, -\boldsymbol{e}_n), \tag{6.3.12}$$

by (6.3.10) we obtain

$$\begin{aligned} \mathcal{C}_1(x; \boldsymbol{\ell}) &= \frac{2}{\omega_n} \sup_{y \in \mathbb{R}^{n-1}} \frac{|(\boldsymbol{e}_n - n(\boldsymbol{e}_{xy}, \boldsymbol{e}_n)\boldsymbol{e}_{xy},\ \boldsymbol{\ell})|}{x_n^n} \left(\frac{x_n}{|y-x|} \right)^n \\ &= \frac{2}{\omega_n x_n^n} \sup_{\sigma \in \mathbb{S}^{n-1}_-} \big|(\boldsymbol{e}_n - n(\boldsymbol{e}_\sigma, \boldsymbol{e}_n)\boldsymbol{e}_\sigma,\ \boldsymbol{\ell})\big| \big(\boldsymbol{e}_\sigma, -\boldsymbol{e}_n\big)^n. \end{aligned}$$

Replacing here $\boldsymbol{e}_\sigma$ by $-\boldsymbol{e}_\sigma$, we arrive at (6.3.2) for $p = 1$ with the sharp constant (6.3.3).

Let $1 < p \leq \infty$. Using (6.3.12) and the equality

$$\frac{1}{|y-x|^{nq}} = \frac{1}{x_n^{nq-n+1}} \left(\frac{x_n}{|y-x|} \right)^{n(q-1)} \frac{x_n}{|y-x|^n},$$

and replacing q by $p/(p-1)$ in (6.3.11), we conclude that (6.3.2) holds with the sharp constant

$$C_p(\boldsymbol{\ell}) = \frac{2}{\omega_n} \left\{ \int_{\mathbb{S}_-^{n-1}} \big|(\boldsymbol{e}_n - n(\boldsymbol{e}_\sigma, \boldsymbol{e}_n)\boldsymbol{e}_\sigma,\ \boldsymbol{\ell})\big|^{p/(p-1)} \big(\boldsymbol{e}_\sigma, -\boldsymbol{e}_n\big)^{n/(p-1)} d\sigma \right\}^{(p-1)/p},$$

where $\mathbb{S}_-^{n-1} = \{\sigma \in \mathbb{S}^{n-1} : (\boldsymbol{e}_\sigma, \boldsymbol{e}_n) < 0\}$. Replacing here $\boldsymbol{e}_\sigma$ by $-\boldsymbol{e}_\sigma$, we arrive at (6.3.4) for $1 < p < \infty$ and at (6.3.5) for $p = \infty$.

By (6.3.9) we have

$$\big|\nabla u(x)\big| = \frac{2}{\omega_n} \sup_{|\boldsymbol{\ell}|=1} \int_{\mathbb{R}^{n-1}} \frac{(\boldsymbol{e}_n - n(\boldsymbol{e}_{xy}, \boldsymbol{e}_n)\boldsymbol{e}_{xy},\ \boldsymbol{\ell})}{|y-x|^n} u(y')dy'.$$

Hence, the sharp coefficient $\mathcal{C}_p(x)$ in (6.3.6) can be written as

$$\mathcal{C}_p(x) = \frac{2}{\omega_n} \sup_{\|u\|_p \leq 1} \sup_{|\boldsymbol{\ell}|=1} \int_{\mathbb{R}^{n-1}} \frac{(\boldsymbol{e}_n - n(\boldsymbol{e}_{xy}, \boldsymbol{e}_n)\boldsymbol{e}_{xy}, \boldsymbol{\ell})}{|y-x|^n} u(y')dy'.$$

Permutating the suprema and using (6.3.11), (6.3.10) and (6.3.2), we find

$$\begin{aligned} \mathcal{C}_p(x) &= \frac{2}{\omega_n} \sup_{|\boldsymbol{\ell}|=1} \left\{ \int_{\mathbb{R}^{n-1}} \frac{\big|(\boldsymbol{e}_n - n(\boldsymbol{e}_{xy}, \boldsymbol{e}_n)\boldsymbol{e}_{xy}, \boldsymbol{\ell})\big|^q}{|y-x|^{nq}} dy' \right\}^{1/q} \\ &= \sup_{|\boldsymbol{\ell}|=1} \mathcal{C}_p(x; \boldsymbol{\ell}) = \sup_{|\boldsymbol{\ell}|=1} C_p(\boldsymbol{\ell}) x_n^{(1-n-p)/p} \qquad (6.3.13) \end{aligned}$$

for $1 < p \leq \infty$, and

$$\begin{aligned} \mathcal{C}_1(x) &= \frac{2}{\omega_n} \sup_{|\boldsymbol{\ell}|=1} \sup_{y \in \mathbb{R}^{n-1}} \frac{\big|(\boldsymbol{e}_n - n(\boldsymbol{e}_{xy}, \boldsymbol{e}_n)\boldsymbol{e}_{xy},\ \boldsymbol{\ell})\big|}{|y-x|^n} \\ &= \sup_{|\boldsymbol{\ell}|=1} \mathcal{C}_1(x; \boldsymbol{\ell}) = \sup_{|\boldsymbol{\ell}|=1} C_1(\boldsymbol{\ell}) x_n^{-n}. \qquad (6.3.14) \end{aligned}$$

Using the notation (6.3.8) in (6.3.13) and (6.3.14), we arrive at (6.3.7). □

REMARK 6.5. Formula (6.3.4) for the coefficient $C_p(\boldsymbol{\ell}), 1 < p \leq \infty$, can be written with the integral over the whole sphere $\mathbb{S}^{n-1}$ in $\mathbb{R}^n$,

$$C_p(\boldsymbol{\ell}) = \frac{2^{1/p}}{\omega_n} \left\{ \int_{\mathbb{S}^{n-1}} \big|(\boldsymbol{e}_n - n(\boldsymbol{e}_\sigma, \boldsymbol{e}_n)\boldsymbol{e}_\sigma,\ \boldsymbol{\ell})\big|^{p/(p-1)} \big|(\boldsymbol{e}_\sigma, \boldsymbol{e}_n)\big|^{n/(p-1)} d\sigma \right\}^{(p-1)/p}.$$

A similar remark relates (6.3.5) as well as formula (6.3.3):

$$C_1(\boldsymbol{\ell}) = \frac{2}{\omega_n} \sup_{\sigma \in \mathbb{S}^{n-1}} \big|(\boldsymbol{e}_n - n(\boldsymbol{e}_\sigma, \boldsymbol{e}_n)\boldsymbol{e}_\sigma,\ \boldsymbol{\ell})(\boldsymbol{e}_\sigma, \boldsymbol{e}_n)^n\big|.$$

6.3.2. The case $1 \le p < \infty$

The next assertion is based on the representation for $C_p(\boldsymbol{\ell})$, obtained in Lemma 6.1.

PROPOSITION 6.3. *Let $u \in h^p(\mathbb{R}^n_+)$, and let x be an arbitrary point in $\mathbb{R}^n_+$. The sharp coefficient $\mathcal{C}_p(x;\boldsymbol{\ell})$ in*

$$|(\nabla u(x), \boldsymbol{\ell})| \le \mathcal{C}_p(x;\boldsymbol{\ell})\|u\|_p \,, \tag{6.3.15}$$

$1 \le p < \infty$, is given by

$$\mathcal{C}_p(x;\boldsymbol{\ell}) = C_p(\boldsymbol{\ell}) x_n^{(1-n-p)/p} \,, \tag{6.3.16}$$

where

$$C_p(\boldsymbol{\ell}) = \frac{2(\omega_{n-2})^{(p-1)/p}}{\omega_n} \left\{ \int_0^\pi d\varphi \int_0^{\pi/2} \mathcal{F}_{n,p}(\varphi,\vartheta;\ell_\nu,\ell_\tau)\, d\vartheta \right\}^{(p-1)/p} . \tag{6.3.17}$$

Here

$$\mathcal{F}_{n,p}(\varphi,\vartheta;\ell_\nu,\ell_\tau) = \big|\mathcal{G}_n(\varphi,\vartheta;\ell_\nu,\ell_\tau)\big|^{p/(p-1)} \cos^{n/(p-1)}\vartheta \sin^{n-2}\vartheta \sin^{n-3}\varphi \tag{6.3.18}$$

with

$$\mathcal{G}_n(\varphi,\vartheta;\ell_\nu,\ell_\tau) = (n\cos^2\vartheta - 1)\ell_\nu + n\ell_\tau \cos\vartheta \sin\vartheta \cos\varphi \,. \tag{6.3.19}$$

In particular, the sharp coefficient $\mathcal{C}_p(x)$ in

$$|\nabla u(x)| \le \mathcal{C}_p(x)\|u\|_p \tag{6.3.20}$$

is given by

$$\mathcal{C}_p(x) = C_p x_n^{(1-n-p)/p} \,, \tag{6.3.21}$$

where

$$C_p = \frac{2(\omega_{n-2})^{(p-1)/p}}{\omega_n} \sup_{\gamma \ge 0} \frac{1}{\sqrt{1+\gamma^2}} \left\{ \int_0^\pi d\varphi \int_0^{\pi/2} \mathcal{F}_{n,p}(\varphi,\vartheta;1,\gamma) d\vartheta \right\}^{(p-1)/p} . \tag{6.3.22}$$

PROOF. With $\sigma = (\sigma_1, \dots, \sigma_{n-1}, \sigma_n) \in \mathbb{S}^{n-1}_+$, we associate the vector $\boldsymbol{\sigma}' = \boldsymbol{e}_\sigma - \sigma_n \boldsymbol{e}_n$. Using the equalities $(\boldsymbol{\sigma}', \boldsymbol{e}_n) = 0$, $\sigma_n = \sqrt{1-|\boldsymbol{\sigma}'|^2}$ and $(\boldsymbol{\tau}, \boldsymbol{e}_n) = 0$, we find an expression for $(\boldsymbol{e}_n - n(\boldsymbol{e}_\sigma, \boldsymbol{e}_n)\boldsymbol{e}_\sigma,\ \boldsymbol{\ell})$ as a function of $\boldsymbol{\sigma}'$:

$$\begin{aligned}
(\boldsymbol{e}_n - n(\boldsymbol{e}_\sigma, \boldsymbol{e}_n)\boldsymbol{e}_\sigma,\ \boldsymbol{\ell}) &= \ell_\nu - n\sigma_n(\boldsymbol{e}_\sigma, \boldsymbol{\ell}) = \ell_\nu - n\sigma_n(\boldsymbol{\sigma}' + \sigma_n \boldsymbol{e}_n,\ \ell_\nu \boldsymbol{\nu} + \ell_\tau \boldsymbol{\tau}) \\
&= \ell_\nu - n\sigma_n\big[(\boldsymbol{\sigma}', \boldsymbol{\tau})\ell_\tau + \sigma_n \ell_\nu\big] \\
&= -\big[n(1-|\boldsymbol{\sigma}'|^2) - 1\big]\ell_\nu - n\sqrt{1-|\boldsymbol{\sigma}'|^2}\,(\boldsymbol{\sigma}', \boldsymbol{\tau})\ell_\tau.
\end{aligned} \tag{6.3.23}$$

Let $\mathbb{B}' = \{x' = (x_1, \dots, x_{n-1}) \in \mathbb{R}^{n-1} : |x'| < 1\}$. By (6.3.4) and (6.3.23), taking into account that $d\sigma = d\sigma'/\sqrt{1-|\boldsymbol{\sigma}'|^2}$, we may write (6.3.4) as

$$\begin{aligned}
C_p(\boldsymbol{\ell}) &= \frac{2}{\omega_n} \left\{ \int_{\mathbb{B}'} \frac{\mathcal{H}_{n,p}\big(|\boldsymbol{\sigma}'|, (\boldsymbol{\sigma}', \boldsymbol{\tau})\ell_\tau\big)\big(1-|\boldsymbol{\sigma}'|^2\big)^{n/2(p-1)}}{\sqrt{1-|\boldsymbol{\sigma}'|^2}} d\sigma' \right\}^{(p-1)/p} \\
&= \frac{2}{\omega_n} \left\{ \int_{\mathbb{B}'} \mathcal{H}_{n,p}\big(|\boldsymbol{\sigma}'|, (\boldsymbol{\sigma}', \boldsymbol{\tau})\ell_\tau\big)\big(1-|\boldsymbol{\sigma}'|^2\big)^{(n+1-p)/2(p-1)} d\sigma' \right\}^{(p-1)/p} ,
\end{aligned} \tag{6.3.24}$$

where

$$\mathcal{H}_{n,p}\big(|\boldsymbol{\sigma}'|, (\boldsymbol{\sigma}', \boldsymbol{\tau})\ell_\tau\big) = \Big|\big[n(1-|\boldsymbol{\sigma}'|^2) - 1\big]\ell_\nu + n\sqrt{1-|\boldsymbol{\sigma}'|^2}\,(\boldsymbol{\sigma}', \boldsymbol{\tau})\ell_\tau\Big|^{p/(p-1)} . \tag{6.3.25}$$

Let $\mathbb{B} = \{x \in \mathbb{R}^n : |x| < 1\}$. Using the well known formula (see, e.g., Prudnikov, Brychkov and Marichev [**PBM1**], **3.3.2(3)**),

$$\int_{\mathbb{B}} g(|\boldsymbol{x}|, (\boldsymbol{a}, \boldsymbol{x}))dx = \omega_{n-1} \int_0^1 r^{n-1}dr \int_0^\pi g(r, |\boldsymbol{a}|r\cos\varphi)\sin^{n-2}\varphi\, d\varphi\,,$$

we obtain

$$\begin{aligned}(6.3.26)\quad &\frac{1}{\omega_{n-2}}\int_{\mathbb{B}'} \mathcal{H}_{n,p}\big(|\boldsymbol{\sigma}'|, (\boldsymbol{\sigma}', \boldsymbol{\tau})\ell_\tau\big)\big(1-|\boldsymbol{\sigma}'|^2\big)^{(n+1-p)/2(p-1)}\, d\sigma' \\ &= \int_0^1 r^{n-2}\big(1-r^2\big)^{(n+1-p)/2(p-1)}dr \int_0^\pi \mathcal{H}_{n,p}\big(r, \ell_\tau r\cos\varphi\big)\sin^{n-3}\varphi d\varphi\,.\end{aligned}$$

Making the change of variable $r = \sin\vartheta$ in (6.3.26), we find

$$\begin{aligned}(6.3.27)\quad &\frac{1}{\omega_{n-2}}\int_{\mathbb{B}'} \mathcal{H}_{n,p}\big(|\boldsymbol{\sigma}'|, (\boldsymbol{\sigma}', \boldsymbol{\tau})\ell_\tau\big)\big(1-|\boldsymbol{\sigma}'|^2\big)^{(n+1-p)/2(p-1)}\, d\sigma' \\ &= \int_0^\pi \sin^{n-3}\varphi d\varphi \int_0^{\pi/2} \mathcal{H}_{n,p}\big(\sin\vartheta,\ \ell_\tau\sin\vartheta\cos\varphi\big)\sin^{n-2}\vartheta\cos^{n/(p-1)}\vartheta d\vartheta\,,\end{aligned}$$

where, by (6.3.25),

$$\mathcal{H}_{n,p}\big(\sin\vartheta,\ \ell_\tau\sin\vartheta\cos\varphi\big) = \Big|(n\cos^2\vartheta - 1)\ell_\nu + n\ell_\tau\cos\vartheta\sin\vartheta\cos\varphi\Big|^{p/(p-1)}.$$

Now, (6.3.17) follows from (6.3.24) and (6.3.27).

Introducing in (6.3.17) the parameter $\gamma = \ell_\tau/\ell_\nu$ and using the equality $\ell_\tau^2 + \ell_\nu^2 = 1$, we obtain

$$(6.3.28)\quad C_p(\boldsymbol{\ell}) = \frac{2(\omega_{n-2})^{(p-1)/p}}{\omega_n}\frac{1}{\sqrt{1+\gamma^2}}\left\{\int_0^\pi d\varphi\int_0^{\pi/2}\mathcal{F}_{n,p}(\varphi, \vartheta; 1, \gamma)d\vartheta\right\}^{(p-1)/p}.$$

Since the integrand in (6.3.4) does not change when $\boldsymbol{\ell} \in \mathbb{S}^{n-1}$ is replaced by $-\boldsymbol{\ell}$, we may assume that $\ell_\nu = (\boldsymbol{\ell}, \boldsymbol{\nu}) > 0$ in (6.3.8), and hence $\gamma > 0$ in (6.3.28). By (6.3.28), taking into account (6.3.8), we arrive at (6.3.22). □

COROLLARY 6.3. *Let $u \in h^1(\mathbb{R}^n_+)$, and let x be an arbitrary point in $\mathbb{R}^n_+$. The sharp coefficient $\mathcal{C}_1(x; \boldsymbol{\ell})$ in*

$$(6.3.29)\quad |(\nabla u(x), \boldsymbol{\ell})| \le \mathcal{C}_1(x; \boldsymbol{\ell})\|u\|_1$$

is given by

$$(6.3.30)\quad \mathcal{C}_1(x; \boldsymbol{\ell}) = C_1(\boldsymbol{\ell})\, x_n^{-n},$$

where

$$\begin{aligned}(6.3.31)\quad C_1(\boldsymbol{\ell}) &= \frac{2}{\omega_n}\sup_{\sigma\in\mathbb{S}^{n-1}_+}\big|\big(\boldsymbol{e}_n - n(\boldsymbol{e}_\sigma, \boldsymbol{e}_n)\boldsymbol{e}_\sigma,\ \boldsymbol{\ell}\big)\big|\big(\boldsymbol{e}_\sigma, \boldsymbol{e}_n\big)^n \\ &= \frac{2}{\omega_n}\sup_{(\varphi,\vartheta)\in\mathbb{S}^2_+}\Big|(n\cos^2\vartheta - 1)\ell_\nu + n\ell_\tau\cos\vartheta\sin\vartheta\cos\varphi\Big|\cos^n\vartheta\,.\end{aligned}$$

In particular,

$$(6.3.32)\quad C_1(\boldsymbol{\nu}) = \frac{2(n-1)}{\omega_n},$$

and

$$C_1(\boldsymbol{\tau}) = \frac{2n(n+1)^{(n+1)/2}}{\omega_n (n+2)^{(n+2)/2}} . \tag{6.3.33}$$

The sharp constant C_1 in

$$|\nabla u(x)| \le C_1 x_n^{-n} \|u\|_1 \tag{6.3.34}$$

is given by

$$C_1 = C_1(\boldsymbol{\nu}) = \frac{2(n-1)}{\omega_n} . \tag{6.3.35}$$

PROOF. The equality (6.3.30) and the first equality in (6.3.31) were proved in Lemma 6.1. The second equality in (6.3.31) follows from (6.3.17) as $p \to 1$. Setting $\ell_\nu = 1, \ell_\tau = 0$ i $\ell_\nu = 0, \ell_\tau = 1$ in (6.3.31) and finding the maximal value of the function obtained, we arrive at (6.3.32) and (6.3.33), respectively.

Using (6.3.31), (6.3.8) and the permutability of two suprema, we find

$$\begin{aligned} C_1 &= \frac{2}{\omega_n} \sup_{|\boldsymbol{\ell}|=1} \sup_{\sigma \in \mathbb{S}^{n-1}_+} \big| \big(\boldsymbol{e}_n - n(\boldsymbol{e}_\sigma, \boldsymbol{e}_n)\boldsymbol{e}_\sigma ,\ \boldsymbol{\ell} \big) \big| \big(\boldsymbol{e}_\sigma, \boldsymbol{e}_n \big)^n \\ &= \frac{2}{\omega_n} \sup_{\sigma \in \mathbb{S}^{n-1}_+} \big| \boldsymbol{e}_n - n(\boldsymbol{e}_\sigma, \boldsymbol{e}_n)\boldsymbol{e}_\sigma \big| \big(\boldsymbol{e}_\sigma, \boldsymbol{e}_n \big)^n . \end{aligned} \tag{6.3.36}$$

Taking into account the equality

$$\begin{aligned} |\boldsymbol{e}_n - n(\boldsymbol{e}_\sigma, \boldsymbol{e}_n)\boldsymbol{e}_\sigma| &= \Big(\boldsymbol{e}_n - n(\boldsymbol{e}_\sigma, \boldsymbol{e}_n)\boldsymbol{e}_\sigma ,\ \boldsymbol{e}_n - n(\boldsymbol{e}_\sigma, \boldsymbol{e}_n)\boldsymbol{e}_\sigma \Big)^{1/2} \\ &= \Big(1 + (n^2 - 2n)(\boldsymbol{e}_\sigma,\ \boldsymbol{e}_n)^2 \Big)^{1/2} , \end{aligned}$$

and using (6.3.36), we arrive at the sharp constant

$$C_1 = \frac{2(n-1)}{\omega_n} \tag{6.3.37}$$

in inequality (6.3.34), which together with (6.3.32) completes the proof. □

COROLLARY 6.4. *Let $u \in h^2(\mathbb{R}^n_+)$, and let x be an arbitrary point in $\mathbb{R}^n_+$. The sharp coefficient $\mathcal{C}_2(x, \boldsymbol{\ell})$ in*

$$|(\nabla u(x), \boldsymbol{\ell})| \le \mathcal{C}_2(x; \boldsymbol{\ell}) \|u\|_2 \tag{6.3.38}$$

is given by

$$\mathcal{C}_2(x; \boldsymbol{\ell}) = C_2(\boldsymbol{\ell})\, x_n^{-(n+1)/2} , \tag{6.3.39}$$

where

$$C_2(\boldsymbol{\ell}) = \sqrt{\frac{n(n-1)\ell_\nu^2 + n\ell_\tau^2}{2^n \omega_n}} . \tag{6.3.40}$$

In particular,

$$C_2(\boldsymbol{\nu}) = \sqrt{\frac{n(n-1)}{2^n \omega_n}} , \tag{6.3.41}$$

and

$$C_2(\boldsymbol{\tau}) = \sqrt{\frac{n}{2^n \omega_n}}\,. \tag{6.3.42}$$

The sharp constant C_2 in

$$|\nabla u(x)| \leq C_2 x_n^{-(n+1)/2} \|u\|_2$$

is given by

$$C_2 = C_2(\boldsymbol{\nu}) = \sqrt{\frac{n(n-1)}{2^n \omega_n}}. \tag{6.3.43}$$

PROOF. The equality (6.3.39) was proved in Lemma 6.1. By (6.3.17), (6.3.18) and (6.3.19),

$$C_2(\boldsymbol{\ell}) = \frac{2\sqrt{\omega_{n-2}}}{\omega_n} \left\{ \int_0^\pi d\varphi \int_0^{\pi/2} \mathcal{F}_{n,2}(\varphi, \vartheta; \ell_\nu, \ell_\tau)\, d\vartheta \right\}^{1/2}, \tag{6.3.44}$$

where

$$\mathcal{F}_{n,2}(\varphi, \vartheta; \ell_\nu, \ell_\tau) = \left[(n\cos^2\vartheta - 1)\ell_\nu + n\ell_\tau \cos\vartheta \sin\vartheta \cos\varphi\right]^2 \cos^n\vartheta \sin^{n-2}\vartheta \sin^{n-3}\varphi.$$

The equality (6.3.44) implies

$$C_2(\boldsymbol{\ell}) = \frac{2\sqrt{\omega_{n-2}}}{\omega_n} \left\{ \mathcal{I}_1 \ell_\nu^2 + \mathcal{I}_2 \ell_\tau^2 \right\}^{1/2}, \tag{6.3.45}$$

where

$$\begin{aligned} \mathcal{I}_1 &= \int_0^\pi \sin^{n-3}\varphi\, d\varphi \int_0^{\pi/2} (n\cos^2\vartheta - 1)^2 \sin^{n-2}\vartheta \cos^n\vartheta\, d\vartheta \\ &= \frac{\sqrt{\pi}\, n(n-1)\Gamma\left(\frac{n-2}{2}\right)\Gamma\left(\frac{n+1}{2}\right)}{8(n-1)!}, \end{aligned} \tag{6.3.46}$$

$$\begin{aligned} \mathcal{I}_2 &= n^2 \int_0^\pi \sin^{n-3}\varphi \cos^2\varphi\, d\varphi \int_0^{\pi/2} \sin^n\vartheta \cos^{n+2}\vartheta\, d\vartheta \\ &= \frac{\sqrt{\pi}\, n\Gamma\left(\frac{n-2}{2}\right)\Gamma\left(\frac{n+1}{2}\right)}{8(n-1)!}. \end{aligned} \tag{6.3.47}$$

By (6.3.45), (6.3.46) and (6.3.47) we arrive at (6.3.40). Now (6.3.41) and (6.3.42) are direct corollaries of (6.3.40).

By (6.3.40) we have

$$C_2(\boldsymbol{\nu}) \leq C_2 \leq \sqrt{\frac{n}{2^n \omega_n}} \max_{|\boldsymbol{\ell}|=1} \left\{ (n-1)\ell_\nu^2 + \ell_\tau^2 \right\}^{1/2} \leq \sqrt{\frac{n(n-1)}{2^n \omega_n}} = C_2(\boldsymbol{\nu}),$$

which proves (6.3.43). □

6.3.3. An auxiliary algebraic inequality

Here we prove an algebraic inequality to be used later for deriving an explicit formula for the sharp constant in the estimate for the modulus of gradient in the case $p = \infty$.

LEMMA 6.2. *For all $x \geq 0$ and any $\mu \geq 1$ the inequality holds*

$$\left(\frac{\mu+1}{\mu+x}\right)^{\mu-1} + \left(\frac{\mu+1}{1+\mu x}\right)^{\mu-1} x^{\mu+1} \leq 2x + \frac{\mu(3\mu+1)}{(\mu+1)^2}(1-x)^2. \tag{6.3.48}$$

The equality sign takes place only for $\mu = 1$ or $x = 1$.

PROOF. Clearly, inequality (6.3.48) becomes equality for $\mu = 1$ or $x = 1$. Suppose that $\mu \in (1, \infty)$.

(i) *The case* $0 \leq x < 1$. Let us write (6.3.48) in the form

$$\left(\frac{\mu+1}{\mu+x}\right)^{\mu-1} + \left(\frac{\mu x+x}{1+\mu x}\right)^{\mu-1} x^2 \leq 2x + \frac{\mu(3\mu+1)}{(\mu+1)^2}(1-x)^2.$$

Introducing the notation

$$F(x) = \left(\frac{\mu+1}{\mu+x}\right)^{\mu-1} + \left(\frac{\mu x+x}{1+\mu x}\right)^{\mu-1} x^2, \tag{6.3.49}$$

we find

$$F'(x) = -\frac{\mu-1}{\mu+1}\left(\frac{\mu+1}{\mu+x}\right)^{\mu} + x\left(2+\frac{\mu-1}{1+\mu x}\right)\left(\frac{\mu x+x}{1+\mu x}\right)^{\mu-1},$$

$$F''(x) = \frac{\mu(\mu-1)}{(\mu+1)^2}\left(\frac{\mu+1}{\mu+x}\right)^{\mu+1} + \mu\left[\frac{2(x+1)}{1+\mu x} + \frac{\mu-1}{(1+\mu x)^2}\right]\left(\frac{\mu x+x}{1+\mu x}\right)^{\mu-1}, \tag{6.3.50}$$

and

$$F'''(x) = -\frac{\mu(\mu-1)}{(\mu+1)^2}\left(\frac{\mu+1}{\mu+x}\right)^{\mu+2} + \frac{\mu(\mu^2-1)}{x(1+\mu x)^3}\left(\frac{\mu x+x}{1+\mu x}\right)^{\mu-1}. \tag{6.3.51}$$

By Taylor's formula with Lagrange's remainder term,

$$F(x)=F(1)+F'(1)(x-1)+\frac{1}{2}F''(t)\,(x-1)^2 = 2+2(x-1)+\frac{1}{2}F''(t)\,(x-1)^2, \tag{6.3.52}$$

where $x \in [0,1)$ and $t \in (x,1)$.

Note that $F''(0) < F''(1)$. In fact, by (6.3.50),

$$F''(0) = \frac{\mu-1}{\mu+1}\left(1+\frac{1}{\mu}\right)^{\mu}, \qquad F''(1) = \frac{2\mu(3\mu+1)}{(\mu+1)^2},$$

which together with the obvious inequality

$$\frac{\mu-1}{\mu+1}e < \frac{2\mu(3\mu+1)}{(\mu+1)^2}$$

implies $F''(0) < F''(1)$.

Next we show that

$$F''(t) < \max\{F''(0),\ F''(1)\} = \frac{2\mu(3\mu+1)}{(\mu+1)^2} \tag{6.3.53}$$

for any $t \in (0,1)$.

Suppose the opposite assertion holds, i.e. there exists a point $t \in (0,1)$ at which (6.3.53) fails. Hence $F''(t)$ attains its maximum value on $[0,1]$ at an inner point τ, i.e.,

$$F''(\tau) = \max_{t\in[0;1]} F''(t) \geq \max\left\{F''(0),\ F''(1)\right\} = \frac{2\mu(3\mu+1)}{(\mu+1)^2}. \tag{6.3.54}$$

Taking into account (6.3.51), we have

$$F'''(\tau) = -\frac{\mu(\mu-1)}{(\mu+1)^2}\left(\frac{\mu+1}{\mu+\tau}\right)^{\mu+2} + \frac{\mu(\mu^2-1)}{\tau(1+\mu\tau)^3}\left(\frac{\mu\tau+\tau}{1+\mu\tau}\right)^{\mu-1} = 0,$$

which is equivalent to

$$\left(\frac{\mu\tau+\tau}{1+\mu\tau}\right)^{\mu-1} = \frac{\tau(1+\mu\tau)^3}{(\mu+1)^3}\left(\frac{\mu+1}{\mu+\tau}\right)^{\mu+2}.$$

Combined with (6.3.50), this implies

$$\begin{aligned} F''(\tau) &= \frac{\mu(\mu-1)}{(\mu+1)^2}\left(\frac{\mu+1}{\mu+\tau}\right)^{\mu+1}\left\{1+\tau(1+\mu\tau)\frac{2(1+\tau)(1+\mu\tau)+\mu-1}{(\mu-1)(\mu+\tau)}\right\} \\ &< \frac{\mu(\mu-1)}{(\mu+1)^2}\left(\frac{\mu+1}{\mu+\tau}\right)^{\mu+1}\left\{2+\frac{2(1+\tau)(1+\mu\tau)}{\mu-1}\right\}. \end{aligned}$$

Therefore,

$$\begin{aligned} F''(\tau) &< \frac{2\mu}{(\mu+1)^2}\left(\frac{\mu+1}{\mu+\tau}\right)^{\mu+1}\left\{(\mu-1)+(1+\tau)(1+\mu\tau)\right\} \\ &= 2\mu(\mu+1)^{\mu-1}\frac{\mu\tau^2+(\mu+1)\tau+\mu}{(\mu+\tau)^{\mu+1}}. \end{aligned} \tag{6.3.55}$$

Setting

$$\eta(\tau) = \frac{\mu\tau^2+(\mu+1)\tau+\mu}{(\mu+\tau)^{\mu+1}}, \tag{6.3.56}$$

we write (6.3.55) as

$$F''(\tau) < 2\mu(\mu+1)^{\mu-1}\eta(\tau). \tag{6.3.57}$$

Noting that

$$\eta'(\tau) = \frac{\mu\tau(\mu-1)(1-\tau)}{(\mu+\tau)^{\mu+2}} > 0$$

for $0<\tau<1$, by (6.3.56) and (6.3.57), we find

$$\max_{t\in[0;1]} F''(t) = F''(\tau) < 2\mu(\mu+1)^{\mu-1}\eta(1) = \frac{2\mu(3\mu+1)}{(\mu+1)^2}.$$

The latter contradicts (6.3.54) which proves (6.3.53) for all $t \in (0,1)$. Thus, it follows from (6.3.49), (6.3.52) and (6.3.53) that

$$\left(\frac{\mu+1}{\mu+x}\right)^{\mu-1} + \left(\frac{\mu x+x}{1+\mu x}\right)^{\mu-1} x^2 < 2x + \frac{\mu(3\mu+1)}{(\mu+1)^2}(x-1)^2$$

for all $0 \leq x < 1$. This means that for $\mu > 1$ and $0 \leq x < 1$ the strict inequality (6.3.48) holds.

(ii) *The case $x > 1$.* Since the function

$$G(x) = \left(\frac{\mu+1}{\mu+x}\right)^{\mu-1} + \left(\frac{\mu x + x}{1+\mu x}\right)^{\mu-1} x^2 - 2x - \frac{\mu(3\mu+1)}{(\mu+1)^2}(x-1)^2$$

satisfies the equality

$$G\left(\frac{1}{x}\right) = \frac{1}{x^2}\, G(x),$$

we have by part (i) that $G(x) < 0$ for $0 \le x < 1$ and hence $G(x) < 0$ for $x > 1$. Thus, the strict inequality (6.3.48) holds for $\mu > 1$ and $x > 1$. □

COROLLARY 6.5. *For all $y \ge 0$ and any integer $n \ge 2$ the inequality holds*

$$\mathcal{P}_n^2(y) + \mathcal{P}_n^2(-y) \le \frac{2n^2 + 4(n-1)(3n-2)y^2}{n^n}, \tag{6.3.58}$$

where

$$\mathcal{P}_n(y) = \frac{\left(\sqrt{1+y^2}+y\right)^{n-1}}{\left(1+(n-1)\left(\sqrt{1+y^2}+y\right)^2\right)^{(n-2)/2}}. \tag{6.3.59}$$

PROOF. Suppose that $0 < x \le 1$. We introduce the new variable

$$y = \frac{1-x}{2\sqrt{x}} \in [0, \infty). \tag{6.3.60}$$

Solving the equation $y = (x^{-1/2} - x^{1/2})/2$ in $\sqrt{x}$, we find $\sqrt{x} = \sqrt{1+y^2} - y$, i.e.,

$$x = \left(\sqrt{1+y^2} - y\right)^2 = \frac{1}{\left(\sqrt{1+y^2}+y\right)^2},$$

Putting $\mu = n-1$, we write (6.3.48) as

$$\frac{1}{(n-1+x)^{n-2}} + \frac{x^n}{(1+(n-1)x)^{n-2}} \le \frac{2n^2 x + (n-1)(3n-2)(1-x)^2}{n^n}. \tag{6.3.61}$$

By (6.3.60) we have $(1-x)^2 = 4y^2 x$, hence (6.3.61) can be rewritten in the form

$$\frac{1}{x(n-1+x)^{n-2}} + \frac{x^{n-1}}{(1+(n-1)x)^{n-2}} \le \frac{2n^2 + 4(n-1)(3n-2)y^2}{n^n}. \tag{6.3.62}$$

Setting $x = \left(\sqrt{1+y^2}+y\right)^{-2}$ in the first term on the left-hand side of (6.3.62), we obtain

$$\frac{1}{x(n-1+x)^{n-2}} = \frac{\left(\sqrt{1+y^2}+y\right)^{2n-2}}{\left(1+(n-1)\left(\sqrt{1+y^2}+y\right)^2\right)^{n-2}} = \mathcal{P}_n^2(y). \tag{6.3.63}$$

Similarly, putting $x=\left(\sqrt{1+y^2}-y\right)^2$ in the second term on the left-hand side of (6.3.62), we find

$$\text{(6.3.64)}\qquad \frac{x^{n-1}}{\left(1+(n-1)x\right)^{n-2}}=\frac{\left(\sqrt{1+y^2}-y\right)^{2n-2}}{\left(1+(n-1)\left(\sqrt{1+y^2}-y\right)^2\right)^{n-2}}=\mathcal{P}_n^2(-y).$$

Using (6.3.63) and (6.3.64), we can rewrite (6.3.62) as (6.3.58). □

We give one more corollary of Lemma 6.2 containing an alternative form of (6.3.48) with an integer $\mu\geq 2$. However, we are not going to use it henceforth.

COROLLARY 6.6. *For all $x\geq 0$ and any integer $n\geq 2$ the inequality holds*

$$\text{(6.3.65)}\qquad \sum_{k=3}^{n+1}\binom{n+1}{k}\left\{\frac{1}{\left(n+x\right)^{k-2}}+\frac{(-1)^k}{\left(1+nx\right)^{k-2}}\right\}(1-x)^k\leq 0\,.$$

The equality sign takes place only for $x=1$.

PROOF. We set $\mu=n, n\geq 2$, in (6.3.48):

$$\left(\frac{n+1}{n+x}\right)^{n-1}+\left(\frac{n+1}{1+nx}\right)^{n-1}x^{n+1}\leq 2x+\frac{n(3n+1)}{(n+1)^2}(1-x)^2.$$

Multiplying the last inequality by $(n+1)^2$, we state it as

$$\text{(6.3.66)}\qquad \frac{(n+1)^{n+1}}{\left(n+x\right)^{n-1}}+\frac{\left((n+1)x\right)^{n+1}}{\left(1+nx\right)^{n-1}}-2(n+1)^2x-n(3n+1)(1-x)^2\leq 0.$$

We rewrite the first term in (6.3.66):

$$\frac{(n+1)^{n+1}}{\left(n+x\right)^{n-1}}=\frac{[(n+x)+(1-x)]^{n+1}}{\left(n+x\right)^{n-1}}=(n+x)^2+(n+1)(n+x)(1-x)$$

$$\text{(6.3.67)}\qquad +\frac{n(n+1)}{2}(1-x)^2+\sum_{k=3}^{n+1}\binom{n+1}{k}\frac{(1-x)^k}{(n+x)^{k-2}}\,.$$

Similarly, the second term in (6.3.66) can be written as

$$\frac{\left((n+1)x\right)^{n+1}}{\left(1+nx\right)^{n-1}}=\frac{[(1+nx)-(1-x)]^{n+1}}{\left(1+nx\right)^{n-1}}=\left(1+nx\right)^2-(n+1)\left(1+nx\right)(1-x)$$

$$\text{(6.3.68)}\qquad +\frac{n(n+1)}{2}(1-x)^2+\sum_{k=3}^{n+1}(-1)^k\binom{n+1}{k}\frac{(1-x)^k}{\left(1+nx\right)^{k-2}}\,.$$

Using (6.3.67) and (6.3.68) in (6.3.66), we arrive at (6.3.65) with the left-hand side as the sum of rational functions. □

6.3.4. The case $p=\infty$

The next assertion is the main theorem of this section. It is based on the representation of the sharp coefficient $\mathcal{C}_p(x;\boldsymbol{\ell})$, $1<p<\infty$, obtained in Proposition 6.3. To find an explicit formula for C_∞ we solve an extremal problem with a scalar parameter in the integrand of a double integral.

THEOREM 6.1. *Let $u \in h^\infty(\mathbb{R}^n_+)$, and let x be an arbitrary point in $\mathbb{R}^n_+$. The sharp coefficient $\mathcal{C}_\infty(x;\boldsymbol{\ell})$ in*

$$|(\nabla u(x), \boldsymbol{\ell})| \le \mathcal{C}_\infty(x;\boldsymbol{\ell}) \|u\|_\infty \tag{6.3.69}$$

is given by

$$\mathcal{C}_\infty(x;\boldsymbol{\ell}) = C_\infty(\boldsymbol{\ell}) x_n^{-1} , \tag{6.3.70}$$

where

$$C_\infty(\boldsymbol{\ell}) = \frac{2\omega_{n-2}}{\omega_n} \int_0^\pi \sin^{n-3}\varphi \, d\varphi \int_0^{\pi/2} \big|\mathcal{G}_n(\varphi,\vartheta;\ell_\nu,\ell_\tau)\big| \sin^{n-2}\vartheta \, d\vartheta . \tag{6.3.71}$$

Here

$$\mathcal{G}_n(\varphi,\vartheta;\ell_\nu,\ell_\tau) = (n\cos^2\vartheta - 1)\ell_\nu + n\ell_\tau \cos\vartheta \sin\vartheta \cos\varphi . \tag{6.3.72}$$

In particular,

$$C_\infty(\boldsymbol{\nu}) = \frac{4(n-1)^{(n-1)/2}\,\omega_{n-1}}{n^{n/2}\,\omega_n} , \tag{6.3.73}$$

$$C_\infty(\boldsymbol{\tau}) = \frac{4\omega_{n-2}}{(n-2)\,\omega_n} . \tag{6.3.74}$$

The sharp constant C_∞ in

$$|\nabla u(x)| \le C_\infty x_n^{-1} \|u\|_\infty \tag{6.3.75}$$

is given by

$$C_\infty = C_\infty(\boldsymbol{\nu}) = \frac{4(n-1)^{(n-1)/2}\,\omega_{n-1}}{n^{n/2}\,\omega_n} . \tag{6.3.76}$$

PROOF. We pass to the limit as $p \to \infty$ in (6.3.16) and (6.3.17). This results in (6.3.70) and (6.3.71). Formulas (6.3.73) and (6.3.74) follow from (6.3.71) for $\ell_\nu = 1, \ell_\tau = 0$ and $\ell_\nu = 0, \ell_\tau = 1$, respectively. Similarly, passing to the limit in (6.3.22) as $p \to \infty$, we obtain

$$C_\infty = \sup_{\gamma \ge 0} \frac{2\omega_{n-2}}{\omega_n\sqrt{1+\gamma^2}} \int_0^\pi \sin^{n-3}\varphi \, d\varphi \int_0^{\pi/2} \big|\mathcal{G}_n(\varphi,\vartheta;1,\gamma)\big| \sin^{n-2}\vartheta \, d\vartheta , \tag{6.3.77}$$

where

$$\mathcal{G}_n(\varphi,\vartheta;1,\gamma) = (n\cos^2\vartheta - 1) + n\gamma\cos\vartheta\sin\vartheta\cos\varphi.$$

We are looking for a solution of the equation

$$(n\cos^2\vartheta - 1) + n\gamma\cos\vartheta\sin\vartheta\cos\varphi = 0 \tag{6.3.78}$$

as a function ϑ of φ. We can rewrite (6.3.78) as

$$\tan^2\vartheta - n\gamma\cos\varphi\tan\vartheta + 1 - n = 0.$$

Since $0 \le \vartheta \le \pi/2$, we find that the nonnegative root of this equation is

$$\vartheta_\gamma(\varphi) = \arctan\frac{h_\gamma(\varphi)}{2} , \tag{6.3.79}$$

where

$$h_\gamma(\varphi) = n\gamma\cos\varphi + \Big(4(n-1) + n^2\gamma^2\cos^2\varphi\Big)^{1/2} . \tag{6.3.80}$$

Taking into account that the function $\mathcal{G}_n(\varphi, \vartheta; 1, \gamma)$ is nonnegative for $0 \le \vartheta \le \vartheta_\gamma(\varphi)$, $0 \le \varphi \le \pi$, and using the equalities

$$\int_0^\vartheta \mathcal{G}_n(\varphi, \vartheta; 1, \gamma) \sin^{n-2} \vartheta d\vartheta = \left[\cos\vartheta + \gamma \cos\varphi \sin\vartheta\right] \sin^{n-1}\vartheta \ ,$$

$$\int_0^\pi \sin^{n-3}\varphi \, d\varphi \int_0^{\pi/2} \mathcal{G}_n(\varphi, \vartheta; 1, \gamma) \sin^{n-2}\vartheta \, d\vartheta = \gamma \int_0^\pi \sin^{n-3}\varphi \cos\varphi \, d\varphi = 0,$$

we write (6.3.77) as

$$C_\infty = \sup_{\gamma \ge 0} \frac{4\omega_{n-2}}{\omega_n \sqrt{1+\gamma^2}} \int_0^\pi \sin^{n-3}\varphi \, d\varphi \int_0^{\vartheta_\gamma(\varphi)} \mathcal{G}_n(\varphi, \vartheta; 1, \gamma) \sin^{n-2}\vartheta d\vartheta$$

(6.3.81)

$$= \sup_{\gamma \ge 0} \frac{4\omega_{n-2}}{\omega_n \sqrt{1+\gamma^2}} \int_0^\pi \left[\cos\vartheta_\gamma(\varphi) + \gamma \cos\varphi \sin\vartheta_\gamma(\varphi)\right] \sin^{n-1}\vartheta_\gamma(\varphi) \sin^{n-3}\varphi d\varphi.$$

By (6.3.79),

$$\sin\vartheta_\gamma(\varphi) = \frac{h_\gamma(\varphi)}{\sqrt{4 + h_\gamma^2(\varphi)}} \ , \tag{6.3.82}$$

$$\cos\vartheta_\gamma(\varphi) = \frac{2}{\sqrt{4 + h_\gamma^2(\varphi)}} \ . \tag{6.3.83}$$

By (6.3.82) and (6.3.83), we find

$$\cos\vartheta_\gamma(\varphi) + \gamma \cos\varphi \sin\vartheta_\gamma(\varphi) = \frac{2 + \gamma h_\gamma(\varphi)\cos\varphi}{\sqrt{4 + h_\gamma^2(\varphi)}} \ . \tag{6.3.84}$$

Using (6.3.82), (6.3.84), and the identity

$$2 + \gamma h_\gamma(\varphi) \cos\varphi = \frac{4 + h_\gamma^2(\varphi)}{2n} \ ,$$

where $h_\gamma(\varphi)$ is defined by (6.3.80), we can write (6.3.81) as

$$C_\infty = \sup_{\gamma \ge 0} \frac{2\omega_{n-2}}{n\omega_n \sqrt{1+\gamma^2}} \int_0^\pi \frac{h_\gamma^{n-1}(\varphi)}{\left(4 + h_\gamma^2(\varphi)\right)^{(n-2)/2}} \sin^{n-3}\varphi \, d\varphi.$$

Introducing the parameter

$$\alpha = \frac{n\gamma}{2\sqrt{n-1}},$$

and taking into account (6.3.80), we obtain

$$C_\infty = \sup_{\alpha \ge 0} \frac{4\omega_{n-2}(n-1)^{(n-1)/2}}{\omega_n \sqrt{n^2 + 4(n-1)\alpha^2}} \int_0^\pi \mathcal{P}_n(\alpha\cos\varphi) \sin^{n-3}\varphi \, d\varphi \tag{6.3.85}$$

with $\mathcal{P}_n$ defined by (6.3.59). The change of variable $t = \cos\varphi$ in (6.3.85) implies

$$C_\infty = \sup_{\alpha \ge 0} \frac{4\omega_{n-2}(n-1)^{(n-1)/2}}{\omega_n \sqrt{n^2 + 4(n-1)\alpha^2}} \int_{-1}^1 \mathcal{P}_n(\alpha t)(1-t^2)^{(n-4)/2} \, dt. \tag{6.3.86}$$

Integrating in (6.3.86) over $(-1,0)$ and $(0,1)$, we have

$$C_\infty = \sup_{\alpha\geq 0} \frac{4\omega_{n-2}(n-1)^{(n-1)/2}}{\omega_n\sqrt{n^2+4(n-1)\alpha^2}} \int_0^1 \frac{\mathcal{P}_n(\alpha t)+\mathcal{P}_n(-\alpha t)}{(1-t^2)^{(4-n)/2}}\, dt.$$

Applying the Cauchy inequality, we see that

$$C_\infty \leq \sup_{\alpha\geq 0} \frac{A_n}{\sqrt{n^2+4(n-1)\alpha^2}} \left\{ \int_0^1 \frac{\big(\mathcal{P}_n(\alpha t)+\mathcal{P}_n(-\alpha t)\big)^2}{(1-t^2)^{(4-n)/2}}\, dt \right\}^{1/2}, \tag{6.3.87}$$

where

$$A_n = \frac{4\omega_{n-2}(n-1)^{(n-1)/2}}{\omega_n} \left\{ \int_0^1 \frac{dt}{(1-t^2)^{(4-n)/2}} \right\}^{1/2}. \tag{6.3.88}$$

By (6.3.59),

$$\mathcal{P}_n(y)\mathcal{P}_n(-y) = \Big(4(n-1)y^2+n^2\Big)^{(2-n)/2},$$

which implies

$$\mathcal{P}_n(y)\mathcal{P}_n(-y) \leq n^{2-n}. \tag{6.3.89}$$

Combining (6.3.89) and (6.3.58), we obtain

$$\big(\mathcal{P}_n(y)+\mathcal{P}_n(-y)\big)^2 \leq \mathcal{P}_n^2(y)+\mathcal{P}_n^2(-y)+2n^{2-n} \leq \frac{2n^2+4(n-1)(3n-2)y^2}{n^n}+\frac{2}{n^{n-2}}.$$

Therefore,

$$\big(\mathcal{P}_n(\alpha t)+\mathcal{P}_n(-\alpha t)\big)^2 \leq \frac{4}{n^{n-2}}\left(1+\frac{(n-1)(3n-2)}{n^2}\,\alpha^2 t^2\right). \tag{6.3.90}$$

By (6.3.87), (6.3.88), (6.3.90) and by

$$\int_0^1 (1-t^2)^{(n-4)/2}\, dt = \frac{\sqrt{\pi}\,\Gamma\left(\frac{n-2}{2}\right)}{2\Gamma\left(\frac{n-1}{2}\right)},$$

$$\int_0^1 t^2(1-t^2)^{(n-4)/2}\, dt = \frac{\sqrt{\pi}\,\Gamma\left(\frac{n-2}{2}\right)}{2(n-1)\Gamma\left(\frac{n-1}{2}\right)},$$

we find

$$C_\infty \leq \frac{4\omega_{n-2}\sqrt{\pi}\,(n-1)^{(n-1)/2}\Gamma\left(\frac{n-2}{2}\right)}{\omega_n n^{n/2}\Gamma\left(\frac{n-1}{2}\right)} \sup_{\alpha\geq 0}\left(\frac{n^2+(3n-2)\alpha^2}{n^2+4(n-1)\alpha^2}\right)^{1/2}. \tag{6.3.91}$$

Note that

$$\frac{d}{d\alpha}\left(\frac{n^2+(3n-2)\alpha^2}{n^2+4(n-1)\alpha^2}\right) = -\frac{2\alpha(n-2)n^2}{\big(n^2+4(n-1)\alpha^2\big)^2} < 0 \ \text{ for } \ \alpha>0,$$

therefore the supremum in α on the right-hand side of (6.3.91) is attained for $\alpha=0$. Thus,

$$C_\infty \leq \frac{4\omega_{n-2}\sqrt{\pi}\,(n-1)^{(n-1)/2}\Gamma\left(\frac{n-2}{2}\right)}{\omega_n n^{n/2}\Gamma\left(\frac{n-1}{2}\right)} = \frac{4(n-1)^{(n-1)/2}}{n^{n/2}}\frac{\omega_{n-1}}{\omega_n}. \tag{6.3.92}$$

Furthermore, in view of (6.3.85) and (6.3.59),

$$\begin{aligned} C_\infty &\geq \frac{4\omega_{n-2}(n-1)^{(n-1)/2}}{n\omega_n}\int_0^\pi \mathcal{P}_n(0)\sin^{n-3}\varphi\, d\varphi \\ &= \frac{4\omega_{n-2}(n-1)^{(n-1)/2}}{n\omega_n}\int_0^\pi \frac{\sin^{n-3}\varphi}{n^{(n-2)/2}}\, d\varphi \\ &= \frac{4\omega_{n-2}\sqrt{\pi}\,(n-1)^{(n-1)/2}\Gamma\left(\frac{n-2}{2}\right)}{\omega_n n^{n/2}\Gamma\left(\frac{n-1}{2}\right)} = \frac{4(n-1)^{(n-1)/2}\,\omega_{n-1}}{n^{n/2}\,\omega_n}, \end{aligned}$$

which together with (6.3.73) and (6.3.92) proves (6.3.76). □

REMARK 6.6. Inequality (6.3.75) can be written in the form

$$|\nabla u(x)| \leq \mathcal{C}_\infty(x) \sup_{y\in\mathbb{R}^n_+} \big|u(y)\big|, \tag{6.3.93}$$

where $\mathcal{C}_\infty(x) = C_\infty/x_n$. Using (6.3.76), we arrive at the explicit sharp inequality

$$|\nabla u(x)| \leq \frac{4(n-1)^{(n-1)/2}\,\omega_{n-1}}{n^{n/2}\,\omega_n x_n} \sup_{y\in\mathbb{R}^n_+} \big|u(y)\big|\,,$$

which generalizes the real value analog (6.6.12) of (6.6.10) to harmonic functions in the n-dimensional half-space.

From (6.3.93) it follows that

$$|\nabla u(x)| \leq \mathcal{C}_\infty(x) \sup_{y\in\mathbb{R}^n_+} \big|u(y)-\omega\big| \tag{6.3.94}$$

with an arbitrary constant ω. Minimizing (6.3.94) in ω, we obtain

$$|\nabla u(x)| \leq \frac{\mathcal{C}_\infty(x)}{2}\,\mathrm{osc}_{\mathbb{R}^n_+}(u), \tag{6.3.95}$$

where $\mathrm{osc}_{\mathbb{R}^n_+}(u)$ is the oscillation of u on $\mathbb{R}^n_+$.

Relations (6.3.76) and (6.3.95) imply the sharp estimate

$$|\nabla u(x)| \leq \frac{2(n-1)^{(n-1)/2}\,\omega_{n-1}}{n^{n/2}\,\omega_n x_n}\,\mathrm{osc}_{\mathbb{R}^n_+}(u),$$

which is an analogue of (6.6.11) for harmonic functions in $\mathbb{R}^n_+$.

6.4. Sharp estimates for directional derivatives and Khavinson's type extremal problem in a ball with boundary data from L^p

6.4.1. Representation for $\mathcal{K}_p(x;\boldsymbol{\ell})$ as an integral over $\mathbb{S}^{n-1}$

We introduce some notation used henceforth. Let $|\cdot|$ be the length of a vector in the n-dimensional Euclidean space $\mathbb{R}^n$. Let $\mathbb{B} = \big\{x\in\mathbb{R}^n : |x|<1\big\}$, and $\mathbb{S}^{n-1} = \{x\in\mathbb{R}^n : |x|=1\}$.

By $\|\cdot\|_p$ we denote the norm in the space $L^p(\mathbb{S}^{n-1})$, that is

$$\|f\|_p = \left\{\int_{\mathbb{S}^{n-1}} |f(x)|^p d\sigma_x\right\}^{1/p},$$

if $1\leq p<\infty$, and

$$\|f\|_\infty = \mathrm{ess\ sup}\{|f(x)| : x\in\mathbb{S}^{n-1}\}.$$

Next, by $h^p(\mathbb{B})$ we denote the Hardy space of harmonic functions on $\mathbb{B}$, which can be represented as the Poisson integral

$$u(x) = \frac{1}{\omega_n} \int_{\mathbb{S}^{n-1}} \frac{1-r^2}{|y-x|^n} u(y) d\sigma_y \tag{6.4.1}$$

with boundary values in $L^p(\mathbb{S}^{n-1})$, $1 \le p \le \infty$, where $r = |x| < 1$.

In what follows, we assume that the Cartesian coordinates with origin $\mathcal{O}$ at the center of the ball are chosen in such a way that $x = r\boldsymbol{e}_n$. By $\boldsymbol{\ell}$ we denote an arbitrary unit vector in $\mathbb{R}^n$ and by $\boldsymbol{\nu}_x$ we mean the unit vector of exterior normal to the sphere $|x| = r$ at a point x. Let $\boldsymbol{\ell}_\tau$ be the orthogonal projection of $\boldsymbol{\ell}$ on the tangent hyperplane to the sphere $|x| = r$ at x. If $\boldsymbol{\ell}_\tau \ne \mathbf{0}$, we set $\boldsymbol{\tau}_x = \boldsymbol{\ell}_\tau / |\boldsymbol{\ell}_\tau|$, otherwise $\boldsymbol{\tau}_x$ is an arbitrary unit vector tangent to the sphere $|x| = r$ at x. Hence $\boldsymbol{\ell} = \ell_\tau \boldsymbol{\tau}_x + \ell_\nu \boldsymbol{\nu}_x$, where $\ell_\tau = |\boldsymbol{\ell}_\tau|$ and $\ell_\nu = (\boldsymbol{\ell}, \boldsymbol{\nu}_x)$.

Now, we find a representation for the best coefficient $\mathcal{K}_p(x; \boldsymbol{\ell})$ in the inequality for the absolute value of derivative of $u \in h^p(\mathbb{B})$ in an arbitrary direction $\boldsymbol{\ell}$ at a point $x \in \mathbb{B}$. In particular, we obtain a formula for the constant in a similar inequality for the modulus of the gradient.

THEOREM 6.2. *Let $u \in h^p(\mathbb{B})$, and let x be an arbitrary point in $\mathbb{B}$. The sharp coefficient $\mathcal{K}_p(x; \boldsymbol{\ell})$ in*

$$|(\nabla u(x), \boldsymbol{\ell})| \le \mathcal{K}_p(x; \boldsymbol{\ell}) \|u\|_p \tag{6.4.2}$$

is given by

$$\mathcal{K}_p(x; \boldsymbol{\ell}) = \frac{N(r)}{\omega_n} \left\{ \int_{\mathbb{S}^{n-1}} \frac{\left|\left(\mathcal{A}(r)\left(y - \lambda(r)x\right), \ \boldsymbol{\ell}\right)\right|^q}{|y-x|^{(n+2)q}} d\sigma_y \right\}^{1/q} . \tag{6.4.3}$$

Here $1/p + 1/q = 1$,

$$N(r) = n\left(1 - r^2\right) + 4r^2, \qquad \lambda(r) = 1 + 2d(r), \tag{6.4.4}$$

and $\mathcal{A}(r)$ is the $(n \times n)$-matrix of the form

$$\mathcal{A}(r) = \operatorname{diag}\{nd(r), \ \dots, nd(r), \ 1\} , \tag{6.4.5}$$

where

$$d(r) = \frac{1-r^2}{n\left(1-r^2\right) + 4r^2} . \tag{6.4.6}$$

In particular, the sharp coefficient $\mathcal{K}_p(x)$ in

$$|\nabla u(x)| \le \mathcal{K}_p(x) \|u\|_p \tag{6.4.7}$$

is given by

$$\mathcal{K}_p(x) = \max_{|\boldsymbol{\ell}|=1} \mathcal{K}_p(x; \boldsymbol{\ell}) . \tag{6.4.8}$$

PROOF. Fix a point $x \in \mathbb{B}$. By (6.4.1) we have

$$\frac{\partial u}{\partial x_i} = \frac{1}{\omega_n} \int_{\mathbb{S}^{n-1}} \left[\frac{-2x_i}{|y-x|^n} + \frac{n\left(1-r^2\right)\left(y_i - x_i\right)}{|y-x|^{n+2}} \right] u(y) d\sigma_y,$$

that is

$$\nabla u(x) = \frac{1}{\omega_n} \int_{\mathbb{S}^{n-1}} \frac{n\left(1-r^2\right)(y-x) - 2|y-x|^2 x}{|y-x|^{n+2}} u(y) d\sigma_y. \tag{6.4.9}$$

Since $y-x=y-r\boldsymbol{e}_n=\big(y-y_n\boldsymbol{e}_n\big)+\big(y_n-r\big)\boldsymbol{e}_n$, we can express the numerator of the integrand in (6.4.9) as the sum

$$\begin{aligned} n\left(1-r^2\right)(y-x)-2|y-x|^2x &= n\left(1-r^2\right)\left(y-y_n\boldsymbol{e}_n\right)\\ &+ \left(n\left(1-r^2\right)\left(y_n-r\right)-2r|y-x|^2\right)\boldsymbol{e}_n\,. \end{aligned} \tag{6.4.10}$$

Introducing the notation

$$T(r)=n\left(1-r^2\right), \qquad N(r)=n\left(1-r^2\right)+4r^2\,, \tag{6.4.11}$$

and

$$\lambda(r)=\frac{n\left(1-r^2\right)+2(1+r^2)}{n\left(1-r^2\right)+4r^2}\,, \tag{6.4.12}$$

using (6.4.10) and the equality

$$n\left(1-r^2\right)\left(y_n-r\right)-2r|y-x|^2=N(r)\big(y_n-\lambda(r)r\big)\,,$$

we write (6.4.9) as

$$\nabla u(x)=\frac{1}{\omega_n}\int_{\mathbb{S}^{n-1}}\frac{T(r)\big(y-y_n\boldsymbol{e}_n\big)+N(r)\big(y_n-\lambda(r)r\big)\boldsymbol{e}_n}{|y-x|^{n+2}}\,u(y)d\sigma_y. \tag{6.4.13}$$

Since the vectors $y-y_n\boldsymbol{e}_n$ and $\boldsymbol{e}_n$ are orthogonal and $x=r\boldsymbol{e}_n$, it follows that

$$y-y_n\boldsymbol{e}_n=\sum_{i=1}^{n-1}\big(y-y_n\boldsymbol{e}_n,\ \boldsymbol{e}_i\big)\boldsymbol{e}_i=\sum_{i=1}^{n-1}\big(y-\lambda(r)x,\ \boldsymbol{e}_i\big)\boldsymbol{e}_i\,,$$

which together with

$$\big(y_n-\lambda(r)r\big)\boldsymbol{e}_n=\big(y-\lambda(r)x,\ \boldsymbol{e}_n\big)\boldsymbol{e}_n$$

implies (6.4.13) written in the form

$$\nabla u(x)=\frac{1}{\omega_n}\int_{\mathbb{S}^{n-1}}\frac{\Lambda(r)\big(y-\lambda(r)x\big)}{|y-x|^{n+2}}\,u(y)d\sigma_y\,, \tag{6.4.14}$$

where the $(n\times n)$-matrix $\Lambda(r)$ is defined by

$$\Lambda(r)=\operatorname{diag}\{T(r),\ \ldots,\ T(r),\ N(r)\}\,. \tag{6.4.15}$$

Using (6.4.11), (6.4.15) as well as notation introduced in the theorem, we can write (6.4.14) as

$$\nabla u(x)=\frac{N(r)}{\omega_n}\int_{\mathbb{S}^{n-1}}\frac{\mathcal{A}(r)\big(y-\lambda(r)x\big)}{|y-x|^{n+2}}\,u(y)d\sigma_y\,. \tag{6.4.16}$$

Thus,

$$(\nabla u(x),\boldsymbol{\ell})=\frac{N(r)}{\omega_n}\int_{\mathbb{S}^{n-1}}\frac{\big(\mathcal{A}(r)\big(y-\lambda(r)x\big),\ \boldsymbol{\ell}\big)}{|y-x|^{n+2}}\,u(y)d\sigma_y\,, \tag{6.4.17}$$

and (6.4.3) follows.

By (6.4.17) we have

$$\big|\nabla u(x)\big|=\frac{N(r)}{\omega_n}\max_{|\boldsymbol{\ell}|=1}\int_{\mathbb{S}^{n-1}}\frac{\big(\mathcal{A}(r)\big(y-\lambda(r)x\big),\ \boldsymbol{\ell}\big)}{|y-x|^{n+2}}\,u(y)d\sigma_y\,.$$

Since the suprema commute, this implies that the sharp constant in (6.4.7) has the form

$$\mathcal{K}_p(x) = \frac{N(r)}{\omega_n} \max_{|\boldsymbol{\ell}|=1} \left\{ \int_{\mathbb{S}^{n-1}} \frac{\left|\left(\mathcal{A}(r)\big(y-\lambda(r)x\big),\ \boldsymbol{\ell}\right)\right|^q}{|y-x|^{(n+2)q}}\, d\sigma_y \right\}^{1/q}. \tag{6.4.18}$$

Taking into account (6.4.3) and (6.4.18), we arrive at (6.4.8). □

6.4.2. The case $p = 1$ and the formula for $\mathcal{K}_p(0)$

The next assertion concerns sharp coefficient in inequality (6.4.2) with $p = 1$ for the cases $\boldsymbol{\ell} = \boldsymbol{\nu}_x$ and $\boldsymbol{\ell} = \boldsymbol{\tau}_x$. It is shown that $\mathcal{K}_1(x) = \mathcal{K}_1(x;\boldsymbol{\nu}_x)$, that is the radial direction is extremal for $p = 1$.

COROLLARY 6.7. *Let $u \in h^1(\mathbb{B})$, and let x be an arbitrary point in $\mathbb{B}$. The sharp coefficient $\mathcal{K}_1(x;\boldsymbol{\ell})$ in*

$$|(\nabla u(x), \boldsymbol{\ell})| \le \mathcal{K}_1(x;\boldsymbol{\ell})\|u\|_1 \tag{6.4.19}$$

is given by

$$\mathcal{K}_1(x;\boldsymbol{\ell}) = \frac{N(r)}{\omega_n} \max_{\alpha} \frac{\left|nd(r)\ell_\tau \sin\alpha + \big(\cos\alpha - \lambda(r)r\big)\ell_\nu\right|}{\big(1-2r\cos\alpha + r^2\big)^{(n+2)/2}}, \tag{6.4.20}$$

where $\boldsymbol{\ell} = \ell_\tau \boldsymbol{\tau}_x + \ell_\nu \boldsymbol{\nu}_x$.

In particular,

$$\mathcal{K}_1(x) = \mathcal{K}_1(x;\boldsymbol{\nu}_x) = \frac{n+(n-2)r}{\omega_n(1-r)^n}, \tag{6.4.21}$$

and

$$\mathcal{K}_1(x;\boldsymbol{\tau}_x) = \frac{n(1-r^2)(1-\rho^2(r))^{1/2}}{\omega_n(1-2r\rho(r)+r^2)^{(n+2)/2}}, \tag{6.4.22}$$

where

$$\rho(r) = \frac{2(n+2)r}{1+r^2+\big((1+r^2)^2+4n(n+2)r^2\big)^{1/2}}.$$

PROOF. Passing to the limit as $q \to \infty$ in the right-hand side of (6.4.18), and using the diagonality of the matrix $\mathcal{A}$, we find

$$\begin{aligned} \mathcal{K}_1(x,\ \boldsymbol{\ell}) &= \frac{N(r)}{\omega_n} \max_{|y|=1} \frac{\left|\mathcal{A}(r)\big(y-\lambda(r)x\big),\ \boldsymbol{\ell}\right)\right|}{|y-x|^{n+2}} \\ &= \frac{N(r)}{\omega_n} \max_{|y|=1} \frac{\left|\big(y-\lambda(r)x\big),\ \mathcal{A}(r)\boldsymbol{\ell}\right)\right|}{|y-x|^{n+2}}. \end{aligned} \tag{6.4.23}$$

Taking into account (6.4.5) and the choice of the coordinate system, we obtain

$$\mathcal{A}(r)\boldsymbol{\ell} = \mathcal{A}(r)\big(\ell_\tau\boldsymbol{\tau}_x + \ell_\nu\boldsymbol{\nu}_x\big) = nd(r)\ell_\tau\boldsymbol{\tau}_x + \ell_\nu\boldsymbol{\nu}_x, \tag{6.4.24}$$

which together with (6.4.23) and the orthogonality of $x = r\boldsymbol{e}_n$ and $\boldsymbol{\tau}_x$ implies

$$\mathcal{K}_1(x;\boldsymbol{\ell}) = \frac{N(r)}{\omega_n} \max_{|y|=1} \frac{\left|nd(r)\big(y,\ \boldsymbol{\tau}_x\big)\ell_\tau + \big((y,\ \boldsymbol{\nu}_x) - \lambda(r)r\big)\ell_\nu\right|}{|y-x|^{n+2}}. \tag{6.4.25}$$

Let α stand for the angle between y and $\boldsymbol{\nu}_x$, and let $y_\tau = y - y_n\boldsymbol{e}_n$. By β we denote the angle between y_τ and $\boldsymbol{\tau}_x$. We have

$$(y, \boldsymbol{\tau}_x) = (y_\tau, \boldsymbol{\tau}_x) = |y_\tau|\cos\beta = \sin\alpha\cos\beta,$$

which allows to write (6.4.25) as

$$\mathcal{K}_1(x;\boldsymbol{\ell})=\frac{N(r)}{\omega_n}\max_{\alpha\in[0,\pi]}\max_{\beta\in[0,2\pi]}\frac{\big|nd(r)\ell_\tau\sin\alpha\cos\beta+\big(\cos\alpha-\lambda(r)r\big)\ell_\nu\big|}{\big(1-2r\cos\alpha+r^2\big)^{(n+2)/2}}\,.$$

Since the maximum in β is attained either at $\beta=0$ or at $\beta=\pi$, we arrive at

$$\mathcal{K}_1(x;\boldsymbol{\ell})=\frac{N(r)}{\omega_n}\max_{\alpha\in[0,2\pi]}\frac{\big|nd(r)\ell_\tau\sin\alpha+\big(\cos\alpha-\lambda(r)r\big)\ell_\nu\big|}{\big(1-2r\cos\alpha+r^2\big)^{(n+2)/2}}\,,$$

which proves (6.4.20).

Next we prove (6.4.21). Using (6.4.20) together with (6.4.4) and (6.4.6), we find

$$\mathcal{K}_1(x)\ \ge\ \mathcal{K}_1(x;\,\boldsymbol{\nu}_x)=\frac{N(r)}{\omega_n}\max_{\alpha\in[0,2\pi]}\frac{\big|\cos\alpha-\lambda(r)r\big|}{\big(1-2r\cos\alpha+r^2\big)^{(n+2)/2}}$$

$$\ge\ \frac{N(r)}{\omega_n}\frac{\big|1-\lambda(r)r\big|}{\big(1-r\big)^{n+2}}=\frac{n+(n-2)r}{\omega_n(1-r)^n}\,. \tag{6.4.26}$$

Now we obtain an upper estimate for $\mathcal{K}_1(x)$. Taking into account (6.4.20) and the equality $\ell_\tau^2+\ell_\nu^2=1$, we obtain the estimate

$$\mathcal{K}_1(x;\boldsymbol{\ell})\le\frac{N(r)}{\omega_n}\max_{\alpha\in[0,2\pi]}\left\{\frac{n^2d^2(r)\sin^2\alpha+(\cos\alpha-\lambda(r)r)^2}{\big(1-2r\cos\alpha+r^2\big)^{n+2}}\right\}^{1/2},$$

which is the same as

$$\mathcal{K}_1(x;\boldsymbol{\ell})\le\frac{N(r)}{\omega_n}\max_{|t|\le1}\left\{\frac{n^2d^2(r)(1-t^2)+(t-\lambda(r)r)^2}{\big(1-2rt+r^2\big)^{n+2}}\right\}^{1/2}$$

as $t=\cos\alpha$. Thus,

$$\mathcal{K}_1(x;\boldsymbol{\ell})\le\frac{N(r)}{\omega_n(1-r)^n}\max_{|t|\le1}\left\{\frac{n^2d^2(r)(1-t^2)+(t-\lambda(r)r)^2}{\big(1-2rt+r^2\big)^2}\right\}^{1/2}. \tag{6.4.27}$$

Let

$$f(t)=\frac{n^2d^2(r)\big(1-t^2\big)+\big(t-\lambda(r)r\big)^2}{\big(1-2rt+r^2\big)^2}\,.$$

Since

$$f'(t)=\frac{2n(n-2)r(1-r^2)^2}{\big(n\,(1-r^2)+4r^2\big)^2\big(1-2rt+r^2\big)^2}\ge0\,,$$

we have

$$\max_{|t|\le1}f(t)=f(1)=\frac{\big(n+(n-2)r\big)^2}{\big(n\,(1-r^2)+4r^2\big)^2}=\frac{\big(n+(n-2)r\big)^2}{N^2(r)}.$$

This and (6.4.27) imply

$$\mathcal{K}_1(x)\le\frac{n+(n-2)r}{\omega_n(1-r)^n}\,,$$

which together with (6.4.26) proves (6.4.21).

Now, by (6.4.4), (6.4.6) and (6.4.20) we obtain

$$\mathcal{K}_1(x; \boldsymbol{\tau}_x) = \frac{n(1-r^2)}{\omega_n} \max_{|t|\le 1} \frac{\sqrt{1-t^2}}{(1-2rt+r^2)^{(n+2)/2}} .$$

Since the maximum in t is attained at

$$t = \frac{2(n+2)r}{1+r^2+\left((1+r^2)^2+4n(n+2)r^2\right)^{1/2}} ,$$

we arrive at (6.4.22). □

COROLLARY 6.8. *Let $u \in h^p(\mathbb{B})$. The sharp coefficient $\mathcal{K}_p(0)$ in*

$$|\nabla u(0)| \le \mathcal{K}_p(0) \|u\|_p$$

is given by

$$\mathcal{K}_1(0) = \frac{n}{\omega_n} , \tag{6.4.28}$$

and

$$\mathcal{K}_p(0) = \frac{n}{\omega_n} \left\{ \frac{2\pi^{(n-1)/2}\Gamma\left(\frac{q+1}{2}\right)}{\Gamma\left(\frac{n+q}{2}\right)} \right\}^{1/q} \tag{6.4.29}$$

for $1 < p \le \infty$.

In particular,

$$\mathcal{K}_2(0) = \left(\frac{n}{\omega_n}\right)^{1/2}, \qquad \mathcal{K}_\infty(0) = \frac{2n\omega_{n-1}}{(n-1)\omega_n} .$$

PROOF. Equality (6.4.28) follows from (6.4.21) with $r = 0$. By (6.4.3) and (6.4.8),

$$\mathcal{K}_p(0) = \frac{n}{\omega_n} \max_{|\boldsymbol{z}|=1} \left\{ \int_{\mathbb{S}^{n-1}} |(y, \boldsymbol{\ell})|^q \, d\sigma_y \right\}^{1/q}$$

with $1 < p \le \infty$. Since the integral does not depend on $\boldsymbol{\ell}$, we find

$$\mathcal{K}_p(0) = \frac{n}{\omega_n} \left\{ \int_{\mathbb{S}^{n-1}} |(y, \boldsymbol{e}_n)|^q \, d\sigma_y \right\}^{1/q} . \tag{6.4.30}$$

Using spherical coordinates in (6.4.30), we obtain

$$\begin{aligned}
\mathcal{K}_p(0) &= \frac{n}{\omega_n} \left\{ \int_0^\pi \cdots \int_0^\pi \int_0^{2\pi} |\cos\vartheta_1|^q \sin^{n-2}\vartheta_1 \ldots \sin\vartheta_{n-2} d\vartheta_1 \ldots d\vartheta_{n-2} d\vartheta_{n-1} \right\}^{1/q} \\
&= \frac{n(2\omega_{n-1})^{1/q}}{\omega_n} \left\{ \int_0^{\pi/2} \cos^q\vartheta_1 \sin^{n-2}\vartheta_1 \, d\vartheta \right\}^{1/q} \\
&= \frac{n(\omega_{n-1})^{1/q}}{\omega_n} B^{1/q}\left(\frac{q+1}{2}, \frac{n-1}{2}\right) ,
\end{aligned}$$

which implies (6.4.29). □

6.4.3. Representations for $\mathcal{K}_p(x;\boldsymbol{\ell})$ as double integrals

In the next assertion we reduce the evaluation of $\mathcal{K}_p(x;\boldsymbol{\ell})$, found as an integral over $\mathbb{S}^{n-1}$, to a double integral.

PROPOSITION 6.4. *Let $u \in h^p(\mathbb{B})$, $1 < p \le \infty$, and let x be an arbitrary point in $\mathbb{B}$. The sharp coefficient $\mathcal{K}_p(x;\boldsymbol{\ell})$ in the inequality*

$$|(\nabla u(x),\boldsymbol{\ell})| \le \mathcal{K}_p(x;\boldsymbol{\ell})\|u\|_p \tag{6.4.31}$$

is given by

$$\mathcal{K}_p(x;\boldsymbol{\ell}) = \frac{N(r)(\omega_{n-2})^{1/q}}{\omega_n}\left\{\int_0^\pi \sin^{n-2}\vartheta d\vartheta \int_0^\pi \mathcal{F}_{n,q}(\vartheta,\varphi;\boldsymbol{\ell},r)\sin^{n-3}\varphi d\varphi\right\}^{1/q},$$

where $1/p+1/q=1$, $\boldsymbol{\ell} = \ell_\tau \boldsymbol{\tau}_x + \ell_\nu \boldsymbol{\nu}_x$,

$$\mathcal{F}_{n,q}(\vartheta,\varphi;r,\boldsymbol{\ell}) = \frac{\left|nd(r)\ell_\tau \sin\vartheta\cos\varphi + (\cos\vartheta - \lambda(r)r)\ell_\nu\right|^q}{\left(1-2r\cos\vartheta + r^2\right)^{(n+2)q/2}}, \tag{6.4.32}$$

with $N(r)$, $\lambda(r) = 1+2d(r)$ and $d(r)$ defined by (6.4.4) and (6.4.6).

In particular,

$$\mathcal{K}_p(x;\boldsymbol{\nu}_x) = \frac{N(r)(\omega_{n-1})^{1/q}}{\omega_n}\left\{\int_0^\pi \frac{|\cos\vartheta - \lambda(r)r|^q \sin^{n-2}\vartheta}{\left(1-2r\cos\vartheta + r^2\right)^{(n+2)q/2}} d\vartheta\right\}^{1/q} \tag{6.4.33}$$

and

$$\mathcal{K}_p(x;\boldsymbol{\tau}_x) = \mathcal{K}_p(0)(1-r^2)\left\{F\left(\frac{n(q-1)+q+2}{2}, \frac{(n+2)q}{2}; \frac{n+q}{2}; r^2\right)\right\}^{1/q}. \tag{6.4.34}$$

Here $F(a,b;c;x)$ is the hypergeometric Gauss function, and $\mathcal{K}_p(0)$ is defined by (6.4.29).

PROOF. Using (6.4.24) and diagonality of the matrix $\mathcal{A}$, we write (6.4.3) as

$$\mathcal{K}_p(x;\boldsymbol{\ell}) = \frac{N(r)}{\omega_n}\left\{\int_{\mathbb{S}^{n-1}} \frac{\left|nd(r)(y,\ \boldsymbol{\tau}_x)\ell_\tau + ((y,\ \boldsymbol{\nu}_x) - \lambda(r)r)\ell_\nu\right|^q}{\left|y-x\right|^{(n+2)q/2}} d\sigma_y\right\}^{1/q}$$

$$= \frac{N(r)}{\omega_n}\left\{\int_{\mathbb{S}^{n-1}} \frac{\left|nd(r)(y_\tau,\ \boldsymbol{\tau}_x)\ell_\tau + (y_n - \lambda(r)r)\ell_\nu\right|^q}{\left(1-2ry_n+r^2\right)^{(n+2)q/2}} d\sigma_y\right\}^{1/q}, \tag{6.4.35}$$

where $y_\tau = (y_1,\dots,y_{n-1},0)$.

Introducing the function

$$\mathcal{H}_{n,q}(s,t;r,\boldsymbol{\ell}) = \frac{\left|nd(r)s\ell_\tau + (t-\lambda(r)r)\ell_\nu\right|^q}{\left(1-2rt+r^2\right)^{(n+2)q/2}}, \tag{6.4.36}$$

we write the integral in (6.4.35) as the sum

$$\int_{\mathbb{S}_+^{n-1}} \mathcal{H}_{n,q}((y_\tau,\boldsymbol{\tau}_x),y_n;r,\boldsymbol{\ell})\,d\sigma_y + \int_{\mathbb{S}_-^{n-1}} \mathcal{H}_{n,q}((y_\tau,\boldsymbol{\tau}_x),y_n;r,\boldsymbol{\ell})\,d\sigma_y, \tag{6.4.37}$$

where $\mathbb{S}_+^{n-1} = \{y\in\mathbb{S}^{n-1} : (y,\boldsymbol{e}_n) > 0\}$, $\mathbb{S}_-^{n-1} = \{y\in\mathbb{S}^{n-1} : (y,\boldsymbol{e}_n) < 0\}$.

Let $y' = (y_1, \dots, y_{n-1}) \in \mathbb{B}' = \{y' \in \mathbb{R}^{n-1} : |y'| < 1\}$. We put

$$\boldsymbol{\tau}_x' = \sum_{i=1}^{n-1} (\boldsymbol{\tau}_x, \boldsymbol{e}_i)\boldsymbol{e}_i.$$

Since $y_n = \sqrt{1-|y'|^2}$ for $y \in \mathbb{S}_+^{n-1}$ and $y_n = -\sqrt{1-|y'|^2}$ for $y \in \mathbb{S}_-^{n-1}$ and since $d\sigma_y = dy'/\sqrt{1-|y'|^2}$, each of integrals in (6.4.37) can be written as

$$\int_{\mathbb{S}_+^{n-1}} \mathcal{H}_{n,q}((y_\tau, \boldsymbol{\tau}_x), y_n; r, \boldsymbol{\ell}) d\sigma_y = \int_{\mathbb{B}'} \frac{\mathcal{H}_{n,q}\Big((y', \boldsymbol{\tau}_x'), \sqrt{1-|y'|^2}; r, \boldsymbol{\ell}\Big)}{\sqrt{1-|y'|^2}} dy', \tag{6.4.38}$$

$$\int_{\mathbb{S}_-^{n-1}} \mathcal{H}_{n,q}((y_\tau, \boldsymbol{\tau}_x), y_n; r, \boldsymbol{\ell}) d\sigma_y = \int_{\mathbb{B}'} \frac{\mathcal{H}_{n,q}\Big((y', \boldsymbol{\tau}_x'), -\sqrt{1-|y'|^2}; r, \boldsymbol{\ell}\Big)}{\sqrt{1-|y'|^2}} dy'. \tag{6.4.39}$$

Putting

$$\mathcal{M}_{n,q}(s, t; r, \boldsymbol{\ell}) = \mathcal{H}_{n,q}(s, t; r, \boldsymbol{\ell}) + \mathcal{H}_{n,q}(s, -t; r, \boldsymbol{\ell}), \tag{6.4.40}$$

and using (6.4.36)-(6.4.39), we write (6.4.35) as

$$\mathcal{K}_p(x; \boldsymbol{\ell}) = \frac{N(r)}{\omega_n} \left\{ \int_{\mathbb{B}'} \frac{\mathcal{M}_{n,q}\Big((y', \boldsymbol{\tau}_x'), \sqrt{1-|y'|^2}; r, \boldsymbol{\ell}\Big)}{\sqrt{1-|y'|^2}} dy' \right\}^{1/q}. \tag{6.4.41}$$

By the identity

$$\int_{\mathbb{B}} g((\boldsymbol{y}, \boldsymbol{\xi}),\ |\boldsymbol{y}|) dy = \omega_{n-1} \int_0^1 \rho^{n-1} d\rho \int_0^\pi g(|\boldsymbol{\xi}|\rho\cos\varphi,\ \rho) \sin^{n-2}\varphi\, d\varphi$$

(see, e.g., Prudnikov, Brychkov and Marichev [**PBM1**], **3.3.2(3)**), we express the integral in (6.4.41) as

$$\begin{aligned} &\int_{\mathbb{B}'} \frac{\mathcal{M}_{n,q}\Big((y', \boldsymbol{\tau}_x'), \sqrt{1-|y'|^2}; r, \boldsymbol{\ell}\Big)}{\sqrt{1-|y'|^2}} dy' \\ &= \omega_{n-2} \int_0^1 \frac{\rho^{n-2}}{\sqrt{1-\rho^2}} d\rho \int_0^\pi \mathcal{M}_{n,q}\Big(\rho\cos\varphi,\ \sqrt{1-\rho^2}; r, \boldsymbol{\ell}\Big) \sin^{n-3}\varphi d\varphi. \end{aligned} \tag{6.4.42}$$

The change of variable $\rho = \sin\vartheta$ in (6.4.42) gives

$$\begin{aligned} &\int_{\mathbb{B}'} \frac{\mathcal{M}_{n,q}\Big((y', \boldsymbol{\tau}_x'), \sqrt{1-|y'|^2}; r, \boldsymbol{\ell}\Big)}{\sqrt{1-|y'|^2}} dy' \\ &= \omega_{n-2} \int_0^{\pi/2} \sin^{n-2}\vartheta d\vartheta \int_0^\pi \mathcal{M}_{n,q}\Big(\sin\vartheta\cos\varphi,\ \cos\vartheta; r, \boldsymbol{\ell}\Big) \sin^{n-3}\varphi\, d\varphi\,. \end{aligned} \tag{6.4.43}$$

By (6.4.36), (6.4.40) and introducing the notation

$$\begin{aligned} \mathcal{F}_{n,q}(\vartheta, \varphi; r, \boldsymbol{\ell}) &= \mathcal{H}_{n,q}\Big(\sin\vartheta\cos\varphi,\ \cos\vartheta; r, \boldsymbol{\ell}\Big) \\ &= \frac{\big|nd(r)\ell_\tau \sin\vartheta\cos\varphi + \big(\cos\vartheta - \lambda(r)r\big)\ell_\nu\big|^q}{\big(1 - 2r\cos\vartheta + r^2\big)^{(n+2)q/2}}, \end{aligned}$$

we write (6.4.43) as

$$\int_{\mathbb{B}'} \frac{\mathcal{M}_{n,q}\Big((y', \boldsymbol{\tau}_x'), \sqrt{1-|y'|^2}; r, \boldsymbol{\ell}\Big)}{\sqrt{1-|y'|^2}}\, dy' \tag{6.4.44}$$

$$= \omega_{n-2}\int_0^{\pi/2} \sin^{n-2}\vartheta d\vartheta \int_0^{\pi}\Big(\mathcal{F}_{n,q}(\vartheta,\varphi;r,\boldsymbol{\ell}) + \mathcal{F}_{n,p}(\pi-\vartheta,\varphi;r,\boldsymbol{\ell})\Big)\sin^{n-3}\varphi\, d\varphi.$$

Changing the variable $\psi = \pi - \vartheta$, we obtain

$$\int_0^{\pi/2} \sin^{n-2}\vartheta\, d\vartheta \int_0^{\pi} \mathcal{F}_{n,q}(\pi-\vartheta,\varphi;r,\boldsymbol{\ell})\sin^{n-3}\varphi\, d\varphi$$

$$= \int_{\pi/2}^{\pi} \sin^{n-2}\psi\, d\psi \int_0^{\pi} \mathcal{F}_{n,q}(\psi,\varphi;r,\boldsymbol{\ell})\sin^{n-3}\varphi\, d\varphi,$$

which together with (6.4.44) leads to the representation of (6.4.41) as

$$\mathcal{K}_p(x;\boldsymbol{\ell}) = S_{n,q}(r)\left\{\int_0^{\pi} \sin^{n-2}\vartheta\, d\vartheta \int_0^{\pi} \mathcal{F}_{n,q}(\vartheta,\varphi;r,\boldsymbol{\ell})\sin^{n-3}\varphi\, d\varphi\right\}^{1/q}, \tag{6.4.45}$$

where

$$S_{n,q}(r) = \frac{N(r)(\omega_{n-2})^{1/q}}{\omega_n}. \tag{6.4.46}$$

Equality (6.4.45) proves the representation for $\mathcal{K}_p(x;\boldsymbol{\ell})$, given in the present Proposition.

Formula (6.4.33) follows directly from (6.4.32) and (6.4.45) for $\boldsymbol{\ell} = \boldsymbol{\nu}_x$. Similarly, (6.4.34) results from (6.4.32), (6.4.45) and the equality

$$\int_0^{\pi} \frac{\sin^{\mu-1} x}{(1-2r\cos x + r^2)^{\nu}}\, dx = B\left(\frac{\mu}{2}, \frac{1}{2}\right) F\left(\nu,\ \nu + \frac{1-\mu}{2};\ \frac{1+\mu}{2};\ r^2\right)$$

(see, e.g., Prudnikov, Brychkov and Marichev [**PBM1**], **2.5.16(43)**), where $B(u,v)$ is the Beta-function, and $F(a,b;c;x)$ is the hypergeometric Gauss function. □

REMARK 6.7. Since the function (6.4.32) is even and 2π-periodic in φ, (6.4.45) can be written as

$$\mathcal{K}_p(x;\boldsymbol{\ell}) = \frac{S_{n,q}(r)}{2^{1/q}}\left\{\int_0^{\pi} \sin^{n-2}\vartheta d\vartheta \int_0^{2\pi} \mathcal{F}_{n,q}(\vartheta,\varphi;r,\boldsymbol{\ell})|\sin^{n-3}\varphi| d\varphi\right\}^{1/q}. \tag{6.4.47}$$

We introduce a Cartesian coordinate system with origin $\mathcal{O}$ with orthonormal vectors $\boldsymbol{i}, \boldsymbol{j}, \boldsymbol{k}$ of the axes $\mathcal{O}x_1, \mathcal{O}x_2, \mathcal{O}x_3$, respectively. We also use spherical coordinates ϑ and φ on the unit sphere $\mathbb{S}^2$, measuring the angle ϑ from the axis $\mathcal{O}x_3$ and the angle φ from the plane $x_1\mathcal{O}x_3$.

The double integral in (6.4.47) can be now written as the integral over $\mathbb{S}^2$:

$$\mathcal{K}_p(x;\boldsymbol{\ell}) = \frac{S_{n,q}(r)}{2^{1/q}}\left\{\int_{\mathbb{S}^2} \frac{\big|(\boldsymbol{\eta} - \lambda(r)r\boldsymbol{k},\ nd(r)\ell_\tau \boldsymbol{i} + \ell_\nu \boldsymbol{k}\ )\big|^q}{|\boldsymbol{\eta} - r\boldsymbol{k}|^{(n+2)q}}\, |(\boldsymbol{\eta}, \boldsymbol{j})|^{n-3} d\sigma_\eta\right\}^{1/q}.$$

In the next assertion, the integral in ϑ in (6.4.45) is reduced to an integral with algebraic integrand. The expression for $\mathcal{K}_p(x;\boldsymbol{\ell})$, given below, contains two factors one of which is an explicitely given function increasing as $r \to 1$ to infinity and the second factor (the double integral) is a bounded function for $r \in [0,\ 1]$.

LEMMA 6.3. *Let $u \in h^p(\mathbb{B})$, $1 < p \le \infty$, and let x be an arbitrary point in $\mathbb{B}$. The sharp coefficient $\mathcal{K}_p(x;\boldsymbol{\ell})$ in the inequality*

$$|(\nabla u(x), \boldsymbol{\ell})| \le \mathcal{K}_p(x;\boldsymbol{\ell})\|u\|_p \tag{6.4.48}$$

is given by

$$\mathcal{K}_p(x;\boldsymbol{\ell}) = \mathcal{B}_{n,q}(r)\left\{\int_0^\pi \sin^{n-3}\varphi\, d\varphi \int_0^\infty \mathcal{T}_{n,q}(t,\varphi;r,\boldsymbol{\ell})dt\right\}^{1/q}, \tag{6.4.49}$$

where $1/p + 1/q = 1$,

$$\mathcal{B}_{n,q}(r) = \frac{n+(n-2)r}{\omega_n}\left\{\frac{2^{n-1}\omega_{n-2}\big(a(r)\big)^{n-1}}{(1+r)^{n-1}(1-r)^{n(q-1)+1}}\right\}^{1/q}, \tag{6.4.50}$$

and

$$\mathcal{T}_{n,q}(t,\varphi;r,\boldsymbol{\ell}) = \frac{\big|2c(r)\ell_\tau t\cos\varphi + (1-t^2)\ell_\nu\big|^q t^{n-2}}{[1+a^2(r)t^2]^{(n+2)q/2}[1+b^2(r)t^2]^{n-1-(nq)/2}}. \tag{6.4.51}$$

Here $\boldsymbol{\ell} = \ell_\tau \boldsymbol{\tau}_x + \ell_\nu \boldsymbol{\nu}_x$,

$$a(r) = \left(\frac{n+(n-2)r}{n-(n-2)r}\right)^{1/2}, \qquad b(r) = \frac{1-r}{1+r}\left(\frac{n+(n-2)r}{n-(n-2)r}\right)^{1/2}, \tag{6.4.52}$$

and

$$c(r) = \frac{n}{\big(n^2-(n-2)^2r^2\big)^{1/2}}. \tag{6.4.53}$$

In particular,

$$\mathcal{K}_p(x) = \sup_{\gamma\ge 0}\frac{\mathcal{C}_{n,q}(r)}{\sqrt{c^2(r)+\gamma^2}}\left\{\int_0^\pi \sin^{n-3}\varphi\, d\varphi\int_0^\infty \mathcal{P}_{n,q}(t,\varphi;r,\gamma)\,dt\right\}^{1/q}, \tag{6.4.54}$$

where

$$\mathcal{C}_{n,q}(r) = \frac{n}{\omega_n}\left\{\frac{2^{n-1}\omega_{n-2}\big(a(r)\big)^{n+q-1}}{(1+r)^{n-1}(1-r)^{n(q-1)+1}}\right\}^{1/q}, \tag{6.4.55}$$

and

$$\mathcal{P}_{n,q}(t,\varphi;r,\gamma) = \frac{\big|1+2\gamma t\cos\varphi - t^2\big|^q t^{n-2}}{[1+a^2(r)t^2]^{(n+2)q/2}[1+b^2(r)t^2]^{n-1-(nq)/2}}. \tag{6.4.56}$$

PROOF. Changing the order of integration in (6.4.45), we obtain

$$\mathcal{K}_p(x;\boldsymbol{\ell}) = S_{n,q}(r)\left\{\int_0^\pi \sin^{n-3}\varphi\, d\varphi\int_0^\pi \mathcal{F}_{n,q}(\vartheta,\varphi;r,\boldsymbol{\ell})\sin^{n-2}\vartheta\, d\vartheta\right\}^{1/q} \tag{6.4.57}$$

with the constant $S_{n,q}(r)$ defined by (6.4.46) and the function $\mathcal{F}_{n,q}(\vartheta,\varphi;r,\boldsymbol{\ell})$ given by (6.4.32). Making the change of variable

$$\vartheta = 2\arctan\big(b(r)t\big)$$

in (6.4.57), we arrive at (6.4.49).

Since the integrand in (6.4.3) does not change when the unit vector $\boldsymbol{\ell}$ is replaced by $-\boldsymbol{\ell}$, we may assume that $\ell_\nu = (\boldsymbol{\ell},\boldsymbol{\nu}_x) > 0$ in (6.4.8). Introducing the parameter $\gamma = c(r)\ell_\tau/\ell_\nu$ in (6.4.49) and using the equality $\ell_\tau^2 + \ell_\nu^2 = 1$ together with (6.4.8), we arrive at (6.4.54). □

6.4.4. The case $p = 2$

The next assertion gives expressions for $\mathcal{K}_2(x;\boldsymbol{\ell})$ for any direction $\boldsymbol{\ell}$, in particular, the sharp values for $\mathcal{K}_2(x;\boldsymbol{\nu}_x)$ and $\mathcal{K}_2(x;\boldsymbol{\tau}_x)$. It is also shown that the value of $\mathcal{K}_2(x;\boldsymbol{\ell})$ is squeezed between the minimal value $\mathcal{K}_2(x;\boldsymbol{\tau}_x)$ and the maximal value of $\mathcal{K}_2(x) = \mathcal{K}_2(x;\boldsymbol{\nu}_x)$.

COROLLARY 6.9. *Let $u \in h^2(\mathbb{B})$, and x be an arbitrary point in $\mathbb{B}$. The sharp coefficient $\mathcal{K}_2(x;\boldsymbol{\ell})$ in the inequality*

$$|(\nabla u(x),\boldsymbol{\ell})| \le \mathcal{K}_2(x;\boldsymbol{\ell})\|u\|_2 \tag{6.4.58}$$

is given by

$$\mathcal{K}_2(x;\boldsymbol{\ell}) = \left\{\frac{n(1+r^2)\ell_\tau^2 + \left[n+(n^2+n-4)r^2+(n-2)^2r^4\right]\ell_\nu^2}{\omega_n(1-r^2)^{n+1}}\right\}^{1/2}. \tag{6.4.59}$$

In particular,

$$\mathcal{K}_2(x;\boldsymbol{\tau}_x) \le \mathcal{K}_2(x;\boldsymbol{\ell}) \le \mathcal{K}_2(x;\boldsymbol{\nu}_x), \tag{6.4.60}$$

where

$$\mathcal{K}_2(x;\boldsymbol{\nu}_x) = \left\{\frac{n+(n^2+n-4)r^2+(n-2)^2r^4}{\omega_n(1-r^2)^{n+1}}\right\}^{1/2}, \tag{6.4.61}$$

and

$$\mathcal{K}_2(x;\boldsymbol{\tau}_x) = \left\{\frac{n(1+r^2)}{\omega_n(1-r^2)^{n+1}}\right\}^{1/2}. \tag{6.4.62}$$

The equality sign on the left-hand side of (6.4.60) *is attained for $\boldsymbol{\ell} = \boldsymbol{\tau}_x$ and that on the right-hand side for $\boldsymbol{\ell} = \boldsymbol{\nu}_x$.*

PROOF. By (6.4.49)-(6.4.51),

$$\mathcal{K}_2(x;\boldsymbol{\ell}) = \mathcal{B}_{n,2}(r)\left\{\int_0^\pi \sin^{n-3}\varphi\, d\varphi \int_0^\infty \mathcal{T}_{n,2}(t,\varphi;r,\boldsymbol{\ell})dt\right\}^{1/2}, \tag{6.4.63}$$

where

$$\mathcal{B}_{n,2}(r) = \frac{n+(n-2)r}{\omega_n}\left\{\frac{2^{n-1}\omega_{n-2}\big(a(r)\big)^{n-1}}{(1+r)^{n-1}(1-r)^{n+1}}\right\}^{1/2}, \tag{6.4.64}$$

and

$$\mathcal{T}_{n,2}(t,\varphi;r,\boldsymbol{\ell}) = \frac{\left[2c(r)\ell_\tau t\cos\varphi + (1-t^2)\ell_\nu\right]^2[1+b^2(r)t^2]t^{n-2}}{[1+a^2(r)t^2]^{n+2}[1+b^2(r)t^2]^{-1}}. \tag{6.4.65}$$

Using the equalities

$$\int_0^\pi \sin^{n-3}\varphi\cos\varphi\, d\varphi = 0,$$

$$\int_0^\pi \sin^{n-3}\varphi\, d\varphi = \frac{\sqrt{\pi}\,\Gamma\left(\frac{n-2}{2}\right)}{\Gamma\left(\frac{n-1}{2}\right)}, \qquad \int_0^\pi \sin^{n-3}\varphi\cos^2\varphi\, d\varphi = \frac{\sqrt{\pi}\,\Gamma\left(\frac{n-2}{2}\right)}{(n-1)\Gamma\left(\frac{n-1}{2}\right)},$$

we write (6.4.63) as

$$\mathcal{K}_2(x;\boldsymbol{\ell}) = \mathcal{D}_n(r)\left\{\int_0^\infty \mathcal{G}_n(t;r,\boldsymbol{\ell})\, dt\right\}^{1/2}, \tag{6.4.66}$$

where

$$\mathcal{D}_n(r) = \mathcal{B}_{n,2}(r) \left(\frac{\sqrt{\pi}\, \Gamma\left(\frac{n-2}{2}\right)}{(n-1)\Gamma\left(\frac{n-1}{2}\right)} \right)^{1/2}, \tag{6.4.67}$$

and

$$\mathcal{G}_n(t; r, \boldsymbol{\ell}) = \frac{\left[4c^2(r)t^2\ell_\tau^2 + (n-1)(1-t^2)^2\ell_\nu^2\right]\left[1+b^2(r)t^2\right]t^{n-2}}{\left[1+a^2(r)t^2\right]^{n+2}}. \tag{6.4.68}$$

Introducing the notations

$$\mathcal{J}_1(r) = 4c^2(r) \int_0^\infty \frac{\left[1+b^2(r)t^2\right]t^n}{\left[1+a^2(r)t^2\right]^{n+2}}\, dt\,, \tag{6.4.69}$$

$$\mathcal{J}_2(r) = (n-1) \int_0^\infty \frac{(1-t^2)^2\left[1+b^2(r)t^2\right]t^{n-2}}{\left[1+a^2(r)t^2\right]^{n+2}}\, dt\,, \tag{6.4.70}$$

by (6.4.66) and (6.4.68) we have

$$\mathcal{K}_2(x; \boldsymbol{\ell}) = \mathcal{D}_n(r) \left\{ \mathcal{J}_1(r)\ell_\tau^2 + \mathcal{J}_2(r)\ell_\nu^2 \right\}^{1/2}. \tag{6.4.71}$$

In view of the formula

$$\int_0^\infty \frac{t^{k-1}}{(1+p^2t^2)^m}\, dt = \frac{\Gamma\left(\frac{k}{2}\right)\Gamma\left(\frac{2m-k}{2}\right)}{2p^k(m-1)!}$$

(see, e.g., Gradshtein and Ryzhik [**GR**], **3.241(4)**), we find that (6.4.69), (6.4.70) become

$$\mathcal{J}_1(r) = \frac{2n^2\Gamma^2\left(\frac{n+1}{2}\right)(1+r^2)}{n!(1+r)^2\big(a(r)\big)^{n+1}\left[n^2-(n-2)^2r^2\right]}, \tag{6.4.72}$$

$$\mathcal{J}_2(r) = \frac{2n\Gamma^2\left(\frac{n+1}{2}\right)\left[n+(n^2+n-4)r^2+(n-2)r^4\right]}{n!(1+r)^2\big(a(r)\big)^{n+1}\left[n^2-(n-2)^2r^2\right]}. \tag{6.4.73}$$

Using (6.4.72), (6.4.73) in (6.4.71), taking into account (6.4.64) and (6.4.67), as well as the identity

$$2^{2z-1}\Gamma(z)\Gamma\left(z+\frac{1}{2}\right) = \sqrt{\pi}\,\Gamma(2z)$$

with $z = (n-1)/2$, we arrive at (6.4.59).

Formulas (6.4.61) and (6.4.62) follow directly from (6.4.59) for $\ell_\nu = 1, \ell_\tau = 0$ and $\ell_\tau = 1, \ell_\nu = 0$, respectively.

Further, since $\ell_\tau^2 + \ell_\nu^2 = 1$ and

$$n(1+r^2) \le n + (n^2+n-4)r^2 + (n-2)^2r^4$$

for $n \ge 2$, it follows from (6.4.59) that

$$\left\{ \frac{n(1+r^2)}{\omega_n(1-r^2)^{n+1}} \right\}^{1/2} \le \mathcal{K}_2(x; \boldsymbol{\ell}) \le \left\{ \frac{n+(n^2+n-4)r^2+(n-2)^2r^4}{\omega_n(1-r^2)^{n+1}} \right\}^{1/2},$$

i.e. $\mathcal{K}_2(x; \boldsymbol{\tau}_x) \le \mathcal{K}_2(x; \boldsymbol{\ell}) \le \mathcal{K}_2(x; \boldsymbol{\nu}_x)$, which proves (6.4.60).

The monotonicity of $\mathcal{K}_2(x; \boldsymbol{\ell})$ as a function of ℓ_ν follows from (6.4.59) and the equality $\ell_\tau^2 + \ell_\nu^2 = 1$. □

6.4.5. The case $p = \infty$

The next assertion concerns particular cases of representations for $\mathcal{K}_p(x; \boldsymbol{\nu}_x)$ and $\mathcal{K}_p(x; \boldsymbol{\tau}_x)$ for $p = \infty$.

COROLLARY 6.10. *Let $u \in h^\infty(\mathbb{B})$, and let x be an arbitrary point in $\mathbb{B}$. The sharp coefficient $\mathcal{K}_\infty(x; \boldsymbol{\nu}_x)$ in the inequality*

$$|(\nabla u(x),\ \boldsymbol{\nu}_x)| \le \mathcal{K}_\infty(x; \boldsymbol{\nu}_x)\|u\|_\infty \tag{6.4.74}$$

is given by

$$\mathcal{K}_\infty(x; \boldsymbol{\nu}_x) = \mathcal{D}_n(r) \int_0^1 \frac{(1-t^2)t^{n-2}}{\left[1+a^2(r)t^2\right]^{(n+2)/2}\left[1+b^2(r)t^2\right]^{(n-2)/2}}\, dt\ , \tag{6.4.75}$$

where

$$\mathcal{D}_n(r) = \frac{2^n \omega_{n-1}\left[n+(n-2)r\right]\left(a(r)\right)^{n-1}}{\omega_n (1+r)^{n-1}(1-r)}\ ,$$

and $a(r)$, $b(r)$ are defined by (6.4.52).

In particular,

$$\mathcal{K}_\infty(x; \boldsymbol{\nu}_x) = \frac{1}{r^2}\left(\frac{\left(1+r^2/3\right)^{3/2}}{1-r^2} - 1\right) \tag{6.4.76}$$

for $n = 3$, and

$$\mathcal{K}_\infty(x; \boldsymbol{\nu}_x) = \frac{1}{\pi r^3}\left(\frac{r(2+r^2)(4-r^2)^{1/2}}{(1-r^2)} - 2\arccos\frac{(2-r^2)^2-2}{2}\right) \tag{6.4.77}$$

for $n = 4$.

The sharp coefficient $\mathcal{K}_\infty(x; \boldsymbol{\tau}_x)$ in the inequality

$$|(\nabla u(x),\ \boldsymbol{\tau}_x)| \le \mathcal{K}_\infty(x; \boldsymbol{\tau}_x)\|u\|_\infty \tag{6.4.78}$$

is given by

$$\mathcal{K}_\infty(x; \boldsymbol{\tau}_x) = \frac{2n\omega_{n-1}(1-r^2)}{(n-1)\omega_n}\, F\left(\frac{3}{2}, \frac{n+2}{2}; \frac{n+1}{2}; r^2\right), \tag{6.4.79}$$

where $F(a, b; c; x)$ is the hypergeometric Gauss function.

In particular,

$$\mathcal{K}_\infty(x; \boldsymbol{\tau}_x) = \frac{2}{\pi(1-r^2)}\left[2\mathbf{K}(r) - (1+r^2)\mathbf{D}(r)\right] \tag{6.4.80}$$

for $n = 3$, and

$$\mathcal{K}_\infty(x; \boldsymbol{\tau}_x) = \frac{8(1-r^2)}{\pi r^2}\left[\frac{1+r^2}{(1-r^2)^2} - \frac{1}{2r}\log\frac{1+r}{1-r}\right] \tag{6.4.81}$$

for $n = 4$. Here $\mathbf{K}(k)$ is the complete elliptic integral of the first kind, and $\mathbf{D}(k)$ is the complete elliptic integral.

PROOF. We put $\ell_\tau = 0, \ell_\nu = 1$ in (6.4.51) and we pass to the limit as $p \to \infty$ ($q \to 1$) in (6.4.49). This results in

$$\mathcal{K}_\infty(x; \boldsymbol{\nu}_x) = \mathcal{B}_{n,1}(r) \int_0^\pi \sin^{n-3}\varphi\, d\varphi \int_0^\infty \mathcal{T}_{n,1}(t, \varphi; r, \boldsymbol{\nu}_x) dt\ , \tag{6.4.82}$$

where

$$\mathcal{B}_{n,1}(r) = \frac{2^{n-1}\omega_{n-2}\big(n+(n-2)r\big)\big(a(r)\big)^{n-1}}{\omega_n(1+r)^{n-1}(1-r)}\,, \tag{6.4.83}$$

and

$$\mathcal{T}_{n,1}(t,\varphi;r,\boldsymbol{\nu}_x) = \frac{\big|1-t^2\big|t^{n-2}}{[1+a^2(r)t^2]^{(n+2)/2}[1+b^2(r)t^2]^{(n-2)/2}}\,. \tag{6.4.84}$$

Note that

$$\int_0^\infty \frac{(1-t^2)t^{n-2}}{(1+a^2(r)t^2)^{(n+2)/2}(1+b^2(r)t^2)^{(n-2)/2}}\,dt = 0\,. \tag{6.4.85}$$

In fact, setting $u \equiv 1$, $\boldsymbol{\ell} = \boldsymbol{\nu}_x$ in (6.4.17), we have

$$\int_{\mathbb{S}^{n-1}} \frac{(y,\ \boldsymbol{\nu}_x) - \lambda(r)r}{|y-x|^{n+2}}\,d\sigma_y = 0.$$

Writing the last integral in the same manner as we did in the proof of Proposition 6.4, we find

$$\int_0^\pi \frac{\cos\vartheta - \lambda(r)r}{\big(1-2r\cos\vartheta + r^2\big)^{n+2}}\sin^{n-2}d\vartheta = 0. \tag{6.4.86}$$

Making the change of variable $\vartheta = 2\arctan\big(b(r)t\big)$ in (6.4.86) as in Lemma 6.3, we arrive at (6.4.85).

By (6.4.85)

$$\int_0^1 \frac{(1-t^2)t^{n-2}}{(1+a^2(r)t^2)^{(n+2)/2}(1+b^2(r)t^2)^{(n-2)/2}}\,dt$$
$$= -\int_1^\infty \frac{(1-t^2)t^{n-2}}{(1+a^2(r)t^2)^{(n+2)/2}(1+b^2(r)t^2)^{(n-2)/2}}\,dt\,.$$

Therefore, (6.4.82) implies

$$\mathcal{K}_\infty(x;\boldsymbol{\nu}_x) = 2\mathcal{B}_{n,1}(r)\int_0^\pi \sin^{n-3}\varphi\,d\varphi\int_0^1 \mathcal{T}_{n,1}(t,\varphi;r,\boldsymbol{\nu}_x)dt.$$

Evaluating the integral in φ, we obtain (6.4.75).

Integrating in (6.4.75), we arrive at (6.4.76) and (6.4.77) for $n=3$ and $n=4$, respectively.

Equality (6.4.79) results from (6.4.34) with $q = 1$. Formulas (6.4.80) and (6.4.81) follow from (6.4.79) with $n = 3, 4$ and the table of particular values of the hypergeometric Gauss function (Prudnikov, Brychkov and Marichev [**PBM2**], Section **7.3.2**, items 217 and 225). □

REMARK 6.8. Note that the change of variable $t = \tan\vartheta/a(r)$ in the inner integral in (6.4.54) and the change of parameter $\gamma = c(r)\beta$ leads to the representation

$$\mathcal{K}_\infty(x) = \frac{2^{n-2}(n-2)}{\pi(1+r)^{n-1}(1-r)}\sup_{\beta\geq 0}\frac{1}{\sqrt{1+\beta^2}}\int_0^\pi \sin^{n-3}\varphi\,d\varphi\int_0^{\pi/2} G_n(\vartheta,\varphi;r,\beta)\,d\vartheta\,,$$

where

$$G_n(\vartheta,\varphi;r,\beta)=\frac{\left|n\cos 2\vartheta+n\beta\sin 2\vartheta\cos\varphi+(n-2)r\right|}{\left[1+\left(\frac{1-r}{1+r}\right)^2\tan^2\vartheta\right]^{(n-2)/2}}\sin^{n-2}\vartheta\ .$$

6.5. Sharp estimates for the gradient of a solution of the Neumann problem in a half-space

Let u be a harmonic function in $\mathbb{R}^n_+$ given as the single layer potential

$$u(x)=\frac{2}{(n-2)\omega_n}\int_{\mathbb{R}^{n-1}}\frac{1}{|y-x|^{n-2}}\frac{\partial u}{\partial \boldsymbol{\nu}}(y',0)\ )dy' \tag{6.5.1}$$

with the boundary values of $\partial u/\partial\boldsymbol{\nu}$ on $\partial\mathbb{R}^n_+$ in $L^p(\mathbb{R}^{n-1}), 1\le p<\infty$. Then $\partial u/\partial\boldsymbol{\nu}\in h^p(\mathbb{R}^n_+)$.

Now, we find a representation for the best coefficient $\mathcal{C}_{n,p}(x;\boldsymbol{\ell})$ in the inequality for the absolute value of the derivative $\partial u/\partial\boldsymbol{\ell}$ in an arbitrary direction $\boldsymbol{\ell}\in\mathbb{S}^{n-1}$ with $\|\partial u/\partial\boldsymbol{\nu}\|_p$ on the right-hand side.

PROPOSITION 6.5. *Let u be a solution of the Neumann problem for the Laplace equation in $\mathbb{R}^n_+$ with $\partial u/\partial\boldsymbol{\nu}\in h^p(\mathbb{R}^n_+), 1\le p<\infty$, and let x be an arbitrary point in $\mathbb{R}^n_+$. The sharp coefficient $\mathcal{C}_{n,p}(x;\boldsymbol{\ell})$ in the inequality*

$$|(\nabla u(x),\boldsymbol{\ell})|\le\mathcal{C}_{n,p}(x;\boldsymbol{\ell})\left|\left|\frac{\partial u}{\partial\boldsymbol{\nu}}\right|\right|_p \tag{6.5.2}$$

is given by

$$\mathcal{C}_{n,p}(x;\boldsymbol{\ell})=\frac{C_{n,p}(\boldsymbol{\ell})}{x_n^{(n-1)/p}}\ , \tag{6.5.3}$$

where

$$C_{n,1}(\boldsymbol{\ell})=\frac{2}{\omega_n}\max_{\sigma\in\mathbb{S}^{n-1}}|(\boldsymbol{e}_\sigma,\boldsymbol{\ell})||(\boldsymbol{e}_\sigma,\boldsymbol{e}_n)|^{n-1}\ , \tag{6.5.4}$$

and

$$C_{n,p}(\boldsymbol{\ell})=\frac{2^{1/p}}{\omega_n}\left\{\int_{\mathbb{S}^{n-1}}|(\boldsymbol{e}_\sigma,\boldsymbol{\ell})|^{p/(p-1)}|(\boldsymbol{e}_\sigma,\boldsymbol{e}_n)|^{(n-p)/(p-1)}d\sigma\right\}^{1-\frac{1}{p}} \tag{6.5.5}$$

for $1<p<\infty$.

In particular, the sharp coefficient $\mathcal{C}_{n,p}(x)$ in the inequality

$$|\nabla u(x)|\le\mathcal{C}_{n,p}(x)\left|\left|\frac{\partial u}{\partial\boldsymbol{\nu}}\right|\right|_p \tag{6.5.6}$$

is given by

$$\mathcal{C}_{n,p}(x)=\frac{C_{n,p}}{x_n^{(n-1)/p}}\ , \tag{6.5.7}$$

where

$$C_{n,p}=\max_{|\boldsymbol{\ell}|=1}C_{n,p}(\boldsymbol{\ell})\ . \tag{6.5.8}$$

PROOF. By (6.5.1),

$$\nabla u(x) = \frac{2}{\omega_n} \int_{\mathbb{R}^{n-1}} \frac{\boldsymbol{e}_{xy}}{|y-x|^{n-1}} \frac{\partial u}{\partial \boldsymbol{\nu}}(y', 0)\,)dy' ,$$

where $\boldsymbol{e}_{xy} = (y-x)/|y-x|$. Hence, the sharp coefficient $\mathcal{C}_{n,p}(x; \boldsymbol{\ell})$ in (6.5.2) has the form

$$\mathcal{C}_{n,p}(x; \boldsymbol{\ell}) = \frac{2}{\omega_n} \left\{ \int_{\mathbb{R}^{n-1}} \frac{|(\boldsymbol{e}_{xy}, \boldsymbol{\ell})|^q}{|y-x|^{(n-1)q}}\, dy' \right\}^{1/q}, \tag{6.5.9}$$

where $1/p + 1/q = 1$. Since

$$\frac{1}{|y-x|^{(n-1)q}} = \frac{1}{x_n^{(n-1)(q-1)}} \left(\frac{x_n}{|y-x|} \right)^{(n-1)q-n} \frac{x_n}{|y-x|^n},$$

it follows that (6.5.9) can be written in the form (6.5.3), where

$$\begin{aligned} C_{n,p}(\boldsymbol{\ell}) &= \frac{2}{\omega_n} \left\{ \int_{\mathbb{S}^{n-1}_-} |(\boldsymbol{e}_\sigma, \boldsymbol{\ell})|^q (\boldsymbol{e}_\sigma, -\boldsymbol{e}_n)^{(n-1)q-n} d\sigma \right\}^{1/q} \\ &= \frac{2}{\omega_n} \left\{ \frac{1}{2} \int_{\mathbb{S}^{n-1}} |(\boldsymbol{e}_\sigma, \boldsymbol{\ell})|^q |(\boldsymbol{e}_\sigma, \boldsymbol{e}_n)|^{(n-1)q-n} d\sigma \right\}^{1/q}. \end{aligned} \tag{6.5.10}$$

The last equality is equivalent to (6.5.5). Formula (6.5.4) is the limit case of (6.5.10) as $q \to \infty$.

Formulas (6.5.7) and (6.5.8) for the best coefficient in (6.5.6) are direct consequences of (6.5.2) and (6.5.3). □

The next assertion provides a solution of the optimization problem (6.5.8) on the unit sphere in $\mathbb{R}^n$ for $p \in [1, n]$.

THEOREM 6.3. *Let u be a solution of the Neumann problem for the Laplace equation in $\mathbb{R}^n_+$ with $\partial u / \partial \boldsymbol{\nu} \in h^p(\mathbb{R}^n_+)$, $p \in [1, n]$, and let x be an arbitrary point in $\mathbb{R}^n_+$. The sharp coefficient $\mathcal{C}_{n,p}$ in the inequality*

$$|\nabla u(x)| \le \frac{\mathcal{C}_{n,p}}{x_n^{(n-1)/p}} \left\| \frac{\partial u}{\partial \boldsymbol{\nu}} \right\|_p \tag{6.5.11}$$

is given by

$$\mathcal{C}_{n,p}(x; \boldsymbol{\ell}) = \frac{C_{n,p}(\boldsymbol{\ell})}{x_n^{(n-1)/p}}, \tag{6.5.12}$$

where

$$C_{n,1} = \frac{2}{\omega_n}, \tag{6.5.13}$$

and

$$C_{n,p} = \frac{2^{1/p}}{\omega_n} \left\{ \frac{2\pi^{(n-1)/2} \Gamma\left(\frac{n+p-1}{2p-2} \right)}{\Gamma\left(\frac{np}{2p-2} \right)} \right\}^{1-\frac{1}{p}} \tag{6.5.14}$$

for $1 < p \le n$.

The maximum in $\boldsymbol{\ell}$ in (6.5.8) is attained for $\boldsymbol{\ell} = \boldsymbol{e}_n$ or for $\boldsymbol{\ell} = -\boldsymbol{e}_n$ if $1 \le p < n$. The coefficient $C_{n,n}(\boldsymbol{\ell})$ is independent of $\boldsymbol{\ell}$.

PROOF. First, let $p = 1$. It follows from (6.5.4) and (6.5.8) that

$$C_{n,1} = \frac{2}{\omega_n} \max_{|\boldsymbol{\ell}|=1} \max_{\sigma \in \mathbb{S}^{n-1}} |(\boldsymbol{e}_\sigma, \boldsymbol{\ell})||(\boldsymbol{e}_\sigma, \boldsymbol{e}_n)|^{n-1} . \tag{6.5.15}$$

Interchanging the order of maxima, we obtain

$$C_{n,1} = \frac{2}{\omega_n} \max_{\sigma \in \mathbb{S}^{n-1}} \max_{|\boldsymbol{\ell}|=1} |(\boldsymbol{e}_\sigma, \boldsymbol{\ell})||(\boldsymbol{e}_\sigma, \boldsymbol{e}_n)|^{n-1} = \frac{2}{\omega_n} \max_{\sigma \in \mathbb{S}^{n-1}} |(\boldsymbol{e}_\sigma, \boldsymbol{e}_n)|^{n-1} = \frac{2}{\omega_n} ,$$

which proves (6.5.13). The maximum in $\boldsymbol{\ell}$ in (6.5.15) is attained at $\boldsymbol{\ell} = \pm \boldsymbol{e}_n$.

The equality (6.5.5) with $p = n$ takes the form

$$C_{n,n}(\boldsymbol{\ell}) = \frac{2^{1/n}}{\omega_n} \left\{ \int_{\mathbb{S}^{n-1}} |(\boldsymbol{e}_\sigma, \boldsymbol{\ell})|^{n/(n-1)} d\sigma \right\}^{1-\frac{1}{n}} . \tag{6.5.16}$$

Thus, $C_{n,n}(\boldsymbol{\ell})$ is independent of $\boldsymbol{\ell}$.

Next, let $p \in (1, n)$. By (6.5.5) and (6.5.8),

$$C_{n,p} = \max_{|\boldsymbol{\ell}|=1} C_{n,p}(\boldsymbol{\ell}) \geq C_{n,p}(\boldsymbol{e}_n) = \frac{2^{1/p}}{\omega_n} \left\{ \int_{\mathbb{S}^{n-1}} |(\boldsymbol{e}_\sigma, \boldsymbol{e}_n)|^{n/(p-1)} d\sigma \right\}^{1-\frac{1}{p}} . \tag{6.5.17}$$

We show that the estimate opposite to (6.5.17) also holds.

Let

$$s = \frac{n}{p} , \quad t = \frac{n}{n-p} . \tag{6.5.18}$$

Then

$$\frac{1}{s} + \frac{1}{t} = \frac{p}{n} + \frac{n-p}{n} = 1 . \tag{6.5.19}$$

Using Hölder's inequality and (6.5.18), we obtain

$$\begin{aligned}
&\int_{\mathbb{S}^{n-1}} |(\boldsymbol{e}_\sigma, \boldsymbol{\ell})|^{\frac{p}{p-1}} |(\boldsymbol{e}_\sigma, \boldsymbol{e}_n)|^{\frac{n-p}{p-1}} d\sigma \\
\leq\ & \left\{ \int_{\mathbb{S}^{n-1}} |(\boldsymbol{e}_\sigma, \boldsymbol{\ell})|^{\frac{ps}{p-1}} d\sigma \right\}^{1/s} \left\{ \int_{\mathbb{S}^{n-1}} |(\boldsymbol{e}_\sigma, \boldsymbol{e}_n)|^{\frac{(n-p)t}{p-1}} d\sigma \right\}^{1/t} \\
=\ & \left\{ \int_{\mathbb{S}^{n-1}} |(\boldsymbol{e}_\sigma, \boldsymbol{\ell})|^{\frac{n}{p-1}} d\sigma \right\}^{1/s} \left\{ \int_{\mathbb{S}^{n-1}} |(\boldsymbol{e}_\sigma, \boldsymbol{e}_n)|^{\frac{n}{p-1}} d\sigma \right\}^{1/t} .
\end{aligned} \tag{6.5.20}$$

Since the first integral in the last equality is independent of $\boldsymbol{\ell}$, it follows that

$$\int_{\mathbb{S}^{n-1}} |(\boldsymbol{e}_\sigma, \boldsymbol{\ell})|^{\frac{n}{p-1}} d\sigma = \int_{\mathbb{S}^{n-1}} |(\boldsymbol{e}_\sigma, \boldsymbol{e}_n)|^{\frac{n}{p-1}} d\sigma ,$$

which, in view of (6.5.19), allows to write (6.5.20) as

$$\int_{\mathbb{S}^{n-1}} |(\boldsymbol{e}_\sigma, \boldsymbol{\ell})|^{\frac{p}{p-1}} |(\boldsymbol{e}_\sigma, \boldsymbol{e}_n)|^{\frac{n-p}{p-1}} d\sigma \leq \int_{\mathbb{S}^{n-1}} |(\boldsymbol{e}_\sigma, \boldsymbol{e}_n)|^{\frac{n}{p-1}} |d\sigma .$$

This together with (6.5.5) and (6.5.8) imply

$$C_{n,p} \leq \frac{2^{1/p}}{\omega_n} \left\{ \int_{\mathbb{S}^{n-1}} |(\boldsymbol{e}_\sigma, \boldsymbol{e}_n)|^{n/(p-1)} d\sigma \right\}^{1-\frac{1}{p}} .$$

By this estimate and (6.5.17) we get

$$C_{n,p} = \frac{2^{1/p}}{\omega_n} \left\{ \int_{\mathbb{S}^{n-1}} |(\boldsymbol{e}_\sigma, \boldsymbol{e}_n)|^{n/(p-1)} d\sigma \right\}^{1-\frac{1}{p}} . \tag{6.5.21}$$

Besides, the maximum in $\boldsymbol{\ell}$ in (6.5.8) is attained at $\boldsymbol{\ell} = \pm \boldsymbol{e}_n$.

Evaluating the integral in (6.5.21), we find

$$\int_{\mathbb{S}^{n-1}} |(\boldsymbol{e}_\sigma, \boldsymbol{e}_n)|^{n/(p-1)} d\sigma = 2\omega_{n-1} \int_0^{\pi/2} \cos^{n/(p-1)} \vartheta \sin^{n-2} \vartheta d\vartheta$$

$$= \omega_{n-1} B\left(\frac{n+p-1}{2(p-1)}, \frac{n-1}{2}\right) = \frac{2\pi^{(n-1)/2}\Gamma\left(\frac{n+p-1}{2(p-1)}\right)}{\Gamma\left(\frac{np}{2(p-1)}\right)},$$

which together with (6.5.21) and (6.5.16) proves (6.5.14). □

6.6. Comments to Chapter 6

1. A number of sharp estimates for derivatives of harmonic functions are known; we mention only a few. The best constant (6.4.76) in inequality (6.4.74) was found by Khavinson [**Kh**]. The following estimate for the absolute value of the gradient of a harmonic function is borrowed from the book by Protter and Weinberger [**PW**]

$$|\nabla u(x)| \leq \frac{n\omega_{n-1}}{(n-1)\omega_n d_x} \operatorname{osc}_{\mathcal{D}}(u) \,. \tag{6.6.1}$$

Here u is harmonic function in $\mathcal{D} \subset \mathbb{R}^n$, $n \geq 2$, ω_n is the area of the unit sphere $\mathbb{S}^{n-1}$ in $\mathbb{R}^n$, $\operatorname{osc}_{\mathcal{D}}(u)$ is the oscillation of u on $\mathcal{D}$, and d_x is the distance from $x \in \mathcal{D}$ to $\partial\mathcal{D}$. Inequality (6.6.1) is a corollary of the estimate for the modulus of the gradient of a harmonic function at the center of the ball $\mathbb{B}_R = \{x \in \mathbb{R}^n : |x| < R\}$:

$$|\nabla u(0)| \leq \frac{n\omega_{n-1}}{(n-1)\omega_n R} \operatorname{osc}_{\mathbb{B}_R}(u) \,.$$

The last estimate is equivalent to the inequality

$$|\nabla u(0)| \leq \frac{2n\omega_{n-1}}{(n-1)\omega_n R} \sup_{|x|<R} |u(x)| \tag{6.6.2}$$

with the best possible constant (see Khavinson [**Kh**], Burgeth [**Burg**]). Corollary 6.8 is a generalization of the last estimate to the case of boundary data from L^p.

2. Estimates obtained in Sections 6.2, 6.3 and 6.4 generalize some classical sharp inequalities in the theory of analytic functions with different characteristics of the functions real part as a majorant. Various inequalities for analytic functions of such a type are called *real-part theorems* in reference to the first assertion of such a kind, the celebrated Hadamard real-part theorem

$$|f(z)| \leq \frac{C|z|}{1-|z|} \max_{|\zeta|=1} \Re f(\zeta).$$

Here $|z| < 1$ and f is an analytic function on the closure $\overline{\mathbb{D}}$ of the unit disk $\mathbb{D} = \{z : |z| < 1\}$ vanishing at $z = 0$. This inequality was first obtained by Hadamard with $C = 4$ in 1892 [**Had**]. The following refinement of Hadamard's real-part theorem due to Borel [**Bo1, Bo2**], Carathéodory [**Land1, Land2**] and Lindelöf [**Li**]

$$|f(z) - f(0)| \leq \frac{2|z|}{1-|z|} \sup_{|\zeta|<1} \Re\{f(\zeta) - f(0)\}, \tag{6.6.3}$$

and corollaries of the last sharp estimate are often called the Borel-Carathéodory inequalities (see, for instance, Titchmarsh [**Ti**], Cartwright [**Ca**], Zalcman [**Za**]).

Sometimes, (6.6.3) is called Hadamard-Borel-Carathéodory inequality (see, e.g., Burckel [**Burc**]). We note that an extension of Hadamard's real-part theorem for holomorphic functions in a subdomain of a complex manifold was established by Aizenberg, Aytuna and Djakov [**AAD**].

Real-part theorems embrace a family of sharp estimates for the first derivative of an arbitrary analytic function. Among them, there is the Lindelöf inequality [**Li**] (see also Jensen [**Je**])

$$|f'(z)| \le \frac{2}{1-|z|^2} \sup_{|\zeta|<1} \{\Re f(\zeta) - \Re f(z)\}, \tag{6.6.4}$$

where z is an arbitrary point of the unit disk $\mathbb{D}$ and f is an analytic function on $\mathbb{D}$ with real part bounded from above. There exists also the analogous sharp estimate (see, e.g., Ingham [**Ing**], Rajagopal [**Raj**]), containing $\Re f(0)$ on the right-hand side instead of $\Re f(z)$:

$$|f'(z)| \le \frac{2}{(1-|z|)^2} \sup_{|\zeta|<1} \{\Re f(\zeta) - \Re f(0)\}. \tag{6.6.5}$$

Furthermore, we mention the Ruscheweyh inequality [**Rusc**]

$$|f'(z)| \le \frac{2}{1-|z|^2} \Re f(z) \tag{6.6.6}$$

for analytic functions with positive real part on $\mathbb{D}$.

Besides, we mention two equivalent inequalities for the first derivative (see Khavinson [**Kh**])

$$|f'(z)| \le \frac{4}{\pi(1-|z|^2)} \sup_{|\zeta|<1} |\Re f(\zeta)| \tag{6.6.7}$$

and

$$|f'(z)| \le \frac{2}{\pi(1-|z|^2)} \operatorname{osc}_{\mathbb{D}}(\Re f), \tag{6.6.8}$$

where z is an arbitrary point in $\mathbb{D}$.

The sharp pointwise estimate for the modulus of the derivative of odd order for a bounded analytic function in a disk was established by Szász [**Szá**]. Sharp estimates for $|f^{(n)}(z)|$ with the $L^p(\partial\mathbb{D})$-norm of $|f|$ in the right-hand side for analytic functions in the unit disk $\mathbb{D}$ are due to Makintyre and Rogosinski [**MaRo**].

There is a series of sharp estimates for derivative of functions f analytic in the upper half-plane $\mathbb{C}_+ = \{z \in \mathbb{C} : \Im z > 0\}$ with different characteristics of the real part of f in the majorant part. In particular, it is the Lindelöf inequality in the half-plane

$$|f'(z)| \le \frac{1}{\Im z} \sup_{\zeta \in \mathbb{C}_+} \{\Re f(\zeta) - \Re f(z)\} \tag{6.6.9}$$

and two equivalent inequalities

$$|f'(z)| \le \frac{2}{\pi \Im z} \sup_{\zeta \in \mathbb{C}_+} |\Re f(\zeta)| \tag{6.6.10}$$

and

$$|f'(z)| \le \frac{1}{\pi \Im z} \operatorname{osc}_{\mathbb{C}_+}(\Re f) \tag{6.6.11}$$

where $\operatorname{osc}_{\mathbb{C}_+}(\Re f)$ is the oscillation of $\Re f$ on $\mathbb{C}_+$, and z is an arbitrary point in $\mathbb{C}_+$.

Estimates for derivatives of analytic functions in $\mathbb{C}_+$ with boundary values of their real parts in $L^p(\partial\mathbb{C}_+)$ were obtained in Kresin and Maz'ya [**KM14, KM15**]. In particular, the inequality with the sharp coefficient

$$|f^{(2m+1)}(z)| \le \frac{2}{\pi}\,\frac{[(2m+1)!!]^2}{(2m+1)(\Im z)^{2m+1}}\,\|\Re f\|_\infty\,,$$

derived in [**KM15**], generalizes (6.6.10).

The inequalities for the first derivative of an analytic function just mentioned can be restated as estimates for the gradient of a harmonic function. For example, Lindelöf's inequality (6.6.4) can be written in the form

$$|\nabla u(z)| \le \frac{2}{1-|z|^2}\sup_{|\zeta|<1}\{u(\zeta)-u(z)\}, \tag{6.6.12}$$

where u is a harmonic function in the unit disk.

The collection of real-part theorems and related assertions is rather broad. It involves assertions of various form (see, e.g., Kresin and Maz'ya [**KM11**] and the bibliography collected there).

We mention another type of sharp estimates for functions which are harmonic in the disk $\mathbb{D}$, different from estimates of the type (6.6.12) (see Kresin and Maz'ya [**KM11**], Corollary 4.7):

$$\left|\frac{\partial u(z)}{\partial \boldsymbol{l}_{\alpha+\beta}}\right| \le \mathcal{C}_p(|z|,\alpha)\left\|\frac{\partial u}{\partial \boldsymbol{l}_\beta}\right\|_p\,, \tag{6.6.13}$$

where $1\le p\le\infty$, $\boldsymbol{l}_\beta$ is a unit vector such that the angle β between $\boldsymbol{l}_\beta$ and the radial direction is constant.

In particular,

$$\mathcal{C}_1(r,\alpha)=\frac{1}{\pi(1-r^2)}\,,\qquad \mathcal{C}_2(r,\alpha)=\frac{1}{\sqrt{\pi(1-r^2)}}\,.$$

Unlike the cases $p=1$ and $p=2$, the coefficient $\mathcal{C}_\infty(r,\alpha)$ depends on α and

$$\max_\alpha \mathcal{C}_\infty(r,\alpha)=\mathcal{C}_\infty(r,\pi/2)=\frac{2}{\pi r}\log\left(\frac{1+r}{1-r}\right)\,, \tag{6.6.14}$$

$$\min_\alpha \mathcal{C}_\infty(r,\alpha)=\mathcal{C}_\infty(r,0)=\frac{4}{\pi r}\arctan r\,.$$

Suppose that $|\partial u(\zeta)/\partial \boldsymbol{l}_\beta|\le 1$ for $|\zeta|=1$. If, in view of (6.6.14), we put $\beta=0, p=\infty$ in (6.6.13), we conclude that the maximum modulus of the directional derivative is attained at the direction which is tangent to the circumference $|z|=r$. Inequality (6.6.13) with $\beta=0, p=\infty$ and $\alpha=\pi/2$ shows that the maximum value in the Khavinson's type extremal problem is not always attained at the radial direction.

Putting $\beta=\pi/2, p=\infty$ in (6.6.13), we find by (6.6.14) that the maximum modulus of the directional derivative is attained at the direction which is normal to the circumference $|z|=r$. In connection with this we mention the paper by Hile and Stanoyevitch [**HS**] which proves that $|\nabla u(x)|$ has the logarithmic growth as x approaches the smooth boundary ∂G of a bounded domain $G\in\mathbb{R}^n$ under the assumption that the boundary values of a harmonic function are Lipschitz.

CHAPTER 7

The Norm and the Essential Norm for Double Layer Vector-Valued Potentials

7.1. Introduction

When studying boundary integral equations on irregular curves and surfaces, one encounters difficulties connected with the change of functional-analytic properties of boundary integral operators. For sufficiently smooth boundaries these operators are compact whereas they lose compactness if the boundary contains a singularity. However, for a rather large class of irregular curves and surfaces boundary integral operators remain bounded. While studying solvability of boundary integral equations of the theory of logarithmic potentials, Radon introduced a notion of the essential norm of a bounded operator (see Radon [**Rad2**]). We give its definition.

Let L be a bounded linear operator acting on a Banach space $\mathfrak{B}$. The quantity

$$\text{ess}\|L\| = \inf\|L - \mathcal{K}\|_{\mathfrak{B}}, \tag{7.1.1}$$

where the infimum is taken over the set $\{\mathcal{K}\}$ of all compact operators in $\mathfrak{B}$, is called the essential norm of the operator L.

In this chapter we study a class of matrix-valued integral operators which appear as limit values of the vector-valued potentials of the double layer type in multidimensional domains with irregular boundaries. We find general representations for the norms and for the essential norms of such operators in the space of continuous vector valued functions. These representations are specified for boundary integral operators of linear isotropic elasticity theory and hydrodynamics of viscous incompressible fluid under the assumption that there are an angle point on the boundary of a plane domain and a conic vertex or an edge on the boundary of a three-dimensional domain.

Here is a layout of this chapter.

Section 7.2 is auxiliary. For reader's convenience we introduce the notion and properties of the perimeter $P(\mathcal{B})$ of a Borel set $\mathcal{B} \subset \mathbb{R}^n$ in the sense of Caccioppoli and De Giorgi, of the solid angle $\omega_{\mathcal{D}}(p, \mathcal{B})$ at which the set $\mathcal{B} \cap \partial\mathcal{D}$ is seen from the point p, of the normal in the sense of Federer, of the reduced boundary $\partial^*\mathcal{D}$ of the set $\mathcal{D}$.

In Section 7.3 we study matrix-valued integral operators of the double layer potential type. Similarly to Burago and Maz'ya [**BuMaz**] and Král [**Kral**], we assume that $\mathcal{D}$ is an open set in $\mathbb{R}^n$ with the compact closure $\overline{\mathcal{D}}$, $\mathcal{D}$ has a finite perimeter $P(\mathcal{D})$, $\partial\mathcal{D} = \partial(\mathbb{R}^n \backslash \overline{\mathcal{D}})$ and that

$$\sup_{p \in \partial\mathcal{D}} |\,\omega_{\mathcal{D}}|(p,\, \partial\mathcal{D}\backslash p) < \infty\,. \tag{7.1.2}$$

This condition is equivalent to

$$\sup_{p \notin \partial\mathcal{D}} |\,\omega_{\mathcal{D}}|(p,\ \partial\mathcal{D}) < \infty \tag{7.1.3}$$

(see Burago and Maz'ya [**BuMaz**]).

In Subsection 7.3.1, to each point $p \in \mathbb{R}^n$ we assign the matrix-valued set function

$$\Psi_{\mathcal{D}}(p, \mathcal{B}) = \int_{\mathcal{B}\cap\partial\mathcal{D}} \mathcal{M}(\boldsymbol{e}_{px})\omega_{\mathcal{D}}(p, dx), \qquad \boldsymbol{e}_{px} = (x-p)|x-p|^{-1},$$

initially defined on Borel sets $\mathcal{B}$ whose closures do not contain the point p. Here $\mathcal{M}$ is a homogeneous of degree 0 even $(m \times m)$-matrix-valued function continuous on $\mathbb{S}^{n-1}$, normalized by the condition

$$\int_{\mathbb{S}^{n-1}} \mathcal{M}(\sigma)d\sigma = I, \tag{7.1.4}$$

where I is the unit matrix of order m. After that, the function $\Psi_{\mathcal{D}}(p, \mathcal{B})$ is defined for all Borel subsets of $\mathbb{R}^n$ and its basic properties are established.

Further, we introduce vector-valued double layer potentials of the form

$$\boldsymbol{W}(q) = \int_{\partial\mathcal{D}} \mathcal{M}(\boldsymbol{e}_{qx})\boldsymbol{\varphi}(x)\omega_{\mathcal{D}}(q, dx), \qquad q \notin \partial\mathcal{D}, \tag{7.1.5}$$

where $\boldsymbol{\varphi}$ is an element of the space $[\mathrm{C}(\partial\mathcal{D})]^m$, i.e. $\boldsymbol{\varphi}$ is a continuous m-component vector-valued function on $\partial\mathcal{D}$ with the norm

$$\|\boldsymbol{\varphi}\| = \sup\big\{|\boldsymbol{\varphi}(x)| : x \in \partial\mathcal{D}\big\}.$$

We prove that the inner limit value $\boldsymbol{W}^{(i)}(p)$ and the outer limit value $\boldsymbol{W}^{(e)}(p)$ of the vector-valued potential $\boldsymbol{W}(q)$ satisfy the relations

$$2\boldsymbol{W}^{(i)}(p) = \boldsymbol{\varphi} + T\boldsymbol{\varphi},$$

$$2\boldsymbol{W}^{(e)}(p) = -\boldsymbol{\varphi} + T\boldsymbol{\varphi}.$$

Here

$$(T\boldsymbol{\varphi})(p) = 2\int_{\partial\mathcal{D}} \Psi_{\mathcal{D}}(p, dx)\boldsymbol{\varphi}(x), \tag{7.1.6}$$

where the last integral is the value of the potential $\boldsymbol{W}(p)$ on $\partial\mathcal{D}$.

In Subsection 7.3.2 we find representations for the norm and for the essential norm of the linear continuous operator T in $[\mathrm{C}(\partial\mathcal{D})]^m$. Our starting point is the following expression for the norm of the matrix-valued integral operator (7.1.6) in $[\mathrm{C}(\partial\mathcal{D})]^m$.

THEOREM 1. *Let $\mathcal{M}^*$ denote the adjoint matrix of $\mathcal{M}$. Then*

$$\|T\| = 2 \sup_{p\in\partial^*\mathcal{D}} \sup_{|\boldsymbol{z}|=1} \int_{\partial\mathcal{D}} \big|\mathcal{M}^*(\boldsymbol{e}_{px})\boldsymbol{z}\big| |\omega_{\mathcal{D}}|(p, dx).$$

This expression is simplified in the case of a convex surface.

COROLLARY 1. *Let $\mathcal{D}$ be a bounded convex domain in $\mathbb{R}^n$. Then*

$$\|T\| = \sup_{|\boldsymbol{z}|=1} \int_{\mathbb{S}^{n-1}} \big|\mathcal{M}^*(\sigma)\boldsymbol{z}\big| d\sigma.$$

The following assertion extends formula (7.5.5) to operator (7.1.6).

THEOREM 2. *The essential norm of T in $[\mathrm{C}(\partial\mathcal{D})]^m$ is equal to*

$$2\lim_{\delta\to 0}\sup_{p\in\partial^*\mathcal{D}}\sup_{|\boldsymbol{z}|=1}\int_{\partial\mathcal{D}\cap\mathbb{B}_{\delta(p)}}\big|\mathcal{M}^*(\boldsymbol{e}_{px})\boldsymbol{z}\big||\omega_{\mathcal{D}}|(p,dx).$$

In Section 7.4 we turn to the boundary integral operators of elasticity and hydrodynamics which are special cases of (7.1.6). We assume that the boundary contains an angle point for $n=2$ and either an edge or a conic vertex for $n=3$.

Vector-valued potentials similar to double layer potentials are introduced in Subsection 7.4.3 for the Lamé system

$$\mu\Delta\boldsymbol{u}+(\lambda+\mu)\text{grad div }\boldsymbol{u}=\boldsymbol{0}$$

and for the Stokes system

$$\nu\Delta\boldsymbol{u}-\text{grad }p=\boldsymbol{0},\ \text{div }\boldsymbol{u}=0.$$

They are defined as follows. Let $H_\kappa^{(n)}(\boldsymbol{e})$ be the matrix with the elements

$$\frac{1}{\omega_n}\big[(1-\kappa)\delta_{ij}+n\kappa(\boldsymbol{e},\boldsymbol{e}_i)(\boldsymbol{e},\boldsymbol{e}_j)\big],\ i,j=1,\dots,n,$$

where $\boldsymbol{e}_j$ is the unit vector directed along the axis Ox_j. For $\kappa=(\lambda+\mu)(\lambda+3\mu)^{-1}$ and for $\kappa=1$ the matrix $\mathcal{M}(\boldsymbol{e})=H_\kappa^{(n)}(\boldsymbol{e})$ generates the elastic and hydrodynamic potentials $\boldsymbol{W}_\kappa^{(n)}(q)$ by (7.1.5). The corresponding matrix-valued integral operators of the type (7.1.6) considered in the space $[\mathrm{C}(\partial\mathcal{D})]^m$ with $m=n$ will be referred as $T_\kappa^{(n)}$.

In Subsection 7.4.4 we give the following explicit formulae for the norm of $T_\kappa^{(n)}$ in the case of a convex domain.

THEOREM 3. *Let $\mathcal{D}$ be convex. Then*

$$\|T_\kappa^{(n)}\|=\frac{2\Gamma(\frac{n}{2})}{\sqrt{\pi}\Gamma(\frac{n-1}{2})}\int_0^{\pi/2}\Big[(1-\kappa)^2+n\kappa(n\kappa-2\kappa+2)\cos^2\vartheta\Big]^{1/2}\sin^{n-2}\vartheta d\vartheta.$$

In particular,

$$\begin{aligned}\|T_\kappa^{(2)}\| &= \frac{2}{\pi}(1+\kappa)\mathbf{E}\left(\frac{2\sqrt{\kappa}}{1+\kappa}\right)\\ &= 1+\frac{1}{2^2}\,\kappa^2+\frac{1}{2^24^2}\,\kappa^4+\cdots+\left[\frac{(2m-3)!!}{2^m m!}\right]^2\kappa^{2m}+\dots,\end{aligned}$$

where $\mathbf{E}(k)$ is the complete elliptic integral of the second kind, and

$$\|T_\kappa^{(3)}\|=\frac{1}{2}\left[(1+2\kappa)+\frac{(1-\kappa)^2}{\sqrt{3\kappa(\kappa+2)}}\log\frac{1+2\kappa+\sqrt{3\kappa(\kappa+2)}}{1-\kappa}\right].$$

For the hydrodynamic double layer potential we have

$$\|T_1^{(n)}\|=\frac{2}{\sqrt{\pi}}\frac{\Gamma\left(\frac{n}{2}+1\right)}{\Gamma\left(\frac{n+1}{2}\right)},$$

so that

$$\|T_1^{(2)}\|=4/\pi,\ \ \|T_1^{(3)}\|=3/2.$$

In Subsection 7.4.5 we obtain the formula for the essential norm of the operator $T_\kappa^{(2)}$ assuming that there is an angle point at the boundary of the plane domain:

$$\text{ess}\|T_\kappa^{(2)}\| = \frac{2}{\pi}(1+\kappa)E\left(\frac{\pi-\alpha}{2}, \frac{2\sqrt{\kappa}}{1+\kappa}\right), \tag{7.1.7}$$

where $\alpha = \min\{\beta, 2\pi-\beta\}$ with β being the opening of the angle on the side of $\mathcal{D}$, and

$$E(\varphi, k) = \int_0^\varphi \sqrt{1-k^2\sin^2\varphi}\, d\varphi$$

is the elliptic integral of the second kind.

In particular,

$$\text{ess}\|T_1^{(2)}\| = \frac{4}{\pi}\cos\left(\frac{\alpha}{2}\right).$$

Expressions for $\text{ess}\|T_\kappa^{(3)}\|$ similar to (7.1.7) are found in Subsections 7.4.6 and 7.4.7 for 3-dimensional domains with either conic vertex or edge, respectively.

For example, in Subsection 7.4.6 it is shown that for a domain $\mathcal{D}$ with vertex of a rotational cone at the boundary one has

$$\text{ess}\|T_1^{(3)}\| = \frac{3}{2\pi}\big(\pi - \alpha + \sin\alpha\big),$$

where $\alpha = \min\big\{\beta, 2\pi-\beta\big\}$ with β being the opening of the cone on the side of $\mathcal{D}$.

Subsection 7.4.7 concerns the cases when the boundary of domain contains a smooth edge E. Let

$$\alpha = \min_{\zeta\in E}\big\{\beta(\zeta),\ 2\pi-\beta(\zeta)\big\},$$

where $\beta(\zeta)$ is the opening of the dihedral angle at the point $\zeta \in E$ on the side of $\mathcal{D}$. Then the essential norm of the hydrodynamic operator $T_1^{(3)}$ is given by

$$\text{ess}\|T_1^{(3)}\| = \frac{3}{2}\cos\left(\frac{\alpha}{2}\right).$$

Below we use the following notation: $\mathcal{B}$ is an arbitrary Borel set in $\mathbb{R}^n$ with the interior int $\mathcal{B}$, the boundary $\partial\mathcal{B}$ and the complement $\mathbf{C}\mathcal{B} = \mathbb{R}^n\backslash\mathcal{B}$; mes_n stands for the Lebesgue measure in $\mathbb{R}^n$; H_k is the k-dimensional Hausdorff measure in $\mathbb{R}^n$; $\mathbb{B}_\rho = \big\{x\in\mathbb{R}^n : |x| < \rho\big\}$; $\mathbb{B} = \mathbb{B}_1$; $v_n = \text{mes}_n\mathbb{B}$; $|\cdot|$ is the Euclidean length of a vector; $\boldsymbol{r}_{xy} = y - x$; $r_{xy} = |y-x|$; $\text{dist}(p, \mathcal{B})$ is the distance from a point p to a set $\mathcal{B}$.

If M is an $(m\times m)$-matrix, then by $\|M\|$ we denote the matrix norm of M induced by the Euclidean norm in $\mathbb{R}^m$.

Henceforth by c with sub-and super-indices we mean constants whose exact values are insignificant.

7.2. Definition and certain properties of a solid angle

We assume in this section that the dimension of the space is greater than 2; in the case $n = 2$ the arguments merely become simpler.

Let G be an open set in $\mathbb{R}^n$. By $\mathrm{C}_0(G)$ and $\mathrm{C}_0^\infty(G)$ we denote the spaces of functions from, respectively, $\mathrm{C}(G)$ and $\mathrm{C}^\infty(G)$ with compact support in G.

Let E be a bounded Borel subset of $\mathbb{R}^n$ and let p be a point in $\mathbb{R}^n$. Let us consider the distribution $\omega_E(p,\varphi)$ on the space $\mathrm{C}_0^\infty(\mathbb{R}^n\backslash p)$ defined by the equality

$$(n-2)\omega_E(p,\varphi) = -\int_E \nabla\varphi(x)\nabla r_{px}^{2-n}\,dx\,. \tag{7.2.1}$$

It is clear that supp $\omega_E(p,\varphi) \subset \partial E \backslash p$.

Let us assume that $\omega_E(p,\varphi)$ is a signed measure (charge) in $\mathbb{R}^n \backslash p$, i.e., that functional (7.2.1) admits an extension up to a linear, continuous functional on $\mathrm{C}_0(\mathbb{R}^n \backslash p)$.

The set function $\omega_E(p,\mathcal{B})$ is defined and finite for all Borel sets $\mathcal{B}$ such that $\overline{\mathcal{B}} \subset \mathbb{R}^n \backslash p$.

DEFINITION 7.1. Let $\mathcal{B} \subset \mathbb{R}^n$ be a Borel set and suppose its characteristic function $\chi_{\mathcal{B}}$ belongs to the space $BV(\mathbb{R}^n)$ of all locally integrable functions on $\mathbb{R}^n$ whose gradients (in the distributional sense) are finite vector-valued charges on $\mathbb{R}^n$. The total variation of the vector-valued charge $\nabla\chi_{\mathcal{B}}$ is called the perimeter of the set $\mathcal{B}$ (in the sense of Caccioppoli and De Giorgi) and is denoted by $P(\mathcal{B})$.

LEMMA 7.1. *If $P(E) < \infty$, then for any point $p \in \mathbb{R}^n$ the distribution $\omega_E(p,\varphi)$ is a charge and, moreover, for any Borel set $\mathcal{B}$ such that $p \notin \overline{\mathcal{B}}$,*

$$(n-2)\omega_E(p,\mathcal{B}) = \int_{\mathcal{B}} \nabla r_{px}^{2-n} \nabla\chi_E(dx)\,. \tag{7.2.2}$$

PROOF. Let $\varphi \in \mathrm{C}_0^\infty(\mathbb{R}^n \backslash p)$. Then

$$(n-2)\omega_E(p,\varphi) = -\int_E \nabla\varphi(x)\nabla r_{px}^{2-n}dx = -\int_E \operatorname{div}\boldsymbol{\psi}(x)dx\,,$$

where $\boldsymbol{\psi}(x) = \varphi(x)\nabla r_{px}^{2-n} \in \mathrm{C}^\infty(\mathbb{R}^n)$. Therefore,

$$(n-2)\omega_E(p,\varphi) = \int_{\mathbb{R}^n} \boldsymbol{\psi}(x)\nabla\chi_E(dx) = \int_{\mathbb{R}^n} \varphi(x)\nabla r_{px}^{2-n}\nabla\chi_E(dx)\,,$$

which proves the lemma. □

COROLLARY 7.1. *If $P(E) < \infty$ and $\operatorname{dist}(p,\mathcal{B}) > 0$, then*

$$|\,\omega_E|(p,\mathcal{B}) \le \big(\operatorname{dist}(p,\mathcal{B})\big)^{1-n}|\nabla\chi_E|(B)\,. \tag{7.2.3}$$

LEMMA 7.2. (De Giorgi [**DeG**]). *If $P(E) < \infty$ then for any point p and for almost all $\varepsilon > 0$,*

$$\nabla\chi_{E\cap\mathbb{B}_\varepsilon(p)}(\mathcal{B}) = \nabla\chi_E(\mathcal{B}\cap\mathbb{B}_\varepsilon(p)) + \nabla\chi_{\mathbb{B}_\varepsilon(p)}(\mathcal{B}\cap E)\,. \tag{7.2.4}$$

The next corollary follows from Lemmas 7.1 and 7.2.

COROLLARY 7.2. *If $P(E) < \infty$ then for any point p and for almost all $\varepsilon > 0$,*

$$\omega_{E\cap\mathbb{B}_\varepsilon(p)}(q,\mathcal{B}) = \omega_E(q,\mathcal{B}\cap\mathbb{B}_\varepsilon(p)) + \omega_{\mathbb{B}_\varepsilon(p)}(q,\mathcal{B}\cap E)\,, \tag{7.2.5}$$

$$\omega_{E\cap C\mathbb{B}_\varepsilon(p)}(q,\mathcal{B}) = \omega_E(q,\mathcal{B}\cap \mathbf{C}\mathbb{B}_\varepsilon(p)) - \omega_{\mathbb{B}_\varepsilon(p)}(q,\mathcal{B}\cap E) \tag{7.2.6}$$

for all sets $\mathcal{B}$ and points q such that $\operatorname{dist}(q,\mathcal{B}) > 0$.

COROLLARY 7.3. *If $P(E) < \infty$ then*

$$\text{(a) } \omega_E(p,\partial E) = \omega_n \quad \text{for } p \in \operatorname{int} E\,,$$
$$\text{(b) } \omega_E(p,\partial E) = 0 \quad \text{for } p \in \mathbf{C}\overline{E}\,.$$

PROOF. (a) Let $f \in \mathrm{C}_0^\infty(\mathbb{R}^n \backslash p)$. Then from equalities (7.2.1) and (7.2.6) we get that for almost all ε,

(7.2.7)
$$(2-n)\int_{E\cap\mathbf{C}\mathbb{B}_\varepsilon(p)} \nabla f(x)\nabla r_{px}^{2-n}dx = \int_{\mathbf{C}\mathbb{B}_\varepsilon(p)} f(x)\omega_E(p,dx) - \varepsilon^{1-n}\int_{E\cap\partial\mathbb{B}_\varepsilon(p)} f(\sigma)d\sigma\,,$$

where $d\sigma$ is an element of area of the sphere $\partial\mathbb{B}_\varepsilon(p)$. We choose f and ε such that $f = 1$ in the neighborhood of $E \cap \mathbf{C}\mathbb{B}_\varepsilon(p)$ and $\overline{\mathbb{B}_\varepsilon(p)} \subset \mathrm{int}E$. Then $\omega_E(p, \partial E) = \omega_n$ follows from (7.2.7).

The assertion (b) follows immediately from equality (7.2.1) when $\varphi \in C_0^\infty(\mathbb{R}^n\backslash p)$, $\varphi = 1$, in the neighborhood of E. □

We now assume that

$$|\,\omega_E|(p,\ \mathcal{B}) \le c(p) < \infty\,, \tag{7.2.8}$$

if $\overline{\mathcal{B}} \subset \mathbb{R}^n\backslash p$. Then the charge $\omega_E(p, \mathcal{B})$ can be extended by additivity onto all Borel subsets in $\mathbb{R}^n\backslash p$.

We define the set function $\omega_E(p, \mathcal{B})$ at the point p. By definition, let

$$\omega_E(p,p) \quad = \quad \begin{cases} \frac{\omega_n}{2} - \omega_E(p, \mathbb{R}^n\backslash p) & \text{for } p \in \partial E\,, \\ 0 & \text{for } p \notin \partial E\,. \end{cases} \tag{7.2.9}$$

We assume, further, that

$$\omega_E(p, \mathcal{B}) = \omega_E(p, \mathcal{B}\backslash p) + \omega_E(p,p)$$

when $p \in \mathcal{B}$. Now, the charge $\omega_E(p, \mathcal{B})$ is defined and is finite for all Borel subsets in $\mathbb{R}^n$.

From equality (7.2.9) and Corollary 7.3, we deduce

$$\omega_E(p, \mathbb{R}^n) = \begin{cases} \omega_n & \text{for } p \in \mathrm{int}E\,, \\ \frac{\omega_n}{2} & \text{for } p \in \partial E\,, \\ 0 & \text{for } p \in \mathbf{C}\overline{E}\,. \end{cases} \tag{7.2.10}$$

In the following lemma we establish a relation of $\omega_E(p, \mathbb{R}^n\backslash p)$ with the volume density $\delta_E(p)$ of the set E at the point p.

LEMMA 7.3. *If $p \in \partial E$ and condition (7.2.8) is fulfilled, then the volume density $\delta_E(p)$ exists and, moreover,*

$$\omega_E(p, \mathbb{R}^n\backslash p) = \omega_n \delta_E(p)\,.$$

PROOF. Setting $f = 1$ in the neighborhood of $E \cap \mathbf{C}\mathbb{B}_\varepsilon(p)$ in equality (7.2.7) for almost all $\varepsilon > 0$, we get

$$\omega_E(p, \mathbf{C}\mathbb{B}_\varepsilon) = \varepsilon^{1-n} H_{n-1}(E \cap \partial\mathbb{B}_\varepsilon)\,.$$

Consequently,

$$\frac{1}{r^n}\int_0^r \omega_E(p, \mathbf{C}\mathbb{B}_\varepsilon)\varepsilon^{n-1} d\varepsilon = \frac{1}{r^n}\int_0^r H_{n-1}(E \cap \partial\mathbb{B}_\varepsilon) d\varepsilon = \frac{1}{r^n}\mathrm{mes}_n(\mathbb{B}_r \cap E)\,. \tag{7.2.11}$$

The expression on the left-hand side of (7.2.11) equals

$$\frac{1}{n}\omega_E(p, \mathbb{R}^n\backslash p) - \frac{1}{r^n}\int_0^r \omega_E(p, \mathbb{B}_\varepsilon\backslash p)\varepsilon^{n-1} d\varepsilon\,. \tag{7.2.12}$$

By (7.2.8),

$$\lim_{\tau \to 0} |\,\omega_E|(p,\ \mathbb{B}_r\backslash p) = 0\,.$$

Consequently, the integral in (7.2.12) tends to zero. Now, from (7.2.11) and (7.2.12), we obtain

$$\frac{1}{\omega_n}\omega_E(p,\mathbb{R}^n\backslash p) = \lim_{r\to 0}\frac{\mathrm{mes}_n(\mathbb{B}_r\cap E)}{\mathrm{mes}_n(\mathbb{B}_r)} = \delta_E(p)\,.$$

The lemma is proved. □

DEFINITION 7.2. A unit vector $\boldsymbol{\nu}(p)$ is referred to as the (outer) normal (in the sense of Federer) to the set E at the point p if

$$\lim_{\varepsilon\to 0}\varepsilon^{-n}\mathrm{mes}_n\big\{y:\ y\in E\cap\mathbb{B}_\varepsilon(p),\ \big(\boldsymbol{\nu}(p),\boldsymbol{r}_{py}\big)>0\big\}=0, \tag{7.2.13}$$

$$\lim_{\varepsilon\to 0}\varepsilon^{-n}\mathrm{mes}_n\big\{y:\ y\in E\cap\mathbb{B}_\varepsilon(p),\ \big(\boldsymbol{\nu}(p),\boldsymbol{r}_{py}\big)<0\big\}=\frac{v_n}{2}, \tag{7.2.14}$$

where v_n is the volume of the n-dimensional unit ball. If the normal does not exist we put formally $\boldsymbol{\nu}(p)=\mathbf{0}$.

It is known (see De Giorgi [**DeG**], Federer [**Fe**]) that

$$\nabla\chi_{\mathcal{D}}(\mathcal{B}) = \int_{\mathcal{B}\cap\partial\mathcal{D}}\boldsymbol{\nu}(x)H_{n-1}(dx). \tag{7.2.15}$$

This allows us to rewrite formula (7.2.2) in the "classical" form

$$\omega_E(p,\mathcal{B}) = \frac{1}{2-n}\int_{\mathcal{B}\cap\partial E}\boldsymbol{\nu}(x)\nabla r_{px}^{2-n}H_{n-1}(dx)\,,\ p\notin\overline{\mathcal{B}}\,.$$

DEFINITION 7.3. The set ∂^*E of all points $p\in\partial E$ at which a normal to E (in the sense of Federer) exists is called the reduced boundary of E.

As is well known (see De Giorgi [**DeG**], Federer [**Fe**]),

$$|\nabla\chi_E|(\mathbb{R}^n\backslash\partial^*E)=0\,. \tag{7.2.16}$$

LEMMA 7.4. *Let $\partial\mathrm{int}E=\partial\mathbf{C}\overline{E}$ and $P(E)<\infty$. Then the set ∂^*E is everywhere dense on ∂E.*

PROOF. We first prove that for any point $s\in\partial E$ and any $\varepsilon>0$,

$$|\nabla\chi_E|(\mathbb{B}_\varepsilon(s))>0\,. \tag{7.2.17}$$

We assume that for all $\varepsilon\le\varepsilon_0$,

$$|\nabla\chi_E|(\mathbb{B}_\varepsilon(s))=0\,. \tag{7.2.18}$$

According to Corollary 7.2, the equalities

$$\omega_{E\cap\mathbb{B}_\varepsilon(s)}(q,\partial(\mathbb{B}_\varepsilon(s)\cap E))=\omega_E(q,\mathbb{B}_\varepsilon(s))+\omega_{\mathbb{B}_\varepsilon(s)}(q,E)\,, \tag{7.2.19}$$

$$\omega_{\mathbf{C}E\cap\mathbb{B}_\varepsilon(s)}(q,\partial(\mathbb{B}_\varepsilon(s)\cap\mathbf{C}E))=\omega_E(q,\mathbb{B}_\varepsilon(s))+\omega_{\mathbb{B}_\varepsilon(s)}(q,\mathbf{C}E) \tag{7.2.20}$$

are satisfied for almost every $\varepsilon>0$ and for all $q\notin\partial\mathbb{B}_\varepsilon(s)\cap\partial E$.

By lemma's hypothesis, the sets $\mathbb{B}_\varepsilon(s)\cap\mathrm{int}E$ and $\mathbb{B}_\varepsilon(s)\cap\mathbf{C}\overline{E}$ are nonempty for all $\varepsilon>0$. We choose $q\in\mathbb{B}_\varepsilon(s)\cap\mathrm{int}E$. Then, the left-hand side of (7.2.19) equals ω_n (Corollary 7.3). Furthermore, estimate (7.2.3) shows that the equality $\omega_E(q,\mathbb{B}_\varepsilon(s))=0$ follows from (7.2.18). Therefore, (7.2.19) yields

$$\omega_{\mathbb{B}_\varepsilon(s)}(q,E)=\omega_n\,.$$

Consequently,

$$H_{n-1}(\partial\mathbb{B}_\varepsilon(s))\cap E)=H_{n-1}(\partial\mathbb{B}_\varepsilon(s))\,.$$

Analogously, by choosing $q \in \mathbb{B}_\varepsilon(s) \cap \mathbf{C}\overline{E}$, from (7.2.20) we get

$$H_{n-1}(\partial\mathbb{B}_\varepsilon(s)) \cap \mathbf{C}E) = H_{n-1}(\partial\mathbb{B}_\varepsilon(s)),$$

which contradicts the previous equality. Consequently, inequality (7.2.17) is valid.

We now assume that the lemma's assertion is false. Then we can find a point $s \in \partial E$ and a number $\varepsilon > 0$ such that $\mathbb{B}_\varepsilon(s) \cap \partial^* E = \emptyset$. Since the charge $|\nabla\chi_E|$ is concentrated on $\partial^* E$, it follows that

$$|\nabla\chi_E|(\mathbb{B}_\varepsilon(s)) = |\nabla\chi_E|(\mathbb{B}_\varepsilon(s) \cap \partial^* E) = 0,$$

which contradicts (7.2.17).

The lemma is proved. □

COROLLARY 7.4. *If* $\partial\text{int}E = \partial\mathbf{C}\overline{E}$, $P(E) < \infty$, *and condition* (7.2.8) *is fulfilled for all* $p \in \partial E$, *then the set of points* $p \in \partial E$, *for which* $\omega_E(p,p) = 0$, *is everywhere dense on* ∂E.

PROOF. By Lemma 7.4, it is enough to show that $\omega_E(p,p) = 0$ for all $p \in \partial^* E$. If $p \in \partial^* E$, then equalities (7.2.13), (7.2.14) are fulfilled at the point p. Adding them we get that the volume density of the set E at the point p equals $1/2$ and, consequently, by Lemma 7.3, $\omega_E(p,p) = 0$. □

REMARK 7.1. If $P(E) < \infty$, the charge $\omega_E(q,\mathcal{B})$ does not have point loadings on $\mathbb{R}^n \backslash q$.

Indeed, in (7.2.5) we set $\mathcal{B} = \mathbb{B}_\varepsilon(p)$, where $p \in \partial E \backslash q$ and ε is so small that $q \notin \overline{\mathbb{B}_\varepsilon(p)}$. Then, using (7.2.10), we obtain

$$\omega_E(q, \mathbb{B}_\varepsilon(p)) = -\omega_{\mathbb{B}_\varepsilon(p)}(q, E).$$

Hence, as $\varepsilon \to 0$ we get $\omega_E(q,p) = 0$.

LEMMA 7.5. *Let* $P(E) < \infty$. *Then for any Borel set* $\mathcal{B}$ *and for any points* $p, q \notin \overline{\mathcal{B}} \cap \partial E$,

$$|\omega_E(p,\mathcal{B}) - \omega_E(q,\mathcal{B})| \le c|\nabla\chi_E|(\mathcal{B}) r^{-n} r_{pq}, \tag{7.2.21}$$

where r *is the smallest of the distances from* p *and* q *to* $\overline{\mathcal{B}} \cap \partial E$ *and* c *is a constant depending only on* n.

PROOF. We set $\mu(\mathcal{B}) = \omega_E(p,\mathcal{B}) - \omega_E(q,\mathcal{B})$. By (7.2.2),

$$\mu(\mathcal{B}) = \frac{1}{n-2}\int_{\mathcal{B}} \nabla_x (r_{px}^{2-n} - r_{qx}^{2-n}) \nabla\chi_E(dx).$$

Consequently,

$$|\mu|(\mathcal{B}) \le \int_{\mathcal{B}} |\nabla_x (r_{px}^{2-n} - r_{qx}^{2-n})| |\nabla\chi_E|(dx) \le \max_{x \in \overline{\mathcal{B}} \cap \partial E} |\nabla_x (r_{px}^{2-n} - r_{qx}^{2-n})| |\nabla\chi_E|(\mathcal{B}).$$

Obviously,

$$\begin{aligned} |\nabla_x (r_{px}^{2-n} - r_{qx}^{2-n})| &\le (n-2)\big(|\boldsymbol{r}_{px} - \boldsymbol{r}_{qx}| r_{px}^{-n} + |r_{px}^n - r_{qx}^n| r_{px}^{-n} r_{qx}^{1-n}\big) \\ &\le (n-2) r_{pq} \sum_{j=0}^{n} r_{px}^{-j} r_{qx}^{j-n} \le (n+1)(n-2) r^{-n} r_{pq}. \end{aligned}$$

The lemma is proved. □

COROLLARY 7.5. *Under the hypothesis of Lemma* 7.4,

$$\big| |\omega_E|(p,B) - |\omega_E|(q,B) \big| \le c|\nabla\chi_E|(\mathcal{B}) r^{-n} r_{pq}\,. \tag{7.2.22}$$

THEOREM 7.1. *If* $\partial \mathrm{int} E = \partial \mathbf{C}\overline{E}$, *then the equality*

$$\sup_{p\notin\partial E} |\omega_E|(p,\mathbb{R}^n) = \sup_{p\in\partial E} |\omega_E|(p,\mathbb{R}^n) + \frac{\omega_n}{2} \tag{7.2.23}$$

is valid.

The assertion of this theorem follows from Lemmas 7.6 and 7.8.

LEMMA 7.6. *If* $\partial \mathrm{int} E = \partial \mathbf{C}\overline{E}$ *and* $p \in \partial E$ *there holds the inequality*

$$|\omega_E|(p,\mathbb{R}^n) + \frac{\omega_n}{2} \le \limsup_{\substack{p_k\to p\\ p_k\notin\partial E}} |\omega_E|(p_k,\mathbb{R}^n)\,. \tag{7.2.24}$$

PROOF. Let us first assume that $\omega_E(p,p) \ge 0$. We consider the sequence of points $p_k \to p$, $p_k \in \mathrm{int} E$. When $r_{pp_k} < \delta/2$, according to (7.2.22) we have

$$\big| |\omega_E|(p,\mathbf{C}\mathbb{B}_\varepsilon(p)) - |\omega_E|(p_k,\mathbf{C}\mathbb{B}_\varepsilon(p)| \le c\Big(\frac{\varepsilon}{2}\Big)^{-n} P(E) r_{pp_k}\,.$$

Consequently,

$$|\omega_E|(p,\mathbf{C}\mathbb{B}_\varepsilon(p)) + |\omega_E(p_k,\mathbf{C}\mathbb{B}_\varepsilon(p))| \le |\omega_E|(p_k,\mathbb{R}^n) + c\Big(\frac{\varepsilon}{2}\Big)^{-n} P(E) r_{pp_k}\,.$$

Note that by virtue of (7.2.10) and Lemma 7.5,

$$\lim_{k\to\infty} |\omega_E(p_k,\mathbb{B}_\varepsilon(p))| = \lim_{k\to\infty} |\omega_n - \omega_E(p_k,\mathbf{C}\mathbb{B}_\varepsilon(p))| = |\omega_n - \omega_E(p,\mathbf{C}\mathbb{B}_\varepsilon(p))|\,.$$

Therefore,

$$|\omega_E|(p,\mathbf{C}\mathbb{B}_\varepsilon(p)) + |\omega_n - \omega_E(p,\mathbf{C}\mathbb{B}_\varepsilon(p))| \le \liminf_{k\to\infty} |\omega_E|(p_k,\mathbb{R}^n)\,.$$

Hence, since ε is arbitrary and $\omega_E(p,p)$ is nonnegative, we find

$$|\omega_E|(p,\mathbb{R}^n) + \frac{\omega_n}{2} = |\omega_E|(p,\mathbb{R}^n\backslash p) + |\omega_n - \omega_E(p,\mathbb{R}^n\backslash p)| \le \liminf_{k\to\infty} |\omega_E|(p_k,\mathbb{R}^n)\,.$$

In the case $\omega_E(p,p) < 0$, we should choose the points p_k lying in $\mathrm{int}\mathbf{C}E$ and repeat the preceding argument. □

LEMMA 7.7. *Let* $p \in \partial E$ *and* $|\omega_E|(p,\mathbb{R}^n) < \infty$. *Then for any function* $f \in \mathrm{C}_0^\infty(\mathbb{R}^n)$ *we have*

$$\frac{1}{2-n}\int_E \nabla f(x)\nabla r_{px}^{2-n}dx = \int_{\mathbb{R}^n} f(x)\omega_E(p,dx) - \frac{\omega_n}{2} f(p)\,, \tag{7.2.25}$$

$$\frac{1}{2-n}\int_{\mathbf{C}E} \nabla f(x)\nabla r_{px}^{2-n}dx = -\int_{\mathbb{R}^n} f(x)\omega(p,dx) + \frac{\omega_n}{2} f(p)\,. \tag{7.2.26}$$

PROOF. First let $f \in \mathrm{C}_0^\infty(\mathbb{R}^n)$ and $f(p) = 0$. Let η_k, $k = 1, 2, \ldots$, denote functions from $C^\infty(\mathbb{R}^n)$, equal to zero in $\mathbb{B}_{\frac{1}{k}}(p)$, to one outside $\mathbb{B}_{\frac{2}{k}}(p)$ and such that $|\nabla\eta_k| \le c_k$, $c_k = \mathrm{const}$. Since $\eta_k f \in \mathrm{C}_0^\infty(\mathbb{R}^n\backslash p)$, by the definition of $\omega_E(p,\varphi)$ we have

$$-\int_E \eta_k\nabla_x f\nabla_x r_{px}^{2-n}dx = \int_E f\nabla_x\eta_k\nabla_x r_{px}^{2-n}dx + (n-2)\int_{\mathbb{R}^n}\eta_k f\omega_E(p,dx)\,.$$

It is clear that

$$\begin{aligned} \left| \int_E f \nabla_x \eta_k \nabla_x r_{px}^{2-n} dx \right| &\leq c' \max_{\mathbb{B}_{\frac{2}{k}}(p)} |f| \, , \\ \left| \int_{\mathbb{R}^n} (1-\eta_k) f \omega_E(p, dx) \right| &\leq c'' \max_{\mathbb{B}_{\frac{2}{k}}(p)} |f| \, , \end{aligned}$$

where $c', c'' = \text{const}$. Therefore, keeping in mind that $f(p) = 0$, and passing to the limit as $k \to \infty$, we obtain

$$\frac{1}{2-n} \int_E \nabla_x f \nabla_x r_{px}^{2-n} dx = \int_{\mathbb{R}^n} f \omega_E(p, dx) \, . \tag{7.2.27}$$

Now let f be an arbitrary function from $\mathrm{C}_0^\infty(\mathbb{R}^n)$. By of (7.2.27) we have

$$\frac{1}{2-n} \int_E \nabla_x \big(f(x) - f(p)\big) \nabla_x r_{px}^{2-n} dx = \int_{\mathbb{R}^n} [f(x) - f(p)] \omega_E(p, dx) \, ,$$

which in view of (7.2.10) is equivalent to (7.2.25).

Equality (7.2.26) is proved analogously. □

LEMMA 7.8. *The inequality*

$$\sup_{p \notin \partial E} |\omega_E|(p, \mathbb{R}^n) \leq \sup_{p \in \partial E} |\omega_E|(p, \mathbb{R}^n) + \frac{\omega_n}{2}$$

is valid.

PROOF. It suffices to consider the case

$$\sup_{p \in \partial E} |\omega_E|(p, \mathbb{R}^n) < \infty \, .$$

We introduce the notation:

$$(n-2) J(p) = \int_{\mathbf{C}E} \nabla f(x) \nabla r_{px}^{2-n} dx \, ,$$

where $f \in \mathrm{C}_0^\infty(\mathbb{R}^n)$ and $p \in \mathrm{int} E$. It is clear that $J(p)$ is a function which is continuous in $\mathbb{R}^n$ and harmonic in $\mathrm{int} E$. Furthermore,

$$J(p) = \int_{\mathbb{R}^n} f(x) \omega_E(p, dx) \, .$$

According to the maximum principle,

$$\left| \int_{\mathbb{R}^n} f(x) \omega_E(p, dx) \right| \leq \max_{s \in \partial E} |J(s)| .$$

Hence, applying (7.2.26), we get

$$\left| \int_{\mathbb{R}^n} f(x) \omega_E(p, dx) \right| \leq \max_{\partial E} |f| \left\{ \sup_{s \in \partial E} |\omega_E|(s, \mathbb{R}^n) + \frac{\omega_n}{2} \right\} .$$

Since $\mathrm{C}_0^\infty(\mathbb{R}^n)$ is dense in $\mathrm{C}(\partial E)$, the latter inequality is true also for all $f \in \mathrm{C}(\partial E)$. Consequently,

$$\sup_{p \in \mathrm{int} E} |\omega_E|(p, \mathbb{R}^n) \leq \sup_{s \in \partial E} |\omega_E|(s, \mathbb{R}^n) + \frac{\omega_n}{2} \, .$$

The inequality

$$\sup_{p \in \mathrm{int} \mathbf{C}E} |\omega_E|(p, \mathbb{R}^n) \leq \sup_{s \in \partial E} |\omega_E|(s, \mathbb{R}^n) + \frac{\omega_n}{2}$$

is proved analogously. The lemma is proved. □

7.3. Matrix-valued integral operators of the double layer potential type

7.3.1. The set function $\Psi_{\mathcal{D}}$

Henceforth we assume that $\mathcal{D}$ is a bounded open set in $\mathbb{R}^n, n \geq 2$, with finite perimeter $P(\mathcal{D})$ and that $\partial\mathcal{D} = \partial(\mathbb{R}^n \backslash \overline{\mathcal{D}})$.

By the previous Section, the solid angle at which a set $\mathcal{B} \cap \partial\mathcal{D}$ is seen from a point $p \notin \overline{\mathcal{B}}$ is the set function

$$\omega_{\mathcal{D}}(p, \mathcal{B}) = \begin{cases} -\displaystyle\int_{\mathcal{B}} \nabla \log r_{px}^{-1} \, \nabla \chi_{\mathcal{D}}(dx) & \text{for } n = 2, \\ \dfrac{1}{2-n} \displaystyle\int_{\mathcal{B}} \nabla r_{px}^{2-n} \nabla \chi_{\mathcal{D}}(dx) & \text{for } n > 2. \end{cases} \tag{7.3.1}$$

In view of (7.2.15) the integrals may also be taken over $\mathcal{B} \cap \partial\mathcal{D}$ and $\chi_{\mathcal{D}}(dx)$ may be replaced by $\boldsymbol{\nu}(x) H_{n-1}(dx)$.

Let $\mathcal{M}$ be a homogeneous of degree 0 even $(m \times m)$-matrix-valued function continuous on $\mathbb{S}^{n-1}$. With each $p \in \mathbb{R}^n$ we connect the matrix-valued set function

$$\Psi_{\mathcal{D}}(p, \mathcal{B}) = \int_{\mathcal{B} \cap \partial\mathcal{D}} \mathcal{M}(\boldsymbol{e}_{px}) \omega_{\mathcal{D}}(p, dx), \tag{7.3.2}$$

first defined for the sets $\mathcal{B}$ whose closures do not contain p.

By (7.2.6) and (7.3.2),

$$\Psi_{\mathcal{D} \cap \mathbf{C}\mathbb{B}_\varepsilon(p)}(q, \mathcal{B}) = \Psi_{\mathcal{D}}\big(q, \mathcal{B} \cap \mathbf{C}\mathbb{B}_\varepsilon(p)\big) - \Psi_{\mathbb{B}_\varepsilon(p)}\big(q, \mathcal{B} \cap \mathcal{D}\big) \tag{7.3.3}$$

for any p, for almost all $\varepsilon > 0$ and for all $\mathcal{B}$ and q with $\operatorname{dist}(q, \mathcal{B}) > 0$.

LEMMA 7.9. *If $p \notin \overline{\mathcal{D}}$ then $\Psi_{\mathcal{D}}(p, \partial\mathcal{D}) = 0$.*

PROOF. Let $n > 2$. The case $n = 2$ is treated in a similar way. If the elements of $\mathcal{M}$ are continuously differentiable on $\mathbb{S}^{n-1}$ then, by the Gauss-Green theorem (see Federer [**Fe**]),

$$0 = \int_{\mathcal{D}} \nabla \mathcal{M}\big(\boldsymbol{e}_{px}\big) \nabla r_{px}^{2-n} dx = \int_{\partial\mathcal{D}} \mathcal{M}\big(\boldsymbol{e}_{px}\big) \boldsymbol{\nu}(x) \nabla r_{px}^{2-n} H_{n-1}(dx)$$

for $p \in \mathbb{R}^n \backslash \overline{\mathcal{D}}$. Hence,

$$\int_{\partial\mathcal{D}} \mathcal{M}\big(\boldsymbol{e}_{px}\big) \omega_{\mathcal{D}}\big(p, dx\big) = 0, \; p \in \mathbb{R}^n \backslash \overline{\mathcal{D}}.$$

Clearly, this equality is valid for continuous matrices $\mathcal{M}$ as well. □

Applying the same arguments to the set $\mathcal{D} \backslash \mathbb{B}_\delta(p)$ and using (7.1.4), we obtain

LEMMA 7.10. *If $p \in \mathcal{D}$, then*

$$\Psi_{\mathcal{D}}(p, \partial\mathcal{D}) = I.$$

Condition (7.1.2) enables one to extend the set function $\Psi_{\mathcal{D}}(p, \mathcal{B})$ to all Borel subsets of $\mathbb{R}^n \backslash p$.

We introduce

$$\Psi_{\mathcal{D}}(p, p) = \begin{cases} 0 & \text{for} \quad p \notin \partial\mathcal{D}, \\ \frac{1}{2} I - \Psi_{\mathcal{D}}(p, \partial\mathcal{D} \backslash p) & \text{for} \quad p \in \partial\mathcal{D}. \end{cases} \tag{7.3.4}$$

For $p \in \mathcal{B}$ we put

$$\Psi_{\mathcal{D}}(p, \mathcal{B}) = \Psi_{\mathcal{D}}(p, \mathcal{B}\backslash p) + \Psi_{\mathcal{D}}(p, p). \tag{7.3.5}$$

Thus the matrix-valued function $\mathcal{B} \to \Psi_{\mathcal{D}}(p, \mathcal{B})$ is defined and is finite for all Borel subsets of $\mathbb{R}^n$.

From Lemmas 7.9, 7.10 and from equalities (7.3.4), (7.3.5) one obtains

$$\Psi_{\mathcal{D}}(p, \partial\mathcal{D}) = \begin{cases} I & \text{for} \quad p \in \mathcal{D}, \\ \frac{1}{2}I & \text{for} \quad p \in \partial\mathcal{D}, \\ 0 & \text{for} \quad p \in \mathbb{R}^n\backslash\overline{\mathcal{D}}. \end{cases} \tag{7.3.6}$$

LEMMA 7.11. *If $p \in \partial^*\mathcal{D}$, then $\Psi_{\mathcal{D}}(p, p) = 0$.*

PROOF. Setting $p = q$ and $\mathcal{B} = \mathbf{C}\mathbb{B}_\varepsilon(p)$ in (7.3.3), we obtain

$$\Psi_{\mathcal{D}\cap\mathbf{C}\mathbb{B}_\varepsilon(p)}\big(p, \mathbf{C}\mathbb{B}_\varepsilon(p)\big) = \Psi_{\mathcal{D}}\big(p, \mathbf{C}\mathbb{B}_\varepsilon(p)\big) - \Psi_{\mathbb{B}_\varepsilon(p)}\big(p, \mathcal{D} \cap \mathbf{C}\mathbb{B}_\varepsilon(p)\big)$$

for almost all $\varepsilon > 0$. Since by Lemma 7.9

$$\Psi_{\mathcal{D}\cap\mathbf{C}\mathbb{B}_\varepsilon(p)}\big(p, \mathbf{C}\mathbb{B}_\varepsilon(p)\big) = 0,$$

it follows that

$$\Psi_{\mathcal{D}}\big(p, \mathbf{C}\mathbb{B}_\varepsilon(p)\big) = \Psi_{\mathbb{B}_\varepsilon(p)}\big(p, \mathcal{D} \cap \mathbf{C}\mathbb{B}_\varepsilon(p)\big) = \varepsilon^{1-n} \int_{\mathcal{D}\cap\partial\mathbb{B}_\varepsilon(p)} \mathcal{M}(\boldsymbol{e}_{px}) d\sigma(x),$$

where $d\sigma$ is a surface element of $\partial\mathbb{B}_\varepsilon(p)$. Consequently,

$$\begin{aligned} \delta^{-n} \int_0^\delta \Psi_{\mathcal{D}}\big(p, \mathbf{C}\mathbb{B}_\varepsilon(p)\big)\varepsilon^{n-1} d\varepsilon &= \delta^{-n} \int_0^\delta \Big[\int_{\mathcal{D}\cap\partial\mathbb{B}_\varepsilon(p)} \mathcal{M}(\boldsymbol{e}_{px}) d\sigma(x)\Big] d\varepsilon \\ &= \delta^{-n} \int_{\mathcal{D}\cap\mathbb{B}_\delta(p)} \mathcal{M}(\boldsymbol{e}_{px}) dx. \end{aligned} \tag{7.3.7}$$

The left-hand side of the last equality can be written as

$$\begin{aligned} &\delta^{-n} \int_0^\delta \Psi_D\big(p, \mathbf{C}\mathbb{B}_\varepsilon(p)\big)\varepsilon^{n-1} d\varepsilon \\ &= \delta^{-n} \int_0^\delta \Big[\Psi_{\mathcal{D}}\big(p, R^n\backslash p\big) - \Psi_{\mathcal{D}}\big(p, \mathbb{B}_\varepsilon(p)\backslash p\big)\Big]\varepsilon^{n-1} d\varepsilon \\ &= \frac{1}{n}\Psi_{\mathcal{D}}\big(p, R^n\backslash p\big) - \delta^{-n} \int_0^\delta \Psi_{\mathcal{D}}\big(p, \mathbb{B}_\varepsilon(p)\backslash p\big)\varepsilon^{n-1} d\varepsilon. \end{aligned}$$

By condition (7.1.2),

$$\lim_{\delta\to 0} |\,\omega_{\mathcal{D}}|(p,\ \mathbb{B}_\delta(p)\backslash p) = 0. \tag{7.3.8}$$

Hence

$$\lim_{\delta\to 0}\ |\Psi_{\mathcal{D}}|(p, \mathbb{B}_\delta(p)\backslash p) = 0.$$

Here and below

$$|\Psi_{\mathcal{D}}|(p, \mathcal{B}) = \big(\big(\,|\,\Psi_{\mathcal{D}}^{(ij)}|(p, \mathcal{B})\big)\big), \quad i, j = 1, 2, \ldots, m.$$

Therefore

$$\lim_{\delta\to 0} \delta^{-n} \int_0^\delta \Psi_{\mathcal{D}}\big(p, \mathbf{C}\mathbb{B}_\varepsilon(p)\big)\varepsilon^{n-1} d\varepsilon = \frac{1}{n}\, \Psi_{\mathcal{D}}(p, \mathbb{R}^n \backslash p)$$

and by (7.3.7)

$$\frac{1}{n}\Psi_{\mathcal{D}}\big(p, \mathbb{R}^n\backslash p\big) = \lim_{\delta\to 0} \delta^{-n} \int_{\mathcal{D}\cap\mathbb{B}_\delta(p)} \mathcal{M}(\boldsymbol{e}_{px}) dx. \tag{7.3.9}$$

We introduce the notation

$$\mathbb{B}_\delta^+(p) = \{x \in \mathbb{B}_\delta(p) : \ (\boldsymbol{\nu}(p),\ \boldsymbol{r}_{px}) > 0\},$$

$$\mathbb{B}_\delta^-(p) = \{x \in \mathbb{B}_\delta(p) : \ (\boldsymbol{\nu}(p),\ \boldsymbol{r}_{px}) < 0\}$$

and divide $\mathcal{D} \cap \mathbb{B}_\delta((p)$ into two sets

$$\mathfrak{M}_\delta^+(p) = \mathcal{D} \cap \mathbb{B}_\delta^+(p),\ \mathfrak{M}_\delta^-(p) = \mathcal{D} \cap \mathbb{B}_\delta^-(p).$$

Using (7.3.9) and the definition of the normal in the sense of Federer, one finds

$$\begin{aligned}
\frac{1}{n}\Psi_{\mathcal{D}}(p, \mathbb{R}^n\backslash p) &= \lim_{\delta\to 0} \delta^{-n} \int_{\mathfrak{M}_\delta^+(p)} \mathcal{M}(\boldsymbol{e}_{px}) dx + \lim_{\delta\to 0} \delta^{-n} \int_{\mathfrak{M}_\delta^-(p)} \mathcal{M}(\boldsymbol{e}_{px}) dx \\
&= \lim_{\delta\to 0} \delta^{-n} \int_{\mathfrak{M}_\delta^-(p)} \mathcal{M}(\boldsymbol{e}_{px}) dx = \frac{1}{2} \lim_{\delta\to 0} \delta^{-n} \int_{\mathbb{B}_\delta(p)} \mathcal{M}(\boldsymbol{e}_{px}) dx \\
&= \frac{1}{2}\cdot\frac{1}{n} \int_{\mathbb{S}^{n-1}} \mathcal{M}(\sigma) d\sigma = \frac{1}{2n} I,
\end{aligned}$$

which together with (7.3.4) completes the proof. □

REMARK 7.2. It is shown in Lemma 7.4 that the set $\partial^*\mathcal{D}$ is everywhere dense on $\partial\mathcal{D}$. Thus, Lemma 7.11 claims that the set of $p \in \partial\mathcal{D}$ with $\Psi_{\mathcal{D}}(p,p) = 0$ is everywhere dense on $\partial\mathcal{D}$. We also note that

$$\Psi_{\mathcal{D}}(p, q) = 0 \quad \text{for} \quad p \neq q, \tag{7.3.10}$$

which follows from the equality $\omega_{\mathcal{D}}(p, q) = 0$, $p \neq q$ (see Remark 7.1) together with definition (7.3.2) of $\Psi_{\mathcal{D}}(p, \mathcal{B})$.

LEMMA 7.12. *The equality*

$$\lim_{p\to q} |\Psi_{\mathcal{D}}(p, \cdot) - \Psi_{\mathcal{D}}(q, \cdot)|(\mathcal{B}) = 0$$

is valid for any Borel set $\mathcal{B}$ *and any* $q \notin \overline{\mathcal{B}} \cap \partial\mathcal{D}$.

PROOF. Let $n > 2$. The case $n = 2$ is treated in the same way. Suppose $p \notin \overline{\mathcal{B}} \cap \partial\mathcal{D}$. By definitions of $\omega_D(p, \mathcal{B})$ and $\Psi_{\mathcal{D}}(p, \mathcal{B})$ one has

$$\begin{aligned}
&\Psi_{\mathcal{D}}^{(ij)}(p, \mathcal{B}) - \Psi_{\mathcal{D}}^{(ij)}(q, \mathcal{B}) \\
&\quad = \frac{1}{n-2}\Big[\int_{\mathcal{B}} \Big(\mathcal{M}_{ij}(\boldsymbol{e}_{px})\nabla r_{px}^{2-n} - \mathcal{M}_{ij}(\boldsymbol{e}_{qx})\nabla r_{qx}^{2-n}\Big)\nabla\chi_{\mathcal{D}}(dx)\Big].
\end{aligned}$$

This implies

$$
\begin{aligned}
(7.3.11)\qquad & |\Psi_{\mathcal{D}}^{(ij)}(p,\cdot) - \Psi_{\mathcal{D}}^{(ij)}(q,\cdot)|(\mathcal{B}) \\
&\leq \frac{1}{n-2}\Big\{ \int_{\mathcal{B}} \Big|\mathcal{M}_{ij}(\boldsymbol{e}_{px}) - \mathcal{M}_{ij}(\boldsymbol{e}_{qx})\Big| |\nabla r_{px}^{2-n}|\ |\nabla\chi_{\mathcal{D}}|(dx) \\
&+ \int_{\mathcal{B}} \Big|\mathcal{M}_{ij}(\boldsymbol{e}_{qx})\Big| \Big|\nabla r_{px}^{2-n} - \nabla_{qx}^{2-n}\Big|\ |\nabla\chi_{\mathcal{D}}|(dx)\Big\}.
\end{aligned}
$$

Estimating the first integral on the right-hand side, we find

$$
\begin{aligned}
(7.3.12)\qquad & \int_{\mathcal{B}} \Big|\mathcal{M}_{ij}(\boldsymbol{e}_{px}) - \mathcal{M}_{ij}(\boldsymbol{e}_{qx})\Big| \Big|\nabla r_{px}^{2-n}\Big|\ |\nabla\chi_{\mathcal{D}}|(dx) \\
&\leq (n-2) \sup_{x\in\overline{\mathcal{B}}\cap\partial\mathcal{D}} \Big|\mathcal{M}_{ij}(\boldsymbol{e}_{px}) - \mathcal{M}_{ij}(\boldsymbol{e}_{qx})\Big| r^{1-n}\ |\nabla\chi_{\mathcal{D}}|(\mathcal{B}),
\end{aligned}
$$

where r is the minimum distance from p and q to $\overline{\mathcal{B}}\cap\partial\mathcal{D}$. Since $x\in\overline{\mathcal{B}}\cap\partial\mathcal{D}$ and $q,p\notin\overline{\mathcal{B}}\cap\partial\mathcal{D}$, then

$$
\begin{aligned}
|\boldsymbol{e}_{px} - \boldsymbol{e}_{qx}| &= |\boldsymbol{r}_{px}r_{px}^{-1} - \boldsymbol{r}_{qx}r_{qx}^{-1}| = |r_{px}^{-1}r_{qx}^{-1}\big[r_{qx}\boldsymbol{r}_{px} - r_{px}\boldsymbol{r}_{qx}\big]| \\
&\leq r_{qx}^{-1}r_{px}^{-1}\big[|r_{qx}\boldsymbol{r}_{px} - r_{qx}\boldsymbol{r}_{qx}| + |r_{qx}\boldsymbol{r}_{qx} - r_{px}\boldsymbol{r}_{qx}|\big] \\
&\leq 2r_{qx}r_{pq}\big(r_{px}r_{qx}\big)^{-1} \leq 2r_{pq}r^{-1}.
\end{aligned}
$$

So estimate (7.3.12) takes the form

$$
\begin{aligned}
(7.3.13)\qquad & \int_{\mathcal{B}} \Big|\mathcal{M}_{ij}(\boldsymbol{e}_{px}) - \mathcal{M}_{ij}(\boldsymbol{e}_{qx})\Big| \Big|\nabla r_{px}^{2-n}\Big|\ |\nabla\chi_{\mathcal{D}}|(dx) \\
&\leq c_1 r^{1-n}\Omega_{\mathcal{M}}\big(2r_{pq}r^{-1}\big)\ |\nabla\chi_{\mathcal{D}}|(\mathcal{B}),
\end{aligned}
$$

where

$$
\Omega_{\mathcal{M}}(\delta) = \max\big\{\|\mathcal{M}(\boldsymbol{e}_2) - \mathcal{M}(\boldsymbol{e}_1)\|:\ |\boldsymbol{e}_2 - \boldsymbol{e}_1| \leq \delta\big\}.
$$

The second integral on the right-hand side in (7.3.11) is majorized by

$$
\begin{aligned}
(7.3.14)\qquad & \int_{\mathcal{B}} \Big|\mathcal{M}_{ij}(\boldsymbol{e}_{qx})\Big| \Big|\nabla r_{px}^{2-n} - \nabla r_{qx}^{2-n}\Big|\ |\nabla\chi_{\mathcal{D}}|(dx) \\
&\leq c_1 \sup_{x\in\overline{\mathcal{B}}\cap\partial\mathcal{D}} \Big|\mathcal{M}_{ij}(\boldsymbol{e}_{qx})\Big| \Big(r_{pq}r_{px}^{-n} + |r_{px}^{n} - r_{qx}^{n}|r_{px}^{-n}r_{qx}^{1-n}\Big)\ |\nabla\chi_{\mathcal{D}}|(\mathcal{B}) \\
&\leq c_2 \max_{|\boldsymbol{e}|=1}\|\mathcal{M}(\boldsymbol{e})\| r^{-n}r_{pq}\ |\nabla\chi_{\mathcal{D}}|(\mathcal{B}).
\end{aligned}
$$

Summing up estimates (7.3.13), (7.3.14) and taking into account (7.3.11), we obtain

$$
\begin{aligned}
(7.3.15)\qquad & |\Psi_{\mathcal{D}}^{(ij)}(p,\cdot) - \Psi_{\mathcal{D}}^{(ij)}(q,\cdot)|(\mathcal{B}) \\
&\leq c\Big(r^{-n}r_{pq} + \Omega_{\mathcal{M}}\big(2r_{pq}r^{-1}\big)r^{1-n}\Big)\ |\nabla\chi_{\mathcal{D}}|(\mathcal{B}),
\end{aligned}
$$

which by continuity of $\mathcal{M}$ on $\mathbb{S}^{n-1}$ completes the proof. □

Let $\varphi \in [\mathrm{C}(\partial\mathcal{D})]^m$. We introduce the vector-valued double layer potential

$$\boldsymbol{W}(q) = \int_{\partial\mathcal{D}} \mathcal{M}(\boldsymbol{e}_{qx})\boldsymbol{\varphi}(x)\omega_{\mathcal{D}}(q, dx),$$

where $q \notin \partial\mathcal{D}$.

THEOREM 7.2. *The inner and the outer limit values of* $\boldsymbol{W}(q)$ *are*

$$\boldsymbol{W}^{(i)}(p) = \frac{1}{2}\boldsymbol{\varphi}(p) + \int_{\partial\mathcal{D}} \Psi_{\mathcal{D}}(p, dx)\boldsymbol{\varphi}(x), \tag{7.3.16}$$

$$\boldsymbol{W}^{(e)}(p) = -\frac{1}{2}\boldsymbol{\varphi}(p) + \int_{\partial\mathcal{D}} \Psi_{\mathcal{D}}(p, dx)\boldsymbol{\varphi}(x). \tag{7.3.17}$$

PROOF. Consider, for example, the case $q \in \mathcal{D}$, $q \to p \in \partial\mathcal{D}$. By (7.3.6),

$$\begin{aligned} \boldsymbol{W}(q) &= \int_{\partial\mathcal{D}\cap\mathbf{C}\mathbb{B}_\delta(p)} \Psi_{\mathcal{D}}(q, dx)\Big(\boldsymbol{\varphi}(x) - \boldsymbol{\varphi}(p)\Big) \\ &+ \int_{\partial\mathcal{D}\cap\mathbb{B}_\delta(p)} \Psi_{\mathcal{D}}(q, dx)\Big(\boldsymbol{\varphi}(x) - \boldsymbol{\varphi}(p)\Big) + \boldsymbol{\varphi}(p). \end{aligned} \tag{7.3.18}$$

By continuity of φ one can choose δ so that $|\boldsymbol{\varphi}(x) - \boldsymbol{\varphi}(p)| < \varepsilon$ for a given ε and for $x \in \partial\mathcal{D} \cap \mathbb{B}_\delta(p)$. Then by (7.3.2), (7.3.4) and (7.3.5) we find

$$\begin{aligned} &\left|\int_{\partial\mathcal{D}\cap\mathbb{B}_\delta(p)} \Psi_{\mathcal{D}}(q, dx)\Big(\boldsymbol{\varphi}(x) - \boldsymbol{\varphi}(p)\Big)\right| \\ &= \left|\int_{\partial\mathcal{D}\cap\mathbb{B}_\delta(p)} \mathcal{M}(\boldsymbol{e}_{qx})\Big(\boldsymbol{\varphi}(x) - \boldsymbol{\varphi}(p)\Big)\omega_{\mathcal{D}}(q, dx)\right| < c\,\varepsilon. \end{aligned}$$

Passing to the limit in (7.3.18) as $q \to p$ and using the last inequality together with Lemma 7.12, we obtain

$$\left|\limsup_{q\to p} \boldsymbol{W}(q) - \int_{\partial\mathcal{D}\cap\mathbf{C}\mathbb{B}_\delta(p)} \Psi_{\mathcal{D}}(p, dx)\Big(\boldsymbol{\varphi}(x) - \boldsymbol{\varphi}(p)\Big) - \boldsymbol{\varphi}(p)\right| < c\,\varepsilon,$$

$$\left|\liminf_{q\to p} \boldsymbol{W}(q) - \int_{\partial\mathcal{D}\cap\mathbf{C}\mathbb{B}_\delta(p)} \Psi_{\mathcal{D}}(p, dx)\Big(\boldsymbol{\varphi}(x) - \boldsymbol{\varphi}(p)\Big) - \boldsymbol{\varphi}(p)\right| < c\,\varepsilon.$$

Passing to the limit in these inequalities as $\delta \to 0$, using the arbitrariness of ε and taking into account (7.3.6) we conclude

$$\begin{aligned} \boldsymbol{W}^{(i)}(p) &= \limsup_{q\to p} \boldsymbol{W}(q) = \liminf_{q\to p} \boldsymbol{W}(q) \\ &= \boldsymbol{\varphi}(p) + \int_{\partial\mathcal{D}\setminus p} \Psi_{\mathcal{D}}(p, dx)\Big(\boldsymbol{\varphi}(x) - \boldsymbol{\varphi}(p)\Big) \\ &= \int_{\mathbb{S}^{n-1}} \mathcal{M}(\sigma)d\sigma\boldsymbol{\varphi}(p) + \int_{\partial D} \Psi_{\mathcal{D}}(p, dx)\Big(\boldsymbol{\varphi}(x) - \boldsymbol{\varphi}(p)\Big) \\ &= \frac{1}{2}\boldsymbol{\varphi}(p) + \int_{\partial\mathcal{D}} \Psi_{\mathcal{D}}(p, dx)\boldsymbol{\varphi}(x). \end{aligned}$$

In a similar way one can prove the existence of outer limit values of $\boldsymbol{W}(q)$. □

7.3.2. The norm and the essential norm of the operator T

In this subsection we find representations for the norm and for the essential norm of the operator T, defined by

$$(T\varphi)(p) = 2\int_{\partial\mathcal{D}} \Psi_{\mathcal{D}}(p, dx)\varphi(x)$$

in the space $[\mathrm{C}(\partial\mathcal{D})]^m$.

Theorem 7.2 implies the continuity of T in $[\mathrm{C}(\partial\mathcal{D})]^m$.

THEOREM 7.3. *Let $\mathcal{M}^*$ denote the adjoint matrix of $\mathcal{M}$. Then*

$$\|T\| = 2\sup_{p\in\partial^*\mathcal{D}}\sup_{|\boldsymbol{z}|=1}\int_{\partial\mathcal{D}} |\mathcal{M}^*(\boldsymbol{e}_{px})\boldsymbol{z}||\omega_{\mathcal{D}}|(p, dx).$$

PROOF. Since T acts in $[\mathrm{C}(\partial\mathcal{D})]^m$ and since $\partial^*\mathcal{D}$ is everywhere dense in $\partial\mathcal{D}$ by Lemma 7.4 we find after changing the order of two suprema

$$\begin{aligned}\|T\| &= \sup_{\|\varphi\|=1}\sup_{p\in\partial\mathcal{D}}\Big|(T\varphi)(p)\Big| = \sup_{\|\varphi\|=1}\sup_{p\in\partial^*\mathcal{D}}\Big|(T\varphi)(p)\Big| \\ &= \sup_{p\in\partial^*\mathcal{D}}\sup_{\|\varphi\|=1}\Big|(T\varphi)(p)\Big|.\end{aligned}$$

By (7.3.2), (7.3.5) together with Lemma 7.11,

$$\begin{aligned}\|T\| &= 2\sup_{p\in\partial^*\mathcal{D}}\sup_{\|\varphi\|=1}\left|\int_{\partial\mathcal{D}\backslash p} \Psi_{\mathcal{D}}(p, dx)\varphi(x) + \Psi_{\mathcal{D}}(p,p)\varphi(p)\right| \\ &= \sup_{p\in\partial^*\mathcal{D}}\sup_{\|\varphi\|=1}\left|\int_{\partial\mathcal{D}\backslash p} \mathcal{M}(\boldsymbol{e}_{px})\varphi(x)\omega_{\mathcal{D}}(p, dx)\right|.\end{aligned}$$

Hence

$$\begin{aligned}\|T\| &= 2\sup_{p\in\partial^*\mathcal{D}}\sup_{\|\varphi\|=1}\sup_{|\boldsymbol{z}|=1}\left(\boldsymbol{z}, \int_{\partial\mathcal{D}\backslash p} \mathcal{M}(\boldsymbol{e}_{px})\varphi(x)\omega_{\mathcal{D}}(p, dx)\right) \\ &= 2\sup_{p\in\partial^*\mathcal{D}}\sup_{|\boldsymbol{z}|=1}\sup_{\|\varphi\|=1}\int_{\partial\mathcal{D}\backslash p}\Big(\mathcal{M}^*(\boldsymbol{e}_{px})\boldsymbol{z},\ \varphi(x)\Big)\omega_{\mathcal{D}}(p, dx) \\ &= 2\sup_{p\in\partial^*\mathcal{D}}\sup_{|\boldsymbol{z}|=1}\int_{\partial\mathcal{D}\backslash p} |\mathcal{M}^*(\boldsymbol{e}_{px})\boldsymbol{z}||\omega_{\mathcal{D}}|(p, dx).\end{aligned}$$

Since $\omega_{\mathcal{D}}(p,p) = 0$ for $p \in \partial^*\mathcal{D}$ (see proof of Corollary 7.4), one can replace the integration over $\partial\mathcal{D}\backslash p$ by integration over $\partial\mathcal{D}$. □

COROLLARY 7.6. *Let $\mathcal{D}$ be a bounded convex domain in $\mathbb{R}^n$. Then*

$$\|T\| = \sup_{|\boldsymbol{z}|=1}\int_{\mathbb{S}^{n-1}} |\mathcal{M}^*(\sigma)\boldsymbol{z}|d\sigma.$$

PROOF. By Definition 7.2, the normal in the sense of Federer coincides with the classical normal in the case of a convex domain. Since $\mathcal{D}$ lays to one side of a hyperplane which is tangent to the boundary at $p \in \partial^*\mathcal{D}$ and since $\mathcal{M}$ is even on $\mathbb{S}^{n-1}$, then

$$\|T\| = 2\sup_{p\in\partial^*\mathcal{D}}\sup_{|\boldsymbol{z}|=1}\int_{\partial\mathcal{D}} |\mathcal{M}^*(\boldsymbol{e}_{px})\boldsymbol{z}|\omega_{\mathcal{D}}(p, dx) = \sup_{|\boldsymbol{z}|=1}\int_{\mathbb{S}^{n-1}} |\mathcal{M}^*(\sigma)\boldsymbol{z}|d\sigma,$$

which completes the proof. □

We introduce the operator A_δ as

$$(A_\delta \boldsymbol{\varphi})(p) = \int_{\partial\mathcal{D}} \big[1 - \alpha_\delta(x-p)\big] \Psi_{\mathcal{D}}(p, dx) \boldsymbol{\varphi}(x),$$

where $\alpha_\delta \in C_0^\infty(\mathbb{B}_\delta)$, $\alpha_\delta(x) = 1$ for $x \in \mathbb{B}_{\delta/2}$, and $0 \le \alpha_\delta(x) \le 1$.

LEMMA 7.13. *The operator A_δ is compact in $[\mathrm{C}(\partial\mathcal{D})]^m$.*

PROOF. It is sufficient to show that the set of vector-valued functions $A_\delta \boldsymbol{\varphi}$ with $\|\boldsymbol{\varphi}\| \le 1$ is equicontinuous. Let $r_{pq} < \delta/10$. Then

$$\begin{aligned} |A_\delta \boldsymbol{\varphi})(p) - (A_\delta \boldsymbol{\varphi})(q)| \;\le\; & \left| \int_{\partial\mathcal{D}} \big[\alpha_\delta(x-q) - \alpha_\delta(x-p)\big] \mathcal{M}(\boldsymbol{e}_{px}) \boldsymbol{\varphi}(x) \omega_{\mathcal{D}}(p, dx) \right| \\ & + \left| \int_{\partial\mathcal{D}} \big[1 - \alpha_\delta(x-q)\big] \big[\mathcal{M}(\boldsymbol{e}_{px}) - \mathcal{M}(\boldsymbol{e}_{qx})\big] \boldsymbol{\varphi}(x) \omega_{\mathcal{D}}(p, dx) \right| \\ & + \left| \int_{\partial\mathcal{D}} \big[1 - \alpha_\delta(x-q)\big] \mathcal{M}(\boldsymbol{e}_{qx}) \boldsymbol{\varphi}(x) \big[\omega_{\mathcal{D}}(p, dx) - \omega_{\mathcal{D}}(q, dx)\big] \right|. \end{aligned}$$

Using the same arguments as in the proof of Lemma 7.12, we estimate the sum on the right by

$$c(\delta)\big(r_{pq} + \Omega_{\mathcal{M}}(5 r_{pq} \delta^{-1})\big),$$

where $\Omega_{\mathcal{M}}$ is the continuity modulus of the matrix $\mathcal{M}$. □

The next assertion extends representation (7.5.5) for the essential norm of the harmonic double layer potential to the vector case.

THEOREM 7.4. *The essential norm of T in $[\mathrm{C}(\partial\mathcal{D})]^m$ is given by*

$$2 \lim_{\delta \to 0} \sup_{p \in \partial^* \mathcal{D}} \sup_{|\boldsymbol{z}|=1} \int_{\partial\mathcal{D} \cap \mathbb{B}_\delta(p)} |\mathcal{M}^*(\boldsymbol{e}_{px}) \boldsymbol{z}| |\omega_{\mathcal{D}}|(p, dx).$$

We preface the proof of this theorem by two auxiliary assertions.

LEMMA 7.14. *The estimate*

$$\mathrm{ess}\|T\| \le 2 \lim_{\delta \to 0} \sup_{p \in \partial^* \mathcal{D}} \sup_{|\boldsymbol{z}|=1} \int_{\partial\mathcal{D} \cap \mathbb{B}_\delta(p)} |\mathcal{M}^*(\boldsymbol{e}_{px}) \boldsymbol{z}| |\omega_{\mathcal{D}}|(p, dx)$$

is valid.

PROOF. According to Lemma 7.13, the operator

$$(T_\delta \boldsymbol{\varphi})(p) = 2 \int_{\partial\mathcal{D}} \big[1 - \alpha_\delta(x-p)\big] \Psi_{\mathcal{D}}(p, dx) \boldsymbol{\varphi}(x)$$

is compact in $[\mathrm{C}(\partial\mathcal{D})]^m$. Using the same arguments as in the proof of Theorem 7.3, one obtains

$$\begin{aligned} \mathrm{ess}\|T\| \;\le\; & \|T - T_\delta\| = 2 \sup_{\|\boldsymbol{\varphi}\|=1} \sup_{p \in \partial\mathcal{D}} \left| \int_{\partial\mathcal{D}} \alpha_\delta(x-p) \Psi_{\mathcal{D}}(p, dx) \boldsymbol{\varphi}(x) \right| \\ \le\; & 2 \sup_{p \in \partial^* \mathcal{D}} \sup_{|\boldsymbol{z}|=1} \int_{\partial\mathcal{D} \cap \mathbb{B}_\delta(p)} |\mathcal{M}^*(\boldsymbol{e}_{px}) \boldsymbol{z}| |\omega_{\mathcal{D}}|(p, dx). \end{aligned}$$

□

By $\mathfrak{Q}$ we denote the set of finite dimensional operators Q' in $[\mathrm{C}(\partial\mathcal{D})]^m$ defined by

$$(Q'\phi)(p) = \sum_{j=1}^{k} f_j(p) \int_{\partial\mathcal{D}} G'_j(ds)\phi(s), \tag{7.3.19}$$

for which all $(m \times m)$-matrix-valued charges G'_j do not have point loadings. Here $f_j \in \mathrm{C}(\partial\mathcal{D})$.

LEMMA 7.15. *The equality*

$$\mathrm{ess}\|T\| = \inf\{\|T - Q'\| : \ Q' \in \mathfrak{Q}\}$$

is valid.

PROOF. Since

$$\mathrm{ess}\|T\| = \inf\|T - \mathcal{K}\|,$$

where the infimum is taken over the set of all finite dimensional operators, the lemma results from

$$\mathrm{ess}\|T\| \geq \inf\{\|T - Q'\| : \ Q' \in \mathfrak{Q}\}$$

The last inequality holds if for any finite dimensional Q there exists $Q' \in \mathfrak{Q}$ such that

$$\|T - Q\| \geq \|T - Q'\|. \tag{7.3.20}$$

We represent a finite dimensional operator in $[\mathrm{C}(\partial\mathcal{D})]^m$ in the form

$$(Q\phi)(p) = \sum_{j=1}^{k} f_j(p) \int_{\partial\mathcal{D}} G_j(ds)\phi(s).$$

Consider the decomposition $G_j = G'_j + G''_j$, where the matrix-valued charge G'_j does not have point loadings while G''_j consists only of point loadings of the charge G_j. Let Q' be the operator defined by (7.3.19). We put $Q'' = Q' - Q$ and we denote the kernels of the operators $Q'', T-Q, T-Q'$ by $O''(p,B), O(p,B), O'(p,B)$, respectively. Since the operators $T - Q'$, $T - Q$ act in $[\mathrm{C}(\partial\mathcal{D})]^m$ and since $\partial^*\mathcal{D}$ is everywhere dense on $\partial\mathcal{D}$, it follows that

$$\|T - Q\| = \sup_{\|\varphi\|=1} \sup_{p\in\partial\mathcal{D}} |(T - Q)\varphi(p)| = \sup_{p\in\partial^*\mathcal{D}} \sup_{\|\varphi\|=1} |(T - Q)\varphi(p)|,$$

$$\|T - Q'\| = \sup_{\|\varphi\|=1} \sup_{p\in\partial\mathcal{D}} |(T - Q')\varphi(p)| = \sup_{p\in\partial^*\mathcal{D}} \sup_{\|\varphi\|=1} |(T - Q')\varphi(p)|. \tag{7.3.21}$$

Therefore, if for all $p \in \partial^*\mathcal{D}$ the inequality

$$\sup_{\|\varphi\|=1} \left| \int_{\partial\mathcal{D}} O(p,ds)\varphi(s) \right| \geq \sup_{\|\varphi\|=1} \left| \int_{\partial\mathcal{D}} O'(p,ds)\varphi(s) \right| \tag{7.3.22}$$

holds, then (7.3.20) holds.

According to Lemma 7.11 and Remark 7.2, $\Psi_{\mathcal{D}}(p,q) = 0$ for $p \in \partial^*\mathcal{D}$ and for all $q \in \partial\mathcal{D}$. Besides, one may assume that

$$\mathrm{supp}\ G'_i \cap \mathrm{supp}\ G''_j = \emptyset,\ i,j = 1,\dots,k.$$

Therefore, $\operatorname{supp} O'(p,\cdot)\cap \operatorname{supp} O''(p,\cdot)=\emptyset$. Consequently, for $p\in\partial^*\mathcal{D}$,

$$\begin{aligned}\sup_{\|\varphi\|=1}\left|\int_{\partial\mathcal{D}} O(p,ds)\varphi(s)\right| &= \sup_{\|\varphi\|=1}\left|\int_{\partial\mathcal{D}}\Big[O'(p,ds)+O''(p,ds)\Big]\varphi(s)\right|\\ &\geq \sup_{\|\varphi\|=1}\left|\int_{\partial\mathcal{D}} O'(p,ds)\varphi(s)\right|.\end{aligned}$$

Inequality (7.3.22) is proved, and hence we arrive at the assertion of the lemma. □

PROOF OF THEOREM 7.4. Let $Q'\in\mathfrak{Q}$. From (7.3.21) we obtain

$$\begin{aligned}\|T-Q'\| &= \sup_{\|\varphi\|=1}\sup_{p\in\partial\mathcal{D}}\left|\int_{\partial\mathcal{D}} O'(p,ds)\varphi(s)\right|\\
&\geq \lim_{\delta\to 0}\sup_{p\in\partial^*\mathcal{D}}\sup_{\|\varphi\|=1}\left|\int_{\partial\mathcal{D}\cap\mathbb{B}_\delta(p)} O'(p,ds)\varphi(s)\right|\\
&= \liminf_{\delta\to 0}\sup_{p\in\partial^*\mathcal{D}}\sup_{\|\varphi\|=1}\left|2\int_{\partial\mathcal{D}\cap\mathbb{B}_\delta(p)}\Psi_{\mathcal{D}}(p,ds)\varphi(s)\right.\\
&\qquad\left.-\sum_{j=1}^{k} f_j(p)\int_{\partial\mathcal{D}\cap\mathbb{B}_\delta(p)} G_j'(p,ds)\varphi(s)\right|\\
&\geq \liminf_{\delta\to 0}\left\{\sup_{p\in\partial^*\mathcal{D}}\sup_{\|\varphi\|=1}\left|2\int_{\partial\mathcal{D}\cap\mathbb{B}_\delta(p)}\Psi_{\mathcal{D}}(p,ds)\varphi(s)\right|\right.\\
&\qquad\left.-\sup_{p\in\partial^*\mathcal{D}}\sup_{\|\varphi\|=1}\left|\sum_{j=1}^{k} f_j(p)\int_{\partial\mathcal{D}\cap\mathbb{B}_\delta(p)} G_j'(p,ds)\varphi(s)\right|\right\}\\
&= 2\lim_{\delta\to 0}\sup_{p\in\partial^*\mathcal{D}}\sup_{|\boldsymbol{z}|=1}\int_{\partial\mathcal{D}\cap\mathbb{B}_\delta(p)}|\mathcal{M}^*(\boldsymbol{e}_{px})\boldsymbol{z}||\omega_{\mathcal{D}}|(p,dx).\end{aligned}$$

The last equality follows from the fact that the charges G_j' do not have point loadings and from Theorem 7.3. Thus

$$\|T-Q'\|\geq 2\lim_{\delta\to 0}\sup_{p\in\partial^*\mathcal{D}}\sup_{|\boldsymbol{z}|=1}\int_{\partial\mathcal{D}\cap\mathbb{B}_\delta(p)}|\mathcal{M}^*(\boldsymbol{e}_{px})\boldsymbol{z}||\omega_{\mathcal{D}}|(p,dx)$$

for $Q'\in\mathfrak{Q}$. This inequality together with Lemmas 7.14, 7.15 completes the proof.

REMARK 7.3. It was shown in the proof of Lemma 7.11 that

$$\Psi_{\mathcal{D}}\Big(p,\mathbf{C}\mathbb{B}_\varepsilon(p)\Big)=\varepsilon^{1-n}\int_{\mathcal{D}\cap\partial\mathbb{B}_\varepsilon(p)}\mathcal{M}(\boldsymbol{e}_{px})d\sigma(x)$$

for any p and for almost all $\varepsilon>0$.

Suppose that at $p\in\partial\mathcal{D}$ there exists a topological limit $\mathcal{E}(p)$ of the projections $\mathcal{E}_\varepsilon(p)$ of the sets $\mathcal{D}\cap\partial\mathbb{B}_\varepsilon(p)$ onto the sphere $\partial\mathbb{B}(p)$ as $\varepsilon\to 0$. Then the last equality implies

$$\Psi_{\mathcal{D}}(p,\partial\mathcal{D}\backslash p)=\int_{\mathcal{E}}\mathcal{M}(\sigma)d\sigma. \tag{7.3.23}$$

Hence, according to (7.3.4),

$$\Psi_{\mathcal{D}}(p,p)=\frac{1}{2}I-\int_{\mathcal{E}(p)}\mathcal{M}(\sigma)d\sigma. \tag{7.3.24}$$

The next assertion is a corollary of the last theorem.

THEOREM 7.5. *Let $\mathcal{D}$ be a bounded convex domain. Then*

$$\text{ess}\|T\| = \sup_{p\in\partial\mathcal{D}\backslash\partial^*\mathcal{D}} \sup_{|\boldsymbol{z}|=1} \left[\int_{\mathbb{S}^{n-1}} \big|\mathcal{M}^*(\sigma)\boldsymbol{z}\big| d\sigma - 2\int_{\mathcal{E}(p)} \big|\mathcal{M}^*(\sigma)\boldsymbol{z}\big| d\sigma \right]$$

if $\partial\mathcal{D} \neq \partial^*\mathcal{D}$ *and* $\text{ess}\|T\| = 0$ *if* $\partial\mathcal{D} = \partial^*\mathcal{D}$.

The proof follows from Lemmas 7.16, 7.17.

LEMMA 7.16. *Let $\partial\mathcal{D} \neq \partial^*\mathcal{D}$ and let the limit $\mathcal{E}(p)$ exists at any $p \in \partial\mathcal{D}\backslash\partial^*\mathcal{D}$. Then*

$$\text{ess}\|T\| \geq \sup_{p\in\partial\mathcal{D}\backslash\partial^*\mathcal{D}} \sup_{|\boldsymbol{z}|=1} \left| \int_{\mathbb{S}^{n-1}} \big|\mathcal{M}^*(\sigma)\boldsymbol{z}\big| d\sigma - 2\int_{\mathcal{E}(p)} \big|\mathcal{M}^*(\sigma)\boldsymbol{z}\big| d\sigma \right|.$$

PROOF. In the same way as in Section 7.3.1 we introduce the set function

$$\Phi_{\mathcal{D}}(\boldsymbol{z}; p, \mathcal{B}) = \int_{\mathcal{B}\cap\partial\mathcal{D}} |\mathcal{M}^*(\boldsymbol{e}_{px})\boldsymbol{z}| \omega_{\mathcal{D}}(p, dx), \qquad \boldsymbol{z} \in \mathbb{R}^m, \tag{7.3.25}$$

defining it, first, for $p \in \mathbb{R}^n$ and for all Borel sets $\mathcal{B}$ in $\mathbb{R}^n$ which do not contain p. For $p \in \mathcal{B}$ we put

$$\Phi_{\mathcal{D}}(\boldsymbol{z}; p, \mathcal{B}) = \Phi_{\mathcal{D}}(\boldsymbol{z}; p, \mathcal{B}\backslash p) + \Phi_{\mathcal{D}}(\boldsymbol{z}; p, p), \tag{7.3.26}$$

where

$$\Phi_{\mathcal{D}}(\boldsymbol{z}; p, p) = \begin{cases} \dfrac{1}{2}\displaystyle\int_{S^{n-1}} \big|\mathcal{M}^*(\sigma)\boldsymbol{z}\big| d\sigma - \Phi_{\mathcal{D}}(\boldsymbol{z}; p, \partial\mathcal{D}\backslash p) & \text{for} \quad p \in \partial\mathcal{D}, \\ 0 & \text{for} \quad p \in \mathbb{R}^n\backslash\partial\mathcal{D}. \end{cases} \tag{7.3.27}$$

Now the function $\mathcal{B} \to \Phi_{\mathcal{D}}(\boldsymbol{z}; p, \mathcal{B})$ is defined and is finite for all Borel subsets of $\mathbb{R}^n$ and for $\boldsymbol{z} \in \mathbb{R}^m$.

For any fixed m-component vector $\boldsymbol{z}$ the operator

$$(G_{\boldsymbol{z}} f)(p) = \int_{\partial\mathcal{D}} f(x)\alpha_\delta(x-p)\Phi_{\mathcal{D}}(\boldsymbol{z}; p, dx)$$

acts in $\mathrm{C}(\partial\mathcal{D})$ and is continuous. Since by Lemma 7.11, $\Phi_{\mathcal{D}}(\boldsymbol{z};p,p)=0$ for all $p\in\partial^*\mathcal{D}$ and since $\partial^*\mathcal{D}$ is everywhere dense on $\partial\mathcal{D}$, it follows that

$$\begin{aligned}
\mathrm{ess}\|T\| &= 2\lim_{\delta\to 0}\sup_{|\boldsymbol{z}|=1}\sup_{p\in\partial^*\mathcal{D}}\int_{\partial\mathcal{D}\cap\mathbb{B}_\delta(p)}|\mathcal{M}^*(\boldsymbol{e}_{px})\boldsymbol{z}|\,|\omega_{\mathcal{D}}|(p,dx)\\
&\geq 2\liminf_{\delta\to 0}\sup_{|\boldsymbol{z}|=1}\sup_{p\in\partial^*\mathcal{D}}\int_{\partial\mathcal{D}\backslash p}\alpha_\delta(x-p)|\mathcal{M}^*(\boldsymbol{e}_{px})\boldsymbol{z}|\,|\omega_{\mathcal{D}}|(p,dx)\\
&\geq 2\liminf_{\delta\to 0}\sup_{|\boldsymbol{z}|=1}\sup_{p\in\partial^*\mathcal{D}}\left|\int_{\partial\mathcal{D}\backslash p}\alpha_\delta(x-p)|\mathcal{M}^*(\boldsymbol{e}_{px})\boldsymbol{z}|\omega_{\mathcal{D}}(p,dx)\right|\\
&= 2\liminf_{\delta\to 0}\sup_{|\boldsymbol{z}|=1}\sup_{p\in\partial^*\mathcal{D}}\left|\int_{\partial\mathcal{D}\backslash p}\alpha_\delta(x-p)|\mathcal{M}^*(\boldsymbol{e}_{px})\boldsymbol{z}|\omega_{\mathcal{D}}(p,dx)+\Phi_{\mathcal{D}}(\boldsymbol{z};p,p)\right|\\
&= 2\liminf_{\delta\to 0}\sup_{|\boldsymbol{z}|=1}\sup_{p\in\partial\mathcal{D}}\left|\int_{\partial\mathcal{D}}\alpha_\delta(x-p)\Phi_{\mathcal{D}}(\boldsymbol{z};p,dx)\right|\\
&\geq 2\sup_{|\boldsymbol{z}|=1}\sup_{p\in\partial\mathcal{D}}\liminf_{\delta\to 0}\left|\int_{\partial\mathcal{D}}\alpha_\delta(x-p)\Phi_{\mathcal{D}}(\boldsymbol{z};p,dx)\right|.
\end{aligned}$$

This and (7.3.8) imply

$$\mathrm{ess}\|T\|\geq 2\sup_{|\boldsymbol{z}|=1}\sup_{p\in\partial\mathcal{D}}|\Phi_{\mathcal{D}}(\boldsymbol{z};p,p)|.$$

Taking into account that $\Phi_{\mathcal{D}}(\boldsymbol{z};p,p)=0$ for $p\in\partial^*\mathcal{D}$, we obtain

$$\mathrm{ess}\|T\|\geq 2\sup_{|\boldsymbol{z}|=1}\sup_{p\in\partial\mathcal{D}\backslash\partial^*\mathcal{D}}\left|\Phi_{\mathcal{D}}(\boldsymbol{z};p,p)\right|=2\sup_{p\in\partial\mathcal{D}\backslash\partial^*\mathcal{D}}\sup_{|\boldsymbol{z}|=1}|\Phi_{\mathcal{D}}(\boldsymbol{z};p,p)|. \tag{7.3.28}$$

Making use of (7.3.25), (7.3.27) and applying the arguments from Remark 7.3 to the function $\Phi_{\mathcal{D}}(\boldsymbol{z};p,\mathcal{B})$, we find

$$\Phi_{\mathcal{D}}(\boldsymbol{z};p,p)=\frac{1}{2}\int_{\mathbb{S}^{n-1}}|\mathcal{M}^*(\sigma)\boldsymbol{z}|d\sigma-\int_{\mathcal{E}(p)}|\mathcal{M}^*(\sigma)\boldsymbol{z}|d\sigma, \tag{7.3.29}$$

where $p\in\partial\mathcal{D}\backslash\partial^*\mathcal{D}$. This together with (7.3.28) completes the proof. □

LEMMA 7.17. *Let $\mathcal{D}$ be a bounded convex domain. Then*

$$\mathrm{ess}\|T\|\leq\sup_{p\in\partial\mathcal{D}\backslash\partial^*\mathcal{D}}\sup_{|\boldsymbol{z}|=1}\left[\int_{\mathbb{S}^{n-1}}|\mathcal{M}^*(\sigma)\boldsymbol{z}|d\sigma-2\int_{\mathcal{E}(p)}|\mathcal{M}^*(\sigma)\boldsymbol{z}|d\sigma\right]$$

if $\partial\mathcal{D}\neq\partial^*\mathcal{D}$ and $\mathrm{ess}\|T\|=0$ *if $\partial\mathcal{D}=\partial^*\mathcal{D}$.*

PROOF. We introduce the notation

$$\mathcal{G}_\delta(\boldsymbol{z};p)=\int_{\partial\mathcal{D}}\alpha_\delta(x-p)\Phi_{\mathcal{D}}(\boldsymbol{z};p,dx), \tag{7.3.30}$$

where the set function $\Phi_{\mathcal{D}}(\boldsymbol{z};p,\mathcal{B})$ is defined by (7.3.25)-(7.3.27).

In the following discussion of this lemma we use the notation $\alpha_{(k)}(x)$, $\mathcal{G}_{(k)}(\boldsymbol{z};p)$ and $\mathbb{B}_{(k)}(p)$ instead of $\alpha_{2^{-k}}(x)$, $\mathcal{G}_{2^{-k}}(\boldsymbol{z};p)$ and $\mathbb{B}_{2^{-k}}(p)$.

Since $\omega_{\mathcal{D}}(p,\mathcal{B}) \geq 0$ and since $\omega_{\mathcal{D}}(p,\mathcal{B})$ does not have point loadings, we have by the monotonicity of the sequence $\{\alpha_{(k)}\}_{k\geq 1}$

$$\begin{aligned}
\mathrm{ess}\|T\| &= 2\lim_{\delta\to 0}\sup_{|\boldsymbol{z}|=1}\sup_{p\in\partial^*\mathcal{D}}\int_{\partial\mathcal{D}\cap\mathbb{B}_\delta(p)}|\mathcal{M}^*(\boldsymbol{e}_{px})\boldsymbol{z}|\omega_{\mathcal{D}}(p,dx)\\
&= 2\lim_{k\to\infty}\sup_{|\boldsymbol{z}|=1}\sup_{p\in\partial^*\mathcal{D}}\int_{\partial\mathcal{D}\cap\mathbb{B}_{(k)}(p)}|\mathcal{M}^*(\boldsymbol{e}_{px})\boldsymbol{z}|\omega_{\mathcal{D}}(p,dx)\\
&\leq 2\lim_{k\to\infty}\sup_{|\boldsymbol{z}|=1}\sup_{p\in\partial^*\mathcal{D}}\int_{\partial\mathcal{D}\backslash p}\alpha_{(k-1)}(x-p)|\mathcal{M}^*(\boldsymbol{e}_{px})\boldsymbol{z}|\omega_{\mathcal{D}}(p,dx).
\end{aligned}$$

Proceeding in the same way as in Lemma 7.16, we obtain

$$\begin{aligned}
\mathrm{ess}\|T\| &\leq 2\lim_{k\to\infty}\sup_{|\boldsymbol{z}|=1}\sup_{p\in\partial^*\mathcal{D}}\left[\int_{\partial\mathcal{D}\backslash p}\alpha_{(k)}(x-p)|\mathcal{M}^*(\boldsymbol{e}_{px})\boldsymbol{z}|\omega_{\mathcal{D}}(p,dx)+\Phi_{\mathcal{D}}(\boldsymbol{z};p,p)\right]\\
&= 2\lim_{k\to\infty}\sup_{|\boldsymbol{z}|=1}\sup_{p\in\partial\mathcal{D}}\int_{\partial\mathcal{D}}\alpha_{(k)}(x-p)\Phi_{\mathcal{D}}(\boldsymbol{z};p,dx) \qquad (7.3.31)\\
&= 2\lim_{k\to\infty}\sup_{|\boldsymbol{z}|=1}\sup_{p\in\partial\mathcal{D}}\mathcal{G}_{(k)}(\boldsymbol{z};p).
\end{aligned}$$

The continuity of the function (7.3.30) on $\mathbb{S}^{n-1}\times\partial\mathcal{D}$ for any fixed $\delta>0$ follows from

$$\begin{aligned}
\big|\mathcal{G}_\delta(\boldsymbol{z}_2;p)-\mathcal{G}_\delta(\boldsymbol{z}_1;p)\big| &\leq \frac{1}{2}\int_{\mathbb{S}^{n-1}}|\mathcal{M}^*(\sigma)(\boldsymbol{z}_2-\boldsymbol{z}_1)|d\sigma\\
&+ \int_{\partial\mathcal{D}\backslash p}\Big[1+\alpha_\delta(x-p)\Big]|\mathcal{M}^*(\boldsymbol{e}_{px})(\boldsymbol{z}_2-\boldsymbol{z}_1)|\omega_{\mathcal{D}}(p,dx)\\
&\leq c|\boldsymbol{z}_2-\boldsymbol{z}_1|
\end{aligned}$$

and from the continuity of the operator $G_{\boldsymbol{z}}$, defined in Lemma 7.16, in $\mathrm{C}(\partial\mathcal{D})$ for any fixed vector $\boldsymbol{z}$.

Since $\mathbb{S}^{m-1}\times\partial\mathcal{D}$ is a compact subset of $\mathbb{R}^m\times\mathbb{R}^n$ and $\mathcal{G}_{(k)}\in\mathrm{C}(\mathbb{S}^{m-1}\times\partial\mathcal{D})$, there exists a vector $\boldsymbol{z}_k\in\mathbb{S}^{m-1}$ and a point $p_k\in\partial\mathcal{D}$ such that

$$\sup_{|\boldsymbol{z}|=1}\sup_{p\in\partial\mathcal{D}}\mathcal{G}_{(k)}(\boldsymbol{z};p)=\mathcal{G}_{(k)}(\boldsymbol{z}_k;p_k).$$

Using (7.3.31), one obtains

$$\mathrm{ess}\|T\|\leq 2\lim_{k\to\infty}\mathcal{G}_{(k)}(\boldsymbol{z}_k;p_k). \qquad (7.3.32)$$

We select from the sequence $\{\boldsymbol{z}_k;p_k\}_{k\geq 1}$ a convergent subsequence $\{\boldsymbol{z}_{k_i};p_{k_i}\}_{i\geq 1}$. Let $(\boldsymbol{\varsigma};q)$ be its limit. By the monotonicity of the sequence $\{\mathcal{G}_{(k)}\}_{k\geq 1}$ and the continuity of the functions $\mathcal{G}_{(k)}$ on $\mathbb{S}^{m-1}\times\partial\mathcal{D}$ we find

$$\lim_{i\to\infty}\mathcal{G}_{(k_i)}(\boldsymbol{z}_{k_i};p_{k_i})\leq\lim_{i\to\infty}\mathcal{G}_{(k_j)}(\boldsymbol{z}_{k_i};p_{k_i})=\mathcal{G}_{(k_j)}(\boldsymbol{\varsigma},q),$$

where k_j is an arbitrary element of the sequence $\{k_i\}_{i\geq 1}$.

This together with (7.3.32) implies

$$\begin{aligned}\text{ess}\|T\| &\le 2\lim_{j\to\infty}\mathcal{G}_{(k_j)}(\varsigma;q)=2\lim_{j\to\infty}\int_{\partial\mathcal{D}}\alpha_{(k_j)}(x-q)\Phi_{\mathcal{D}}(\varsigma;q,dx)\\ &= 2\lim_{j\to\infty}\Big[\int_{\partial\mathcal{D}\backslash q}\alpha_{(k_j)}(x-q)|\mathcal{M}^*(\boldsymbol{e}_{qx})\varsigma|\omega_{\mathcal{D}}(q,dx)+\Phi_{\mathcal{D}}(\varsigma;q,q)\Big].\end{aligned}$$

Hence

$$\text{ess}\|T\|\le 2\Phi_{\mathcal{D}}(\varsigma;q,q), \tag{7.3.33}$$

where (ς,q) is a point of the set $\mathbb{S}^{m-1}\times\partial\mathcal{D}$. If $\partial\mathcal{D}=\partial^*\mathcal{D}$, then by Lemma 7.11, $\Phi_{\mathcal{D}}(\varsigma;q,q)=0$ for all $q\in\partial\mathcal{D}$ and so $\text{ess}\|T\|=0$. On the contrary, if $\partial\mathcal{D}\neq\partial^*\mathcal{D}$ then from (7.3.33) we find

$$\text{ess}\|T\|\le 2\sup_{p\in\partial\mathcal{D}\backslash\partial^*\mathcal{D}}\ \sup_{|\boldsymbol{z}|=1}\Phi_{\mathcal{D}}(\boldsymbol{z};p,p). \tag{7.3.34}$$

The existence of the topological limit $\mathcal{E}(p)$ for any $p\in\partial\mathcal{D}$ follows from the monotonicity of the family $\{\mathcal{E}_\varepsilon(p)\}$ as $\varepsilon\to 0$, i.e. $\mathcal{E}_{\varepsilon_2}(p)\supset\mathcal{E}_{\varepsilon_1}(p)$ if $\varepsilon_2<\varepsilon_1$. The last remark together with (7.3.34) and (7.3.29) completes the proof. □

PROPOSITION 7.1. *Let $\mathcal{D}$ be a bounded convex domain in $\mathbb{R}^n$ and let $\partial\mathcal{D}\neq\partial^*\mathcal{D}$. Then for any $q\in\partial\mathcal{D}\backslash\partial^*\mathcal{D}$ and for any arbitrary $\varepsilon>0$*

$$\begin{aligned}&\lim_{\delta\to 0}\sup_{p\in\partial^*\mathcal{D}\cap\overline{\mathbb{B}_\varepsilon(q)}}\ \sup_{|\boldsymbol{z}|=1}\int_{\partial\mathcal{D}\cap\mathbb{B}_\delta(p)}|\mathcal{M}^*(\boldsymbol{e}_{px})\boldsymbol{z}||\omega_{\mathcal{D}}|(p,dx)\\ &=\sup_{p\in(\partial\mathcal{D}\backslash\partial^*\mathcal{D})\cap\overline{\mathbb{B}_\varepsilon(q)}}\ \sup_{|\boldsymbol{z}|=1}\Big[\frac{1}{2}\int_{\mathbb{S}^{n-1}}|\mathcal{M}^*(\sigma)\boldsymbol{z}|d\sigma-\int_{\mathcal{E}(p)}|\mathcal{M}^*(\sigma)\boldsymbol{z}|d\sigma\Big].\end{aligned}$$

The proof is the same as that of Theorem 7.5.

7.4. Boundary integral operators of elasticity and hydrodynamics

7.4.1. Matrices of fundamental solutions

Consider the internal and the external Dirichlet problems for the Lamé system

$$\mu\Delta\boldsymbol{u}+(\lambda+\mu)\text{grad div}\,\boldsymbol{u}=\boldsymbol{0}\quad\text{in}\quad\mathcal{D},\qquad \boldsymbol{u}\big|_{\partial\mathcal{D}}=\boldsymbol{f}, \tag{7.4.1}$$

$$\mu\Delta\boldsymbol{u}+(\lambda+\mu)\text{grad div}\,\boldsymbol{u}=\boldsymbol{0}\quad\text{in}\quad\mathbb{R}^n\backslash\overline{\mathcal{D}},\quad \boldsymbol{u}\big|_{\partial\mathcal{D}}=\boldsymbol{f}, \tag{7.4.2}$$

where λ and μ are the Lamé constants, $\boldsymbol{u}=(u_1,\dots,u_n)$ is the displacement vector and $\boldsymbol{f}\in[\mathrm{C}(\partial\mathcal{D})]^n$.

In elasticity theory one assumes that $\mu>0, 3\lambda+2\mu>0$; specific strain energy, in this case, is a positive definite quadratic form (see Kupradze, Gegelia, Basheleishvili and Burchuladze [**KGBB**]). We impose a slightly weaker condition $\lambda+\mu>0$ for $\mu>0$.

REMARK 7.4. The characteristic matrix $A(\boldsymbol{\sigma})$ of the Lamé system has the elements

$$a_{jk}(\boldsymbol{\sigma})=\mu|\boldsymbol{\sigma}|^2\delta_{jk}+(\lambda+\mu)\sigma_j\sigma_k,\qquad j,k=1,2,\dots,n.$$

For any vector $\boldsymbol{\xi} \in \mathbb{R}^n$ we have

$$\begin{aligned}(A(\boldsymbol{\sigma})\boldsymbol{\xi}, \boldsymbol{\xi}) &= \sum_{j=1}^{n} \left(\sum_{k=1}^{n} a_{jk}(\boldsymbol{\sigma})\xi_k \right) \xi_j = \sum_{j=1}^{n} \left[\sum_{k=1}^{n} \left(\mu|\boldsymbol{\sigma}|^2\delta_{jk} + (\lambda+\mu)\sigma_j\sigma_k\right) \xi_k \right] \xi_j \\ &= \sum_{j=1}^{n} \left[\mu|\boldsymbol{\sigma}|^2\xi_j + (\lambda+\mu)\sigma_j \left(\sum_{k=1}^{n} \sigma_k\xi_k \right) \right] \xi_j \,,\end{aligned}$$

that is

$$\big(A(\boldsymbol{\sigma})\boldsymbol{\xi}, \boldsymbol{\xi}\big) = \mu|\boldsymbol{\sigma}|^2|\boldsymbol{\xi}|^2 + (\lambda+\mu)\big(\boldsymbol{\sigma}, \boldsymbol{\xi}\big)^2.$$

Hence, for $\mu > 0$ and $\lambda + \mu > -\mu$ the inequality

$$\big(A(\boldsymbol{\sigma})\boldsymbol{\xi}, \boldsymbol{\xi}\big) \geq \delta|\boldsymbol{\sigma}|^2|\boldsymbol{\xi}|^2$$

holds with $\delta > 0$. Thus the Lamé system is strongly elliptic for $\mu > 0$ and $\lambda+2\mu > 0$.

When solving the problems (7.4.1) and (7.4.2) by methods of potential theory, one uses the fundamental matrix of the Lamé system. According to the definition of the fundamental solution of a linear partial differential operator with constant coefficients (see, for example, Hörmander [**Hör**]), the matrix of fundamental solutions $\Gamma(x) = \big(\big(\Gamma_{jk}(x)\big)\big)_{n\times n}$ of the Lamé system consists of elements of n-component vector-valued functions

$$\boldsymbol{\Gamma}_1(x),\ \boldsymbol{\Gamma}_2(x), \ldots,\ \boldsymbol{\Gamma}_n(x),$$

for which

$$\mu\Delta\boldsymbol{\Gamma}_k + (\lambda+\mu)\text{grad div}\,\boldsymbol{\Gamma}_k = \delta(x)\boldsymbol{e}_k, \tag{7.4.3}$$

with δ standing for the Dirac function. For $n \geq 3$ system (7.4.3) is supplied with the condition of vanishing of $\boldsymbol{\Gamma}_k(x)$ as $|x| \to \infty$.

For $n = 2$ (the Boussinesq matrix) and for $n = 3$ (the Kelvin-Somigliana matrix) the matrices $\Gamma(x)$ are well known (see, for example, Parton and Perlin [**PP1**], Kupradze, Gegelia, Basheleishvili and Burchuladze [**KGBB**], McLean [**McL**]). In particular,

$$\Gamma_{jk}(x) = -\frac{1}{4\pi\mu(\lambda+2\mu)} \left[(\lambda+3\mu)\delta_{jk}\log\frac{1}{|x|} + (\lambda+\mu)\frac{x_j x_k}{|x|^2} \right] \tag{7.4.4}$$

for $n = 2$, where δ_{jk} is the Kronecker delta.

Consider now the internal and the external Dirichlet problems for the Stokes system

$$\mu\Delta\boldsymbol{u} - \text{grad}\,p = \boldsymbol{0}, \quad \text{div}\,\boldsymbol{u} = 0 \quad \text{in} \quad \mathcal{D}, \qquad \boldsymbol{u}\big|_{\partial\mathcal{D}} = \boldsymbol{f}, \tag{7.4.5}$$

$$\mu\Delta\boldsymbol{u} - \text{grad}\,p = \boldsymbol{0}, \quad \text{div}\,\boldsymbol{u} = 0 \quad \text{in} \quad \mathbb{R}^n\backslash\overline{\mathcal{D}}, \quad \boldsymbol{u}\big|_{\partial\mathcal{D}} = \boldsymbol{f}, \tag{7.4.6}$$

where μ is the coefficient of kinematic viscosity, $\boldsymbol{u} = (u_1, \ldots u_n)$ is the velocity vector of a fluid and p is the pressure.

Similarly to elasticity theory, the fundamental solution of the Stokes system plays an important role in the theory of hydrodynamic potentials. Let the vector-valued functions $\mathring{\boldsymbol{\Gamma}}_k$ with n-components and scalar function $\mathring{P}_k(x)$ be solutions of the system

$$\mu\Delta\mathring{\boldsymbol{\Gamma}}_k - \text{grad}\,\mathring{P}_k = \delta(x)\boldsymbol{e}_k, \quad \text{div}\,\mathring{\boldsymbol{\Gamma}}_k = 0 \tag{7.4.7}$$

in $\mathbb{R}^n$. The matrix of fundamental solutions to the Stokes system

$$\mathring{\Phi}(x) = \big((\mathring{\Phi}_{jk}(x))\big)_{(n+1)\times n}$$

consists of vector-valued functions with $n+1$ components

$$\mathring{\mathbf{\Phi}}_1(x) = \begin{pmatrix} \mathring{\mathbf{\Gamma}}_1(x) \\ \mathring{P}_1(x) \end{pmatrix}, \ \ldots, \ \mathring{\mathbf{\Phi}}_n(x) = \begin{pmatrix} \mathring{\mathbf{\Gamma}}_n(x) \\ \mathring{P}_n(x) \end{pmatrix}.$$

For $n \geq 3$ it is assumed that the components of every vector-valued function $\mathring{\mathbf{\Phi}}_k(x)$ vanish as $|x| \to \infty$.

For $n = 2$ and $n = 3$ the matrices $\mathring{\Phi}(x)$ can be found, for example, in the book [**Lad**] by Ladyzhenskaya. In particular, for $n = 2$

$$\mathring{\Gamma}_{jk}(x) = -\frac{1}{4\pi\mu}\left[\delta_{jk}\log\frac{1}{|x|} + \frac{x_j x_k}{|x|^2}\right], \qquad \mathring{P}_k(x) = -\frac{x_k}{2\pi|x|^2}\,. \tag{7.4.8}$$

In the next assertion we derive formulas for the matrix of fundamental solutions of the Lamé and Stokes systems for $(n \geq 3)$ using the Fourier transform. As usual, the summation over repeated indices is meant. We consider the system

$$\mu\Delta \boldsymbol{u} - \operatorname{grad} p = \mathbf{0}, \qquad p = -(\lambda+\mu)\operatorname{div}\boldsymbol{u}, \tag{7.4.9}$$

which is equivalent to the Lamé system. By the matrix of fundamental solutions of (7.4.9) we mean the matrix

$$\Phi(x) = \big((\Phi_{jk}(x))\big)_{(n+1)\times n},$$

whose elements are vector-valued functions with $n+1$ components

$$\mathbf{\Phi}_1(x) = \begin{pmatrix} \mathbf{\Gamma}_1(x) \\ P_1(x) \end{pmatrix}, \ \ldots, \ \mathbf{\Phi}_n(x) = \begin{pmatrix} \mathbf{\Gamma}_n(x) \\ P_n(x) \end{pmatrix},$$

where $\{\mathbf{\Gamma}_k(x),\ P_k(x)\}$ are solutions of the system

$$\mu\Delta\mathbf{\Gamma}_k - \operatorname{grad} P_k = \delta(x)\boldsymbol{e}_k, \qquad P_k = -(\lambda+\mu)\operatorname{div}\Gamma_k. \tag{7.4.10}$$

LEMMA 7.18. *The components of the matrix of fundamental solutions of system* (7.4.9), $n \geq 3$, *are given by*

$$\Gamma_{jk}(x) = -\frac{1}{2\omega_n\mu(\lambda+2\mu)}\left[\frac{(\lambda+3\mu)\delta_{jk}}{(n-2)}\frac{1}{|x|^{n-2}} + (\lambda+\mu)\frac{x_j x_k}{|x|^n}\right], \tag{7.4.11}$$

$$P_k(x) = -\frac{(\lambda+\mu)x_k}{(\lambda+2\mu)\omega_n|x|^n}\,, \tag{7.4.12}$$

where ω_n is the area of the $(n-1)$-dimensional unit sphere.

The components of the matrix of fundamental solutions of the n-dimensional Lamé system, $n \geq 3$, are given by (7.4.11).

The limit case of (7.4.11) *and* (7.4.12) *as $\lambda \to \infty$, provides the components of the matrix of fundamental solutions of the Stokes system in $\mathbb{R}^n, n \geq 3$:*

$$\mathring{\Gamma}_{jk}(x) = -\frac{1}{2\omega_n\mu}\left[\frac{\delta_{jk}}{(n-2)}\frac{1}{|x|^{n-2}} + \frac{x_j x_k}{|x|^n}\right], \tag{7.4.13}$$

$$\mathring{P}_k(x) = -\frac{x_k}{\omega_n|x|^n}. \tag{7.4.14}$$

PROOF. We make the Fourier transform in (7.4.10) and use the notations

$$F[\varphi] = \frac{1}{(2\pi)^{n/2}} \int_{\mathbb{R}^n} e^{-i(x,\xi)} \varphi(x) dx, \qquad F^{-1}[\psi] = \frac{1}{(2\pi)^{n/2}} \int_{\mathbb{R}^n} e^{i(x,\xi)} \psi(\xi) d\xi.$$

Then

$$-\mu|\xi|^2 F[\Gamma_{jk}] - i\xi_j F[P_k] = \frac{\delta_{jk}}{(2\pi)^{n/2}}, \tag{7.4.15}$$

$$F[P_k] = -i(\lambda+\mu)\xi_j F[\Gamma_{jk}]. \tag{7.4.16}$$

Multiplying (7.4.15) by ξ_j and summing over j, we obtain

$$-\mu|\xi|^2 \xi_j F[\Gamma_{jk}] - i|\xi|^2 F[P_k] = \frac{\xi_k}{(2\pi)^{n/2}}. \tag{7.4.17}$$

This and (7.4.16) imply

$$F[P_k] = \frac{\lambda+\mu}{(2\pi)^{n/2}(\lambda+2\mu)} \frac{i\xi_k}{|\xi|^2}. \tag{7.4.18}$$

Using (7.4.18) in (7.4.15), we find

$$F[\Gamma_{jk}] = \frac{1}{(2\pi)^{n/2}\mu} \left(-\frac{\delta_{jk}}{|\xi|^2} + \frac{\lambda+\mu}{\lambda+2\mu} \frac{\xi_j \xi_k}{|\xi|^4} \right). \tag{7.4.19}$$

In order to get Γ_{jk} and P_k we use the fundamental solutions of the operators Δ i Δ^2. Since

$$\Delta \left(-\frac{1}{(n-2)\omega_n |x|^{n-2}} \right) = \delta(x),$$

we have

$$|\xi|^2 F\left[\frac{1}{(n-2)\omega_n |x|^{n-2}} \right] = \frac{1}{(2\pi)^{n/2}},$$

that is

$$F\left[\frac{(2\pi)^{n/2}}{(n-2)\omega_n |x|^{n-2}} \right] = \frac{1}{|\xi|^2}. \tag{7.4.20}$$

Thus,

$$F^{-1}\left[\frac{1}{|\xi|^2} \right] = \frac{(2\pi)^{n/2}}{(n-2)\omega_n |x|^{n-2}}. \tag{7.4.21}$$

Using (7.4.20) and the equality

$$F[D^\alpha \varphi] = (i\xi)^\alpha F[\varphi], \tag{7.4.22}$$

we find

$$F^{-1}\left[\frac{i\xi_k}{|\xi|^2} \right] = \frac{\partial}{\partial x_k} \left(\frac{(2\pi)^{n/2}}{(n-2)\omega_n |x|^{n-2}} \right) = -\frac{(2\pi)^{n/2} x_k}{\omega_n |x|^n},$$

which together with (7.4.18), proves (7.4.12).

The fundamental solution $\mathcal{H}_n$ of the biharmonic operator has the form (see, for example, Shilov [**Shi**], §7, sect. 9)

$$\mathcal{H}_n(x) = \begin{cases} \dfrac{|x|^{4-n}}{2\omega_n (n-2)(n-4)} & \text{for } n=3 \text{ and } n>4, \\[2ex] -\dfrac{1}{8\pi^2} \log|x| & \text{for } n=4. \end{cases} \tag{7.4.23}$$

Applying the Fourier transform to the equation $\Delta^2\mathcal{H}_n(x) = \delta(x)$, we obtain

$$|\xi|^4 F[\mathcal{H}_n] = \frac{1}{(2\pi)^{n/2}}\,,$$

that is

$$F\big[(2\pi)^{n/2}\mathcal{H}_n\big] = \frac{1}{|\xi|^4}\,.$$

This together with (7.4.22) implies

$$F^{-1}\left[\frac{\xi_j\xi_k}{|\xi|^4}\right] = -(2\pi)^{n/2}\frac{\partial^2\mathcal{H}_n}{\partial x_j\partial x_k}.$$

This identity and (7.4.23) result in

$$F^{-1}\left[\frac{\xi_j\xi_k}{|\xi|^4}\right] = \frac{(2\pi)^{n/2}}{2(n-2)\omega_n}\left[\frac{\delta_{jk}}{|x|^{n-2}} + (2-n)\frac{x_jx_k}{|x|^n}\right] \tag{7.4.24}$$

for all $n \geq 3$.

Combining (7.4.19), (7.4.21), and (7.4.24), we arrive at (7.4.11). Next, by(7.4.12),

$$\frac{P_k}{\lambda+\mu} = -\frac{x_k}{(\lambda+2\mu)\omega_n|x|^n} \to 0 \text{ as } \lambda\to\infty.$$

Besides, by (7.4.11) and (7.4.12) one concludes that the limits of $\mathbf{\Gamma}_k$ and P_k as $\lambda \to \infty$ exist. Therefore, the solution of (7.4.7) is the limit of the solution of (7.4.10) as $\lambda \to \infty$. Now, passing to the limit in (7.4.11) and (7.4.12) as $\lambda \to \infty$, we arrive at (7.4.13) and (7.4.14). □

7.4.2. Integral representations of solutions

Let $\{\boldsymbol{u}, p\}$ be a solution of (7.4.9). We introduce the tensor (see, for example, Parton and Perlin [**PP1**], Kupradze, Gegelia, Basheleishvili and Burchuladze [**KGBB**]) with the elements

$$\sigma_{ij}(\boldsymbol{u}) = (\lambda+\mu-\gamma)\delta_{ij}\text{div }\boldsymbol{u} + \gamma\frac{\partial u_j}{\partial x_i} + \mu\frac{\partial u_i}{\partial x_j}\,, \tag{7.4.25}$$

corresponding to the field $\boldsymbol{u}$. For $\gamma = \mu$ this tensor becomes the stress tensor, while for $\gamma = \mu(\lambda+\mu)/(\lambda+3\mu)$ it is called the pseudostress tensor.

In what follows we are interested only in the pseudostress tensor which will be denoted by $\tilde{\sigma}_{ij}$:

$$\begin{aligned}\tilde{\sigma}_{ij}(\boldsymbol{u}) &= \frac{(\lambda+\mu)(\lambda+2\mu)}{\lambda+3\mu}\,\delta_{ij}\text{ div }\boldsymbol{u} + \frac{\mu(\lambda+\mu)}{\lambda+3\mu}\frac{\partial u_j}{\partial x_i} + \mu\frac{\partial u_i}{\partial x_j} \\ &= -\,\frac{1+\kappa}{2}\,\delta_{ij}p + \mu\kappa\frac{\partial u_j}{\partial x_i} + \mu\frac{\partial u_i}{\partial x_j}\,,\end{aligned} \tag{7.4.26}$$

where $\kappa = (\lambda+\mu)/(\lambda+3\mu)$ and, by (7.4.9), $p = -(\lambda+\mu)\text{ div }\boldsymbol{u}$.

The pseudotraction vector

$$\boldsymbol{N}(\boldsymbol{u};\boldsymbol{\nu}) = \tilde{\sigma}_{ij}(\boldsymbol{u})\nu_j\boldsymbol{e}_i, \tag{7.4.27}$$

corresponds to the pseudostress tensor, applied to a surface with the normal $\boldsymbol{\nu}$. Thus, by (7.4.26),

$$\begin{aligned}\boldsymbol{N}(\boldsymbol{u};\boldsymbol{\nu}) &= -\frac{1+\kappa}{2}\,\boldsymbol{\nu} p + \mu\kappa\nu_j\boldsymbol{e}_i\frac{\partial u_j}{\partial x_i} + \mu\nu_j\boldsymbol{e}_i\frac{\partial u_i}{\partial x_j}\\ &= -\frac{1+\kappa}{2}\,\boldsymbol{\nu} p + \mu\kappa\nu_j \text{ grad } u_j + \mu\frac{\partial \boldsymbol{u}}{\partial \boldsymbol{\nu}}\,,\end{aligned} \tag{7.4.28}$$

where $p = -(\lambda+\mu) \text{ div } \boldsymbol{u}$.

Let Ω be bounded domain in $\mathbb{R}^n$ with piecewise smooth boundary. We introduce the notation

$$\boldsymbol{L}(\boldsymbol{u};p) = \mu\Delta\boldsymbol{u} - \text{grad } p. \tag{7.4.29}$$

The next assertion concerns the first, the second and the third Betti identities for pseudostresses (see Parton and Perlin [**PP1**], § 14) for the operator $L(\boldsymbol{u};p)$ with $p = -(\lambda+\mu) \text{ div } \boldsymbol{u}$.

LEMMA 7.19. *For all vector-valued functions $\boldsymbol{u},\boldsymbol{v}$ from the class $[\mathrm{C}^2(\overline{\Omega})]^n$ and the functions $p = -(\lambda+\mu) \text{ div } \boldsymbol{u}, q = -(\lambda+\mu) \text{ div } \boldsymbol{v}$ the Betti identities hold*

$$\int_\Omega (\boldsymbol{L}(\boldsymbol{v};q),\ \boldsymbol{u})dx = \int_{\partial\Omega} (\boldsymbol{N}(\boldsymbol{v};\boldsymbol{\nu}(x)),\boldsymbol{u})d\sigma_x - \int_\Omega E(\boldsymbol{v},\boldsymbol{u})dx, \tag{7.4.30}$$

$$\int_\Omega (\boldsymbol{L}(\boldsymbol{u};p),\ \boldsymbol{u})dx = \int_{\partial\Omega} (\boldsymbol{N}(\boldsymbol{u};\boldsymbol{\nu}(x)),\boldsymbol{u})d\sigma_x - \int_\Omega E(\boldsymbol{u},\boldsymbol{u})dx, \tag{7.4.31}$$

$$\int_\Omega\big\{(\boldsymbol{L}(\boldsymbol{v};q),\boldsymbol{u})-(\boldsymbol{L}(\boldsymbol{u};p),\boldsymbol{v})\big\}dx=\int_{\partial\Omega}\big\{(\boldsymbol{N}(\boldsymbol{v};\boldsymbol{\nu}(x)),\boldsymbol{u})-(\boldsymbol{N}(\boldsymbol{u};\boldsymbol{\nu}(x)),\boldsymbol{v})\big\}d\sigma_x, \tag{7.4.32}$$

where $\boldsymbol{\nu}(x)$ is the outward normal to $\partial\Omega$ at a point x and

$$E(\boldsymbol{v},\boldsymbol{u})=\frac{1+\kappa}{2(\lambda+\mu)}p\,q+\frac{\mu\kappa}{2}\left(\frac{\partial v_j}{\partial x_i}+\frac{\partial v_i}{\partial x_j}\right)\left(\frac{\partial u_j}{\partial x_i}+\frac{\partial u_i}{\partial x_j}\right)+\mu(1-\kappa)\frac{\partial v_i}{\partial x_j}\frac{\partial u_i}{\partial x_j}\,. \tag{7.4.33}$$

PROOF. In correspondence with (7.4.27), we introduce the scalar product

$$\big(\boldsymbol{N}(\boldsymbol{v};\boldsymbol{\nu}),\boldsymbol{u}\big) = \tilde\sigma_{ij}(\boldsymbol{v})\nu_j u_i = \big(\boldsymbol{Q}(\boldsymbol{v};\boldsymbol{u}),\ \boldsymbol{\nu}\big), \tag{7.4.34}$$

where

$$\boldsymbol{Q}(\boldsymbol{v};\boldsymbol{u}) = \tilde\sigma_{ij}(\boldsymbol{v})u_i\boldsymbol{e}_j.$$

Therefore,

$$\text{div } \boldsymbol{Q}(\boldsymbol{v};\boldsymbol{u}) = \frac{\partial\tilde\sigma_{ij}(\boldsymbol{v})}{\partial x_j}\,u_i + \tilde\sigma_{ij}(\boldsymbol{v})\,\frac{\partial u_i}{\partial x_j}. \tag{7.4.35}$$

Using (7.4.26) and the equality $q = -(\lambda+\mu)\text{div } \boldsymbol{v}$ together with the notation $\kappa = (\lambda+\mu)/(\lambda+3\mu)$, we obtain

$$\begin{aligned}\frac{\partial\tilde\sigma_{ij}(\boldsymbol{v})}{\partial x_j} &= \frac{\partial}{\partial x_j}\left\{-\frac{1+\kappa}{2}\,\delta_{ij}q + \mu\kappa\frac{\partial v_j}{\partial x_i} + \mu\frac{\partial v_i}{\partial x_j}\right\}\\ &= -\frac{1+\kappa}{2}\frac{\partial q}{\partial x_i} + \mu\kappa\frac{\partial}{\partial x_i}\text{ div } \boldsymbol{v} + \mu\Delta v_i = \mu\Delta v_i - \frac{\partial q}{\partial x_i}.\end{aligned} \tag{7.4.36}$$

Let the second term in (7.4.35) be denoted by $E(\boldsymbol{v}, \boldsymbol{u})$. Taking into account (7.4.26) and the equality $p = -(\lambda+\mu)\text{div}\, \boldsymbol{u}$, we find

$$
\begin{aligned}
E(\boldsymbol{v}, \boldsymbol{u}) = \tilde{\sigma}_{ij}(\boldsymbol{v})\,\frac{\partial u_i}{\partial x_j} &= \Big(-\frac{1+\kappa}{2}\,\delta_{ij} q + \mu\kappa\frac{\partial v_j}{\partial x_i} + \mu\frac{\partial v_i}{\partial x_j}\Big)\frac{\partial u_i}{\partial x_j} \\
&= -\frac{1+\kappa}{2}\, q \,\text{div}\, \boldsymbol{u} + \mu\kappa\frac{\partial v_j}{\partial x_i}\frac{\partial u_i}{\partial x_j} + \mu\frac{\partial v_i}{\partial x_j}\frac{\partial u_i}{\partial x_j} \qquad (7.4.37)\\
&= \frac{1+\kappa}{2(\lambda+\mu)} p\, q + \mu\kappa\frac{\partial v_j}{\partial x_i}\frac{\partial u_i}{\partial x_j} + \mu\frac{\partial v_i}{\partial x_j}\frac{\partial u_i}{\partial x_j}\,.
\end{aligned}
$$

It is easily seen that the bilinear form $E(\boldsymbol{v}, \boldsymbol{u})$ can be written in the form (7.4.33), hence $E(\boldsymbol{v}, \boldsymbol{u}) = E(\boldsymbol{u}, \boldsymbol{v})$ and $E(\boldsymbol{u}, \boldsymbol{u}) \geq 0$.

Thus, by (7.4.35)-(7.4.37),

$$
\text{div}\, \boldsymbol{Q}(\boldsymbol{v}; \boldsymbol{u}) = \big(\mu\Delta\boldsymbol{v} - \text{grad}\, q,\ \boldsymbol{u}\big) + E(\boldsymbol{v}, \boldsymbol{u}), \tag{7.4.38}
$$

where $E(\boldsymbol{v}, \boldsymbol{u}) = E(\boldsymbol{u}, \boldsymbol{v})$. Integrating (7.4.38) over Ω, using the Gauss-Green theorem and (7.4.34), we arrive at the first Betti formula for pseudostresses

$$
\int_\Omega (\mu\Delta\boldsymbol{v} - \text{grad}\, q,\ \boldsymbol{u}) dx = \int_{\partial\Omega} (\boldsymbol{N}(\boldsymbol{v}; \boldsymbol{\nu}(x)), \boldsymbol{u}) d\sigma_x - \int_\Omega E(\boldsymbol{v}, \boldsymbol{u}) dx. \tag{7.4.39}
$$

By (7.4.29), the identity (7.4.39) takes the form (7.4.30). Setting $\boldsymbol{u} = \boldsymbol{v}$ in (7.4.30), we arrive at the second Betti formula (7.4.31) for pseudostresses. Interchanging the places of $\boldsymbol{u}$ and $\boldsymbol{v}$ in (7.4.39), we find

$$
\int_\Omega (\mu\Delta\boldsymbol{u} - \text{grad}\, p,\ \boldsymbol{v}) dx = \int_{\partial\Omega} (\boldsymbol{N}(\boldsymbol{u}; \boldsymbol{\nu}(x)), \boldsymbol{v}) d\sigma_x - \int_\Omega E(\boldsymbol{u}, \boldsymbol{v}) dx. \tag{7.4.40}
$$

By (7.4.39), (7.4.40) and the symmetry of $E(\boldsymbol{v}, \boldsymbol{u})$ the third Betti formula (7.4.32) follows. □

REMARK 7.5. The solution $\{\boldsymbol{\mathcal{E}}_k(x, y),\ Q_k(x, y)\}$ of the system

$$
\mu\Delta\boldsymbol{\mathcal{E}}_k - \text{grad}\, Q_k = \delta(x-y)\boldsymbol{e}_k, \qquad Q_k = -(\lambda+\mu)\,\text{div}\, \boldsymbol{\mathcal{E}}_k,
$$

where $y \in \mathbb{R}^n$ plays the role of a parameter, is

$$
\boldsymbol{\mathcal{E}}_k(x, y) = \boldsymbol{\Gamma}_k(x-y), \qquad Q_k(x, y) = P_k(x-y). \tag{7.4.41}
$$

Here Γ_k and P_k are defined by (7.4.11) and (7.4.12).

Thus, $\boldsymbol{\Gamma}_k(x-y)$ is a solution of the Lamé system

$$
\mu\Delta\boldsymbol{\Gamma}_k + (\lambda+\mu)\text{grad}\,\text{div}\, \boldsymbol{\Gamma}_k = \delta(x-y)\boldsymbol{e}_k,
$$

and the vector-valued function

$$
\begin{pmatrix} \mathring{\boldsymbol{\Gamma}}_k(x-y) \\ \mathring{P}_k(x-y) \end{pmatrix},
$$

with $\mathring{\boldsymbol{\Gamma}}_k$ and $\mathring{P}_k$ defined by (7.4.13) and (7.4.14), satisfies the Stokes system

$$
\mu\Delta\mathring{\boldsymbol{\Gamma}}_k - \text{grad}\, \mathring{P}_k = \delta(x-y)\boldsymbol{e}_k, \qquad \text{div}\, \mathring{\boldsymbol{\Gamma}}_k = 0.
$$

Let $\boldsymbol{N}\big(\Gamma(x-y); \boldsymbol{\nu}(x)\big)$ denote the matrix with columns

$$
\boldsymbol{N}\big(\boldsymbol{\Gamma}_1(x-y); \boldsymbol{\nu}(x)\big),\ \dots\ , \boldsymbol{N}\big(\boldsymbol{\Gamma}_n(x-y); \boldsymbol{\nu}(x)\big),
$$

where $\boldsymbol{N}\big(\boldsymbol{u}; \boldsymbol{\nu}\big)$ is defined by (7.4.28), the components $\Gamma_{jk}(x)$ of the vector-valued function $\boldsymbol{\Gamma}_k(x)$ are given by (7.4.11), and y plays the role of a parameter.

We recall that the scalar component P_k of the fundamental solution of system (7.4.9) is defined by (7.4.12). Consider the vector-valued function

$$\boldsymbol{P}(x) = P_k(x)\boldsymbol{e}_k = -\frac{(\lambda+\mu)x}{(\lambda+2\mu)\omega_n|x|^n}\,, \tag{7.4.42}$$

which appears in the next assertion.

PROPOSITION 7.2. *Let $\mu > 0, \lambda+\mu > 0$. The solution $\{\boldsymbol{u}, p\}$ of system (7.4.9) with $\boldsymbol{u} \in [\mathrm{C}^2(\overline{\Omega})]^n$ can be expressed in the form*

$$\boldsymbol{u}(y) = \int_{\partial\Omega} \boldsymbol{N}\big(\Gamma(x-y);\boldsymbol{\nu}(x)\big)\boldsymbol{u}(x)d\sigma_x - \int_{\partial\Omega} \Gamma(x-y)\boldsymbol{N}\big(\boldsymbol{u}(x);\boldsymbol{\nu}(x)\big)d\sigma_x, \tag{7.4.43}$$

$$p(y) = \int_{\partial\Omega}\big(\boldsymbol{P}(x-y), \boldsymbol{N}(\boldsymbol{u}(x);\boldsymbol{\nu}(x))\big)d\sigma_x - \mu(1+\kappa)\int_{\partial\Omega}\left(\frac{\partial \boldsymbol{P}(x-y)}{\partial\boldsymbol{\nu}(x)}, \boldsymbol{u}(x)\right)d\sigma_x, \tag{7.4.44}$$

where $\kappa = (\lambda+\mu)/(\lambda+3\mu)$ and elements of the $(n\times n)$-matrices $\boldsymbol{N}\big(\Gamma(\boldsymbol{\xi});\boldsymbol{\nu}(x)\big)$ and $\Gamma(\boldsymbol{\xi})$ are

$$\boldsymbol{N}\big(\Gamma(\boldsymbol{\xi});\boldsymbol{\nu}(x)\big)_{ik} = \frac{1}{\omega_n}\left((1-\kappa)\delta_{ik} + \frac{n\kappa\xi_i\xi_k}{|\boldsymbol{\xi}|^2}\right)\frac{(\boldsymbol{\xi},\ \boldsymbol{\nu}(x))}{|\boldsymbol{\xi}|^n}\,, \tag{7.4.45}$$

$$\Gamma_{ik}(\boldsymbol{\xi}) = \begin{cases} -\dfrac{\lambda+\mu}{4\pi\mu(\lambda+2\mu)}\left(\dfrac{\delta_{ik}}{\kappa}\log\dfrac{1}{|\boldsymbol{\xi}|} + \dfrac{n\kappa\xi_i\xi_k}{|\boldsymbol{\xi}|^2}\right) & \text{for } n=2, \\[2ex] -\dfrac{\lambda+\mu}{2\omega_n\mu(\lambda+2\mu)}\left(\dfrac{\delta_{ik}}{(n-2)\kappa}\dfrac{1}{|\boldsymbol{\xi}|^{n-2}} + \dfrac{n\kappa\xi_i\xi_k}{|\boldsymbol{\xi}|^n}\right) & \text{for } n\geq 3. \end{cases} \tag{7.4.46}$$

PROOF. Given $y \in \Omega$, let ε be so small that $\overline{\mathbb{B}}_\varepsilon(y) \subset \Omega$. We write the third Betti identity for pseudostresses (7.4.32) in Ω_ε, putting $\boldsymbol{v} = \boldsymbol{\Gamma}_k(x-y), q = P_k(x-y)$, where $\boldsymbol{\Gamma}_k$ and P_k are defined by (7.4.11) and (7.4.12) (see Remark 7.5). Let $\{\boldsymbol{u}, p\}$ be a solution of the system (7.4.9) in Ω with $\boldsymbol{u} \in [\mathrm{C}^2(\overline{\Omega})]^n$.

Taking into account that the volume integrals in (7.4.32) vanish, we obtain

$$\int_{\partial\Omega_\varepsilon}\big\{\big(\boldsymbol{N}(\boldsymbol{\Gamma}_k(x-y);\boldsymbol{\nu}(x)), \boldsymbol{u}(x)\big) - \big(\boldsymbol{N}(\boldsymbol{u}(x);\boldsymbol{\nu}(x)), \boldsymbol{\Gamma}_k(x-y)\big)\big\}d\sigma_x = 0,$$

that is

$$\begin{aligned} &\int_{\partial\mathbb{B}_\varepsilon(y)}\big\{\big(\boldsymbol{N}(\boldsymbol{u}(x);\boldsymbol{\nu}(x)), \boldsymbol{\Gamma}_k(x-y)\big) - \big(\boldsymbol{N}(\boldsymbol{\Gamma}_k(x-y);\boldsymbol{\nu}(x)), \boldsymbol{u}(x)\big)\big\}d\sigma_x \\ &= \int_{\partial\Omega}\big\{\big(\boldsymbol{N}(\boldsymbol{\Gamma}_k(x-y);\boldsymbol{\nu}(x)), \boldsymbol{u}(x)\big) - \big(\boldsymbol{N}(\boldsymbol{u}(x);\boldsymbol{\nu}(x)), \boldsymbol{\Gamma}_k(x-y)\big)\big\}d\sigma_x. \end{aligned} \tag{7.4.47}$$

By (7.4.4), (7.4.11), for $\boldsymbol{u} \in [\mathrm{C}^2(\overline{\Omega})]^n$ we have

$$\lim_{\varepsilon\to 0}\int_{\partial\mathbb{B}_\varepsilon(y)}\big(\boldsymbol{N}(\boldsymbol{u}(x);\boldsymbol{\nu}(x)), \boldsymbol{\Gamma}_k(x-y)\big)d\sigma_x = 0. \tag{7.4.48}$$

Since $\boldsymbol{u}$ is continuous in $\overline{\Omega}$, the limit, as $\varepsilon \to 0$, of the second integral on the left-hand side of (7.4.47) is reduced to calculation of the integral

$$\int_{\partial\mathbb{B}_\varepsilon(y)}\big(\boldsymbol{N}(\boldsymbol{\Gamma}_k(x-y);\boldsymbol{\nu}(x)), \boldsymbol{u}(y)\big)d\sigma_x. \tag{7.4.49}$$

With this aim in view we find by (7.4.28) that

$$(7.4.50)\quad \boldsymbol{N}\big(\boldsymbol{\Gamma}_k(x);\boldsymbol{\nu}(x)\big) = -\frac{1+\kappa}{2}\,\boldsymbol{\nu}(x)P_k(x) + \mu\kappa\nu_j(x)\boldsymbol{e}_i\frac{\partial\Gamma_{jk}}{\partial x_i} + \mu\nu_j(x)\boldsymbol{e}_i\frac{\partial\Gamma_{ik}}{\partial x_j},$$

with P_k defined by (7.4.12) and Γ_{jk} given by (7.4.4) and (7.4.11). Using (7.4.4), (7.4.11), and (7.4.12), we write $\Gamma_{jk}(x)$ and $P_k(x)$ as

$$(7.4.51)\qquad \Gamma_{jk}(x) = \begin{cases} c_2\left(\dfrac{\delta_{jk}}{\kappa}\log\dfrac{1}{|x|} + \dfrac{x_j x_k}{|x|^2}\right) & \text{for} \quad n=2, \\[2ex] c_n\left(\dfrac{\delta_{jk}}{(n-2)\kappa}\dfrac{1}{|x|^{n-2}} + \dfrac{x_j x_k}{|x|^n}\right) & \text{for} \quad n\ge 3, \end{cases}$$

$$(7.4.52)\qquad P_k(x) = 2c_n\mu\frac{x_k}{|x|^n},$$

where

$$(7.4.53)\qquad c_n = -\frac{\lambda+\mu}{2\omega_n\mu(\lambda+2\mu)}, \qquad n\ge 2.$$

Taking into account (7.4.51) and (7.4.52), we find the terms in (7.4.50):

$$-\frac{1+\kappa}{2}\,\boldsymbol{\nu}(x)P_k(x) = -(1+\kappa)c_n\mu\boldsymbol{\nu}(x)\frac{x_k}{|x|^n},$$

$$\mu\kappa\nu_j(x)\boldsymbol{e}_i\frac{\partial\Gamma_{jk}}{\partial x_i} = c_n\mu\kappa\left(-\frac{\nu_k(x)x}{\kappa|x|^n} + \frac{\boldsymbol{\nu}(x)x_k + \boldsymbol{e}_k(x,\,\boldsymbol{\nu}(x))}{|x|^n} - \frac{n(x,\,\boldsymbol{\nu}(x))x x_k}{|x|^{n+2}}\right),$$

$$\mu\nu_j(x)\boldsymbol{e}_i\frac{\partial\Gamma_{ik}}{\partial x_j} = c_n\mu\left(-\frac{\boldsymbol{e}_k(x,\,\boldsymbol{\nu}(x))}{\kappa|x|^n} + \frac{\boldsymbol{\nu}(x)x_k + \nu_k(x)x}{|x|^n} - \frac{n(x,\,\boldsymbol{\nu}(x))x x_k}{|x|^{n+2}}\right),$$

which together with (7.4.50) and the equality $\kappa = (\lambda+\mu)/(\lambda+3\mu)$ implies

$$(7.4.54)\quad \boldsymbol{N}\big(\boldsymbol{\Gamma}_k(x-y);\boldsymbol{\nu}(x)\big) = \frac{1}{\omega_n}\left[(1-\kappa)\delta_{ik} + \frac{n\kappa(x_i-y_i)(x_k-y_k)}{|x-y|^2}\right]\frac{(x-y,\,\boldsymbol{\nu}(x))}{|x-y|^n}\,\boldsymbol{e}_i.$$

Hence, by (3.3.21) we obtain

$$(7.4.55)\quad \int_{\partial\mathbb{B}_\varepsilon(y)} \big(\boldsymbol{N}(\boldsymbol{\Gamma}_k(x-y);\boldsymbol{\nu}(x)),\boldsymbol{u}(y)\big)d\sigma_x$$

$$= \frac{1}{\omega_n}\int_{\partial\mathbb{B}_\varepsilon(y)} \left[(1-\kappa)\delta_{ik} + \frac{n\kappa(x_i-y_i)(x_k-y_k)}{|x-y|^2}\right]\frac{(x-y,\,\boldsymbol{\nu}(x))}{|x-y|^n}u_i(y)d\sigma_x$$

$$= -\frac{1}{\omega_n}\int_{\mathbb{S}^{n-1}} \left[(1-\kappa)\delta_{ik} + n\kappa(\boldsymbol{e}_\sigma,\boldsymbol{e}_i)(\boldsymbol{e}_\sigma,\boldsymbol{e}_k)\right]u_i(y)d\sigma = -u_k(y).$$

Passing to the limit as $\varepsilon\to 0$ in (7.4.47) and using (7.4.48), (7.4.49), and (7.4.55), we find

$$(7.4.56)\quad u_k(y) = \int_{\partial\Omega}\big\{\big(\boldsymbol{N}(\boldsymbol{\Gamma}_k(x-y);\boldsymbol{\nu}(x)),\ \boldsymbol{u}(x)\big) - \big(\boldsymbol{N}(\boldsymbol{u}(x);\boldsymbol{\nu}(x)),\ \boldsymbol{\Gamma}_k(x-y)\big)\big\}d\sigma_x.$$

By (7.4.11) and (7.4.54) we conclude that the matrices $\Gamma(x)$ and $\boldsymbol{N}(\boldsymbol{\Gamma}_k(x);\boldsymbol{\nu}(x))$ are symmetric. Hence (7.4.56) can be written in the vector form (7.4.43) with (7.4.45) and (7.4.46) resulting from (7.4.54) and (7.4.51).

The expression for p can be found using the equation $p = -(\lambda+\mu)\,\mathrm{div}\,\boldsymbol{u}$ of system (7.4.9) together with (7.4.43), (7.4.45), and (7.4.46).

By (7.4.51) we have

$$\operatorname{div}_y\big[\Gamma(x-y)\boldsymbol{N}(\boldsymbol{u}(x);\boldsymbol{\nu}(x))\big] = \frac{\partial}{\partial y_i}\,\Gamma_{ik}(x-y)\big(\boldsymbol{N}(\boldsymbol{u}(x);\boldsymbol{\nu}(x))\big)_k$$

$$= -\big(\boldsymbol{N}(\boldsymbol{u}(x);\boldsymbol{\nu}(x))\big)_k \frac{\partial}{\partial x_i}\,\Gamma_{ik}(x-y) = \Big(\frac{1}{\kappa}-1\Big)\, c_n\big(\boldsymbol{N}(\boldsymbol{u}(x);\boldsymbol{\nu}(x))\big)_k \frac{x_k-y_k}{|x-y|^n}$$

$$= -\frac{1}{\omega_n(\lambda+2\mu)}\big(\boldsymbol{N}(\boldsymbol{u}(x);\boldsymbol{\nu}(x))\big)_k \frac{x_k-y_k}{|x-y|^n}, \tag{7.4.57}$$

which in view of (7.4.42) and (7.4.53) implies

$$\operatorname{div}_y\big[\Gamma(x-y)\boldsymbol{N}(\boldsymbol{u}(x);\boldsymbol{\nu}(x))\big] = \frac{1}{\lambda+\mu}\big(\boldsymbol{P}(x-y),\boldsymbol{N}(\boldsymbol{u}(x);\boldsymbol{\nu}(x))\big). \tag{7.4.58}$$

Next, by (7.4.45) we obtain

$$\operatorname{div}_y\big[\boldsymbol{N}\big(\Gamma(x-y);\boldsymbol{\nu}(x)\big)\boldsymbol{u}(x)\big] = \frac{\partial}{\partial y_i}\,\boldsymbol{N}\big(\Gamma(x-y);\boldsymbol{\nu}(x)\big)_{ik}u_k(x) \tag{7.4.59}$$

$$= \frac{1}{\omega_n}\frac{\partial}{\partial y_i}\left[(1-\kappa)\delta_{ik}+\frac{n\kappa(x_i-y_i)(x_k-y_k)}{|x-y|^2}\right]\frac{(x-y,\ \boldsymbol{\nu}(x))}{|x-y|^n}\,u_k(x)$$

$$= -\frac{2\mu}{\omega_n(\lambda+3\mu)}\left(\frac{\nu_k(x)}{|x-y|^n}-\frac{n(x-y,\ \boldsymbol{\nu}(x))(x_k-y_k)}{|x-y|^{n+2}}\right)u_k(x).$$

Using (7.4.42), we find

$$\frac{\partial\boldsymbol{P}(x-y)}{\partial\boldsymbol{\nu}(x)} = \nu_i(x)\boldsymbol{e}_k\frac{\partial P_k(x-y)}{\partial x_i} = -\frac{\lambda+\mu}{\omega_n(\lambda+2\mu)}\,\nu_i(x)\boldsymbol{e}_k\frac{\partial}{\partial x_i}\frac{x_k-y_k}{|x-y|^n}$$

$$= -\frac{\lambda+\mu}{\omega_n(\lambda+2\mu)}\left(\frac{\boldsymbol{\nu}(x)}{|x-y|^n}-\frac{n(x-y,\ \boldsymbol{\nu}(x))(x-y)}{|x-y|^{n+2}}\right),$$

which allows to write (7.4.59) in the form

$$\operatorname{div}_y\big[\boldsymbol{N}\big(\Gamma(x-y);\boldsymbol{\nu}(x)\big)\boldsymbol{u}(x)\big] = \frac{\mu(1+\kappa)}{\lambda+\mu}\,\frac{\partial\boldsymbol{P}(x-y)}{\partial\boldsymbol{\nu}(x)}\,. \tag{7.4.60}$$

Taking divergence of (7.4.43), using the equality $p = -(\lambda+\mu)\operatorname{div}\boldsymbol{u}$ and (7.4.58), (7.4.60), we arrive at (7.4.44). □

Since the system (7.4.9) is equivalent to the Lamé system, it follows that the representation for $\boldsymbol{u}$ in Proposition 7.2 is the representation for the solution of the Lamé system.

It follows from (7.4.43), (7.4.57), and (7.4.58) that $\operatorname{div}\boldsymbol{u}\to 0$ as $\lambda\to\infty$ in Ω, where $\boldsymbol{u}$ is the vector component of the solution of (7.4.9) in the class $[\mathrm{C}^2(\overline{\Omega})]^n$. Therefore, the vector component $\boldsymbol{u}$ of the solution of system (7.4.9) becomes the vector component of the solution of the Stokes system as $\lambda\to\infty$ (see Maz'ya and Mikhlin [**MazMik**]). Thus, letting $\lambda\to\infty$ in Proposition 7.2, we arrive at the following assertion.

COROLLARY 7.7. *Any solution* $\{\boldsymbol{u},p\}$ *of the Stokes system with* $\boldsymbol{u}\in[\mathrm{C}^2(\overline{\Omega})]^n$ *and* $p\in\mathrm{C}^1(\overline{\Omega})$ *can be written as*

$$\boldsymbol{u}(y) = \int_{\partial\Omega}\mathring{\boldsymbol{N}}\big(\mathring{\Gamma}(x-y);\boldsymbol{\nu}(x)\big)\boldsymbol{u}(x)d\sigma_x - \int_{\partial\Omega}\mathring{\Gamma}(x-y)\mathring{\boldsymbol{N}}\big(\boldsymbol{u}(x);\boldsymbol{\nu}(x)\big)d\sigma_x, \tag{7.4.61}$$

$$p(y)=\int_{\partial\Omega}\big(\mathring{\boldsymbol{P}}(x-y),\mathring{\boldsymbol{N}}(\boldsymbol{u}(x);\boldsymbol{\nu}(x))\big)d\sigma_x-2\mu\int_{\partial\Omega}\left(\frac{\partial\mathring{\boldsymbol{P}}(x-y)}{\partial\boldsymbol{\nu}(x)},\boldsymbol{u}(x)\right)d\sigma_x, \tag{7.4.62}$$

where

$$\mathring{\boldsymbol{N}}(\boldsymbol{u};\boldsymbol{\nu})=-\boldsymbol{\nu}p+\mu\nu_j\,\mathrm{grad}\,u_j+\mu\frac{\partial\boldsymbol{u}}{\partial\boldsymbol{\nu}}\,,\qquad \mathring{\boldsymbol{P}}(x)=-\frac{x}{\omega_n|x|^n}\,,$$

and elements of the matrices $\mathring{\boldsymbol{N}}\big(\mathring{\Gamma}(\boldsymbol{\xi});\boldsymbol{\nu}(x)\big)$ *and* $\mathring{\Gamma}(\boldsymbol{\xi})$ *are*

$$\mathring{\boldsymbol{N}}\big(\mathring{\Gamma}(\boldsymbol{\xi});\boldsymbol{\nu}(x)\big)_{ik}=\frac{n}{\omega_n}\frac{\xi_i\xi_k\big(\boldsymbol{\xi},\,\boldsymbol{\nu}(x)\big)}{|\boldsymbol{\xi}|^{n+2}}\,, \tag{7.4.63}$$

$$\mathring{\Gamma}_{ik}(\boldsymbol{\xi})=\begin{cases}-\dfrac{1}{4\pi\mu}\left(\delta_{ik}\log\dfrac{1}{|\boldsymbol{\xi}|}+\dfrac{n\xi_i\xi_k}{|\boldsymbol{\xi}|^2}\right) & \text{for } n=2,\\[2ex] -\dfrac{1}{2\mu\omega_n}\left(\dfrac{\delta_{ik}}{n-2}\dfrac{1}{|\boldsymbol{\xi}|^{n-2}}+\dfrac{n\xi_i\xi_k}{|\boldsymbol{\xi}|^n}\right) & \text{for } n\geq 3,\end{cases} \tag{7.4.64}$$

7.4.3. Vector-valued potentials and boundary integral equations

Using the first term in (7.4.43), we introduce the double layer potential

$$\boldsymbol{\mathcal{W}}_\kappa(y)=\frac{1}{\omega_n}\int_{\partial\Omega}\big[(1-\kappa)\boldsymbol{\varphi}(x)+n\kappa\big(\boldsymbol{e}_{yx},\varphi(x)\big)\boldsymbol{e}_{yx}\big]\frac{\big(\boldsymbol{e}_{yx},\boldsymbol{\nu}(x)\big)}{|x-y|^{n-1}}\,d\sigma_x, \tag{7.4.65}$$

where $\boldsymbol{e}_{yx}=(x-y)/|x-y|$, $\kappa=(\lambda+\mu)/(\lambda+3\mu)$ and $y\notin\partial\Omega$. Since we assume that $\mu>0$ and $\lambda+\mu>0$, it follows that $0<\kappa<1$.

Similarly, the first term in (7.4.61) generates the hydrodynamic double layer potential

$$\mathring{\boldsymbol{\mathcal{W}}}(y)=\frac{n}{\omega_n}\int_{\partial\Omega}\frac{\big(\boldsymbol{e}_{yx},\varphi(x)\big)\big(\boldsymbol{e}_{yx},\boldsymbol{\nu}(x)\big)\boldsymbol{e}_{yx}}{|x-y|^{n-1}}\,d\sigma_x, \tag{7.4.66}$$

with $y\notin\partial\Omega$.

If the density $\boldsymbol{\varphi}$ of the vector-valued double layer potentials (7.4.65), (7.4.66) belongs to the space $[\mathrm{C}(\partial\Omega)]^n$ and the boundary of Ω is in the class $\mathrm{C}^{1,\alpha}$ with $0<\alpha\leq 1$, then the boundary value problems (7.4.1), (7.4.2) for the Lamé system and the problems (7.4.5), (7.4.6) for the Stokes system are reduced to systems of regular integral equations for the above mentioned potentials. The regularity of integral equations is understood in the sense that the kernel of the matrix-valued integral operators of the systems for the density of the elastic and hydrodynamic double layer potentials has the weak singularity $O(|x-y|^{1+\alpha-n})$. Therefore, the integral operator is compact and the Fredholm theory applies to the systems of boundary integral equations of the Dirichlet problem for elasticity and hydrodynamics together with the corresponding adjoint systems. The three-dimensional boundary value problems are treated, for example in the book by Kupradze, Gegelia, Basheleishvili, Burchuladze [**KGBB**] and in the book by Ladyzhenskaya [**Lad**].

Since $\mathring{\boldsymbol{\mathcal{W}}}(y)=\boldsymbol{\mathcal{W}}_1(y)$, we can consider the class of potentials $\boldsymbol{\mathcal{W}}_\kappa(y)$ with $0<\kappa\leq 1$, which includes both elastic and hydrodynamic double layer potentials. Note that $\boldsymbol{\mathcal{W}}_1(y)$ coincide with the vector-valued harmonic double layer potential, generated by the representation of the solution to $\Delta\boldsymbol{u}=\boldsymbol{0}$ in Ω.

In what follows we consider the potentials $\mathcal{W}_\kappa(y)$ with $0 \le \kappa \le 1$ in domains with nonsmooth boundary. Namely, assuming that $\mathcal{D}$ has a nonregular boundary, we introduce the vector-valued double layer potential

$$(7.4.67)\quad \boldsymbol{W}_\kappa^{(n)}(q) = \frac{1}{\omega_n}\int_{\partial\Omega}\left[(1-\kappa)\boldsymbol{\varphi}(x) + n\kappa\big(\boldsymbol{e}_{qx}, \varphi(x)\big)\boldsymbol{e}_{qx}\right]\frac{\big(\boldsymbol{e}_{qx}, \boldsymbol{\nu}(x)\big)}{|x-q|^{n-1}} H_{n-1}(dx),$$

where $\boldsymbol{\nu}(x)$ is the normal to $\partial\mathcal{D}$ in the sense of Federer, κ is a parameter with values in $[0,1]$, $\boldsymbol{\varphi} \in [\mathrm{C}(\partial\mathcal{D})]^n$, and $q \notin \partial\mathcal{D}$.

Consider the $(n \times n)$-matrix-valued function $\mathcal{H}_\kappa^{(n)}(\boldsymbol{e})$, defined on the sphere $\mathbb{S}^{n-1}$, with the entries

$$(7.4.68)\qquad \frac{1}{\omega_n}\Big[(1-\kappa)\delta_{ij} + n\kappa(\boldsymbol{e}, \boldsymbol{e}_i)(\boldsymbol{e}, \boldsymbol{e}_j)\Big],$$

$\boldsymbol{e}_j$ being the unit vector directed along the j-th coordinate axis. According to (7.2.15) and (7.3.1), the solid angle is

$$(7.4.69)\qquad \omega_{\mathcal{D}}(q, \mathcal{B}) = \int_{\mathcal{B}\cap\partial\mathcal{D}} \frac{\big(\boldsymbol{r}_{qx}, \boldsymbol{\nu}(x)\big)}{r_{qx}^n} H_{n-1}(dx).$$

Then the potential (7.4.67) can be written as

$$(7.4.70)\qquad \boldsymbol{W}_\kappa^{(n)}(q) = \int_{\partial\mathcal{D}} \mathcal{H}_\kappa^{(n)}(\boldsymbol{e}_{qx})\boldsymbol{\varphi}(x)\omega_{\mathcal{D}}(q, dx).$$

Following (7.3.2), (7.3.4), and (7.3.5), we define the set function $\Psi_{\mathcal{D}}(p, \mathbb{B})$ for $\mathcal{H}_\kappa^{(n)}(\boldsymbol{e})$. Applying Theorem 7.2 to potential (7.4.70), we arrive at the equation

$$\boldsymbol{\varphi}(p) + \big(T_\kappa^{(n)}\boldsymbol{\varphi}\big)(p) = 2\boldsymbol{f}(p)$$

for the interior boundary value problems (7.4.1), (7.4.5) and at the equation

$$-\boldsymbol{\varphi}(p) + \big(T_\kappa^{(n)}\boldsymbol{\varphi}\big)(p) = 2\boldsymbol{f}(p)$$

for the exterior boundary value problems (7.4.2), (7.4.6). The integral operator

$$\big(T_\kappa^{(n)}\boldsymbol{\varphi}\big)(p) = 2\int_{\partial\mathcal{D}} \Psi_{\mathcal{D}}(p, dx)\boldsymbol{\varphi}(x)$$

acts in $[\mathrm{C}(\partial\mathcal{D})]^n$ and is continuous. It can be written in the form

$$(7.4.71)\qquad \big(T_\kappa^{(n)}\boldsymbol{\varphi}\big)(p) = 2\Psi_{\mathcal{D}}(p,p)\boldsymbol{\varphi}(p) + 2\int_{\partial\mathcal{D}\backslash p} \mathcal{H}_\kappa^{(n)}(\boldsymbol{e}_{px})\boldsymbol{\varphi}(x)\omega_{\mathcal{D}}(p, dx),$$

where, by (7.3.4)

$$(7.4.72)\qquad \Psi_{\mathcal{D}}(p,p) = \frac{1}{2}I - \int_{\partial\mathcal{D}\backslash p} \mathcal{H}_\kappa^{(n)}(\boldsymbol{e}_{px})\omega_{\mathcal{D}}(p, dx).$$

Hence, using (7.4.68) we obtain the following representations

$$(7.4.73)\qquad \begin{aligned} \Psi_{\mathcal{D}}^{(ij)}(p,p) &= \frac{1}{2\omega_n}\Big\{\Big[\omega_n + 2(\kappa-1)\omega_{\mathcal{D}}(p, \partial\mathcal{D}\backslash p)\Big]\delta_{ij} \\ &- 2n\kappa\int_{\partial\mathcal{D}\backslash p} (\boldsymbol{e}_{px}, \boldsymbol{e}_i)(\boldsymbol{e}_{px}, \boldsymbol{e}_j)\omega_{\mathcal{D}}(p, dx)\Big\}, \end{aligned}$$

$1 \le i,\ j \le n$, and

$$(7.4.74)\quad (T_\kappa^{(n)}\boldsymbol{\varphi})(p) = \frac{1}{\omega_n}\Big\{\big[\omega_n + 2(\kappa-1)\omega_{\mathcal{D}}(p,\partial\mathcal{D}\backslash p)\big]\boldsymbol{\varphi}(p) - 2n\kappa\int_{\partial\mathcal{D}\backslash p}\boldsymbol{e}_{px}(\boldsymbol{e}_{px},\boldsymbol{\varphi}(p))\omega_{\mathcal{D}}(p,dx)\Big\} + \frac{2}{\omega_n}\int_{\partial\mathcal{D}\backslash p}\Big[(1-\kappa)\boldsymbol{\varphi}(x) + n\kappa\frac{(\boldsymbol{r}_{px},\boldsymbol{\varphi}(x))\boldsymbol{r}_{px}}{r_{px}^2}\Big]\omega_{\mathcal{D}}(p,dx).$$

REMARK 7.6. By Lemma 7.11, $\Psi_{\mathcal{D}}(p,p) = 0$ for $p \in \partial^*\mathcal{D}$. Therefore, the sum in curly brackets in (7.4.74) vanishes for p in $\partial^*\mathcal{D}$ which is everywhere dense on $\partial\mathcal{D}$. We also note that (7.4.74) can be written in the form

$$(T_\kappa^{(n)}\boldsymbol{\varphi})(p) = \boldsymbol{\varphi}(p) + \frac{2}{\omega_n}\int_{\partial\mathcal{D}}\Big[(1-\kappa)\big(\boldsymbol{\varphi}(x)-\boldsymbol{\varphi}(p)\big) + n\kappa\frac{(\boldsymbol{r}_{px},\boldsymbol{\varphi}(x)-\boldsymbol{\varphi}(p))\boldsymbol{r}_{px}}{r_{px}^2}\Big]\omega_{\mathcal{D}}(p,dx).$$

Consider now representations for $\Psi_{\mathcal{D}}(p,p)$ and for the operator $T_\kappa^{(n)}$ for domains of a more narrow class. Suppose at each $p \in \partial\mathcal{D}$ there exists a tangent cone to $\mathcal{D}$ denoted by $\mathcal{V}(p)$ such that $\mathcal{D}\cap\mathcal{V}(p) \ne \emptyset$. Let $\mathcal{E}(p) = \partial\mathbb{B}(p)\cap\mathcal{V}(p)$ and let $\boldsymbol{e}_{p\sigma}$ denote the vector connecting a point $p \in \partial\mathcal{D}$ with $\sigma \in \partial\mathbb{B}(p)$. Taking into account (7.3.2), (7.3.23), (7.4.69) one can write (7.4.73) and (7.4.74) as

$$(7.4.75)\quad \Psi_{\mathcal{D}}^{(ij)}(p,p) = \frac{1}{2\omega_n}\Big\{\Big[\omega_n + 2(\kappa-1)\mathrm{mes}_{n-1}\mathcal{E}(p)\Big]\delta_{ij} - 2n\kappa\int_{\mathcal{E}(p)}(\boldsymbol{e}_{p\sigma},\boldsymbol{e}_i)(\boldsymbol{e}_{p\sigma},\boldsymbol{e}_j)d\sigma\Big\}, 1 \le i,\ j \le n,$$

and

$$(7.4.76)\quad (T_\kappa^{(n)}\boldsymbol{\varphi})(p) = \frac{1}{\omega_n}\Big\{\Big[\omega_n + 2(\kappa-1)\mathrm{mes}_{n-1}\mathcal{E}(p)\Big]\boldsymbol{\varphi}(p) - 2n\kappa\int_{\mathcal{E}(p)}\boldsymbol{e}_{p\sigma}(\boldsymbol{e}_{p\sigma},\boldsymbol{\varphi}(p))d\sigma\Big\} + \frac{2}{\omega_n}\int_{\partial\mathcal{D}\backslash p}\Big[(1-\kappa)\boldsymbol{\varphi}(x) + n\kappa\frac{(\boldsymbol{r}_{px},\boldsymbol{\varphi}(x))\boldsymbol{r}_{px}}{r_{px}^2}\Big]\frac{(\boldsymbol{r}_{px},\boldsymbol{\nu}(x))}{r_{px}^n}H_{n-1}(dx).$$

EXAMPLE 7.1. Let $\mathcal{D}$ be a planar domain with the only angular point p and let β denote the opening of the angle measured within the domain. We introduce a Cartesian coordinate system with origin at p so that the axis px_1 coincides with the bisectrix of the angle and denote by ϑ the angle between the vector $\boldsymbol{e}_{p\sigma}$ and the axis px_1.

Using (7.4.75), we obtain

$$\Psi_{\mathcal{D}}^{(12)}(p,p) = \Psi_{\mathcal{D}}^{(21)}(p,p) = -\frac{\kappa}{\pi}\int_{-\beta/2}^{\beta/2}\sin\vartheta\cos\vartheta\,d\vartheta = 0,$$

$$\Psi_{\mathcal{D}}^{(11)}(p,p) = \frac{1}{2\pi}\big(\pi + (\kappa-1)\beta\big) - \frac{\kappa}{\pi}\int_{-\beta/2}^{\beta/2}\cos^2\vartheta\,d\vartheta = \frac{1}{2\pi}(\pi - \beta - \kappa\sin\,\beta),$$

$$\Psi_{\mathcal{D}}^{(22)}(p,p) = \frac{1}{2\pi}\big(\pi + (\kappa-1)\beta\big) - \frac{\kappa}{\pi}\int_{-\beta/2}^{\beta/2}\sin^3\vartheta\,d\vartheta = \frac{1}{2\pi}(\pi - \beta + \kappa\sin\,\beta).$$

Thus,

$$\Psi_{\mathcal{D}}(p,p) = \frac{1}{2\pi}\text{diag}\,\{\pi-\beta-\kappa\sin\,\beta,\ \pi-\beta+\kappa\sin\,\beta\}. \tag{7.4.77}$$

EXAMPLE 7.2. Let the tangent cone $\mathcal{V}(p)$ with vertex p at the boundary of the domain $\mathcal{D}\subset\mathbb{R}^3$ be a circular cone with opening β. We choose a coordinate system with origin at p in such a way that the axis px_3 passes through the center of the spherical segment $\mathcal{E}(p)$. We introduce the spherical coordinates ϑ and χ, where ϑ is the angle between the planes σpx_3 and x_2px_3 and χ is the angle between the vectors $\boldsymbol{e}_{p\sigma}$ and $\boldsymbol{e}_3$. Using (7.4.75), we obtain

$$\begin{aligned}
\Psi_{\mathcal{D}}^{(ij)}(p,p) &= -\frac{3}{4\pi}\kappa\int_{\mathcal{E}(p)}(\boldsymbol{e}_{p\sigma},\boldsymbol{e}_i)(\boldsymbol{e}_{p\sigma},\boldsymbol{e}_j)d\sigma = 0 \text{ for } i\neq j,\\
\Psi_{\mathcal{D}}^{(11)}(p,p) &= \frac{1}{8\pi}\left\{\left[4\pi+4\pi(\kappa-1)\left(1-\cos\frac{\beta}{2}\right)\right]-6\kappa\int_0^{\beta/2}\int_0^{2\pi}\sin^3\chi\sin^2\vartheta d\chi d\vartheta\right\}\\
&= \frac{1}{4}\cos\frac{\beta}{2}\left(2+\kappa\sin^2\frac{\beta}{2}\right),\\
\Psi_{\mathcal{D}}^{(22)}(p,p) &= \frac{1}{8\pi}\left\{\left[4\pi+4\pi(\kappa-1)\left(1-\cos\frac{\beta}{2}\right)\right]-6\kappa\int_0^{\beta/2}\int_0^{2\pi}\sin^3\chi\cos^2\vartheta d\chi d\vartheta\right\}\\
&= \frac{1}{4}\cos\frac{\beta}{2}\left(2+\kappa\sin^2\frac{\beta}{2}\right),\\
\Psi_{\mathcal{D}}^{(33)}(p,p) &= \frac{1}{8\pi}\left\{\left[4\pi+4\pi(\kappa-1)\left(1-\cos\frac{\beta}{2}\right)\right]-6\kappa\int_0^{\beta/2}\int_0^{2\pi}\cos^2\chi\sin\chi d\chi d\vartheta\right\}\\
&= \frac{1}{2}\cos\frac{\beta}{2}\left(1-\kappa\sin^2\frac{\beta}{2}\right).
\end{aligned}$$

Thus,

$$\begin{aligned}
&\Psi_{\mathcal{D}}(p,p)\\
&= \frac{1}{4}\cos\frac{\beta}{2}\ \text{diag}\left\{\left(2+\kappa\sin^2\frac{\beta}{2}\right),\left(2+\kappa\sin^2\frac{\beta}{2}\right),2\left(1-\kappa\sin^2\frac{\beta}{2}\right)\right\}.
\end{aligned} \tag{7.4.78}$$

EXAMPLE 7.3. Consider now the domain $\mathcal{D}\subset\mathbb{R}^3$ with an edge at the boundary. Let the tangent cone $\mathcal{V}(p)$ with vertex at $p\in\partial\mathcal{D}$ be a dihedral angle with the opening β. Suppose the Cartesian coordinate system with origin at p is chosen in such a way that the axis px_2 passes through the center of the spherical biangle $\mathcal{E}(p)$ and that the vertices of this biangle lay in the plane x_2px_3. Further, let ϑ be the angle between the planes σpx_3 and x_2px_3, $-\pi<\vartheta\leq\pi$, and let χ denote the angle between the vectors $\boldsymbol{e}_{p\sigma}$ and $\boldsymbol{e}_3$.

Using (7.4.75) we find

$$\Psi_{\mathcal{D}}^{(ij)}(p,p) = -\frac{3}{4\pi}\kappa\int_{\mathcal{E}(p)}(\boldsymbol{e}_{p\sigma},\boldsymbol{e}_i)(\boldsymbol{e}_{p\sigma},\boldsymbol{e}_j)d\sigma = 0 \quad \text{for} \quad i\neq j,$$

$$\Psi_{\mathcal{D}}^{(11)}(p,p) = \frac{1}{8\pi}\left\{\Big[4\pi+4\beta(\kappa-1)\Big] - 6\kappa\int_{-\beta/2}^{\beta/2}\int_0^\pi \sin^3\chi\sin^2\vartheta d\chi d\vartheta\right\}$$
$$= \frac{1}{2\pi}(\pi-\beta+\kappa\sin\beta),$$

$$\Psi_{\mathcal{D}}^{(22)}(p,p) = \frac{1}{8\pi}\left\{\Big[4\pi+4\beta(\kappa-1)\Big] - 6\kappa\int_{-\beta/2}^{\beta/2}\int_0^\pi \sin^2\chi\cos^2\vartheta d\chi d\vartheta\right\}$$
$$= \frac{1}{2\pi}(\pi-\beta-\kappa\sin\beta),$$

$$\Psi_{\mathcal{D}}^{(33)}(p,p) = \frac{1}{8\pi}\left\{\Big[4\pi+4\beta(\kappa-1)\Big] - 6\kappa\int_{-\beta/2}^{\beta/2}\int_0^\pi \cos^2\chi\sin\chi d\chi d\vartheta\right\}$$
$$= \frac{1}{2\pi}(\pi-\beta).$$

Thus,

$$\Psi_{\mathcal{D}}(p,p) = \frac{1}{2\pi}\,\mathrm{diag}\,\{\pi-\beta+\kappa\sin\beta,\ \pi-\beta-\kappa\sin\beta,\ \pi-\beta\}. \tag{7.4.79}$$

7.4.4. The norm of the integral operators of elasticity theory and hydrodynamics for a convex domain

In the case of a convex domain $\Psi_{\mathcal{D}}(p,p)$ is calculated by (7.4.75) and the operator $T_\kappa^{(n)}$ can be written in the form (7.4.76).

To calculate the norm of the operator $T_\kappa^{(n)}$ in the space $[\mathrm{C}(\partial\mathcal{D})]^n$ we apply Corollary 7.6. First we note that the matrix $\mathcal{H}_\kappa^{(n)}$ is symmetric and

$$\big|\mathcal{H}_\kappa^{(n)}(\boldsymbol{e})\boldsymbol{z}\big| = \frac{1}{\omega_n}\left\{\sum_{i=1}^n\Big[(1-\kappa)z_i + n\kappa\sum_{j=1}^n(\boldsymbol{e},\boldsymbol{e}_i)(\boldsymbol{e},\boldsymbol{e}_j)z_j\Big]^2\right\}^{1/2}$$
$$= \frac{1}{\omega_n}\Big[(1-\kappa)^2|\boldsymbol{z}|^2 + n\kappa(n\kappa-2\kappa+2)(\boldsymbol{e},\boldsymbol{z})^2\Big]^{1/2}. \tag{7.4.80}$$

THEOREM 7.6. *Let $\mathcal{D}$ be a convex domain. Then*

$$\|T_\kappa^{(n)}\| = \frac{2\Gamma(\frac{n}{2})}{\sqrt{\pi}\Gamma(\frac{n-1}{2})}\int_0^{\pi/2}\big[(1-\kappa)^2+n\kappa(n\kappa-2\kappa+2)\cos^2\vartheta\big]^{1/2}\sin^{n-2}\vartheta d\vartheta.$$

In particular,

$$\|T_\kappa^{(2)}\| = \frac{2}{\pi}(1+\kappa)\mathbf{E}\left(\frac{2\sqrt{\kappa}}{1+\kappa}\right) = 1+\frac{1}{2^2}\kappa^2+\frac{1}{2^24^2}\kappa^4+\dots+\left[\frac{(2m-3)!!}{2^m m!}\right]^2\kappa^{2m}+\dots,$$

where $\mathbf{E}(k)$ is the complete elliptic integral of the second kind, and

$$\|T_\kappa^{(3)}\| = \frac{1}{2}\left[1+2\kappa+\frac{(1-\kappa)^2}{\sqrt{3\kappa(\kappa+2)}}\log\frac{1+2\kappa+\sqrt{3\kappa(\kappa+2)}}{1-\kappa}\right].$$

For the hydrodynamic double layer potential we have

$$\|T_1^{(n)}\| = \frac{2}{\sqrt{\pi}} \frac{\Gamma(\frac{n}{2}+1)}{\Gamma(\frac{n+1}{2})},$$

so that

$$\|T_1^{(2)}\| = 4/\pi, \ \|T_1^{(3)}\| = 3/2.$$

PROOF. By Corollary 7.6 and by (7.4.80)

$$\|T_\kappa^{(n)}\| = \frac{1}{\omega_n} \sup_{|\boldsymbol{z}|=1} \int_{\mathbb{S}^{n-1}} \Big[(1-\kappa)^2 + n\kappa(n\kappa - 2\kappa + 2)(\boldsymbol{e}_\sigma, \boldsymbol{z})^2\Big]^{1/2} d\sigma,$$

where $\boldsymbol{e}_\sigma$ is the vector connecting the origin with $\sigma \in \mathbb{S}^{n-1}$.

Hence

$$\|T_\kappa^{(n)}\| = \frac{1}{\omega_n} \sup_{|\boldsymbol{z}|=1} \int_{\mathbb{S}^{n-1}} \Big[(1-\kappa)^2 + n\kappa(n\kappa - 2\kappa + 2)\cos^2\vartheta(\sigma)\Big]^{1/2} d\sigma,$$

$\vartheta(\sigma)$ being the angle between the vectors $\boldsymbol{e}_\sigma$ and $\boldsymbol{z}$. The integral in the last equality does not depend on z. In the spherical coordinates we obtain

$$\begin{aligned}
\|T_\kappa^{(n)}\| &= \frac{1}{\omega_n} \int_0^\pi \cdots \int_0^\pi \int_0^{2\pi} \Big[(1-\kappa)^2 + n\kappa(n\kappa - 2\kappa + 2)\cos^2\vartheta\Big] \\
&\qquad \times \sin^{n-2}\vartheta \sin^{n-3}\vartheta_1 \dots \sin\vartheta_{n-3} d\vartheta\, d\vartheta_1 \dots d\vartheta_{n-2} \\
&= \frac{2}{\sqrt{\pi}} \frac{\Gamma(\frac{n}{2})}{\Gamma(\frac{n-1}{2})} \int_0^{\pi/2} \Big[(1-\kappa)^2 + n\kappa(n\kappa - 2\kappa + 2)\cos^2\vartheta\Big]^{1/2} \sin^{n-2}\vartheta d\vartheta.
\end{aligned}$$

In particular, for $n = 2$ and $n = 3$,

$$\begin{aligned}
\|T_\kappa^{(2)}\| &= \frac{2}{\pi}(1+\kappa) \int_0^{\pi/2} \Big[1 - \frac{4\kappa}{(1+\kappa)^2}\sin^2\vartheta\Big]^{1/2} d\vartheta, \\
\|T_\kappa^{(3)}\| &= \int_0^{\pi/2} \Big[(1-\kappa)^2 + 3\kappa(\kappa+2)\cos^2\vartheta\Big]^{1/2} \sin\vartheta d\vartheta.
\end{aligned}$$

The proof is complete. □

The next assertion generalizes Theorem 7.6 showing that for a convex domain $\mathcal{D}$ the value $\|T_\kappa^{(n)}\|$ yields the minimum of the norms of the operator $T_\kappa^{(n)}$ in some class of spaces of continuous vector-valued functions on $\partial\mathcal{D}$.

Let $\mathbb{E}^n$ be an n-dimensional Euclidean space endowed additionally with a norm $\|\cdot\|$ and let

$$\|\boldsymbol{z}\|_* = \sup\{(\boldsymbol{z}, \boldsymbol{\zeta}) : \|\boldsymbol{\zeta}\| \le 1\}$$

be the dual norm to $\|\cdot\|$. By $[\mathrm{C}(\partial\mathcal{D}, \|\cdot\|)]^n$ we denote the space of continuous n-component vector-valued functions on $\partial\mathcal{D}$ with the norm

$$\|\boldsymbol{\varphi}\| = \sup\big\{\|\boldsymbol{\varphi}(x)\| : \ x \in \partial\mathcal{D}\big\}.$$

PROPOSITION 7.3. *Let $\mathcal{D}$ be a convex domain. Then*

$$\|T_\kappa^{(n)}\|_{[\mathrm{C}(\partial\mathcal{D},\|\cdot\|)]^n} = \frac{1}{\omega_n} \sup_{\|\boldsymbol{z}\|_*=1} \int_{\mathbb{S}^{n-1}} \|(1-\kappa)\boldsymbol{z} + n\kappa \boldsymbol{e}_\sigma(\boldsymbol{e}_\sigma, \boldsymbol{z})\|_* d\sigma, \tag{7.4.81}$$

Moreover,

$$\|T_\kappa^{(n)}\| = \min_{\|\cdot\|} \|T_\kappa^{(n)}\|_{[\mathrm{C}(\partial\mathcal{D},\|\cdot\|)]^n},$$

where the minimum is taken over the set of all norms in $\mathbb{E}^n$.

PROOF. Taking into account (7.4.69), (7.4.71) and using the same arguments as in the proof of Theorem 7.3, we obtain

$$\|T_\kappa^{(n)}\|_{[C(\partial\mathcal{D},\|\cdot\|)]^n} = \sup_{p\in\partial^*\mathcal{D}} \sup_{\|\varphi\|=1} \|(T_\kappa^{(n)}\varphi)(p)\| \tag{7.4.82}$$

$$= 2 \sup_{p\in\partial^*\mathcal{D}} \sup_{\|\varphi\|=1} \left\| \int_{\partial\mathcal{D}\backslash p} \mathcal{H}_\kappa^{(n)}(\boldsymbol{e}_{px})\boldsymbol{\varphi}(x)\frac{(\boldsymbol{r}_{px},\boldsymbol{\nu}(x))}{r_{px}^n} H_{n-1}(dx)\right\|.$$

By virtue of the equalities

$$\|\boldsymbol{\zeta}\| = \sup\big\{(\boldsymbol{z},\boldsymbol{\zeta}):\ \|\boldsymbol{z}\|_* = 1\big\}, \qquad \|\boldsymbol{z}\|_* = \sup\big\{(\boldsymbol{z},\boldsymbol{\zeta}):\ \|\boldsymbol{\zeta}\| = 1\big\}$$

and the symmetry of the matrix $\mathcal{H}_\kappa^{(n)}$ one has

$$\sup_{\|\varphi\|=1} \left\| \int_{\partial\mathcal{D}\backslash p} \mathcal{H}_\kappa^{(n)}(\boldsymbol{e}_{px})\boldsymbol{\varphi}(x)\frac{(\boldsymbol{r}_{px},\boldsymbol{\nu}(x))}{r_{px}^n} H_{n-1}(dx)\right\|$$

$$= \sup_{\|\varphi\|=1} \sup_{\|\boldsymbol{z}\|_*=1} \left(\boldsymbol{z}, \int_{\partial\mathcal{D}\backslash p} \mathcal{H}_\kappa^{(n)}(\boldsymbol{e}_{px})\boldsymbol{\varphi}(x)\frac{(\boldsymbol{r}_{px},\boldsymbol{\nu}(x))}{r_{px}^n} H_{n-1}(dx)\right)$$

$$= \sup_{\|\varphi\|=1} \sup_{\|\boldsymbol{z}\|_*=1} \int_{\partial\mathcal{D}\backslash p} \left(\boldsymbol{z}, \mathcal{H}_\kappa^{(n)}(\boldsymbol{e}_{px})\boldsymbol{\varphi}(x)\right)\frac{(\boldsymbol{r}_{px},\boldsymbol{\nu}(x))}{r_{px}^n} H_{n-1}(dx)$$

$$= \sup_{\|\boldsymbol{z}\|_*=1} \sup_{\|\varphi\|=1} \int_{\partial\mathcal{D}\backslash p} \left(\mathcal{H}_\kappa^{(n)}(\boldsymbol{e}_{px})\boldsymbol{z}, \boldsymbol{\varphi}(x)\right)\frac{(\boldsymbol{r}_{px},\boldsymbol{\nu}(x))}{r_{px}^n} H_{n-1}(dx)$$

$$= \sup_{\|\boldsymbol{z}\|_*=1} \int_{\partial\mathcal{D}\backslash p} \|\mathcal{H}_\kappa^{n}(\boldsymbol{e}_{px})\boldsymbol{z}\|_*\frac{(\boldsymbol{r}_{px},\boldsymbol{\nu}(x))}{r_{px}^n} H_{n-1}(dx).$$

Therefore, by (7.4.82)

$$\|T_\kappa^{(n)}\|_{[C(\partial\mathcal{D},\|\cdot\|)]^n} = 2 \sup_{p\in\partial^*\mathcal{D}} \sup_{\|\boldsymbol{z}\|_*=1} \int_{\partial\mathcal{D}\backslash p} \|\mathcal{H}_\kappa^{(n)}(\boldsymbol{e}_{px})\boldsymbol{z}\|_*\frac{(\boldsymbol{r}_{px},\boldsymbol{\nu}(x))}{r_{px}^n} H_{n-1}(dx).$$

Proceeding in the same way as in the proof of Corollary 7.6 from this we obtain

$$\|T_\kappa^{(n)}\|_{[C(\partial\mathcal{D},\|\cdot\|)]^n} = \sup_{\|\boldsymbol{z}\|_*=1} \int_{\mathbb{S}^{n-1}} \|\mathcal{H}_\kappa^{(n)}(\sigma)\boldsymbol{z}\|_* d\sigma. \tag{7.4.83}$$

Substituting the elements of the matrix-valued function $\mathcal{H}_\kappa^{(n)}$ specified by (7.4.68) into (7.4.83) we arrive at (7.4.81).

Let $\boldsymbol{z}_0$ be an n-dimensional vector such that

$$|\boldsymbol{z}_0| = \max\{|\boldsymbol{z}|:\ \|\boldsymbol{z}\|_* = 1\}.$$

Clearly, two parallel vectors $(1-\kappa)\boldsymbol{z}_0 + n\kappa\boldsymbol{e}_\sigma(\boldsymbol{e}_\sigma,\boldsymbol{z}_0)$ and $\boldsymbol{Z}(\sigma)$ with $\|\boldsymbol{Z}(\sigma)\|_* = 1$ satisfy

$$\|(1-\kappa)\boldsymbol{z}_0 + n\kappa\boldsymbol{e}_\sigma(\boldsymbol{e}_\sigma,\boldsymbol{z}_0)\|_* = \frac{|(1-\kappa)\boldsymbol{z}_0 + n\kappa\boldsymbol{e}_\sigma(\boldsymbol{e}_\sigma,\boldsymbol{z}_0)|}{|\boldsymbol{Z}(\sigma)|}.$$

So

$$\|(1-\kappa)\boldsymbol{z}_0 + n\kappa\boldsymbol{e}_\sigma(\boldsymbol{e}_\sigma,\boldsymbol{z}_0)\|_* \geq \frac{|(1-\kappa)\boldsymbol{z}_0 + n\kappa\boldsymbol{e}_\sigma(\boldsymbol{e}_\sigma,\boldsymbol{z}_0)|}{|\boldsymbol{z}_0|}.$$

Then, by (7.4.81)

$$\|T_\kappa^{(n)}\|_{[\mathrm{C}(\partial\mathcal{D},\|\cdot\|)]^n} \geq \frac{1}{\omega_n}\int_{\mathbb{S}^{n-1}} \|(1-\kappa)\boldsymbol{z}_0 + n\kappa\boldsymbol{e}_\sigma(\boldsymbol{e}_\sigma,\boldsymbol{z}_0)\|_* d\sigma$$
$$\geq \frac{1}{\omega_n}\int_{\mathbb{S}^{n-1}} \left|(1-\kappa)\frac{\boldsymbol{z}_0}{|\boldsymbol{z}_0|} + n\kappa\boldsymbol{e}_\sigma\left(\boldsymbol{e}_\sigma,\frac{\boldsymbol{z}_0}{|\boldsymbol{z}_0|}\right)\right| d\sigma$$
$$= \frac{1}{\omega_n}\int_{\mathbb{S}^{n-1}} \left[(1-\kappa)^2 + n\kappa(n\kappa-2\kappa+2)\left(\boldsymbol{e}_\sigma,\frac{\boldsymbol{z}_0}{|\boldsymbol{z}_0|}\right)^2\right]^{1/2} d\sigma$$
$$= \frac{1}{\omega_n}\int_{\mathbb{S}^{n-1}} \left[(1-\kappa)^2 + n\kappa(n\kappa-2\kappa+2)\cos^2\vartheta_0(\sigma)\right]^{1/2} d\sigma, \tag{7.4.84}$$

where $\vartheta_0(\sigma)$ is the angle between $\boldsymbol{e}_\sigma$ and $\boldsymbol{z}_0$.

In the proof of Theorem 7.6 it was shown that

$$\|T_\kappa^{(n)}\| = \frac{1}{\omega_n}\int_{\mathbb{S}^{n-1}} \left[(1-\kappa)^2 + n\kappa(n\kappa-2\kappa+2)\cos^2\vartheta(\sigma)\right]^{1/2} d\sigma,$$

where $\vartheta(\sigma)$ is the angle between the vectors $\boldsymbol{e}_\sigma$ and an arbitrary fixed vector $\boldsymbol{z}$. Comparing (7.4.84) with the last equality, we obtain

$$\|T_\kappa^{(n)}\|_{[\mathrm{C}(\partial\mathcal{D},\|\cdot\|)]^n} \geq \|T_\kappa^{(n)}\|,$$

which completes the proof. □

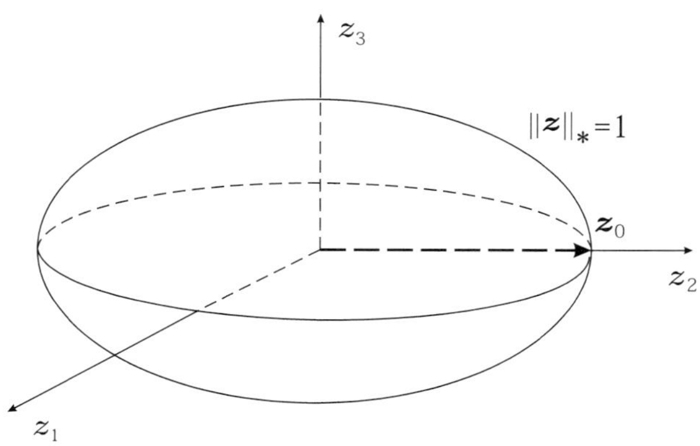

FIGURE 7.1. Choice of the vector $\boldsymbol{z}_0$ in the proof of Proposition 7.3

7.4.5. The essential norm of the operator $T_\kappa^{(2)}$ for a domain with an angle point

Let $\mathcal{D}$ be a bounded domain in $\mathbb{R}^2$ with a smooth boundary except for a single angular point. Let β be the opening of the angle calculated in $\mathcal{D}$, $0 \leq \beta \leq 2\pi$, and let $\alpha = \min\{\beta,\ 2\pi - \beta\}$.

THEOREM 7.7. *The equality*

$$\mathrm{ess}\|T_\kappa^{(2)}\| = \frac{2}{\pi}(1+\kappa)E\Big(\frac{\pi-\alpha}{2}, \frac{2\sqrt{\kappa}}{1+\kappa}\Big) \tag{7.4.85}$$

is valid, where $E(\varphi, k)$ is the elliptic integral of the second kind.

In particular,

$$\text{ess}\|T_1^{(2)}\| = (4/\pi)\cos(\alpha/2).$$

PROOF. Let V_ε denote the angle (calculated in $\mathcal{D}$) between two rays with origin at the angular point q which pass through the intersection points of $\partial\mathcal{D}$ with $\partial\mathbb{B}_\varepsilon(q)$. By V_0 we denote the limit of V_ε as $\varepsilon \to 0$.

By a standard argument we derive from Theorem 7.4 that

$$\text{ess}\|T_\kappa^{(2)}\| = 2\lim_{\delta\to 0}\sup_{p\in\partial V_\varepsilon\setminus q}\sup_{|\boldsymbol{z}|=1}\int_{\partial V_\varepsilon\cap\mathbb{B}_\delta(p)}\left|\mathcal{H}_\kappa^{(2)}(\boldsymbol{e}_{px})\boldsymbol{z}\right|\frac{|(\boldsymbol{r}_{px},\boldsymbol{\nu}(x))|}{r_{px}^2}d_xS + O(\varepsilon).$$

Hence, applying Proposition 7.1 and then passing to the limit as $\varepsilon \to 0$, we obtain

$$\text{ess}\|T_\kappa^{(2)}\| = \sup_{|\boldsymbol{z}|=1}\left[\int_{\mathbb{S}^1}\left|\mathcal{H}_\kappa^{(2)}(\sigma)\boldsymbol{z}\right|d\sigma - 2\int_{\mathcal{C}}\left|\mathcal{H}_\kappa^{(2)}(\sigma)\boldsymbol{z}\right|d\sigma\right], \tag{7.4.86}$$

where $\mathcal{C}$ is that of the sets $\partial\mathbb{B}(q)\cap V_0$, $\partial\mathbb{B}(q)\setminus V_0$ which has the least length. Replacing the integrand in (7.4.86) by (7.4.80), we arrive at

$$\text{ess}\|T_\kappa^{(2)}\| = \frac{1}{\pi}\sup_{|\boldsymbol{z}|=1}\left\{\frac{1}{2}\int_{\mathbb{S}^1}\left[(1-\kappa)^2 + 4\kappa(\boldsymbol{e}_\sigma,\boldsymbol{z})^2\right]^{1/2}d\sigma - \int_{\mathcal{C}}\left[(1-\kappa)^2 + 4\kappa(\boldsymbol{e}_{q\sigma},\boldsymbol{z})\right]^{1/2}d\sigma\right\}.$$

Here the first integral does not depend on $\boldsymbol{z}$. Hence

$$\text{ess}\|T_\kappa^{(2)}\| = \mathfrak{J} - \frac{1}{\pi}\inf_{|\boldsymbol{z}|=1}\int_{\mathcal{C}}\left[(1-\kappa)^2 + 4\kappa(\boldsymbol{e}_{q\sigma},\boldsymbol{z})^2\right]^{1/2}d\sigma, \tag{7.4.87}$$

where

$$\begin{aligned}\mathfrak{J} &= \frac{1}{2\pi}\int_{\mathbb{S}^1}\left[(1-\kappa)^2 + 4\kappa(\boldsymbol{e}_\sigma,\boldsymbol{z})^2\right]^{1/2}d\sigma \\ &= \frac{2}{\pi}(1+\kappa)\int_0^{\pi/2}\left[1 - \frac{4\kappa}{(1+\kappa)^2}\sin^2\gamma\right]^{1/2}d\gamma,\end{aligned} \tag{7.4.88}$$

γ being the value of the angle between the vectors $\boldsymbol{e}_\sigma$ and $\boldsymbol{z}$.

Suppose the Cartesian coordinate system with origin at q is chosen in such a way that the axis qx_1 passes through the point $a \in \mathcal{C}$ equidistant from points of the set $\partial\mathcal{C}$. Let ϑ denote the angle between the vector $\boldsymbol{e}_{q\sigma}$ and the axis qx_1, $-\pi < \vartheta \le \pi$. The value of the angle between the vector $\boldsymbol{z}$ and the axis qx_1 will be denoted by η, $-\pi < \eta \le \pi$. The infimum over $\boldsymbol{z}$ of the integral in (7.4.87) is equal to

$$\inf\left\{\int_{-\alpha/2}^{\alpha/2}\left[(1-\kappa)^2 + 4\kappa\cos^2(\vartheta-\eta)\right]^{1/2}d\vartheta :\ -\pi < \eta \le \pi\right\}. \tag{7.4.89}$$

In order to find this infimum consider the function

$$f(\eta) = \int_{-\alpha/2}^{\alpha/2}\left[(1-\kappa)^2 + 4\kappa\cos^2(\upsilon-\eta)\right]^{1/2}d\upsilon,$$

where $-\pi < \eta \le \pi$. Since $f(0) = f(-\pi) = f(\pi)$, the end points of the interval $[-\pi,\pi]$ can be excluded from the consideration. By the substitution $\xi = \upsilon - \eta$ we represent $f(\eta)$ as

$$\int_{-\alpha/2-\eta}^{\alpha/2-\eta}\left[(1-\kappa)^2 + 4\kappa\cos^2\xi\right]^{1/2}d\xi$$

and find its critical points for $\eta \in (-\pi, \pi)$. We have

$$f'(\eta) = -\Big[(1-\kappa)^2 + 4\kappa\cos^2\Big(\frac{\alpha}{2} - \eta\Big)\Big]^{1/2} + \Big[(1-\kappa)^2 + 4\kappa\cos^2\Big(\frac{\alpha}{2} + \eta\Big)\Big]^{1/2} = 0$$

which can be simplified and reduced to the equation $\sin\alpha\sin 2\eta = 0$. If $\alpha = 0$ or $\alpha = \pi$, then $f(\eta) = \text{const}$. Suppose, $0 < \alpha < \pi$. Then the last equation implies that the critical points of $f(\eta)$ at the interval $(-\pi, \pi)$ are $\eta = 0,\ \pi/2,\ -\pi/2$.

Let us find the sign of the function

$$f''(\eta) = -2\kappa\left\{\frac{\sin(\alpha - 2\eta)}{[(1-\kappa)^2 + 4\kappa\cos^2(\frac{\alpha}{2} - \eta)]^{1/2}} + \frac{\sin(\alpha + 2\eta)}{[(1-\kappa)^2 + 4\kappa\cos^2(\frac{\alpha}{2} + \eta)]^{1/2}}\right\}$$

for $\eta = 0,\ \pi/2,\ -\pi/2$. Since $\kappa \neq 0$, then

$$f''(0) = \frac{-4\kappa\sin\alpha}{\big[(1-\kappa)^2 + 4\kappa\cos^2(\alpha/2)\big]^{1/2}} < 0,$$

$$f''\big(\frac{\pi}{2}\big) = f''\big(-\frac{\pi}{2}\big) = \frac{4\kappa\sin\alpha}{\big[(1-\kappa)^2 + 4\kappa\sin^2(\alpha/2)\big]^{1/2}} > 0.$$

We note also that $f(-\pi/2) = f(\pi/2)$. Therefore the minimum value of $f(\eta)$ is attained for $\eta = \pm\pi/2$ and equals

$$\int_{(\pi-\alpha)/2}^{(\pi+\alpha)/2} \big[(1-\kappa)^2 + 4\kappa\cos^2\xi\big]^{1/2} d\xi,$$

Consequently,

$$\text{(7.4.90)}\quad \begin{aligned} \text{ess}\|T_\kappa^{(2)}\| &= \mathfrak{J} - \frac{1}{\pi}\int_{(\pi-\alpha)/2}^{(\pi+\alpha)/2} \big[(1-\kappa)^2 + 4\kappa\cos^2\xi\big]^{1/2} d\xi \\ &= \mathfrak{J} - \frac{2}{\pi}(1+\kappa)\int_{(\pi-\alpha)/2}^{\pi/2} \Big[1 - \frac{4\kappa}{(1+\kappa)^2}\sin^2\xi\Big]^{1/2} d\xi. \end{aligned}$$

Hence the result follows from (7.4.88). □

REMARK 7.7. The following formula for the Fredholm radius of the operator $T_\kappa^{(2)}$ in the space of continuous vector-valued functions

$$\text{(7.4.91)}\qquad R(T_\kappa^{(2)}) = \frac{\pi}{\pi - \alpha + \kappa\sin\alpha},$$

was obtained by Shelepov [**Shel2**].

By (7.4.88) and (7.4.90),

$$\text{ess}\|T_\kappa^{(2)}\| = \frac{2}{\pi}\int_0^{(\pi-\alpha)/2} \big[(1-\kappa)^2 + 4\kappa\cos^2\vartheta\big]^{1/2} d\vartheta.$$

Taking into account the inequality

$$\big[(1-\kappa)^2 + 4\kappa\cos^2\chi\big]^{1/2} > 1 - \kappa + 2\kappa\cos^2\chi,$$

with $\kappa > 0$, along with $\cos\chi \neq 0,\ \pm 1$, and using (7.4.91), we find

$$\text{(7.4.92)}\qquad R(T_\kappa^{(2)}) > \frac{1}{\text{ess}\|T_\kappa^{(2)}\|}$$

for $\kappa > 0$.

We note that for $\kappa = 0$, i.e. for the logarithmic potential T, the Fredholm radius and the quantity inverse to the the essential norm are equal, as shown by J. Radon [**Rad2**].

7.4.6. The essential norm of the operator $T_\kappa^{(3)}$ for a domain with a conic point

Let $\mathcal{D}$ be a bounded domain in $\mathbb{R}^3$ and let $\partial\mathcal{D}$ be a smooth surface except the only conic point q. Suppose that the tangent cone $V(q)$ is rotational and denote its opening by β, $0 \le \beta \le 2\pi$. We put $\alpha = \min\{\beta, 2\pi - \beta\}$.

THEOREM 7.8. *The equality*

$$\begin{aligned}\operatorname{ess}\|T_\kappa^{(3)}\| &= \frac{(1-\kappa)}{\pi}\left[\sqrt{1+\Gamma^2\sin^2\frac{\alpha}{2}}\, E\left(\frac{\Gamma\sin\frac{\alpha}{2}}{\sqrt{1+\Gamma^2\sin^2\frac{\alpha}{2}}}\right)\cos\frac{\alpha}{2}\right.\\ &\left. + \int_0^{\alpha/2}\frac{1+\Gamma^2\cos^2\vartheta}{\Gamma\cos\vartheta}\arcsin\frac{\Gamma\cos\vartheta\cos(\alpha/2)}{\sqrt{1+\Gamma^2\cos^2\vartheta}}d\vartheta\right]\end{aligned}$$

holds with

$$\Gamma = (1-\kappa)^{-1}\sqrt{3\kappa(\kappa+2)}.$$

In particular,

$$\operatorname{ess}\|T_1^{(3)}\| = \frac{3}{2\pi}(\pi - \alpha + \sin\alpha).$$

PROOF. By a standard argument, from Theorem 7.4 and Proposition 7.1 we derive

$$\operatorname{ess}\|T_\kappa^{(3)}\| = \sup_{|\boldsymbol{z}|=1}\left[\int_{\mathbb{S}^2}\left|\mathcal{H}_\kappa^{(3)}(\sigma)\boldsymbol{z}\right|d\sigma - 2\int_{\mathcal{S}}\left|\mathcal{H}_\kappa^{(3)}(\sigma)\boldsymbol{z}\right|d\sigma\right], \tag{7.4.93}$$

where $\mathcal{S}$ is that of the sets $\partial\mathbb{B}(q)\cap V(q)$, $\partial\mathbb{B}(q)\backslash V(q)$ which is placed in a hemisphere.

Substituting the integrand from (7.4.80) into (7.4.93) we arrive at

$$\begin{aligned}\operatorname{ess}\|T_\kappa^{(3)}\| = \frac{1}{2\pi}\sup_{|\boldsymbol{z}|=1}\Big\{\frac{1}{2}\int_{\mathbb{S}^2}&\left[(1-\kappa)^2 + 3\kappa(\kappa+2)(\boldsymbol{e}_\sigma,\boldsymbol{z})^2\right]^{1/2}d\sigma\\ &- \int_{\mathcal{S}}\left[(1-\kappa)^2 + 3\kappa(\kappa+2)(\boldsymbol{e}_{q\sigma},\boldsymbol{z})^2\right]^{1/2}d\sigma\Big\}.\end{aligned} \tag{7.4.94}$$

The first integral in (7.4.94) does not depend on $\boldsymbol{z}$. Thus

$$\operatorname{ess}\|T_\kappa^{(3)}\| = \mathcal{J} - \frac{1}{2\pi}\inf_{|\boldsymbol{z}|=1}\int_{\mathcal{S}}\left[(1-\kappa)^2 + 3\kappa(\kappa+2)(\boldsymbol{e}_{q\sigma},\boldsymbol{z})^2\right]^{1/2}d\sigma, \tag{7.4.95}$$

where

$$\mathcal{J} = \frac{1}{4\pi}\int_{\mathbb{S}^2}\left[(1-\kappa)^2 + 3\kappa(\kappa+2)(\boldsymbol{e}_\sigma,\boldsymbol{z})^2\right]^{1/2}d\sigma. \tag{7.4.96}$$

Let a denote the center of $\mathcal{S}$. Suppose the Cartesian coordinate system with origin at q is chosen in such a way that the axis qx_1 passes through a and that the vector $\boldsymbol{z}$ lays in the plane x_1qx_2. We introduce the spherical coordinates φ and ϑ, where φ is the angle between the planes σqx_3 and x_1qx_3, $-\pi < \varphi \le \pi$ and ϑ is the angle between the vector $\boldsymbol{e}_{q\sigma}$ and the axis qx_3, $0 \le \vartheta \le \pi$. By η we denote the angle between the vector $\boldsymbol{z}$ and the axis qx_1, $-\pi < \eta \le \pi$.

The set $\overline{\mathcal{S}}$ is described by the inequalities

$$(\pi-\alpha)/2 \le \vartheta \le (\pi+\alpha)/2, \qquad -\gamma(\vartheta) \le \varphi \le \gamma(\vartheta),$$

where $\gamma(\vartheta) = \arccos\,(\cos(\alpha/2)/\sin\vartheta)$. Therefore, the integral in (7.4.95) can be written in the form

$$\int_{(\pi-\alpha)/2}^{(\pi+\alpha)/2} \sin\vartheta d\vartheta \int_{-\gamma(\vartheta)}^{\gamma(\vartheta)} \left[(1-\kappa)^2+3\kappa(\kappa+2)\sin^2\vartheta\cos^2(\varphi-\eta)\right]^{1/2} d\varphi. \tag{7.4.97}$$

Using the same argument as in the proof of Theorem 7.7, we show that the minimum of the inner integral, considered as a function of η, is attained for $\eta = \pm\pi/2$. Thus, the infimum over $\boldsymbol{z}$ of (7.4.95) is attained for $(\boldsymbol{z}, \boldsymbol{e}_{qa}) = 0$. Let $\boldsymbol{z} = \boldsymbol{e}_2$. Then (7.4.95) can be written in the form

$$\operatorname{ess}\|T_\kappa^{(3)}\| = \frac{1}{2\pi}\int_{\partial\mathbb{B}^+(q)\setminus\mathcal{S}} \left[(1-\kappa)^2 + 3\kappa(\kappa+2)(\boldsymbol{e}_{q\sigma}, \boldsymbol{e}_2)^2\right]^{1/2} d\sigma, \tag{7.4.98}$$

where $\partial\mathbb{B}^+(q) = \{\sigma \in \partial\mathbb{B}(q) : (\boldsymbol{e}_{q\sigma}, \boldsymbol{e}_1) \ge 0\}$.

We introduce the coordinates χ and ω, χ being the angle between the planes $\sigma q x_1$ and $x_1 q x_2$, ω being the angle between the vector $\boldsymbol{e}_{q\sigma}$ and the axis qx_1. By (7.4.98) we obtain

$$\operatorname{ess}\|T_\kappa^{(3)}\| = \frac{2}{\pi}\int_{\alpha/2}^{\pi/2} \sin\omega d\omega \int_0^{\pi/2} \left[(1-\kappa)^2 + 3\kappa(\kappa+2)\sin^2\omega\cos^2\chi\right]^{1/2} d\chi.$$

The theorem is proved. □

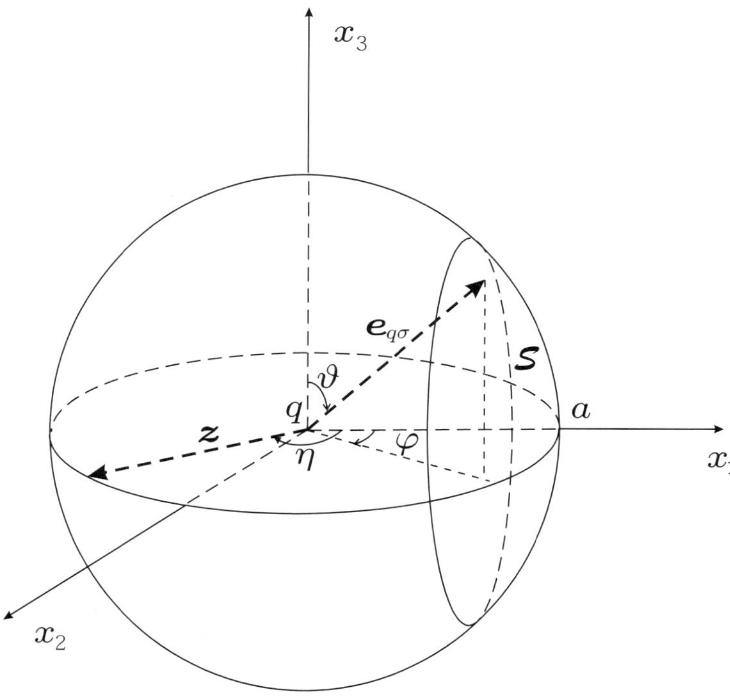

FIGURE 7.2. Spherical segment $\mathcal{S}$ on $\mathbb{S}^2$ in the proof of Theorem 7.8

7.4.7. The essential norm of the operator $T_\kappa^{(3)}$ for a domain with an edge

Let $\mathcal{D}$ be a bounded domain in $\mathbb{R}^3$ whose boundary the union of two smooth surfaces with the common part E. Suppose the curve E is smooth. Then at each $\zeta \in E$ two tangent planes are defined. The angle between them calculated in $\mathcal{D}$, will be denoted by $\beta(\zeta)$, $0 \le \beta(\zeta) \le 2\pi$. Further, we put

$$\alpha = \min_{\zeta \in E}\{\beta(\zeta),\ 2\pi - \beta(\zeta)\}.$$

THEOREM 7.9. *The equality*

$$\text{ess}\|T_\kappa^{(3)}\| = \frac{(1-\kappa)}{\pi}\Big[\frac{\pi-\alpha}{2} + \int_{\alpha/2}^{\pi/2} \frac{1+\Gamma^2\sin^2\vartheta}{\Gamma\sin\vartheta}\arccos\big(1+\Gamma^2\sin^2\vartheta\big)^{-1/2}d\vartheta\Big]$$

is valid with the same constant Γ *as in Theorem* 7.8.

In particular,

$$\text{ess}\|T_1^{(3)}\| = (3/2)\cos(\alpha/2).$$

PROOF. By a standard argument, from Theorem 7.4 and Proposition 7.1 we derive

$$\text{ess}\|T_\kappa^{(3)}\| = \sup_{\zeta\in E}\sup_{|\boldsymbol{z}|=1}\Big[\int_{\mathbb{S}^2}\big|\mathcal{H}_\kappa^{(3)}(\sigma)\boldsymbol{z}\big|d\sigma - 2\int_{\mathcal{R}(\zeta)}\big|\mathcal{H}_\kappa^{(3)}(\sigma)\boldsymbol{z}\big|d\sigma\Big], \tag{7.4.99}$$

where $\mathcal{R}(\zeta)$ is that of the sets $\partial\mathbb{B}(\zeta)\cap V(\zeta)$, $\partial\mathbb{B}(\zeta)\backslash V(\zeta)$ which has the least area.

Substituting the integrand from (7.4.80) into (7.4.99) we arrive at

$$\text{ess}\|T_\kappa^{(3)}\| = \frac{1}{2\pi}\sup_{\zeta\in E}\sup_{|\boldsymbol{z}|=1}\Big\{\frac{1}{2}\int_{\mathbb{S}^2}\big[(1-\kappa)^2+3\kappa(\kappa+2)(\boldsymbol{e}_\sigma,\boldsymbol{z})^2\big]^{1/2}d\sigma$$
$$-\int_{\mathcal{R}(\zeta)}\big[(1-\kappa)^2+3\kappa(\kappa+2)(\boldsymbol{e}_{\zeta\sigma},\boldsymbol{z})^2\big]^{1/2}d\sigma\Big\}.$$

Hence, in the same way as in Theorem 7.8, we obtain

$$\text{ess}\|T_\kappa^{(3)}\| = \mathcal{J} - \frac{1}{2\pi}\inf_{\zeta\in E}\inf_{|\boldsymbol{z}|=1}\int_{\mathcal{R}(\zeta)}F(\boldsymbol{e}_{\zeta\sigma},\boldsymbol{z})d\sigma, \tag{7.4.100}$$

where $\mathcal{J}$ is defined by (7.4.96) and

$$F(\boldsymbol{e}_{\zeta\sigma},\boldsymbol{z}) = \big[(1-\kappa)^2+3\kappa(\kappa+2)(\boldsymbol{e}_{\zeta\sigma},\boldsymbol{z})^2\big]^{1/2}. \tag{7.4.101}$$

We put

$$\mathcal{R}^*(\zeta) = \big\{v\in\partial\mathbb{B}(\zeta):\ \boldsymbol{e}_{\zeta v} = -\boldsymbol{e}_{\zeta\sigma}, \sigma\in\mathcal{R}(\zeta)\big\}.$$

Taking into account that $F(\boldsymbol{e}_{\zeta\sigma},\boldsymbol{z})$ is an even function of $\boldsymbol{e}_{\zeta\sigma}$ on $\partial\mathbb{B}(\zeta)$ we have

$$\int_{\mathcal{R}(\zeta)}F(\boldsymbol{e}_{\zeta\sigma},\boldsymbol{z})d\sigma = \int_{\mathcal{R}^*(\zeta)}F(\boldsymbol{e}_{\zeta\sigma},\boldsymbol{z})d\sigma$$

and, consequently,

$$\int_{\mathcal{R}(\zeta)}F(\boldsymbol{e}_{\zeta\sigma},\boldsymbol{z})d\sigma = \frac{1}{2}\int_{\mathcal{R}(\zeta)\cup\mathcal{R}^*(\zeta)}F(\boldsymbol{e}_{\zeta\sigma},\boldsymbol{z})d\sigma. \tag{7.4.102}$$

Let $\Lambda(\zeta)$ denote the plane passing through ζ and dividing the angle of the spherical biangle $\mathcal{R}(\zeta)$ into two equal parts. Let further p be a point of the circumference $\mathfrak{J}(\zeta) = \mathcal{R}(\zeta) \cap \Lambda(\zeta)$. We set

$$\mathcal{R}_p^+(\zeta) = \{\sigma \in \mathcal{R}(\zeta) \cup \mathcal{R}^*(\zeta) : (\boldsymbol{e}_{\zeta\sigma}, \boldsymbol{e}_{\zeta p}) > 0\},$$

$$\mathcal{R}_p^-(\zeta) = \{\sigma \in \mathcal{R}(\zeta) \cup \mathcal{R}^*(\zeta) : (\boldsymbol{e}_{\zeta\sigma}, \boldsymbol{e}_{\zeta p}) < 0\}.$$

Using the equality

$$\int_{\mathcal{R}_p^+(\zeta)} F(\boldsymbol{e}_{\zeta\sigma}, \boldsymbol{z}) d\sigma = \int_{\mathcal{R}_p^-(\zeta)} F(\boldsymbol{e}_{\zeta\sigma}, \boldsymbol{z}) d\sigma,$$

from (7.4.102) we find

$$\int_{\mathcal{R}(\zeta)} F(\boldsymbol{e}_{\zeta\sigma}, \boldsymbol{z}) d\sigma = \int_{\mathcal{R}_p^+(\zeta)} F(\boldsymbol{e}_{\zeta\sigma}, \boldsymbol{z}) d\sigma. \tag{7.4.103}$$

Let $\boldsymbol{n}(\zeta)$ be the unit normal to $\Lambda(\zeta)$ and let

$$Z(p) = \{\boldsymbol{z} \in \mathbb{R}^3 : \boldsymbol{z} = d_1 \boldsymbol{e}_{\zeta p} + d_2 \boldsymbol{n}(\zeta),\ d_1^2 + d_2^2 = 1\}.$$

Since the set of unit vectors in $\mathbb{R}^3$ can be represented in the form

$$\bigcup_{p \in \mathfrak{J}(\zeta)} Z(p),$$

by (7.4.103) we get

$$\inf_{|\boldsymbol{z}|=1} \int_{\mathcal{R}(\zeta)} F(\boldsymbol{e}_{\zeta\sigma}, \boldsymbol{z}) d\sigma = \inf_{p \in \mathfrak{J}(\zeta)} \inf_{\boldsymbol{z} \in Z(p)} \int_{\mathcal{R}_p^+(\zeta)} F(\boldsymbol{e}_{\zeta\sigma}, \boldsymbol{z}) d\sigma. \tag{7.4.104}$$

Suppose the Cartesian coordinate system with origin at ζ is chosen in such a way that the plane $x_1\zeta x_3$ coincides with $\Lambda(\zeta)$ and that $\boldsymbol{e}_1 = \boldsymbol{e}_{\zeta p}$. The angle between the vector $\boldsymbol{z} \in Z(p)$ and the axis ζx_1 will be denoted by η, $-\pi < \eta \le \pi$. We introduce the spherical coordinates φ and ϑ, where φ is the angle between the planes $\sigma\zeta x_3$ and $x_1\zeta x_3$, $-\pi < \varphi \le \pi$, and ϑ is the angle between the vector $\boldsymbol{e}_{\zeta\sigma}$ and the axis ζx_3, $0 \le \vartheta \le \pi$. The set $\overline{\mathcal{R}_p^+(\zeta)}$ is described by the inequalities

$$0 \le \vartheta \le \pi, \quad -\gamma_p(\vartheta) \le \varphi \le \gamma_p(\vartheta),$$

where $\gamma_p(\vartheta)$ is a certain continuous function. Taking into account (7.4.101) one can write the integral on the right-hand side in (7.4.104) in the form

$$\int_0^\pi \sin\vartheta\, d\vartheta \int_{-\gamma_p(\vartheta)}^{\gamma_p(\vartheta)} \big[(1-\kappa)^2 + 3\kappa(\kappa+2)\sin^2\vartheta \cos^2(\varphi - \eta)\big]^{1/2} d\varphi. \tag{7.4.105}$$

Using the same argument as in the proof of Theorem 7.7 we find that the inner integral in (7.4.105) attains its minimal value for $\eta = \pm\pi/2$. Thus, the infimum over $\boldsymbol{z} \in Z(p)$ of the integral on the right-hand side of (7.4.104) is equal to

$$\int_{\mathcal{R}_p^+(\zeta)} F(\boldsymbol{e}_{\zeta\sigma}, \boldsymbol{e}_2) d\sigma$$

and so, by (7.4.103),

$$\inf_{|\boldsymbol{z}|=1} \int_{\mathcal{R}(\zeta)} F(\boldsymbol{e}_{\zeta\sigma}, \boldsymbol{z}) d\sigma = \int_{\mathcal{R}(\zeta)} F(\boldsymbol{e}_{\zeta\sigma}, \boldsymbol{e}_2) d\sigma. \tag{7.4.106}$$

Now we fix the Cartesian coordinate system assuming that the point p is equidistant from the vertices of biangle $\mathcal{R}(\zeta)$. Taking into account (7.4.96), (7.4.101) and (7.4.106), we can write (7.4.100) in the form

$$\text{ess}\|T_\kappa^{(3)}\| = \frac{1}{2\pi}\sup_{\zeta\in E}\int_{\partial\mathbb{B}^+(\zeta)\setminus\mathcal{R}(\zeta)}\left[(1-\kappa)^2+3\kappa(\kappa+2)(\boldsymbol{e}_{\zeta\sigma},\boldsymbol{e}_2)^2\right]^{1/2}d\sigma,$$

where

$$\partial\mathbb{B}^+(\zeta)=\{\sigma\in\partial\mathbb{B}(\zeta):\ (\boldsymbol{e}_{\zeta\sigma},\boldsymbol{e}_1)\geq 0\}.$$

Hence

$$\text{ess}\|T_\kappa^{(3)}\| = \frac{2}{\pi}\int_0^{\pi/2}\sin\vartheta d\vartheta\int_{\alpha/2}^{\pi/2}\left[(1-\kappa)^2+3\kappa(\kappa+2)\sin^2\vartheta\sin^2\varphi\right]^{1/2}d\varphi,$$

which completes the proof. □

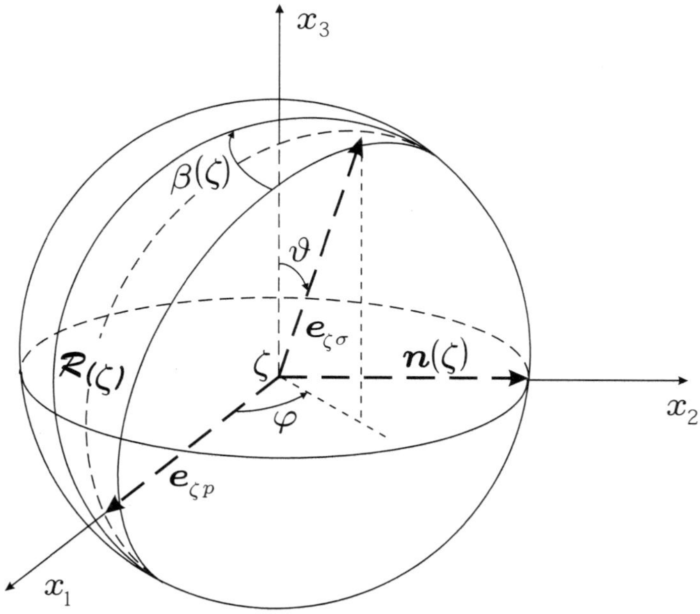

FIGURE 7.3. Biangle $\mathcal{R}(\zeta)$ on $\mathbb{S}^2$ in the proof of Theorem 7.9

7.5. Comments to Chapter 7

1. The notion of the Fredholm radius introduced by Radon in [**Rad1**] plays the central role in the solvability theory for boundary value problems in nonsmooth domains. This notion is introduced as follows.

Let L be a bounded linear operator acting on a Banach space $\mathfrak{B}$, and let I be the identity operator on $\mathfrak{B}$. The Fredholm radius $R(L)$ of the operator L is the radius of the largest circle on the complex λ-plane centered at $\lambda = 0$ inside which $I + \lambda L$ is a Fredholm operator.

Radon also was the first who used the notion of the continuity degree of an operator L. It runs as follows. Let $\mathcal{K}$ be a compact operator on $\mathfrak{B}$, and let $C_L(\mathcal{K})$ be the radius of the largest disk in the λ-plane centered at $\lambda = 0$ in which the series

$$I + \lambda(L - \mathcal{K}) + \lambda^2(L - \mathcal{K})^2 + \cdots$$

converges uniformly. The number $C_L = \sup C_L(\mathcal{K})$, where the supremum is taken over the set of all compact operators $\mathcal{K}$ on $\mathfrak{B}$, is called the continuity degree of the operator. It is shown in [**Rad1**] that

$$R(L) = C_L. \tag{7.5.1}$$

The notion of the essential norm of L, also introduced by Radon [**Rad2**], is closely related to the Fredhom radius. Namely, the quantity

$$\operatorname{ess}\|L\| = \inf \|L - \mathcal{K}\|_{\mathfrak{B}},$$

where the infimum is taken over the set of all linear compact operators in $\mathfrak{B}$, is called the essential norm of the operator L. By (7.5.1), the Fredholm radius of the operator L is related to the essential norm as follows

$$R(L) \geq \frac{1}{\operatorname{ess}\|L\|}. \tag{7.5.2}$$

The essential norm of a bounded operator is a measure of non-compactness of this operator (for measure of non-compactness see Rodkina, Potapov, Kamenskiĭ, Akhmerov, Sadowsky [**RPKAS**] and references therein). The essential norm of boundary integral operators in potential theory can be considered as a quantitative characteristic of the influence of boundary singularities on properties of the boundary integral operator.

Note that the Fredholm radius can be approximated with any accuracy by the reciprocal of the essential norm and the accuracy depends on the choice of the norm in $\mathfrak{B}$. More precisely, it was shown by Gohberg and Marcus [**GM**] that the Fredholm radius of the operator L in a Banach space $\mathfrak{B}$ with the norm $\|\cdot\|_0$ satisfies

$$R(L) = \sup\Big\{ \big(\operatorname{ess}\|L\|\big)^{-1} : \ \|\cdot\| \in \mathcal{N}(\|\cdot\|_0)\Big\}, \tag{7.5.3}$$

where $\operatorname{ess}\|L\|$ is calculated with respect to a norm $\|\cdot\|$ in the set $\mathcal{N}(\|\cdot\|_0)$ of all norms in $\mathfrak{B}$, topologically equivalent to the norm $\|\cdot\|_0$.

The results of Radon layed the foundation of an important part of functional analysis (see works by Nikol'skii [**Nikl**],Atkinson [**At**], Gohberg and Krein [**GK1**]).

2. The notions of the Fredholm radius and the essential norm were applied by Radon [**Rad2**] to the theory of the logarithmic potential for the so-called "curves with bounded rotation", which are defined as follows. Let Γ be a rectifiable curve and let s denote the arc length on Γ $(0 \leq s \leq l)$. If the angle $\vartheta(s)$ made by the positively oriented tangent and the abscissa is a function of bounded variation on [0,1], the curve Γ is said to be a curve with bounded rotation.

Suppose that a curve with bounded rotation Γ is the boundary of a domain $D \subset \mathbb{R}^2$ with compact closure and that u is a function harmonic on D and satisfying the Dirichlet condition

$$u\Big|_{\Gamma} = g,$$

where g is in the space of continuous functions $\mathrm{C}(\Gamma)$. When one represents u as a double layer logarithmic potential with unknown density $\varphi \in \mathrm{C}(\Gamma)$, one arrives at the integral equation

$$\varphi + T\varphi = 2g \quad \text{on} \quad \Gamma.$$

Here T is a continuous but, generally speaking, is not a compact operator in $C(\Gamma)$.

Radon found the formula for the essential norm of T in $C(\Gamma)$

$$\text{ess}\|T\| = \alpha/\pi, \tag{7.5.4}$$

where

$$\alpha = \sup_{0\le s\le l} \big|\theta(s+0) - \theta(s-0)\big|.$$

The following inequality, which is a corollary of (7.5.2) and (7.5.4),

$$R(T) \ge \frac{\pi}{\alpha}$$

is sufficient for applicability of the Fredholm theory to the above integral equation provided $\alpha < \pi$, i.e. there are no cusps on Γ. Moreover, showing that the converse estimate $R(T) \le \pi/\alpha$ holds, Radon proved that in the theory of logarithmic potential on curves with bounded rotation one has

$$R(T) = \frac{1}{\text{ess}\|T\|}.$$

The example of the double layer elastic potential (see (7.4.92)) shows that the above equality in general does not hold for vector-valued potentials.

We note, that three years earlier than Radon Carleman [**Car**] showed that if the boundary Γ of the domain $D \subset \mathbb{R}^2$ is the union of a finite number N of closed arcs $p_j p_{j+1} (p_{N+1} = p_1)$ of class C^2, then the essential norm of T in the space with the norm

$$\sup_j \sup_{x\in L} |x - p_j|^{\kappa_j} |u(x)|, \quad 0 \le \kappa_j \le 1,$$

admits the estimate

$$\text{ess}\|T\| \le \sup_j \frac{\sin(|\pi - \alpha_j|\kappa_j)}{\sin(\pi\kappa_j)},$$

where α_j is the angle between the half-tangents to Γ at p_j. In the case $\kappa_j = 0$, Carleman's estimate coincides with that of Radon.

Carleman also established an analogous result for two-dimensional surfaces S comprised by two surfaces S_1 and S_2 of the class C^2 having a common curved edge E, which is assumed to be a twice continuously differentiable curve. He proved, in particular, that for the essential norm of T on the space $C(S)$ the estimate

$$\text{ess}\|T\| \le \sup_{p\in E} |1 - \alpha(p)/\pi|$$

holds, $\alpha(p)$ being the minimum of the two angles made by the tangent planes to S_1 and S_2 at $p \in E$.

It was shown in the works of Burago and Maz'ya [**BuMaz**] and Král [**Kra1**] that for a multidimensional domain $\mathcal{D}$ with bounded variation of the solid angle $\omega_{\mathcal{D}}(p, \mathcal{B})$ $(p \in \partial\mathcal{D},\ \mathcal{B} \subset \mathbb{R}^n)$ the essential norm of T in $C(\partial\mathcal{D})$ is

$$\frac{2}{\omega_n} \lim_{\delta\to 0} \sup_{p\in\partial\mathcal{D}} |\,\omega_{\mathcal{D}}|(p, \mathbb{B}_\delta(p)), \tag{7.5.5}$$

where $\mathbb{B}_\delta(p) = \big\{x \in \mathbb{R}^n : |x - p| < \delta\big\}$.

A formula expressing the essential norm of the operator T in terms of the volume density of the set was obtained by Netuka [**Ne1**] for convex domains, who also showed that $\text{ess}\|T\| < 1$. A theory of harmonic potentials in domains with irregular boundaries is developed in the book by Král [**Kra2**].

Note that the boundary of a planar domain with bounded variation of the solid angle may be a smooth curve with unbounded rotation (see Burago and Maz'ya [**BuMaz**]).

Another direction in the study of boundary integral operators in various function spaces on domains with irregular boundaries, different from estimates of the Fredholm radius by the essential norm, originated in the paper [**Lo2**] by Lopatinskiĭ. His approach is based on consideration of a "model" integral operator, its reduction to a matrix-valued integral operator of convolution type on the half-axis and the use of results due to Gohberg and Krein [**GK2**] concerning systems of integral equations on the half-axis with kernels depending on the difference of arguments. We also mention the papers [**BaSh**] by Bazaliĭ and Shelepov and [**Shel1**]-[**Shel3**] by Shelepov dedicated to various properties of the harmonic double layer potentials and their vector-valued analogues, in particular, elastostatic potentials, on curves with bounded rotation.

A different approach in the analysis of spectral properties of integral operators on nonsmooth boundaries of planar and three-dimensional domains is based on the Mellin transform techniques. This approach is used, in particular, by Fabes, Jodeit and Lewis [**FJL**] and I. Mitrea [**Mi1**].

Many results on the invertibility in various function spaces and the Fredholm property of boundary integral operators on piecewise smooth boundaries are either presented or referred to in the papers Angell, Kleinman and Král [**AKK**], Costabel [**Cos**], Elschner [**El**], Fabes [**Fa**], Fabes, Sand and Seo [**FSS**], Grachev [**Gr**], Grachev and Maz'ya [**GM1, GM2**], Hansen [**Han**], Král and Medková [**KraMe1, KraMe2, KraMe3**], Král and Wendland [**KW1, KW2**], Maz'ya [**Ma1, Ma2**], Medková [**Me1, Me2, Me3**], Mitrea [**Mi2, IM3**], Mitrea and Tucker [**MT**], Rathsfeld [**Rat1, Rat2**], Rempel and Schmidt [**RS**], Steinbach and Wendland [**SW**], Verchota [**Ve**], Verchota and Vogel [**VV**].

3. In Section 7.2, for reader's convenience, we collect some properties of the solid angle for non-regular domains following the first part of the book by Burago and Maz'ya [**BuMaz**]. A selection of properties of the signed measure (charge) is presented, for example, in the book by Landkof [**Land**]. For a theory of the space $BV(\mathbb{R}^n)$ see, for example, the second part of the book by Burago and Maz'ya [**BuMaz**], and the book by Volpert and Hudyaev [**VH**].

The survey by Maz'ya [**Ma2**] on boundary integral equations presents the history of this topic, numerous results in various directions and an extensive bibliography until 1990. The modern state of the art in the theory of boundary value problems for partial differential equations in domains with piecewise smooth boundaries is described in the books by Kozlov, Maz'ya and Rossmann [**KMR1, KMR2**], Maz'ya and Rossmann [**MR4**].

Part 2

Parabolic Systems

CHAPTER 8

Maximum Modulus Principle for Parabolic Systems

8.1. Introduction

In this chapter we find criteria for validity of the classical maximum modulus principle for solutions of the uniformly parabolic system in the sense of Petrovskiĭ

$$\frac{\partial \boldsymbol{u}}{\partial t} - \sum_{|\beta|\leq 2\ell} \mathcal{A}_\beta(x,t) D_x^\beta \boldsymbol{u} = \boldsymbol{0}. \tag{8.1.1}$$

Here $\boldsymbol{u}$ is an m-component vector-valued function, $\mathcal{A}_\beta$ are real or complex $(m\times m)$-matrix-valued functions, $\beta = (\beta_1, \dots, \beta_n)$ is a multi-index of order $|\beta| = \beta_1 + \cdots + \beta_n$ and $D_x^\beta = \partial^{|\beta|}/\partial x_1^{\beta_1} \dots \partial x_n^{\beta_n}$. For $\ell \geq 1$ the vector-valued function $\boldsymbol{u}$ is defined on the closure $\overline{\mathbb{R}_T^{n+1}}$ of the layer $\mathbb{R}_T^{n+1} = \mathbb{R}^n \times (0, T]$. In the special case $\ell = 1$ it will be defined also on the closure $\overline{Q}_T$ of the cylinder $Q_T = \Omega \times (0, T]$, where Ω is a bounded domain in $\mathbb{R}^n$.

Throughout the chapter we make the following assumptions:

(A) The matrix-valued functions $\mathcal{A}_\beta$ are defined in $\overline{\mathbb{R}_T^{n+1}}$ and have bounded derivatives in x up to the order $|\beta|$ which satisfy the uniform Hölder condition on $\overline{\mathbb{R}_T^{n+1}}$ with exponent α, $0 < \alpha \leq 1$, with respect to the parabolic distance

$$d\,[(x,t),(x',t')] = \left(|x-x'|^2 + |t-t'|^{1/\ell}\right)^{1/2}$$

between the points (x,t) and (x',t') in $\mathbb{R}^{n+1}$;

(B) For any point $(x,t) \in \overline{\mathbb{R}_T^{n+1}}$, the real parts of λ-roots of the equation

$$\det\left((-1)^\ell \sum_{|\beta|=2\ell} \mathcal{A}_\beta(x,t)\boldsymbol{\sigma}^\beta - \lambda I\right) = 0$$

satisfy the inequality $\Re\, \lambda(x,t,\boldsymbol{\sigma}) \leq -\delta|\boldsymbol{\sigma}|^{2\ell}$, where $\delta = \text{const} > 0$ for any $\boldsymbol{\sigma} = (\sigma_1, \dots, \sigma_n) \in \mathbb{R}^n$, I is the identity matrix of order m, and $|\cdot|$ is the Euclidean length of a vector.

In Section 8.2 the Cauchy problem for system (8.1.1) is treated. We obtain an expression for the best constant $\mathcal{K}(\mathbb{R}^n, T)$ in the inequality

$$|\boldsymbol{u}(x,t)| \leq \mathcal{K}(\mathbb{R}^n, T) \sup\{|\boldsymbol{u}(y,0)| : y \in \mathbb{R}^n\},$$

where $(x,t) \in \mathbb{R}_T^{n+1}$. It is shown that $\mathcal{K}(\mathbb{R}^n, T) > 1$ for all $\ell \geq 2$.

Section 8.3 contains criteria for validity of the classical maximum modulus principle for second order parabolic systems. For $\ell = 1$, besides the constant $\mathcal{K}(\mathbb{R}^n, T)$, we study the best constant $\mathcal{K}(\Omega, T)$ in the inequality

$$|\boldsymbol{u}(x,t)| \leq \mathcal{K}(\Omega, T) \sup\{|\boldsymbol{u}(y,\tau)| : (y,\tau) \in \overline{\Gamma}_T\},$$

where $(x,t)\in Q_T, \Gamma_T = \{(x,t)\in \partial Q_T : t<T\}$. The closure $\overline{\Gamma}_T$ of Γ_T is called the parabolic boundary of the domain $\Omega\times(0,T)$.

Then we give separate necessary and sufficient conditions for validity of the classical maximum modulus principle (i.e. $\mathcal{K}(\Omega,T)=1$, $\mathcal{K}(\mathbb{R}^n,T)=1$) for solutions of the parabolic second order system

$$\frac{\partial \boldsymbol{u}}{\partial t} - \sum_{j,k=1}^{n} \mathcal{A}_{jk}(x,t)\frac{\partial^2 \boldsymbol{u}}{\partial x_j \partial x_k} + \sum_{j=1}^{n} \mathcal{A}_j(x,t)\frac{\partial \boldsymbol{u}}{\partial x_j} + \mathcal{A}_0(x,t)\boldsymbol{u} = \mathbf{0}. \tag{8.1.2}$$

Without loss of generality one may assume that $\mathcal{A}_{jk}=\mathcal{A}_{kj}$, $1\le j,\ k\le n$.

If the coefficients of system (8.1.2) do not depend on t, then the above mentioned necessary and sufficient conditions coincide. More precisely, the following statement concerning the system

$$\frac{\partial \boldsymbol{u}}{\partial t} - \sum_{j,k=1}^{n} \mathcal{A}_{jk}(x)\frac{\partial^2 \boldsymbol{u}}{\partial x_j \partial x_k} + \sum_{j=1}^{n} \mathcal{A}_j(x)\frac{\partial \boldsymbol{u}}{\partial x_j} + \mathcal{A}_0(x)\boldsymbol{u} = \mathbf{0} \tag{8.1.3}$$

holds for the case of real coefficients.

THEOREM 1. *The classical maximum modulus principle holds for solutions of system* (8.1.3) *in* Q_T ($\mathbb{R}^{n+1}_T$) *if and only if:*

(i) *for all* $x\in\Omega$ ($x\in\mathbb{R}^n$) *the equalities*

$$\mathcal{A}_{jk}(x) = a_{jk}(x)I,\ 1\le j,\ k\le n,$$

hold, where $((a_{jk}))$ *is a positive-definite* $(n\times n)$*- matrix-valued function;*

(ii) *for all* $x\in\Omega$ ($x\in\mathbb{R}^n$) *and for any* $\boldsymbol{\xi}_j,\ \boldsymbol{\zeta}\in\mathbb{R}^m$, $j=1,\dots,n$, *with* $(\boldsymbol{\xi}_j,\boldsymbol{\zeta})=0$, *the inequality*

$$\sum_{j,k=1}^{n} a_{jk}(x)(\boldsymbol{\xi}_j,\boldsymbol{\xi}_k) + \sum_{j=1}^{n}(\mathcal{A}_j(x)\boldsymbol{\xi}_j,\boldsymbol{\zeta}) + (\mathcal{A}_0(x)\boldsymbol{\zeta},\boldsymbol{\zeta}) \ge 0$$

holds.

The next assertion immediately follows from this theorem.

COROLLARY 1. *The classical maximum modulus principle holds for solutions of system* (8.1.3) *in* Q_T ($\mathbb{R}^{n+1}_T$) *if and only if condition* (i) *of Theorem* 1 *is satisfied and for all* $x\in\Omega$ ($x\in\mathbb{R}^n$) *and any* $\boldsymbol{\zeta}\in\mathbb{R}^m$, $|\boldsymbol{\zeta}|=1$, *the inequality*

$$\sum_{i,j=1}^{n} b_{ij}(x)\big[(\mathcal{A}_i(x)\boldsymbol{\zeta},\boldsymbol{\zeta})(\mathcal{A}_j(x)\boldsymbol{\zeta},\boldsymbol{\zeta}) - (\mathcal{A}_i^*(x)\boldsymbol{\zeta},\ \mathcal{A}_j^*(x)\boldsymbol{\zeta})\big] + 4(\mathcal{A}_0(x)\boldsymbol{\zeta},\boldsymbol{\zeta}) \ge 0$$

is valid. Here $((b_{ij}))$ *is the* $(n\times n)$*-matrix-valued function inverse of* $((a_{ij}))$ *and* $*$ *means the passage to the transposed matrix.*

We demonstrate by an example that the scalar character of the principal part of system (8.1.2) everywhere in the domain is not necessary for validity of the maximum principle when coefficients depend both on x and t.

Finally, it is shown that all the facts concerning the maximum modulus principle for solutions of systems with complex coefficients are corollaries of corresponding assertions for systems with real coefficients.

In particular, for the scalar parabolic equation with complex coefficients

$$\frac{\partial u}{\partial t} - \sum_{j,k=1}^{n} a_{jk}(x)\frac{\partial^2 u}{\partial x_j \partial x_k} + \sum_{j=1}^{n} a_j(x)\frac{\partial u}{\partial x_j} + a_0(x)u = 0$$

we obtain the following assertion.

COROLLARY 2. *The classical maximum modulus principle is valid in* Q_T $(\mathbb{R}^{n+1}_T)$ *if and only if:*

(i) *the* $(n \times n)$*-matrix-valued function* $((a_{jk}))$ *is real and positive-definite;*

(ii) *for all* $x \in \Omega (x \in \mathbb{R}^n)$ *the inequality*

$$4\Re\, a_0(x) \geq \sum_{j,k=1}^{n} b_{jk}(x) \Im\, a_j(x) \Im\, a_k(x)$$

holds.

Section 8.4 concerns the parabolic Lamé system

$$\frac{\partial \boldsymbol{u}}{\partial t} - \Delta\, \boldsymbol{u} - \omega \text{ grad div } \boldsymbol{u} = \boldsymbol{0},$$

where $\omega > -1$. For solutions of this system in the half-space $\mathbb{R}^{n+1}_+ = \{(x,t) \in \mathbb{R}^{n+1} : t > 0\}$, $n \geq 3$, we find an expression for the sharp constant $\mathcal{K}$ in the inequality

$$|\boldsymbol{u}(x,t)| \leq \mathcal{K} \sup\{\, |\boldsymbol{u}(y,0)| : y \in \mathbb{R}^n\}.$$

8.2. The Cauchy problem for systems of order 2ℓ

8.2.1. The case of real coefficients

8.2.1.1. *Some notations.* We introduce the operators

$$\mathfrak{P}(x,t,D_x) = \sum_{|\beta| \leq 2\ell} \mathcal{A}_\beta(x,t) D_x^\beta, \qquad \mathfrak{P}_0(x,t,D_x) = \sum_{|\beta| = 2\ell} \mathcal{A}_\beta(x,t) D_x^\beta,$$

where $D_x = (\partial/\partial x_1, \ldots, \partial/\partial x_n)$, $\mathcal{A}_\beta$ are real $(m \times m)$-matrix-valued functions satisfying in the sequel conditions (A) and (B), formulated in Section 8.1.

Below we use the following notations. Let F be a set in the Euclidean space. By $[\mathrm{C_b}(F)]^m$ we denote the space of continuous and bounded m-component vector-valued functions on F with the norm

$$\|\boldsymbol{u}\| = \sup\,\{|\boldsymbol{u}(q)| :\ q \in F\}.$$

Let $\Pi_T = \mathcal{D} \times (0,T]$, where $\mathcal{D}$ is either a bounded domain Ω in $\mathbb{R}^n$ or $\mathcal{D} = \mathbb{R}^n$ and $0 < T \leq \infty$. By $[\mathrm{C}^{(k,1)}(\Pi_T)]^m$ we mean the space of m-component vector-valued functions on Π_T whose derivatives with respect to x up to order k and first derivative with respect to t are continuous. By $[\mathrm{C}^{k,\alpha}_{\mathrm{b}}(\mathbb{R}^n)]^m$ we denote the space of m-component vector-valued functions on $\mathbb{R}^n$ with continuous and bounded derivatives with respect to x up to order k which satisfy the uniform Hölder condition with exponent $\alpha, 0 < \alpha \leq 1$. Let $\mathbb{R}^{n+1}_T = \mathbb{R}^n \times (0,T]$ and $Q_T = \Omega \times (0,T]$. Finally, let $[\mathrm{C}^{k+\alpha,\alpha/2\ell}_{\mathrm{b}}(\overline{\mathbb{R}^{n+1}_T})]^m$ denote the space of m-component vector-valued functions with derivatives up to order k with respect to x which are bounded in $\overline{\mathbb{R}^{n+1}_T}$ and satisfy the uniform Hölder condition on $\overline{\mathbb{R}^{n+1}_T}$ with exponent α with respect to the parabolic distance

$$d\,[(x,t),(x',t')] = \left(|x-x'|^2 + |t-t'|^{1/\ell}\right)^{1/2}.$$

For the space of $(m \times m)$-matrix-valued functions, defined on $\overline{\mathbb{R}^{n+1}_T}$ and having similar properties, we use the notation $[\mathrm{C}^{k+\alpha,\alpha/2\ell}_{\mathrm{b}}(\overline{\mathbb{R}^{n+1}_T})]^{m\times m}$.

For $\ell \geq 1$ we put

$$\mathcal{K}(\mathbb{R}^n, T) = \sup \frac{\|\boldsymbol{u}\|_{[\mathrm{C_b}(\overline{\mathbb{R}^{n+1}_T})]^m}}{\|\boldsymbol{u}|_{t=0}\|_{[\mathrm{C_b}(\mathbb{R}^n)]^m}}, \tag{8.2.1}$$

where the supremum is taken over all functions in the class

$$[\mathrm{C}^{(2\ell,1)}(\mathbb{R}^{n+1}_T)]^m \cap [\mathrm{C_b}(\overline{\mathbb{R}^{n+1}_T})]^m$$

satisfying the system

$$\frac{\partial \boldsymbol{u}}{\partial t} - \mathfrak{P}(x, t, D_x)\boldsymbol{u} = \boldsymbol{0}. \tag{8.2.2}$$

Let $\mathbb{R}^{n+1}_+ = \{(x,t) \in \mathbb{R}^{n+1} : t > 0\}$. We introduce one more function

$$\mathcal{K}_0(y) = \sup \frac{\|\boldsymbol{u}\|_{[\mathrm{C_b}(\overline{\mathbb{R}^{n+1}_+})]^m}}{\|\boldsymbol{u}|_{t=0}\|_{[\mathrm{C_b}(\mathbb{R}^n)]^m}}, \tag{8.2.3}$$

where the supremum is taken over all functions in the class

$$[\mathrm{C}^{(2\ell,1)}(\mathbb{R}^{n+1}_+)]^m \cap [\mathrm{C_b}(\overline{\mathbb{R}^{n+1}_+})]^m$$

satisfying the system

$$\frac{\partial \boldsymbol{u}}{\partial t} - \mathfrak{P}_0(y, 0, D_x)\boldsymbol{u} = \boldsymbol{0}$$

and $y \in \mathbb{R}^n$ plays the role of a parameter.

8.2.1.2. *Representations for the constants $\mathcal{K}(\mathbb{R}^n, T)$ and $\mathcal{K}_0(y)$.* According to Eidel'man [**Ei1**], there exists one and only one function in the class

$$[\mathrm{C}^{(2\ell,1)}(\mathbb{R}^{n+1}_T)]^m \cap [\mathrm{C_b}(\overline{\mathbb{R}^{n+1}_T})]^m$$

which satisfies the Cauchy problem

$$\frac{\partial \boldsymbol{u}}{\partial t} - \mathfrak{P}(x, t, D_x)\boldsymbol{u} = \boldsymbol{0} \quad \text{in} \quad \mathbb{R}^{n+1}_T, \quad \boldsymbol{u}|_{t=0} = \boldsymbol{\psi}, \tag{8.2.4}$$

with $\boldsymbol{\psi} \in [\mathrm{C_b}(\mathbb{R}^n)]^m$. This solution can be represented in the form

$$\boldsymbol{u}(x,t) = \int_{\mathbb{R}^n} G(t, 0, x, \eta)\boldsymbol{\psi}(\eta) d\eta. \tag{8.2.5}$$

Here $G(t, \tau, x, \eta)$ is the Green matrix (or the fundamental matrix of solutions of the Cauchy problem (8.2.4)). The Green matrix for the system

$$\frac{\partial \boldsymbol{u}}{\partial t} - \mathfrak{P}(y, t, D_x)\boldsymbol{u} = \boldsymbol{0}$$

will be denoted by $\mathcal{G}(t, \tau, x - \eta; y)$.

The Green matrix $\mathcal{G}_0(t - \tau, x - \eta; y)$ for the system

$$\frac{\partial \boldsymbol{u}}{\partial t} - \mathfrak{P}_0(y, 0, D_x)\boldsymbol{u} = \boldsymbol{0}$$

has the representation

$$\mathcal{G}_0(t - \tau, x - \eta; y) = (2\pi)^{-n} \int_{\mathbb{R}^n} \exp\Big[(-1)^\ell \sum_{|\beta|=2\ell} \mathcal{A}_\beta(y, 0)\sigma^\beta (t - \tau)\Big] e^{i(x-\eta,\sigma)} d\sigma,$$

where $\sigma = (\sigma_1, \dots, \sigma_n) \in \mathbb{R}^n$. This implies

$$\mathcal{G}_0(t - \tau, x - \eta; y) = (t - \tau)^{-n/2\ell} P\left(\frac{x - \eta}{(t - \tau)^{1/2\ell}}; y\right) \tag{8.2.6}$$

with

$$P(x;y) = (2\pi)^{-n} \int_{\mathbb{R}^n} e^{i(x,\sigma)} \exp\Big[(-1)]^\ell \sum_{|\beta|=2\ell} \mathcal{A}_\beta(y,0)\sigma^\beta\Big] d\sigma.$$

When discussing system (8.2.2) with coefficients depending only on t we use the notation

$$\mathcal{A}_\alpha(t),\ \mathfrak{P}(t,D_x),\ \mathfrak{P}_0(t,D_x),\ \mathcal{G}(t,\tau,x-\eta),\ \mathcal{G}_0(t-\tau,x-\eta),\ P(x).$$

THEOREM 8.1. *The following formula is valid*

$$\mathcal{K}(\mathbb{R}^n,T) = \sup_{x\in\mathbb{R}^n}\ \sup_{0<t\le T}\ \sup_{|\boldsymbol{z}|=1} \int_{\mathbb{R}^n} |G^*(t,0,x,\eta)\boldsymbol{z}|d\eta, \tag{8.2.7}$$

where the star * *denotes the passage to the transposed matrix.*

In particular,

$$\mathcal{K}_0(y) = \sup_{|\boldsymbol{z}|=1} \int_{\mathbb{R}^n} |P^*(\eta;y)\boldsymbol{z}|d\eta. \tag{8.2.8}$$

PROOF. Let (x,t) be a fixed point in $\mathbb{R}^{n+1}_T$. We find the norm $||\boldsymbol{u}(x,t)||$ of the mapping $[\mathrm{C_b}(\mathbb{R}^n)]^m \ni \boldsymbol{\psi} \to \boldsymbol{u}(x,t) \in \mathbb{R}^m$, where $\boldsymbol{u}$ is defined by (8.2.5). Using the properties of the inner product in $\mathbb{R}^m$ and the fact that the supremum operations commute, we obtain

$$\begin{aligned} ||\boldsymbol{u}(x,t)|| &= \sup_{|\boldsymbol{\psi}|\le 1} \Big| \int_{\mathbb{R}^n} G(t,0,x,\eta)\boldsymbol{\psi}(\eta)d\eta \Big| \\ &= \sup_{|\boldsymbol{\psi}|\le 1}\ \sup_{|\boldsymbol{z}|=1} \Big(\boldsymbol{z}, \int_{\mathbb{R}^n} G(t,0,x,\eta)\boldsymbol{\psi}(\eta)d\eta\Big) \\ &= \sup_{|\boldsymbol{z}|=1}\ \sup_{|\boldsymbol{\psi}|\le 1} \int_{\mathbb{R}^n} (\boldsymbol{z}, G(t,0,x,\eta)\boldsymbol{\psi}(\eta))d\eta \\ &= \sup_{|\boldsymbol{z}|=1}\ \sup_{|\boldsymbol{\psi}|\le 1} \int_{\mathbb{R}^n} (G^*(t,0,x,\eta)\boldsymbol{z}, \boldsymbol{\psi}(\eta))d\eta. \end{aligned} \tag{8.2.9}$$

Let $N_{(x,t)}(\boldsymbol{z})$ denote the set of points $\eta \in \mathbb{R}^n$ for which

$$G^*(t,0,x,\eta)\boldsymbol{z} = 0.$$

By (8.2.9) we have

$$||\boldsymbol{u}(x,t)|| = \sup_{|\boldsymbol{z}|=1}\ \sup_{|\boldsymbol{\psi}|\le 1} \int_{\mathbb{R}^n\setminus N_{(x,t)}(\boldsymbol{z})} (G^*(t,0,x,\eta)\boldsymbol{z}, \boldsymbol{\psi}(\eta))d\eta.$$

Clearly, the inside supremum of the last integral is attained at the vector-valued function

$$\frac{G^*(t,0,x,\eta)\boldsymbol{z}}{|G^*(t,0,x,\eta)\boldsymbol{z}|}.$$

Consequently,

$$||\boldsymbol{u}(x,t)|| = \sup_{|\boldsymbol{z}|=1} \int_{\mathbb{R}^n\setminus N_{(x,t)}(\boldsymbol{z})} |G^*(t,0,x,\eta)\boldsymbol{z}|d\eta = \sup_{|\boldsymbol{z}|=1} \int_{\mathbb{R}^n} |G^*(t,0,x,\eta)\boldsymbol{z}|d\eta.$$

Using definition (8.2.1) of the constant $\mathcal{K}(\mathbb{R}^n, T)$, we obtain

$$\begin{aligned}
\mathcal{K}(\mathbb{R}^n,T) &= \sup\Big\{ \|\boldsymbol{u}\|_{[\mathrm{C}(\overline{\mathbb{R}^{n+1}_T})]^m} :\ \|\boldsymbol{u}|_{t=0}\|_{[\mathrm{C}(\mathbb{R}^n)]^m} \le 1, \\
&\quad \partial \boldsymbol{u}/\partial t - \mathfrak{P}(x,t,D_x)\boldsymbol{u} = \boldsymbol{0} \quad \text{in} \quad \mathbb{R}^{n+1}_T \Big\} \\
&= \sup_{x\in\mathbb{R}^n} \sup_{0<t\le T} \sup\Big\{ |\boldsymbol{u}(x,t)| :\ \|\boldsymbol{u}|_{t=0}\|_{[\mathrm{C}(\mathbb{R}^n)]^m} \le 1, \\
&\quad \partial \boldsymbol{u}/\partial t - \mathfrak{P}(x,t,D_x)\boldsymbol{u} = \boldsymbol{0} \quad \text{in} \quad \mathbb{R}^{n+1}_T \Big\} \\
&= \sup_{x\in\mathbb{R}^n} \sup_{0<t\le T} \|\boldsymbol{u}(x,t)\| \\
&= \sup_{x\in\mathbb{R}^n} \sup_{0<t\le T} \sup_{|\boldsymbol{z}|=1} \int_{\mathbb{R}^n} |G^*(t,0,x,\eta)\boldsymbol{z}|\,d\eta
\end{aligned} \tag{8.2.10}$$

which gives representation (8.2.7).

Substituting $\mathcal{G}_0(t-\tau, x-\eta; y)$ given by (8.2.6) into (8.2.7) in place of $G(t,0,x,\eta)$, we arrive at (8.2.8). □

8.2.1.3. *Necessity of the condition $\ell = 1$ for the maximum modulus principle in* $\mathbb{R}^{n+1}_T$. Henceforth all positive constants with non-significant value will be denoted by c with various indices.

LEMMA 8.1. *The inequality*

$$\mathcal{K}(\mathbb{R}^n,T) \ge \sup\big\{\mathcal{K}_0(y) :\ y \in \mathbb{R}^n\big\} \tag{8.2.11}$$

is valid.

PROOF. From (8.2.7) it follows

$$\mathcal{K}(\mathbb{R}^n,T) \ge \lim_{r\to 0} \sup_{y\in\mathbb{R}^n} \sup_{0<t\le T} \sup_{|\boldsymbol{z}|=1} \int_{\mathbb{B}_r(y)} |G^*(t,0,y,\eta)\boldsymbol{z}|\,d\eta, \tag{8.2.12}$$

where $\mathbb{B}_r(y) = \big\{x \in \mathbb{R}^n :\ |x-y| < r\big\}$.

We denote the norm

$$\sup\big\{|L\boldsymbol{z}| :\ \boldsymbol{z} \in \mathbb{R}^m;\ |\boldsymbol{z}| = 1\big\}$$

of an $(m \times m)$-matrix L by $\|L\|$. Since $\mathcal{A}_\beta \in [\mathrm{C}_{\mathrm{b}}^{\alpha,\alpha/2\ell}(\overline{\mathbb{R}^{n+1}_T})]^{m\times m}$, then, according to estimates given in Eidel'man [**Ei1**], Green's matrix $G(t,\tau,y,\eta)$ for $\tau = 0$ admits the representation

$$\begin{aligned}
G(t,0,y,\eta) &= \mathcal{G}_0(t, y-\eta;\ y) \\
&+ [\mathcal{G}_0(t,y-\ \eta;\ \eta) - \mathcal{G}_0(t,y-\eta;\ y)] + W(t,y,\eta),
\end{aligned} \tag{8.2.13}$$

where

$$\begin{aligned}
&\|\mathcal{G}_0(t,y-\eta;\ \eta) - \mathcal{G}_0(t,y-\eta;\ y)\| \\
&\qquad \le c_1|y-\eta|^\alpha t^{-n/2\ell} \exp\Big[-c_2\Big(\frac{|y-\eta|}{t^{1/2\ell}}\Big)^{2\ell/(2\ell-1)}\Big]
\end{aligned} \tag{8.2.14}$$

and

$$\|W(t,y,\eta)\| \le c_3 t^{-(n-\alpha)/2\ell} \exp\Big[-c_4\Big(\frac{|y-\eta|}{t^{1/2\ell}}\Big)^{2\ell/(2\ell-1)}\Big]. \tag{8.2.15}$$

Using (8.2.13)-(8.2.15) and representation (8.2.6), from (8.2.12) we obtain

$$\begin{aligned}
\mathcal{K}(\mathbb{R}^n, T) &\geq \sup_{y\in\mathbb{R}^n} \sup_{|\boldsymbol{z}|=1} \lim_{r\to 0} \limsup_{t\to+0} \Big\{ \int_{\mathbb{B}_r(y)} |\mathcal{G}_0^*(t, y-\eta; y)| d\eta - c_5 r^\alpha - c_6 t^{\alpha/2\ell} \Big\} \\
&= \sup_{y\in\mathbb{R}^n} \sup_{|\boldsymbol{z}|=1} \lim_{r\to 0} \limsup_{t\to+0} \int_{\mathbb{B}_r(y)} t^{-n/2\ell} \left| P^* \left(\frac{y-\eta}{t^{1/2\ell}}; y \right) \boldsymbol{z} \right| d\eta \\
&= \sup_{y\in\mathbb{R}^n} \sup_{|\boldsymbol{z}|=1} \lim_{r\to 0} \lim_{t\to+0} \int_{\mathbb{B}_{r/t^{1/2\ell}}} |P^*(x; y)\boldsymbol{z}| dx \\
&= \sup_{y\in\mathbb{R}^n} \sup_{|\boldsymbol{z}|=1} \int_{\mathbb{R}^n} |P^*(x; y)\boldsymbol{z}| dx.
\end{aligned}$$

This and (8.2.8) yield the inequality for $\mathcal{K}(\mathbb{R}^n, T)$ given in the statement of the Lemma. □

LEMMA 8.2. *If the classical maximum modulus principle holds for solutions of the system*

$$\frac{\partial \boldsymbol{u}}{\partial t} - \sum_{1\leq|\beta|\leq 2\ell} \mathcal{A}_\beta(t) D_x^\beta \boldsymbol{u} = \boldsymbol{0} \tag{8.2.16}$$

in $\mathbb{R}_T^{n+1}$, *then* $\mathcal{A}_\beta(t) = a_\beta(t) I$, *where* a_β *are scalar functions.*

PROOF. *1. The structure of the matrix* $\mathcal{G}$. From (8.2.9) and (8.2.10) it follows that

$$\mathcal{K}(\mathbb{R}^n, T) \geq \sup_{|\boldsymbol{z}|=1} \sup_{|\boldsymbol{\psi}|\leq 1} \int_{\mathbb{R}^n} (\mathcal{G}^*(t, 0, x-\eta)\boldsymbol{z},\ \boldsymbol{\psi}(\eta))\, d\eta, \tag{8.2.17}$$

where (x, t) is an arbitrary point of the layer $\mathbb{R}_T^{n+1}$.

Let $\boldsymbol{z}$ be a fixed unit vector in $\mathbb{R}^m$. Since

$$\int_{\mathbb{R}^n} \mathcal{G}(t, 0, x-\eta) d\eta = I, \tag{8.2.18}$$

(see Eidel'man [**Ei1**]), the set $\mathbb{R}^n \backslash N_{(x,t)}(\boldsymbol{z})$ has non-zero measure for any fixed $\boldsymbol{z} \in \mathbb{R}^m$, $|\boldsymbol{z}| = 1$, and $(x, t) \in \mathbb{R}_T^{n+1}$.

Suppose $\mathcal{K}(\mathbb{R}^n, T) = 1$ and let there exist a unit m-dimensional vector $\boldsymbol{z}_0$ and a set $M \subset \mathbb{R}^n \backslash N_{(x,t)}(\boldsymbol{z}_0)$, $\text{mes}_n M > 0$, such that for all $\eta \in M$ the inequality

$$\boldsymbol{z}_0 \neq \mathcal{G}^*(t, 0, x-\eta)\boldsymbol{z}_0 / |\mathcal{G}^*(t, 0, x-\eta)\boldsymbol{z}_0|$$

holds. Then, using (8.2.17) and (8.2.18), we obtain

$$\begin{aligned}
1 &= \mathcal{K}(\mathbb{R}^n, T) \geq \sup_{|\boldsymbol{\psi}|\leq 1} \int_{\mathbb{R}^n} (\mathcal{G}^*(t, 0, x-\eta)\boldsymbol{z}_0,\ \boldsymbol{\psi}(\eta))\, d\eta \\
&= \sup_{|\boldsymbol{\psi}|\leq 1} \int_{\mathbb{R}^n \backslash N_{(x,t)}(\boldsymbol{z}_0)} (\mathcal{G}^*(t, 0, x-\eta)\boldsymbol{z}_0,\ \boldsymbol{\psi}(\eta))\, d\eta \\
&= \int_{\mathbb{R}^n \backslash N_{(x,t)}(\boldsymbol{z}_0)} \frac{(\mathcal{G}^*(t, 0, x-\eta)\boldsymbol{z}_0,\ \mathcal{G}^*(t, 0, x-\eta)\boldsymbol{z}_0)}{|\mathcal{G}^*(t, 0, x-\eta)\boldsymbol{z}_0|} d\eta \\
&> \int_{\mathbb{R}^n \backslash N_{(x,t)}(\boldsymbol{z}_0)} (\mathcal{G}^*(t, 0, x-\eta)\boldsymbol{z}_0,\ \boldsymbol{z}_0)\, d\eta \\
&= \int_{\mathbb{R}^n} (\mathcal{G}^*(t, 0, x-\eta)\boldsymbol{z}_0,\ \boldsymbol{z}_0)\, d\eta = 1.
\end{aligned}$$

Consequently, if $\mathcal{K}(\mathbb{R}^n, T) = 1$, then, for all $\boldsymbol{z} \in \mathbb{R}^m$ with $|\boldsymbol{z}| = 1$ and for almost all $\eta \in \mathbb{R}^n \backslash N_{(x,t)}(\boldsymbol{z})$, one has

$$\boldsymbol{z} = \mathcal{G}^*(t, 0, x - \eta)\boldsymbol{z}/|\mathcal{G}^*(t, 0, x - \eta)\boldsymbol{z}|. \tag{8.2.19}$$

Let g_{jk}, $j, k = 1, \dots, m$, denote the elements of the matrix $\mathcal{G}$. Setting $\boldsymbol{z}_1 = (1, 0, \dots, 0)^*, \dots, \boldsymbol{z}_m = (0, 0, \dots, 1)^*$ successively instead of $\boldsymbol{z}$ in (8.2.19) and using the continuity of the Green matrix $\mathcal{G}(t, \tau, x - \eta)$ for $t > \tau$, we find that $g_{jk}(t, 0, x - \eta) = 0$ for $j \neq k$ for all $\eta \in \mathbb{R}^n \backslash N_{(x,t)}(\boldsymbol{z}_k)$.

Since $g_{jk}(t, 0, x - \eta) = 0$ for $j = 1, 2, \dots, m$ and for $\eta \in N_{(x,t)}(\boldsymbol{z}_k)$, $k = 1, 2, \dots, m$, we conclude that $g_{jk}(t, 0, x - \eta) = 0$ for $j \neq k$ and for all $\eta \in \mathbb{R}^n$. Now we put $\boldsymbol{z}' = m^{-1/2}(1, 1, \dots, 1)^*$ instead of $\boldsymbol{z}$ in (8.2.19). Then

$$\left[g_{11}^2(t, 0, x - \eta) + \dots + g_{mm}^2(t, 0, x - \eta)\right]^{1/2} = m^{1/2} g_{jj}(t, 0, x - \eta)$$

for all $j = 1, 2, \dots, m$ and for $\eta \in \mathbb{R}^n \backslash N_{(x,t)}(\boldsymbol{z}')$. Hence making use of the equalities $g_{jj}(t, 0, x - \eta) = 0$ for $j = 1, 2, \dots, m$ and for $\eta \in N_{(x,t)}(\boldsymbol{z}')$, we get

$$g_{11}(t, 0, x - \eta) = g_{22}(t, 0, x - \eta) = \dots = g_{mm}(t, 0, x - \eta)$$

for all $\eta \in \mathbb{R}^n$.

Let $g(t, 0, x - \eta) = g_{jj}(t, 0, x - \eta)$, $1 \le j \le m$, and assume that $\mathcal{K}(\mathbb{R}^n, T) = 1$. Then the solution of the Cauchy problem for system (8.2.16) has the form

$$\boldsymbol{u}(x, t) = \int_{\mathbb{R}^n} g(t, 0, x - \eta)\boldsymbol{\psi}(\eta) d\eta, \tag{8.2.20}$$

where $\boldsymbol{\psi}(\eta) = \boldsymbol{u}(\eta, 0)$.

2. The structure of the operator $\mathfrak{A}$. By ψ_0 we denote a scalar function which is continuous and bounded on $\mathbb{R}^n$. Let

$$u_0(x, t) = \int_{\mathbb{R}^n} g(t, 0, x - \eta)\psi_0(\eta) d\eta. \tag{8.2.21}$$

By (8.2.20), the vector-valued function $\boldsymbol{h}_z(x, t) = u_0(x, t)\boldsymbol{z}$, with $\boldsymbol{z} \in \mathbb{R}^m$, is a solution of the Cauchy problem

$$\frac{\partial \boldsymbol{h}_{\boldsymbol{z}}}{\partial t} - \sum_{1 \le |\beta| \le 2\ell} \mathcal{A}_\beta(t) D_x^\beta \boldsymbol{h}_{\boldsymbol{z}} = \boldsymbol{0} \quad \text{in} \quad \mathbb{R}_T^{n+1}, \quad \boldsymbol{h}_{\boldsymbol{z}}\big|_{t=0} = \psi_0 \boldsymbol{z}. \tag{8.2.22}$$

Setting $\boldsymbol{z}_1 = (1, 0, \dots, 0)^*, \dots, \boldsymbol{z}_m = (0, 0, \dots, 1)^*$ successively instead of $\boldsymbol{z}$ in (8.2.22), we obtain m^2 boundary value problems

$$\delta_{pq} \frac{\partial u_0}{\partial t} - \sum_{1 \le |\beta| \le 2\ell} \mathcal{A}_\beta^{(pq)}(t) D_x^\beta u_0 = 0 \quad \text{in} \quad \mathbb{R}_T^{n+1}, \qquad u_0|_{t=0} = \psi_0. \tag{8.2.23}$$

Here δ_{pq} is the Kronecker symbol, $\mathcal{A}_\beta^{(pq)}$ is the element of the matrix $\mathcal{A}_\beta$ placed at the intersection of the p-th row and the q-th column, $p, q = 1, 2, \dots, m$.

Suppose the initial function $\psi_0(\eta)$ in (8.2.21) has a compact support. Since the Green matrix $g(t, \tau, x - \eta)I$ satisfies the inequality

$$\|g(t, \tau, x - \eta)I\| = |g(t, \tau, x - \eta)| \le c_1(t - \tau)^{-n/2\ell} \exp\left[-c_2\left(\frac{|x - \eta|^{2\ell}}{t - \tau}\right)^{1/(2\ell - 1)}\right]$$

(see Eidel'man [**Ei1**]), (8.2.21) implies the estimate

$$|u_0(x,t)| \le c_3(t)\exp(-c_4(t)|x|^{2\ell/(2\ell-1)})$$

for any fixed $t > 0$.

Applying the Fourier transform with respect to the variables $x_1, \dots, x_n$ to equation (8.2.23), we get

$$\delta_{pq}\frac{dF[u_0]}{dt} - F[u_0]\sum_{1\le|\beta|\le 2\ell} i^{|\beta|}\mathcal{A}_\beta^{(pq)}(t)\sigma^\beta = 0. \tag{8.2.24}$$

Let $p \neq q$. Since the function ψ_0 determining u_0 by (8.2.21) is arbitrary, the last equality yields $\mathcal{A}_\beta^{(pq)}(t) = 0$ for all $p \neq q$ and all multi-indices β, $1 \le |\beta| \le 2\ell$.

Suppose now that $p = q$. Integrating equation (8.2.24) and taking into account the initial condition $F[u_0] = F[\psi_0]$ for $t = 0$, we find

$$F[u_0] = F[\psi_0]\exp\Big[\sum_{1\le|\beta|\le 2\ell} i^{|\beta|}\sigma^\beta\int_0^t \mathcal{A}_\beta^{(pq)}(\tau)d\tau\Big].$$

Therefore,

$$\begin{aligned} &F[\psi_0]\exp\Big[\sum_{1\le|\beta|\le 2\ell} i^{|\beta|}\sigma^\beta\int_0^t \mathcal{A}_\beta^{(pp)}(\tau)d\tau\Big] \\ = \; &F[\psi_0]\exp\Big[\sum_{1\le|\beta|\le 2\ell} i^{|\beta|}\sigma^\beta\int_0^t \mathcal{A}_\beta^{(qq)}(\tau)d\tau\Big], \end{aligned}$$

where $p, q = 1, 2, \dots, m$. Hence $\mathcal{A}_\beta^{(pp)}(t) = \mathcal{A}_\beta^{(qq)}(t)$ for all $p, q = 1, 2, \dots, m$ and for all multi-indices β, $1 \le |\beta| \le 2\ell$. Thus, if $\mathcal{K}(\mathbb{R}^n, T) = 1$, then the operator $\partial/\partial t - \mathfrak{P}(t; D_x)$ on the left-hand side of (8.2.16) satisfies the equality

$$\frac{\partial}{\partial t} - \mathfrak{P}(t, D_x) = \frac{\partial}{\partial t} - \sum_{1\le|\beta|\le 2\ell} a_\beta(t)D_x^\beta,$$

where a_β are scalar functions. □

LEMMA 8.3. *Let the parabolic system have the form*

$$\frac{\partial \boldsymbol{u}}{\partial t} - \sum_{|\beta|=2\ell} a_\beta D_x^\beta \boldsymbol{u} = \boldsymbol{0}, \tag{8.2.25}$$

where a_β are constant scalar coefficients. Then the equality $\mathcal{K}(\mathbb{R}^n, T) = 1$ holds if and only if $\ell = 1$.

PROOF. *Sufficiency* of the condition $\ell = 1$ for the equality

$$\mathcal{K}(\mathbb{R}^n, T) = 1$$

follows from the positivity of the fundamental solution $g_0(t-\tau, x-\eta)$ of the Cauchy problem for the parabolic second order equation with constant real coefficients

$$\frac{\partial u}{\partial t} - \sum_{j,k=1}^n a_{jk}\frac{\partial^2 u}{\partial x_j \partial x_k} = 0$$

and from the formula

$$\boldsymbol{u}(x,t) = \int_{\mathbb{R}^n} g_0(t, x-\eta)\boldsymbol{\psi}(\eta)d\eta \tag{8.2.26}$$

for the solution of the Cauchy problem

$$\frac{\partial \boldsymbol{u}}{\partial t} - \sum_{j,k=1}^{n} a_{jk}\frac{\partial^2 \boldsymbol{u}}{\partial x_j \partial x_k} = \boldsymbol{0} \quad \text{in} \quad \mathbb{R}^{n+1}_T,\ \boldsymbol{u}(x,0) = \boldsymbol{\psi}(x),$$

where $\boldsymbol{u}$ and $\boldsymbol{\psi}$ are m-component vector-valued functions, $\boldsymbol{\psi} \in [\mathrm{C_b}(\mathbb{R}^n)]^m$.

Necessity. The solution of the Cauchy problem for the system (8.2.25) can be expressed by (8.2.26) for any ℓ, where

$$g_0(t-\tau, x-\eta) = (t-\tau)^{-n/2\ell} P\left(\frac{x-\eta}{(t-\tau)^{1/2\ell}}\right)$$

is the fundamental solution of the Cauchy problem for the scalar equation (8.2.25), i.e. for $m = 1$. Consequently, the constant

$$\mathcal{K}(\mathbb{R}^n, T) = \int_{\mathbb{R}^n} |g_0(t, x-\eta)|d\eta = \int_{\mathbb{R}^n} |P(\eta)|d\eta$$

does not depend on m and one can put $m = 1$.

We note that the solution of the Cauchy problem

$$\frac{\partial v}{\partial t} - a_{1\dots 1}\frac{\partial^{2\ell} v}{\partial x_1^{2\ell}} = 0 \quad \text{in} \quad \mathbb{R}^2_T, \tag{8.2.27}$$

$$v(x_1, 0) = \psi(x_1),$$

where sign $a_{1\dots 1} = (-1)^{\ell-1}$ and $v = v(x_1, t)$ is a scalar function, also satisfies the Cauchy problem for the scalar equation (8.2.25) with the initial condition $u(x,0) = \psi(x_1)$. Therefore,

$$\mathcal{K}(\mathbb{R}^n, T) \ge k = \sup\left(\|v\|_{\mathrm{C}(\overline{\mathbb{R}^2_{\mathrm{T}}})} / \|v|_{t=0}\|_{\mathrm{C}(\mathbb{R})}\right),$$

where the supremum is taken over all solutions of (8.2.27) in the class

$$\mathrm{C}^{(2\ell,1)}(\mathbb{R}^2_{\mathrm{T}}) \cap \mathrm{C}(\overline{\mathbb{R}^2_{\mathrm{T}}}).$$

Henceforth in this lemma x_1 is denoted by x. We set

$$a = |a_{1\dots 1}|^{1/2\ell}.$$

The solution of the Cauchy problem

$$\frac{\partial v}{\partial t} + (-1)^s a^{2\ell}\frac{\partial^{2\ell} v}{\partial x^{2\ell}} = 0 \text{ in } \mathbb{R}^2_T,\ v(x,0) = \psi(x),$$

$\psi \in \mathrm{C}(\mathbb{R})$, has the form

$$v(x,t) = t^{-1/2\ell}\int_{-\infty}^{\infty} P\left(\frac{x-\eta}{t^{1/2\ell}}\right)\psi(\eta)d\eta, \tag{8.2.28}$$

where

$$\begin{aligned} P(x) &= \frac{1}{2\pi}\int_{-\infty}^{\infty} e^{ix\sigma}\exp(-a^{2\ell}|\sigma|^{2\ell})d\sigma = \frac{1}{\pi}\int_0^{\infty}\cos(x\sigma)\exp(-a^{2\ell}\sigma^{2\ell})d\sigma \\ &= \frac{1}{\pi a}\int_0^{\infty}\cos\left(\frac{x\theta}{a}\right)\exp(-\theta^{2\ell})d\theta. \end{aligned}$$

From (8.2.28) it follows that

$$k = \int_{-\infty}^{+\infty} |P(\eta)| d\eta.$$

Since

$$t^{-1/2\ell} \int_{-\infty}^{\infty} P\left(\frac{x-\eta}{t^{1/2\ell}}\right) d\eta = \int_{-\infty}^{\infty} P(\eta) d\eta = 1,$$

we have $k > 1$ in case $P(\eta)$ changes sign.

We show that $P(\eta)$ is a function with alternating signs. Put

$$\mathcal{F}_\lambda(r) = \int_0^\infty \cos(r\theta) \exp(-\theta^\lambda) d\theta, \tag{8.2.29}$$

where $r \geq 0$, $\lambda \geq 1$. Then $P(x) = (\pi a)^{-1} \mathcal{F}_{2\ell}(x/a)$. Since $\mathcal{F}_\lambda(0) > 0$, then $\mathcal{F}_\lambda(r) > 0$ with $r \in (0, \varepsilon]$ for some $\varepsilon > 0$. We verify that $\mathcal{F}_\lambda(3\pi/2) < 0$ for $\lambda \geq 4$.

Integrating by parts in (8.2.29), we find

$$\begin{aligned} \mathcal{F}_\lambda(r) &= \frac{1}{r} \int_0^\infty \exp(-\theta^\lambda)\, d\left(\sin(r\theta)\right) \\ &= \frac{1}{r} \exp(-\theta^\lambda) \sin(r\theta)\Big|_0^\infty + \frac{\lambda}{r} \int_0^\infty \theta^{\lambda-1} \exp(-\theta^\lambda) \sin(r\theta) d\theta \\ &= \frac{\lambda}{r} \int_0^\infty \theta^{\lambda-1} \exp(-\theta^\lambda) \sin(r\theta) d\theta = \frac{\lambda}{r}\big(A_\lambda(r) + B_\lambda(r)\big), \end{aligned}$$

where

$$A_\lambda(r) = \int_0^2 \theta^{\lambda-1} \exp(-\theta^\lambda) \sin(r\theta) d\theta,$$

$$B_\lambda(r) = \int_2^\infty \theta^{\lambda-1} \exp(-\theta^\lambda) \sin(r\theta) d\theta.$$

The maximum value of the function $f(\theta) = \theta^{\lambda-1} \exp(-\theta^\lambda)$ is attained at $\theta_0 = ((\lambda-1)/\lambda)^{1/\lambda} < 1$. Hence $B_\lambda(3\pi/2) < 0$ for $\lambda \geq 1$, so it suffices to show that $A_\lambda(3\pi/2) < 0$ when $\lambda \geq 4$.

Estimating each of four integrals in the equality

$$\begin{aligned} A_\lambda(3\pi/2) = &\int_0^{2/3} \theta^{\lambda-1} \exp(-\theta^\lambda) \sin(3\pi\theta/2) d\theta \\ &+ \int_{2/3}^1 \theta^{\lambda-1} \exp(-\theta^\lambda) \sin(3\pi\theta/2) d\theta + \int_1^{4/3} \theta^{\lambda-1} \exp(-\theta^\lambda) \sin(3\pi\theta/2) d\theta \\ &+ \int_{4/3}^2 \theta^{\lambda-1} \exp(-\theta^\lambda) \sin(3\pi\theta/2) d\theta, \end{aligned}$$

we obtain

$$\begin{aligned}
A_\lambda(3\pi/2) \ &\leq \ \int_0^{3/2} \theta^{\lambda-1}\exp(-\theta^\lambda)d\theta + \left(\frac{2}{3}\right)^{\lambda-1}\exp(-(2/3)^\lambda)\int_{2/3}^1 \sin(3\pi\theta/2)d\theta \\
&+ \ \left(\frac{4}{3}\right)^{\lambda-1}\exp(-(4/3)^\lambda)\int_1^{4/3}\sin(3\pi\theta/2)d\theta + \int_{4/3}^2 \theta^{\lambda-1}\exp(-\theta^\lambda)d\theta \\
&= \ \frac{1}{\lambda}\Big(1-\exp(-(2/3)^\lambda)\Big) - (2/3\pi)(2/3)^{\lambda-1}\exp(-(2/3)^\lambda) \\
&- \ (4/3)^{\lambda-1}(2/3\pi)\exp(-(4/3)^\lambda) + \frac{1}{\lambda}(\exp(-(4/3)^\lambda) - \exp(-2^\lambda)) \\
&\leq \ (2/3)^\lambda\left(\frac{1}{\lambda} - \frac{1}{\pi}\exp(-(2/3)^\lambda)\right) \\
&+ \ \left(\frac{1}{\lambda} - (4/3)^{\lambda-1}(2/3\pi)\right)\exp(-(4/3)^\lambda) - \frac{1}{\lambda}\exp(-2^\lambda).
\end{aligned}$$

The functions

$$f_1(\lambda) = \frac{1}{\lambda} - \frac{1}{\pi}\exp(-(2/3)^\lambda), \ f_2(\lambda) = \frac{1}{\lambda} - \left(\frac{2}{3\pi}\right)\left(\frac{4}{3}\right)^{\lambda-1}$$

are decreasing as λ increases. Since $f_1(4) < 0, \ f_2(4) < 0$, then $A_\lambda(3\pi/2) < 0$ for $\lambda \geq 4$. Taking into account that

$$\mathcal{F}_\lambda(r) = (\lambda/r)\big(A_\lambda(r) + B_\lambda(r)\big), \quad P(x) = (\pi a)^{-1}\mathcal{F}_{2s}(x/a),$$

$$B_\lambda(3\pi/2) < 0 \quad \text{for} \quad \lambda \geq 1, \quad \text{and} \quad \mathcal{F}_\lambda(0) > 0,$$

we arrive at the conclusion that P changes sign for $s \geq 2$. Thus, $\mathcal{K}(\mathbb{R}^n, T) \geq k > 1$ for $s \geq 2$. □

THEOREM 8.2. *The classical maximum modulus principle does not hold for solutions of the system*

$$\frac{\partial \boldsymbol{u}}{\partial t} - \mathfrak{P}(x, t, D_x)\boldsymbol{u} = \boldsymbol{0} \quad \text{in} \quad \mathbb{R}^{n+1}_T$$

if $\ell > 1$.

PROOF. Lemma 8.2 implies that the equality $\mathcal{K}_0(y) = 1$ is valid for the system

$$\frac{\partial \boldsymbol{u}}{\partial t} - \sum_{|\beta|=2\ell} \mathcal{A}_\beta(y, 0)D_x^\beta \boldsymbol{u} = \boldsymbol{0}$$

if and only if $\mathcal{A}_\beta(y, 0) = a_\beta(y, 0)I$.

By Lemma 8.3 we have $\mathcal{K}_0(y) > 1$ for the system

$$\frac{\partial \boldsymbol{u}}{\partial t} - \sum_{|\beta|=2\ell} a_\beta(y, 0)D_x^\beta \boldsymbol{u} = \boldsymbol{0},$$

with $\ell > 1$ which together with (8.2.11) completes the proof of the theorem. □

8.2.2. The case of complex coefficients

In this subsection we extend basic results of Subsection 8.2.1 to system (8.2.1) with complex coefficients and with solutions $\boldsymbol{u} = \boldsymbol{v} + i\boldsymbol{w}$, where $\boldsymbol{v}$ and $\boldsymbol{w}$ are m-component vector-valued functions with real-valued components.

For spaces of vector-valued functions with complex components we retain the same notation as in the case of real components but use bold. The same relates notation for the spaces of matrix-valued functions.

We introduce the operators

$$\mathfrak{L}(x,t,D_x) = \sum_{|\beta|\le 2\ell} \mathcal{A}_\beta(x,t)D_x^\beta, \qquad \mathfrak{L}_0(x,t,D_x) = \sum_{|\beta|=2\ell} \mathcal{A}_\beta(x,t)D_x^\beta,$$

where $\mathcal{A}_\beta$ are $(m\times m)$-matrix-valued functions with complex elements satisfying conditions (A) and (B), formulated in Section 8.1.

Let $\mathcal{R}_\beta$ and $\mathcal{H}_\beta$ be real $(m\times m)$-matrix-valued functions defined on $\overline{\mathbb{R}^{n+1}_T}$ such that $\mathcal{A}_\beta(x,t) = \mathcal{R}_\beta(x,t) + i\mathcal{H}_\beta(x,t)$. We use the following notation

$$\mathfrak{R}(x,t,D_x) = \sum_{|\beta|\le 2\ell} \mathcal{R}_\beta(x,t)D_x^\beta, \qquad \mathfrak{H}(x,t;D_x) = \sum_{|\beta|\le 2\ell} \mathcal{H}_\beta(x,t)D_x^\beta,$$

$$\mathfrak{R}_0(x,t,D_x) = \sum_{|\beta|=2\ell} \mathcal{R}_\beta(x,t)D_x^\beta, \qquad \mathfrak{H}_0(x,t,D_x) = \sum_{|\beta|=2\ell} \mathcal{H}_\beta(x,t)D_x^\beta.$$

Separating the real and imaginary parts of the system

$$\frac{\partial \boldsymbol{u}}{\partial t} - \mathfrak{L}(x,t,D_x)\boldsymbol{u} = \boldsymbol{0},$$

we get the system with real coefficients, which like the original system is uniformly parabolic in $\overline{\mathbb{R}^{n+1}_T}$

$$\frac{\partial \boldsymbol{v}}{\partial t} - \mathfrak{R}(x,t,D_x)\boldsymbol{v} + \mathfrak{H}(x,t,D_x)\boldsymbol{w} = \boldsymbol{0},$$
$$\frac{\partial \boldsymbol{w}}{\partial t} - \mathfrak{H}(x,t,D_x)\boldsymbol{v} - \mathfrak{R}(x,t,D_x)\boldsymbol{w} = \boldsymbol{0}.$$

REMARK 8.1. The preservation of the uniform parabolicity in the sense of Petrovskiĭ under the passage from a system with complex coefficients to a system with real coefficients is a corollary of the following simple algebraic property.

Let Q be an $(m\times m)$-matrix with R and H being its real and imaginary parts, respectively. Let the eigenvalues λ of Q satisfy $\Re\lambda < -\delta$, $\delta > 0$. Then for the eigenvalues μ of the matrix

$$\Lambda = \begin{pmatrix} R & -H \\ H & H \end{pmatrix}$$

the inequality $\Re\mu < -\delta$ holds. In fact,

$$\begin{aligned} \det(\Lambda - \mu E) &= \det\begin{pmatrix} R-\mu I & -H \\ H & R-\mu I \end{pmatrix} = \det\begin{pmatrix} Q-\mu I & i(Q-\mu I) \\ H & R-\mu I \end{pmatrix} \\ &= \det\begin{pmatrix} Q-\mu I & 0 \\ H & \overline{Q}-\mu I \end{pmatrix} = \det(Q-\mu I)\ \det(\overline{Q}-\mu I), \end{aligned}$$

where $\overline{Q}$ is a matrix whose elements are complex conjugate of corresponding elements of Q, and E is the identity matrix of order $2m$. Since

$$\det(\overline{Q}-\mu I) = \det(\overline{Q-\overline{\mu}I}) = \overline{\det(Q-\overline{\mu}I)},$$

we have $\Re\mu < -\delta$.

We introduce the matrix differential operators

$$\mathfrak{K}(x,t,D_x) = \begin{pmatrix} \mathfrak{R}(x,t,D_x) & -\mathfrak{H}(x,t,D_x) \\ \mathfrak{H}(x,t;D_x) & \mathfrak{R}(x,t;D_x) \end{pmatrix},$$

$$\mathfrak{K}_0(x,t,D_x) = \begin{pmatrix} \mathfrak{R}_0(x,t,D_x) & -\mathfrak{H}_0(x,t,D_x) \\ \mathfrak{H}_0(x,t,D_x) & \mathfrak{R}_0(x,t,D_x) \end{pmatrix}.$$

Let $G'(t,\tau,x,\eta)$ and $\mathcal{G}'_0(t-\tau,x-\eta;y)$ denote the fundamental matrices of solutions of the Cauchy problem for the systems

$$\left(\frac{\partial}{\partial t} - \mathfrak{K}(x,t,D_x)\right)\begin{pmatrix} \boldsymbol{v} \\ \boldsymbol{w} \end{pmatrix} = \mathbf{0} \quad \text{and} \quad \left(\frac{\partial}{\partial t} - \mathfrak{K}_0(y,0,D_x)\right)\begin{pmatrix} \boldsymbol{v} \\ \boldsymbol{w} \end{pmatrix} = \mathbf{0},$$

respectively. Further, let $P'(x;y)$ be the $(2m \times 2m)$-matrix-valued function in the representation

$$\mathcal{G}'_0(t-\tau,x-\eta;y) = (t-\tau)^{-n/2\ell} P'\left(\frac{x-\eta}{(t-\tau)^{1/2\ell}};y\right).$$

Let F be a set in the Euclidean space. The norm in the space $[\mathbf{C_b}(F)]^m$ of continuous and bounded on F vector-valued functions $\boldsymbol{u} = \boldsymbol{v} + i\boldsymbol{w}$ with m complex components is defined by the equality

$$\|\boldsymbol{u}\| = \sup\Big\{(|\boldsymbol{v}(q)|^2 + |\boldsymbol{w}(q)|^2)^{1/2} : q \in F\Big\}.$$

As in the definition of $\mathcal{K}(\mathbb{R}^n,T)$ we put

$$\mathcal{K}'(\mathbb{R}^n,T) = \sup \frac{\|u\|_{[\mathbf{C_b}(\overline{\mathbb{R}^{n+1}_T})]^m}}{\|u|_{t=0}\|_{[\mathbf{C_b}(\mathbb{R}^n)]^m}},$$

where the supremum is taken over all functions $\boldsymbol{u} = \boldsymbol{v} + i\boldsymbol{w}$ in the class

$$[\mathbf{C}^{(2\ell,1)}(\mathbb{R}^{n+1}_T)]^m \cap [\mathbf{C_b}(\overline{\mathbb{R}^{n+1}_T})]^m$$

satisfying the system

$$\frac{\partial \boldsymbol{u}}{\partial t} - \mathfrak{L}(x,t,D_x)\boldsymbol{u} = \mathbf{0}.$$

We define the function

$$\mathcal{K}'_0(y) = \sup \frac{\|u\|_{[\mathbf{C_b}(\overline{\mathbb{R}^{n+1}_+})]^m}}{\|u|_{t=0}\|_{[\mathbf{C_b}(\mathbb{R}^n)]^m}},$$

where the supremum is taken over all solutions $\boldsymbol{u} = \boldsymbol{v} + i\boldsymbol{w}$ of the system

$$\frac{\partial \boldsymbol{u}}{\partial t} - \mathfrak{L}_0(y,0,D_x)\boldsymbol{u} = \mathbf{0}$$

in the class $[\mathbf{C}^{(2\ell,1)}(\mathbb{R}^{n+1}_+)]^m \cap [\mathbf{C_b}(\overline{\mathbb{R}^{n+1}_+})]^m$ and $y \in \mathbb{R}^n$ plays the role of a parameter.

Clearly, the constant $\mathcal{K}'(\mathbb{R}^n,T)$ for the system

$$\frac{\partial \boldsymbol{u}}{\partial t} - \mathfrak{L}(x,t,D_x)\boldsymbol{u} = \mathbf{0}$$

coincides with the constant $\mathcal{K}(\mathbb{R}^n,T)$ for the system

$$\left(\frac{\partial}{\partial t} - \mathfrak{K}(x,t,D_x)\right)\begin{pmatrix} \boldsymbol{v} \\ \boldsymbol{w} \end{pmatrix} = \mathbf{0}.$$

Therefore, all the assertions concerning $\mathcal{K}'(\mathbb{R}^n, T)$ are immediate corollaries of analogous assertions about $\mathcal{K}(\mathbb{R}^n, T)$. Taking this into account, we obtain the following three assertions from Theorems 8.1, 8.2 and Lemma 8.1.

THEOREM 8.3. *The following formula is valid*

$$\mathcal{K}'(\mathbb{R}^n, T) = \sup_{x\in\mathbb{R}^n} \sup_{0<t\leq T} \sup_{\{\boldsymbol{z}\in\mathbb{R}^{2m}:|\boldsymbol{z}|=1\}} \int_{\mathbb{R}^n} |(G')^*(t,0,x,\eta)\boldsymbol{z}|d\eta,$$

where $*$ *means the passage to the transposed matrix.*

In particular,

$$\mathcal{K}'_0(y) = \sup_{\{\boldsymbol{z}\in\mathbb{R}^{2m}:|\boldsymbol{z}|=1\}} \int_{\mathbb{R}^n} |(P')^*(\eta;y)\boldsymbol{z}|d\eta.$$

LEMMA 8.4. *The inequality*

$$\mathcal{K}'(\mathbb{R}^n, T) \geq \sup\{\mathcal{K}'_0(y) : \ y \in \mathbb{R}^n\}$$

holds.

THEOREM 8.4. *The classical maximum modulus principle fails for solutions of the system*

$$\frac{\partial \boldsymbol{u}}{\partial t} - \mathfrak{L}(x,t,D_x)\boldsymbol{u} = \boldsymbol{0} \quad in \quad \mathbb{R}^{n+1}_T$$

if $\ell > 1$.

8.3. Second order systems

8.3.1. The case of real coefficients

8.3.1.1. *Necessary conditions.* In this subsection we study the validity of the classical maximum modulus principle for system (8.1.2) with coefficients $\mathcal{A}_{jk} = \mathcal{A}_{kj}$, $\mathcal{A}_j$, $\mathcal{A}_0 (1 \leq j, k \leq n)$ that are real $(m \times m)$-matrix-valued functions in the layer $\overline{\mathbb{R}^{n+1}_T}$ and in the cylinder $\overline{Q}_T = \overline{\Omega} \times [0, T]$, where Ω is a bounded domain in $\mathbb{R}^n$. Let $\Gamma_T = \{(x,t) \in \partial Q_T : t < T\}$.

We retain all notations introduced in Section 8.2 for arbitrary $\ell \geq 1$ and introduce the constant

$$\mathcal{K}(\Omega, T) = \sup \frac{\|\boldsymbol{u}\|_{[\mathrm{C}(\overline{Q}_T)]^m}}{\|\boldsymbol{u}|_{\overline{\Gamma}_T}\|_{[\mathrm{C}(\overline{\Gamma}_T)]^m}}, \tag{8.3.1}$$

where the supremum is taken over all solutions of system (8.1.2) in the class $[\mathrm{C}^{(2,1)}(Q_T)]^m \cap [\mathrm{C}(\overline{Q}_T)]^m$.

LEMMA 8.5. *The inequality*

$$\mathcal{K}(\Omega, T) \geq \sup\{\mathcal{K}_0(y) : \ y \in \Omega\} \tag{8.3.2}$$

holds.

PROOF. Let y be an arbitrary point of Ω and let the radius of the ball $\mathbb{B}_r(y)$ be so small that $\overline{\mathbb{B}_r(y)} \subset \Omega$. Further, let

$$\boldsymbol{\psi}_\varepsilon \in [\mathrm{C}_{\mathrm{b}}(\mathbb{R}^n)]^m, \ \ |\boldsymbol{\psi}_\varepsilon(x)| \leq 1, \ \ \operatorname{supp} \boldsymbol{\psi}_\varepsilon \subset \mathbb{B}_\varepsilon(y), \ \ 0 < \varepsilon \leq r/2.$$

The vector-valued function

$$\boldsymbol{u}_\varepsilon(x,t) = \int_{\mathbb{R}^n} G(t,0,x,\eta)\boldsymbol{\psi}_\varepsilon(\eta)d\eta = \int_{\mathbb{B}_\varepsilon(y)} G(t,0,x,\eta)\boldsymbol{\psi}_\varepsilon(\eta)d\eta \tag{8.3.3}$$

satisfies the Cauchy problem

$$\frac{\partial \boldsymbol{u}_\varepsilon}{\partial t} - \sum_{j,k=1}^{n} \mathcal{A}_{jk}(x,t)\frac{\partial^2 \boldsymbol{u}_\varepsilon}{\partial x_j \partial x_k} + \sum_{j=1}^{n} \mathcal{A}_j(x,t)\frac{\partial \boldsymbol{u}_\varepsilon}{\partial x_j} + \mathcal{A}_0(x,t)\boldsymbol{u}_\varepsilon = \boldsymbol{0} \quad \text{in} \quad \mathbb{R}^{n+1}_T,$$

$$\boldsymbol{u}_\varepsilon(x,0) = \boldsymbol{\psi}_\varepsilon(x).$$

Since $\mathcal{A}_{jk}$, $\mathcal{A}_j$, $\mathcal{A}_0 \in [\mathrm{C}_{\mathrm{b}}^{\alpha,\alpha/2}(\overline{\mathbb{R}^{n+1}_T})]^{m\times m}$, it follows by Eidel'man [**Ei1**], that

$$\|G(t,0,x,\eta)\| \leq c_1 t^{-n/2}\exp\left(-c_2\frac{|x-\eta|^2}{t}\right). \tag{8.3.4}$$

The last estimate, used for $(x,t) \in (\overline{\Omega}\backslash\mathbb{B}_r(y)) \times (0,T]$, yields

$$\begin{aligned} ||\boldsymbol{u}_\varepsilon(x,t)|| &\leq \int_{\mathbb{B}_\varepsilon(y)} \|G(t,0,x,\eta)\|d\eta \leq c_1 t^{-n/2}\int_{\mathbb{B}_\varepsilon(y)} \exp\left(-c_2\frac{|x-\eta|^2}{t}\right)d\eta \\ &\leq c_3\varepsilon^n t^{-n/2}\exp\left(-c_4\frac{r^2}{t}\right) \end{aligned}$$

which implies

$$||\boldsymbol{u}_\varepsilon(x,t)|| \leq c_5\varepsilon^n,$$

where $||\boldsymbol{u}_\varepsilon(x,t)||$ is the norm of the mapping $\boldsymbol{\psi}_\varepsilon \to \boldsymbol{u}_\varepsilon(x,t)$. Hence, the following estimate is valid for a sufficiently small ε

$$\sup\{||\boldsymbol{u}_\varepsilon(x,t)|| : (x,t) \in \partial\Omega \times [0,T]\} \leq 1. \tag{8.3.5}$$

From this and definition (8.3.1) of the constant $\mathcal{K}(\Omega,T)$ we obtain

$$\mathcal{K}(\Omega,T) \geq \sup_{y\in\Omega}\ \limsup_{\varepsilon\to 0}\ \limsup_{t\to+0}\ ||\boldsymbol{u}_\varepsilon(y,t)||. \tag{8.3.6}$$

Using (8.3.3) and (8.3.4), in the same way as in the proof of Theorem 8.1, we find

$$||\boldsymbol{u}_\varepsilon(y,t)|| = \sup_{|\boldsymbol{z}|=1}\int_{\mathbb{B}_\varepsilon(y)} |G^*(t,0,y,\eta)\boldsymbol{z}|d\eta.$$

The last equality and (8.3.6) yield

$$\begin{aligned} \mathcal{K}(\Omega,T) &\geq \sup_{y\in\Omega}\ \lim_{\varepsilon\to 0}\ \limsup_{t\to+0}\ \sup_{|\boldsymbol{z}|=1}\int_{\mathbb{B}_\varepsilon(y)} |G^*(t,0,y,\eta)\boldsymbol{z}|d\eta \\ &\geq \sup_{y\in\Omega}\ \sup_{|\boldsymbol{z}|=1}\ \lim_{\varepsilon\to 0}\ \limsup_{t\to+0}\int_{\mathbb{B}_\varepsilon(y)} |G^*(t,0,y,\eta)\boldsymbol{z}|d\eta. \end{aligned} \tag{8.3.7}$$

It was shown in Lemma 8.1 that

$$\begin{aligned} &\sup_{|\boldsymbol{z}|=1}\ \lim_{\varepsilon\to 0}\ \limsup_{t\to+0}\int_{\mathbb{B}_\varepsilon(y)} |G^*(t,0,y,\eta)\boldsymbol{z}|d\eta \\ &\qquad = \sup_{|\boldsymbol{z}|=1}\ \lim_{\varepsilon\to 0}\ \limsup_{t\to+0}\int_{\mathbb{B}_\varepsilon(y)} |\mathcal{G}_0^*(t,y-\eta;y)\boldsymbol{z}|d\eta = \mathcal{K}_0(y) \end{aligned}$$

which together with (8.3.7) gives the lower estimate for $\mathcal{K}(\Omega,T)$ in the statement of the lemma. □

THEOREM 8.5. *The classical maximum modulus principle holds for solutions of the system*

$$\frac{\partial \boldsymbol{u}}{\partial t} - \sum_{j,k=1}^{n} \mathcal{A}_{jk}(t)\frac{\partial^2 \boldsymbol{u}}{\partial x_j \partial x_k} + \sum_{j=1}^{n} \mathcal{A}_j(t)\frac{\partial \boldsymbol{u}}{\partial x_j} = \boldsymbol{0} \tag{8.3.8}$$

in $\mathbb{R}_T^{n+1}$ *if and only if*

$$\mathcal{A}_{jk}(t) = a_{jk}(t)I, \ \ \mathcal{A}_j(t) = a_j(t)I, \ \ 1 \le j, \ k \le n,$$

where $((a_{jk}))$ *is a positive-definite* $(n \times n)$*-matrix-valued function and* a_j *are scalar functions.*

PROOF. The necessity of the above equalities follows from Lemma 8.2. We show that $\mathcal{K}(\mathbb{R}^n, T) = 1$ under the conditions of the theorem.

Consider the Cauchy problem

$$\frac{\partial \boldsymbol{u}}{\partial t} - \sum_{j,k=1}^{n} a_{jk}(t)\frac{\partial^2 \boldsymbol{u}}{\partial x_j \partial x_k} + \sum_{j=1}^{n} a_j(t)\frac{\partial \boldsymbol{u}}{\partial x_j} = \boldsymbol{0} \quad \text{in} \quad \mathbb{R}_T^{n+1}, \tag{8.3.9}$$

$\boldsymbol{u}|_{t=0} = \boldsymbol{\psi}$, where $\boldsymbol{\psi} \in [\mathrm{C_b}(\mathbb{R}^n)]^m$. The solution has the form

$$\boldsymbol{u}(x,t) = \int_{\mathbb{R}^n} g(t,0,x-\eta)\boldsymbol{\psi}(\eta)d\eta,$$

where $g(t,\tau,x-\eta)$ is the fundamental solution of the Cauchy problem

$$\frac{\partial u}{\partial t} - \sum_{j,k=1}^{n} a_{jk}(t)\frac{\partial^2 u}{\partial x_j \partial x_k} + \sum_{j=1}^{n} a_j(t)\frac{\partial u}{\partial x_j} = 0 \quad \text{in} \quad \mathbb{R}_T^{n+1},$$

$u|_{t=0} = \psi$, where $\psi \in \mathrm{C_b}(\mathbb{R}^n)$.

Substituting $g(t,0,x-\eta)I$ in place of $G(t,0,x-\eta)$ in (8.2.7), we obtain

$$\mathcal{K}(\mathbb{R}^n, T) = \sup_{x \in \mathbb{R}^n} \sup_{0<t\le T} \int_{\mathbb{R}^n} |g(t,0,x-\eta)|d\eta = \sup_{0<t\le T} \int_{\mathbb{R}^n} |g(t,0,\xi)|d\xi$$

which means that $\mathcal{K}(\mathbb{R}^n, T)$ does not depend on m. Since $\mathcal{K}(\mathbb{R}^n, T) = 1$ for $m = 1$ we arrive at the sufficiency of conditions of the theorem. □

THEOREM 8.6. *Let the classical maximum modulus principle be valid for solutions of system* (8.1.2) *in* Q_T $(\mathbb{R}_T^{n+1})$. *Then*

(i) *for all* $x \in \Omega$ $(x \in \mathbb{R}^n)$ *the equalities*

$$A_{jk}(x,0) = a_{jk}(x)I, \ 1 \le j,k \le n,$$

hold, where $((a_{jk}))$ *is a positive-definite* $(n \times n)$*-matrix-valued function;*

(ii) *for all* $x \in \Omega$ $(x \in \mathbb{R}^n)$ *and for all* $\boldsymbol{\xi}_j$, $\boldsymbol{\zeta} \in \mathbb{R}^m$, $j = 1, \dots, n$, *with* $(\boldsymbol{\xi}_j, \boldsymbol{\zeta}) = 0$ *the inequality*

$$\sum_{j,k=1}^{n} a_{jk}(x)(\boldsymbol{\xi}_j, \boldsymbol{\xi}_k) + \sum_{j=1}^{n} (\mathcal{A}_j(x,0)\boldsymbol{\xi}_j, \boldsymbol{\zeta}) + (\mathcal{A}_0(x,0)\boldsymbol{\zeta}, \boldsymbol{\zeta}) \ge 0$$

holds.

PROOF. We assume for brevity that Ω is a bounded subdomain of $\mathbb{R}^n$ or $\Omega = \mathbb{R}^n$. From Lemmas 8.1, 8.5 it follows that the equality $\mathcal{K}(\Omega, T) = 1$ is valid only if $\mathcal{K}_0(y) = 1$ for all $y \in \Omega$, where $\mathcal{K}_0(y)$ is the constant (8.2.3) defined for the system

$$\frac{\partial \boldsymbol{u}}{\partial t} - \sum_{j,k=1}^{n} \mathcal{A}_{jk}(y, 0) \frac{\partial^2 \boldsymbol{u}}{\partial x_j \partial x_k} = \boldsymbol{0}.$$

By Theorem 8.5 the equality $\mathcal{K}_0(y) = 1$ takes place if and only if (i) is satisfied.

Now we prove the necessity of (ii). Let y be an arbitrary fixed point of Ω, and let the radius of the ball $\mathbb{B}_r(y)$ be so small that $\overline{\mathbb{B}_r(y)} \subset \Omega$. We introduce the vector-valued function

$$\boldsymbol{v}(x) = \Big(\sum_{j=1}^{n} \boldsymbol{\xi}_j x_j + \boldsymbol{\zeta} \Big) \Big(\Big| \sum_{j=1}^{n} \boldsymbol{\xi}_j x_j \Big|^2 + |\boldsymbol{\zeta}|^2 \Big)^{-1/2},$$

where $\boldsymbol{\xi}_j \in \mathbb{R}^m$, $\boldsymbol{\zeta} \in \mathbb{R}^m \backslash \{\boldsymbol{0}\}$, $(\boldsymbol{\xi}_j, \boldsymbol{\zeta}) = 0$, $j = 1, \dots, n$. Further, let

$$\chi_\varepsilon \in \mathrm{C}^\infty(\mathbb{R}^n), \ \chi_\varepsilon(x) = 1 \quad \text{for} \quad |x| \le \varepsilon/2, \ \chi_\varepsilon(x) = 0 \quad \text{for} \quad |x| \ge \varepsilon$$

and

$$0 \le \chi_\varepsilon(x) \le 1 \quad \text{for all} \quad x \in \mathbb{R}^n,$$

where $\varepsilon \le r$.

The vector-valued function

$$\boldsymbol{u}_\varepsilon(x, t) = \int_{\mathbb{R}^n} G(t, 0, x, \eta) \chi_\varepsilon(\eta - y) \boldsymbol{v}(\eta - y) d\eta, \tag{8.3.10}$$

where $G(t, \tau, x, \eta)$ is the Green matrix of system (8.1.2), $\boldsymbol{u}_\varepsilon$ is the solution of the Cauchy problem for system (8.1.2) with the initial data $\boldsymbol{u}_\varepsilon(x, 0) = \boldsymbol{v}_\varepsilon(x)$, where $\boldsymbol{v}_\varepsilon(x) = \chi_\varepsilon(x - y)\boldsymbol{v}(x - y)$. If Ω is a bounded subdomain of $\mathbb{R}^n$, then by (8.3.5) ε can be chosen so small that

$$\sup\big\{ |\boldsymbol{u}_\varepsilon(x, t)| : \ (x, t) \in \partial\Omega \times [0, T] \big\} \le 1.$$

Then $\|\boldsymbol{u}_\varepsilon|_{\overline{\Gamma}_T}\|_{[C(\overline{\Gamma}_T)]^m} = 1$, and, consequently

$$\mathcal{K}(\Omega, T) \ge \sup_{0<t\le T} |\boldsymbol{u}_\varepsilon(y, t)|. \tag{8.3.11}$$

In the case $\Omega = \mathbb{R}^n$ the last inequality is obvious.

Suppose condition (i) is satisfied. We take the scalar product of system (8.1.2) with $\boldsymbol{u}$ and transform the equality

$$\begin{aligned} \Big(\frac{\partial \boldsymbol{u}}{\partial t}, \boldsymbol{u} \Big) &- \sum_{j,k=1}^{n} a_{jk}(x) \Big(\frac{\partial^2 \boldsymbol{u}}{\partial x_j \partial x_k}, \boldsymbol{u} \Big) + \sum_{j=1}^{n} \Big(\mathcal{A}_j(x, t) \frac{\partial \boldsymbol{u}}{\partial x_j}, \boldsymbol{u} \Big) \\ &+ (\mathcal{A}_0(x, t)\boldsymbol{u}, \boldsymbol{u}) - \sum_{j,k=1}^{n} \Big(\Big[\mathcal{A}_{jk}(x, t) - a_{jk}(x) I \Big] \frac{\partial^2 \boldsymbol{u}}{\partial x_j \partial x_k}, \boldsymbol{u} \Big) = 0, \end{aligned}$$

where $a_{jk}(x)I = \mathcal{A}_{jk}(x,0)$. As a result we find

$$\begin{aligned}
\frac{1}{2}\frac{\partial |\boldsymbol{u}|^2}{\partial t} &= \frac{1}{2}\sum_{j,k=1}^{n} a_{jk}(x)\frac{\partial^2 |\boldsymbol{u}|^2}{\partial x_j \partial x_k} - \sum_{j,k=1}^{n} a_{jk}(x)\left(\frac{\partial \boldsymbol{u}}{\partial x_j}, \frac{\partial \boldsymbol{u}}{\partial x_k}\right)\\
&- \sum_{j=1}^{n}\left(\mathcal{A}_j(x,t)\frac{\partial \boldsymbol{u}}{\partial x_j}, \boldsymbol{u}\right) - (\mathcal{A}_0(x,t)\boldsymbol{u}, \boldsymbol{u}) \qquad (8.3.12)\\
&+ \sum_{j,k=1}^{n}\left(\Big[\mathcal{A}_{jk}(x,t) - a_{jk}(x)I\Big]\frac{\partial^2 \boldsymbol{u}}{\partial x_j \partial x_k}, \boldsymbol{u}\right).
\end{aligned}$$

Since the coefficients of system (8.1.2) belong to the class $[\mathrm{C}_{\mathrm{b}}^{\alpha,\alpha/2}(\overline{\mathbb{R}_T^{n+1}})]^{m\times m}$ and since $\boldsymbol{v}_\varepsilon \in [\mathrm{C}_{\mathrm{b}}^{2,\alpha}(\mathbb{R}^n)]^m$ then, according to Eidel'man [**Ei1**],

$$\boldsymbol{u}_\varepsilon \in [\mathrm{C}_{\mathrm{b}}^{2+\alpha,\alpha/2}(\overline{\mathbb{R}_T^{n+1}})]^m.$$

So after substitution of $\boldsymbol{u}_\varepsilon$ defined by (8.3.10) into (8.3.12) in place of $\boldsymbol{u}$, we obtain

$$\begin{aligned}
\lim_{(x,t)\to(y,+0)} \frac{\partial |\boldsymbol{u}_\varepsilon|^2}{\partial t} = 2\Bigg\{ &\frac{1}{2}\sum_{j,k=1}^{n} a_{jk}(y)\frac{\partial^2 |\boldsymbol{v}_\varepsilon|^2}{\partial x_j \partial x_k} - \sum_{j,k=1}^{n} a_{jk}(y)\left(\frac{\partial \boldsymbol{v}_\varepsilon}{\partial x_j}, \frac{\partial \boldsymbol{v}_\varepsilon}{\partial x_k}\right)\\
&- \sum_{j=1}^{n}\left(\mathcal{A}_j(y,0)\frac{\partial \boldsymbol{v}_\varepsilon}{\partial x_j},\ \boldsymbol{v}_\varepsilon\right) - (\mathcal{A}_0(y,0)\boldsymbol{v}_\varepsilon,\ \boldsymbol{v}_\varepsilon)\Bigg\}\Bigg|_{x=y}.
\end{aligned}$$

Hence, using the equalities

$$\frac{\partial^2 |\boldsymbol{v}_\varepsilon|^2}{\partial x_j \partial x_k}\Big|_{x=y} = 0, \qquad \frac{\partial \boldsymbol{v}_\varepsilon}{\partial x_j}\Big|_{x=y} = \frac{\boldsymbol{\xi}_j}{|\boldsymbol{\zeta}|}, \qquad \boldsymbol{v}_\varepsilon(y) = \frac{\boldsymbol{\zeta}}{|\boldsymbol{\zeta}|},$$

where $j,k = 1,\dots,n$, we find

$$\begin{aligned}
\lim_{t\to+0} \frac{\partial |\boldsymbol{u}_\varepsilon(y,t)|^2}{\partial t} &= -\frac{2}{|\boldsymbol{\zeta}|^2}\Big[\sum_{j,k=1}^{n} a_{jk}(y)(\boldsymbol{\xi}_j, \boldsymbol{\xi}_k) + \sum_{j=1}^{n}(\mathcal{A}_j(y,0)\boldsymbol{\xi}_j, \boldsymbol{\zeta})\\
&+ (\mathcal{A}_0(y,0)\boldsymbol{\zeta}, \boldsymbol{\zeta})\Big]. \qquad (8.3.13)
\end{aligned}$$

The function $|\boldsymbol{u}_\varepsilon(y,t)|^2$ is continuous on $[0,T]$ and differentiable at all points of the interval $(0,T)$, its derivative $\partial |\boldsymbol{u}_\varepsilon(y,t)|^2/\partial t$ tends to a finite limit as $t\to+0$. Therefore, the function $|\boldsymbol{u}_\varepsilon(y,t)|^2$ has the right-hand derivative $\partial_+ |\boldsymbol{u}_\varepsilon(y,t)|^2/\partial t$ at $t=0$, and by (8.3.13) we have

$$\begin{aligned}
\frac{\partial_+ |\boldsymbol{u}_\varepsilon(y,t)|^2}{\partial t}\Big|_{t=0} &= -\frac{2}{|\boldsymbol{\zeta}|^2}\Big[\sum_{j,k=1}^{n} a_{jk}(y)(\boldsymbol{\xi}_j, \boldsymbol{\xi}_k) + \sum_{j=1}^{n}(\mathcal{A}_j(y,0)\boldsymbol{\xi}_j, \boldsymbol{\zeta})\\
&+ (\mathcal{A}_0(y,0)\boldsymbol{\zeta}, \boldsymbol{\zeta})\Big]. \qquad (8.3.14)
\end{aligned}$$

Suppose $\mathcal{K}(\Omega,T) = 1$ and let there exist a point $y\in\Omega$ and vectors $\boldsymbol{\xi}_j \in \mathbb{R}^m$, $\boldsymbol{\zeta}\in\mathbb{R}^m\backslash\{\boldsymbol{0}\}$ with $(\boldsymbol{\xi}_j, \boldsymbol{\zeta}) = 0$, $j = 1,\dots,n$, for which

$$\sum_{j,k=1}^{n} a_{jk}(y)(\boldsymbol{\xi}_j, \boldsymbol{\xi}_k) + \sum_{j=1}^{n}(\mathcal{A}_j(y,0)\boldsymbol{\xi}_j, \boldsymbol{\zeta}) + (\mathcal{A}_0(y,0)\boldsymbol{\zeta}, \boldsymbol{\zeta}) < 0. \qquad (8.3.15)$$

Since $|\boldsymbol{u}_\varepsilon(y,0)| = |\boldsymbol{v}(0)| = 1$, relations (8.3.14) and (8.3.15) imply the existence of a constant $\delta > 0$ such that $|\boldsymbol{u}_\varepsilon(y,t)| > 1$ for $0 < t \leq \delta$. From this and (8.3.11) it follows that $\mathcal{K}(\Omega, T) > 1$ which contradicts our assumption on the validity of the classical maximum modulus principle.

Thus, if $\mathcal{K}(\Omega, T) = 1$, then for all $x \in \Omega$ and for all vectors $\boldsymbol{\xi}_j \in \mathbb{R}^m$, $\boldsymbol{\zeta} \in \mathbb{R}^m \backslash \{\mathbf{0}\}$ with $(\boldsymbol{\xi}_j, \boldsymbol{\zeta}) = 0$, $j = 1, \dots, n$, the inequality

$$\sum_{j,k=1}^{n} a_{jk}(x)(\boldsymbol{\xi}_j, \boldsymbol{\xi}_k) + \sum_{j=1}^{n} (\mathcal{A}_j(x,0)\boldsymbol{\xi}_j, \boldsymbol{\zeta}) + (\mathcal{A}_0(x,0)\boldsymbol{\zeta}, \boldsymbol{\zeta}) \geq 0$$

holds. The condition $\boldsymbol{\zeta} \in \mathbb{R}^m \backslash \{0\}$ can be omitted since the inequality

$$\sum_{j,k=1}^{n} a_{jk}(x)(\boldsymbol{\xi}_j, \boldsymbol{\xi}_k) \geq 0$$

holds by the necessity of condition (i) for validity of the classical maximum modulus principle. □

REMARK 8.2. In what follows we show that conditions (i), (ii) of Theorem 8.6, are necessary and sufficient for validity of the classical maximum modulus principle for second order systems with the coefficients depending only on x. But in general, when the coefficients depend on x and t these conditions are not sufficient.

Consider, for example, the parabolic system

$$\frac{\partial \boldsymbol{u}}{\partial t} - \sum_{j,k=1}^{n} \mathcal{A}_{jk}(x,t) \frac{\partial^2 \boldsymbol{u}}{\partial x_j \partial x_k} + \sum_{j=1}^{n} \mathcal{A}_j(x,t) \frac{\partial \boldsymbol{u}}{\partial x_j} = \mathbf{0} \tag{8.3.16}$$

in $\mathbb{R}^{n+1}_T$, where

$$\mathcal{A}_{jk} \in [\mathrm{C}_{\mathrm{b}}^{2+\alpha,\alpha/2}(\overline{\mathbb{R}^{n+1}_T})]^{m\times m}, \quad \mathcal{A}_j \in [\mathrm{C}_{\mathrm{b}}^{1+\alpha,\alpha/2}(\overline{\mathbb{R}^{n+1}_T})]^{m\times m}.$$

Suppose the coefficients of system (8.3.16) do not depend on x in the layer $\mathbb{R}^{n+1}_\delta$, $0 < \delta < T$, and let

$$\mathcal{A}_{jk}(x,0) = a_{jk} I, \quad \mathcal{A}_j(x,0) = a_j I,$$

where $((a_{jk}))$ is a positive-definite $(n \times n)$-matrix and a_j are scalars. Let the matrix $A_1(x,t)$ be non-diagonal for all $(x,t) \in \mathbb{R}^{n+1}_\delta$. Then, according to Theorem 8.5, the classical maximum modulus principle fails for system (8.3.16) in $\mathbb{R}^{n+1}_\delta$ (the more so in $\mathbb{R}^{n+1}_T$) whereas conditions (i), (ii) of Theorem 8.6 are satisfied.

8.3.1.2. *Sufficient conditions for systems with scalar principal part.* Next we present a theorem on a sufficient condition for validity of the classical maximum modulus principle for second order systems with scalar principal part.

THEOREM 8.7. *Let the coefficients of the parabolic system*

$$\frac{\partial \boldsymbol{u}}{\partial t} - \sum_{j,k=1}^{n} a_{jk}(x,t) \frac{\partial^2 \boldsymbol{u}}{\partial x_j \partial x_k} + \sum_{j=1}^{n} \mathcal{A}_j(x,t) \frac{\partial \boldsymbol{u}}{\partial x_j} + \mathcal{A}_0(x,t)\boldsymbol{u} = \mathbf{0} \tag{8.3.17}$$

satisfy the condition

(i) *for all* $(x,t) \in Q_T$ $\big((x,t) \in \mathbb{R}_T^{n+1}\big)$ *and for all vectors* $\boldsymbol{\xi}_j$, $\boldsymbol{\zeta} \in \mathbb{R}^m$ *with* $(\boldsymbol{\xi}_j, \boldsymbol{\zeta}) = 0$, $j = 1, \dots, n$, *the inequality*

$$\sum_{j,k=1}^{n} a_{jk}(x,t)(\boldsymbol{\xi}_j, \boldsymbol{\xi}_k) + \sum_{j=1}^{n} (\mathcal{A}_j(x,t)\boldsymbol{\xi}_j, \boldsymbol{\zeta}) + (\mathcal{A}_0(x,t)\boldsymbol{\zeta}, \boldsymbol{\zeta}) \geq 0 \tag{8.3.18}$$

holds. Then $\mathcal{K}(\Omega, T) = 1$ $(\mathcal{K}(\mathbb{R}^n, T) = 1)$.

PROOF. Suppose first that for all $(x,t) \in Q_T$ and for all $\boldsymbol{\xi}_j \in \mathbb{R}^m$, $\boldsymbol{\zeta} \in \mathbb{R}^m \backslash \{\mathbf{0}\}$, $j = 1, \dots, n$, with $(\boldsymbol{\xi}_j, \boldsymbol{\zeta}) = 0$ we have

$$\sum_{j,k=1}^{n} a_{jk}(x,t)(\boldsymbol{\xi}_j, \boldsymbol{\xi}_k) + \sum_{j=1}^{n} (\mathcal{A}_j(x,t)\boldsymbol{\xi}_j, \boldsymbol{\zeta}) + (\mathcal{A}_0(x,t)\boldsymbol{\zeta}, \boldsymbol{\zeta}) > 0. \tag{8.3.19}$$

We show that for any non-trivial solution $\boldsymbol{u} \in [\mathrm{C}^{(2,1)}(Q_T)]^m \cap [\mathrm{C}(\overline{Q}_T)]^m$ of system (8.3.17) the function $|\boldsymbol{u}(x,t)|$ cannot attain its global maximum at a point $(x,t) \in Q_T$. This will imply that if system (8.3.17) has a regular solution in Q_T and if (8.3.19) holds, then the function $|\boldsymbol{u}(x,t)|$ takes its maximum value on $\overline{\Gamma}_T$.

From (8.3.17) we have

$$\begin{aligned} \frac{1}{2}\frac{\partial |\boldsymbol{u}|^2}{\partial t} &= \frac{1}{2}\sum_{j,k=1}^{n} a_{jk}(x,t)\frac{\partial^2 |\boldsymbol{u}|^2}{\partial x_j \partial x_k} - \sum_{j=1}^{n} a_{jk}(x,t)\left(\frac{\partial \boldsymbol{u}}{\partial x_j}, \frac{\partial \boldsymbol{u}}{\partial x_k}\right) \\ &- \sum_{j=1}^{n}\left(\mathcal{A}_j(x,t)\frac{\partial \boldsymbol{u}}{\partial x_j}, \boldsymbol{u}\right) - (\mathcal{A}_0(x,t)\boldsymbol{u}, \boldsymbol{u}). \end{aligned} \tag{8.3.20}$$

Suppose the function $|\boldsymbol{u}(x,t)|$ takes its global maximum at a point $(x_0, t_0) \in Q_T$. Then

$$\frac{\partial |\boldsymbol{u}|^2}{\partial x_j}\Big|_{(x_0,t_0)} = 2\left(\frac{\partial \boldsymbol{u}}{\partial x_j}, \boldsymbol{u}\right)\Big|_{(x_0,t_0)} = 0, \tag{8.3.21}$$

$$\frac{\partial |\boldsymbol{u}|^2}{\partial t}\Big|_{(x_0,t_0)} = 2\left(\frac{\partial \boldsymbol{u}}{\partial t}, \boldsymbol{u}\right)\Big|_{(x_0,t_0)} \geq 0, \tag{8.3.22}$$

$$\sum_{j,k=1}^{n} a_{jk}(x_0,t_0)\,\frac{\partial^2 |\boldsymbol{u}|^2}{\partial x_j \partial x_k}\Bigg|_{(x_0,t_0)} \leq 0. \tag{8.3.23}$$

By (8.3.20)-(8.3.23) we have

$$\begin{aligned} \Bigg\{ &\sum_{j,k=1}^{n} a_{jk}(x_0,t_0)\left(\frac{\partial \boldsymbol{u}}{\partial x_j}, \frac{\partial \boldsymbol{u}}{\partial x_k}\right) \\ &+ \sum_{j=1}^{n}\left(\mathcal{A}_j(x_0,t_0)\frac{\partial \boldsymbol{u}}{\partial x_j},\ \boldsymbol{u}\right) + (\mathcal{A}_0(x_0,t_0)\boldsymbol{u}, \boldsymbol{u})\Bigg\}\Bigg|_{(x_0,t_0)} \leq 0, \end{aligned}$$

which contradicts (8.3.19) and hence $|\boldsymbol{u}(x,t)|$ does not attain its global maximum at $(x_0, t_0) \in Q_T$. This implies

$$\|\boldsymbol{u}\|_{[\mathrm{C}(\overline{Q}_T)]^m} = \|\boldsymbol{u}|_{\overline{\Gamma}_T}\|_{[\mathrm{C}(\overline{\Gamma}_T)]^m}.$$

Suppose now that (8.3.18) holds under the conditions in the statement of Theorem. Let $\varepsilon = \text{const} > 0$ and let u be the solution of system (8.3.17) in the class $[\mathrm{C}^{(2,1)}]^m(Q_T) \cap [\mathrm{C}(\overline{Q}_T)]^m$. The vector-valued function $\boldsymbol{v}(x,t) = \boldsymbol{u}(x,t)\exp(-\varepsilon t)$ satisfies the system

$$\frac{\partial \boldsymbol{v}}{\partial t} - \sum_{j,k=1}^{n} a_{jk}(x,t)\frac{\partial^2 \boldsymbol{v}}{\partial x_j \partial x_k} + \sum_{j=1}^{n} \mathcal{A}_j(x,t)\frac{\partial \boldsymbol{v}}{\partial x_j} + \mathcal{A}_0(x,t)\boldsymbol{v} + \varepsilon \boldsymbol{v} = \mathbf{0}.$$

According to our assumption, for all $(x,t) \in Q_T$ and all $\boldsymbol{\xi}_j \in \mathbb{R}^m$, $\boldsymbol{\zeta} \in \mathbb{R}^m \backslash \{\mathbf{0}\}$ with $(\boldsymbol{\xi}_j, \boldsymbol{\zeta}) = 0$, $1 \le j \le n$, we have

$$\sum_{j,k=1}^{n} a_{jk}(x,t)(\boldsymbol{\xi}_j, \boldsymbol{\xi}_k) + \sum_{j=1}^{n} (\mathcal{A}_j(x,t)\boldsymbol{\xi}_j, \boldsymbol{\zeta}) + (\mathcal{A}_0(x,t)\boldsymbol{\zeta}, \boldsymbol{\zeta}) + \varepsilon|\boldsymbol{\zeta}|^2 > 0.$$

By what we proved above

$$\|\boldsymbol{v}\|_{[\mathrm{C}(\overline{Q}_T)]^m} = \|\boldsymbol{v}|_{\overline{\Gamma}_T}\|_{[\mathrm{C}(\overline{\Gamma}_T)]^m},$$

and hence

$$\max_{\overline{Q}_T} |e^{-\varepsilon t}\boldsymbol{u}(x,t)| = \max_{\overline{\Gamma}_T} |e^{-\varepsilon t}\boldsymbol{u}(x,t)|.$$

Consequently,

$$\|\boldsymbol{u}\|_{[\mathrm{C}(\overline{Q}_T)]^m} \le e^{\varepsilon T}\|\boldsymbol{u}|_{\overline{\Gamma}_T}\|_{[\mathrm{C}(\overline{\Gamma}_T)]^m}.$$

Since ε is arbitrary, the best constant in the last inequality is equal to one.

Now we turn to the constant $\mathcal{K}(\mathbb{R}^n, T)$. Contrary to the case of a bounded domain, one cannot immediately conclude that $\mathcal{K}(\mathbb{R}^n, T) = 1$ because of the absence of the global maximum of $|\boldsymbol{u}(x,t)|$ in $\mathbb{R}^{n+1}_T$.

Suppose inequality (8.3.18) holds for all $(x,t) \in \mathbb{R}^{n+1}_T$ and all $\boldsymbol{\xi}_j$, $\boldsymbol{\zeta} \in \mathbb{R}^m$ with $(\boldsymbol{\xi}_j, \boldsymbol{\zeta}) = 0$, $j = 1, \ldots, n$, and that the classical maximum modulus principle fails for solutions of the system (8.3.17) in $\mathbb{R}^{n+1}_T$. Then there exists a point $(x_0, t_0) \in \mathbb{R}^{n+1}_T$ and a vector-valued function $\boldsymbol{\phi} \in [\mathrm{C_b}(\mathbb{R}^n)]^m$, $|\boldsymbol{\phi}(x)| \le 1$ such that

$$|\boldsymbol{u}(x_0, t_0)| = \left| \int_{\mathbb{R}^n} G(t_0, 0, x_0, \eta)\boldsymbol{\phi}(\eta)d\eta \right| > 1. \tag{8.3.24}$$

By (8.3.4) one can assume that $\boldsymbol{\phi}$ has a compact support, supp $\boldsymbol{\phi} \subset \mathbb{B}_\rho(x_0)$. If $|x - x_0| \ge R > \rho$ then (8.3.4) implies

$$\left| \int_{\mathbb{R}^n} G(t, 0, x, \eta)\boldsymbol{\phi}(\eta)d\eta \right| \le c_1 \exp\left(-c_2 \frac{(R-\rho)^2}{T} \right),$$

where $0 < t \le T$, $x \in \mathbb{R}^n \backslash \mathbb{B}_R(x_0)$.

Applying the assertion of the present theorem for a cylinder with a bounded base, we get

$$|\boldsymbol{u}(x,t)| = \left| \int_{\mathbb{R}^n} G(t, 0, x, \eta)\boldsymbol{\phi}(\eta)d\eta \right| \le 1,$$

where $(x,t) \in \overline{\mathbb{B}_R(x_0)} \times [0, T]$ and R is sufficiently large. The last inequality contradicts (8.3.24) which proves the validity of the classical maximum modulus principle in $\mathbb{R}^{n+1}_T$. □

8.3.1.3. *Necessary and sufficient conditions.* Theorems 8.6 and 8.7 immediately imply the following assertion.

THEOREM 8.8. *The classical maximum modulus principle is valid for solutions of system* (8.1.3) *in* Q_T $(\mathbb{R}^{n+1}_T)$ *if and only if:*

(i) *for all* $x \in \Omega$ $(x \in \mathbb{R}^n)$ *the equalities*

$$\mathcal{A}_{jk}(x) = a_{jk}(x)I, \ \ 1 \le j, \ \ k \le m,$$

hold, where $((a_{jk}))$ *is a positive-definite* $(n \times n)$*-matrix-valued function;*

(ii) *for all* $x \in \Omega$ $(x \in \mathbb{R}^n)$ *and all* $\boldsymbol{\xi}_j$, $\boldsymbol{\zeta} \in \mathbb{R}^m$, $j = 1, \dots, n$, *with* $(\boldsymbol{\xi}_j, \boldsymbol{\zeta}) = 0$, *the inequality*

$$\sum_{j,k=1}^{n} a_{jk}(x)(\boldsymbol{\xi}_j, \boldsymbol{\xi}_k) + \sum_{j=1}^{n} (\mathcal{A}_j(x)\boldsymbol{\xi}_j, \boldsymbol{\zeta}) + (\mathcal{A}_0(x)\boldsymbol{\zeta}, \boldsymbol{\zeta}) \ge 0 \tag{8.3.25}$$

is valid.

Theorem 8.8 implies

COROLLARY 8.1. *The classical maximum modulus principle holds for solutions of the system*

$$\frac{\partial \boldsymbol{u}}{\partial t} - \sum_{j,k=1}^{n} \mathcal{A}_{jk}(x) \frac{\partial^2 \boldsymbol{u}}{\partial x_j \partial x_k} + \sum_{j=1}^{n} \mathcal{A}_j(x) \frac{\partial \boldsymbol{u}}{\partial x_j} = \boldsymbol{0}$$

in Q_T $(\mathbb{R}^{n+1}_T)$ *if and only if for all* $x \in \Omega$ $(x \in \mathbb{R}^n)$

$$\mathcal{A}_{jk}(x) = a_{jk}(x)I, \quad \mathcal{A}_j(x) = a_j(x)I,$$

where $((a_{jk}))$ *is a positive-definite* $(n \times n)$*-matrix-valued function and* a_j *are scalar functions.*

PROOF. Putting $\mathcal{A}_0 = 0$ in (8.3.25), we get

$$\sum_{j,k=1}^{n} a_{jk}(x)(\boldsymbol{\xi}_j, \boldsymbol{\xi}_k) + \sum_{j=1}^{n} (\mathcal{A}_j(x)\boldsymbol{\xi}_j, \boldsymbol{\zeta}) \ge 0,$$

which can be valid for all $x \in \Omega$ $(x \in \mathbb{R}^n)$ and for all $\boldsymbol{\xi}_j$, $\boldsymbol{\zeta} \in \mathbb{R}^m$ with $(\boldsymbol{\xi}_j, \boldsymbol{\zeta}) = 0$, $j = 1, \dots, n$, only provided $\big(\mathcal{A}_j(x)\boldsymbol{\xi}_j, \boldsymbol{\zeta}\big) = 0$.

Let $\boldsymbol{e}_1, \dots, \boldsymbol{e}_m$ be an orthonormal basis in $\mathbb{R}^m$. Setting $\boldsymbol{\xi}_j = \boldsymbol{e}_j$, $\boldsymbol{\zeta} = \boldsymbol{e}_k$, $j \ne k$, $1 \le j, \ \ k \le m$ in the last equality, we arrive at the conclusion that the matrix-valued functions $\mathcal{A}_j$ are diagonal.

Thus, condition (ii) implies

$$\mathcal{A}_j^{(1,1)}(x)\xi_j^{(1)}\zeta^{(1)} + \dots + \mathcal{A}_j^{(m,m)}\xi_j^{(m)}\zeta^{(m)} = 0,$$
$$\xi_j^{(1)}\zeta^{(1)} + \dots + \xi_j^{(m)}\zeta^{(m)} = 0,$$

where $\xi_j^{(k)} = (\boldsymbol{\xi}_j, \boldsymbol{e}_k)$, $\zeta^{(k)} = (\boldsymbol{\zeta}, \boldsymbol{e}_k)$, $j = 1, \dots, n$. Putting here

$$\xi_j^{(i)} = \xi_j^{(m)} = \zeta^{(i)} = 1, \quad \zeta^{(m)} = -1, \quad \xi_j^{(k)} = \zeta^{(k)} = 0\ ,$$

for $i \ne m, k \ne i, k \ne m$, we find that $\mathcal{A}_j^{(i,i)}(x) = \mathcal{A}_j^{(m,m)}(x)$ for all $i = 1, \dots, m-1$. Consequently, $\mathcal{A}_j(x) = a_j(x)I$, where

$$a_j(x) = \mathcal{A}_j^{(1,1)}(x) = \mathcal{A}_j^{(2,2)}(x) = \dots = \mathcal{A}_j^{(m,m)}(x), \ \ j = 1, \dots, n. \qquad \square$$

REMARK 8.3. Minimizing the left-hand side of (8.3.25) over

$$\boldsymbol{\xi} = (\boldsymbol{\xi}_1, \dots, \boldsymbol{\xi}_n),\ \boldsymbol{\xi}_j \in \mathbb{R}^m,$$

for a fixed $\boldsymbol{\zeta} \in \mathbb{R}^m$ with $(\boldsymbol{\xi}_j, \boldsymbol{\zeta}) = 0$, $j = 1, \dots, n$, one can write condition (ii) of Theorem 8.8 in another form.

One may assume that $\boldsymbol{\zeta} \in \mathbb{R}^m \backslash \{\mathbf{0}\}$, since the inequality

$$\sum_{j,k=1}^{n} a_{jk}(x)(\boldsymbol{\xi}_j, \boldsymbol{\xi}_k) \geq 0$$

follows from condition (i) of Theorem 8.8 for all $\boldsymbol{\xi}_j \in \mathbb{R}^m$.

Let

$$\mathcal{F}_{\boldsymbol{\zeta}}(\boldsymbol{\xi}_1, \dots, \boldsymbol{\xi}_n) = \sum_{j,k=1}^{n} a_{jk}(x)(\boldsymbol{\xi}_j, \boldsymbol{\xi}_k) + \sum_{j=1}^{n} (\mathcal{A}_j(x)\boldsymbol{\xi}_j, \boldsymbol{\zeta}) + (\mathcal{A}_0(x)\boldsymbol{\zeta}, \boldsymbol{\zeta})$$

or, which is the same,

$$\begin{aligned}\mathcal{F}_{\boldsymbol{\zeta}}(\boldsymbol{\xi}_1, \dots, \boldsymbol{\xi}_n) \quad &= \quad \sum_{j,k=1}^{n} \sum_{i=1}^{m} a_{jk}(x)\xi_j^{(i)}\xi_k^{(i)} + \sum_{j=1}^{n} \sum_{i,k=1}^{m} \mathcal{A}_j^{(i,k)}(x)\xi_j^{(k)}\zeta^{(i)} \\ &+ \quad \sum_{i,k=1}^{m} \mathcal{A}_0^{(i,k)}(x)\zeta^{(k)}\zeta^{(i)},\end{aligned}$$

where $\xi_j^{(k)}$ and $\zeta^{(k)}$ are components of the vectors $\boldsymbol{\xi}_j$ and $\boldsymbol{\zeta}$, respectively. At a point of the constraint extremum of the function $\mathcal{F}_{\boldsymbol{\zeta}}(\boldsymbol{\xi}_1, \dots, \boldsymbol{\xi}_n)$ one has

$$\frac{\partial}{\partial \xi_k^{(i)}} \left(\mathcal{F}_{\boldsymbol{\zeta}}(\boldsymbol{\xi}_1, \dots, \boldsymbol{\xi}_n) - \sum_{j=1}^{n} \sum_{i=1}^{m} \lambda_j \xi_j^{(i)} \zeta^{(i)} \right)$$

$$= 2\sum_{j=1}^{n} a_{jk}(x)\xi_j^{(i)} + \sum_{j=1}^{m} \mathcal{A}_k^{(j,i)}(x)\zeta^{(j)} - \lambda_k \zeta^{(i)} = 0, \tag{8.3.26}$$

where $k = 1, 2, \dots, n$, $i = 1, \dots, m$ and the following constraint relations are valid

$$\sum_{i=1}^{m} \xi_j^{(i)} \zeta^{(i)} = 0, \quad j = 1, 2, \dots, n.$$

Multiplying (8.3.26) by $\zeta^{(i)}$ and summing up over i from 1 to m, we obtain

$$\lambda_k = |\zeta|^{-2} \left(\mathcal{A}_k^*(x)\boldsymbol{\zeta}, \boldsymbol{\zeta} \right),$$

where $*$ means the passage to the transposed matrix. Consequently, conditions (8.3.26) defining $\boldsymbol{\xi}_j$ can be written as

$$2\sum_{j=1}^{n} a_{jk}(x)\xi_j^{(i)} + \sum_{j=1}^{m} \mathcal{A}_k^{(j,i)}(x)\zeta^{(j)} - |\boldsymbol{\zeta}|^{-2} \left(\mathcal{A}_k^*(x)\boldsymbol{\zeta}, \boldsymbol{\zeta} \right) \zeta^{(i)} = 0.$$

Taking into account the symmetry of the matrix $((a_{jk}(x)))$, we find

$$\boldsymbol{\xi}_j = \frac{1}{2} \sum_{k=1}^{n} b_{jk}(x) \left[\, |\boldsymbol{\zeta}|^{-2} \left(\mathcal{A}_k^*(x)\boldsymbol{\zeta}, \boldsymbol{\zeta} \right) \boldsymbol{\zeta} - \mathcal{A}_k^*(x)\boldsymbol{\zeta} \,\right], \tag{8.3.27}$$

where $j = 1, 2, \dots n$ and $((b_{jk}(x)))$ is the inverse matrix of $((a_{jk}(x)))$.

The function $\mathcal{F}_\zeta(\boldsymbol{\xi}_1,\dots,\boldsymbol{\xi}_n)$ attains its constraint minimum at vectors (8.3.27) because of the positive-definiteness of the matrix $((a_{jk}(x)))$. Calculating $\mathcal{F}_\zeta(\boldsymbol{\xi}_1,\dots,\boldsymbol{\xi}_n)$ at vectors (8.3.27), we obtain

$$\begin{aligned}\min\Big\{\mathcal{F}_\zeta(\boldsymbol{\xi}_1,\dots,\boldsymbol{\xi}_n) \;&:\; \boldsymbol{\xi}_1,\dots,\boldsymbol{\xi}_n\in\mathbb{R}^m, (\boldsymbol{\xi}_1,\boldsymbol{\zeta})=0,\dots,(\boldsymbol{\xi}_n,\boldsymbol{\zeta})=0\Big\}\\ &= \frac{1}{4}|\boldsymbol{\zeta}|^{-2}\sum_{i,j=1}^n b_{ij}(x)\,(\mathcal{A}_i(x)\boldsymbol{\zeta},\boldsymbol{\zeta})\,(\mathcal{A}_j(x)\boldsymbol{\zeta},\boldsymbol{\zeta})\\ &- \frac{1}{4}\sum_{i,j=1}^n b_{ij}(x)\,\big(\mathcal{A}_i^*(x)\boldsymbol{\zeta},\mathcal{A}_j^*(x)\boldsymbol{\zeta}\big)+(\mathcal{A}_0(x)\boldsymbol{\zeta},\boldsymbol{\zeta}).\end{aligned}$$

Thus Theorem 8.8 implies the following assertion.

COROLLARY 8.2. *The classical maximum modulus principle is valid for solutions of system* (8.1.3) *in* Q_T $(\mathbb{R}^{n+1}_T)$ *if and only if condition* (i) *of Theorem* 8.8 *is satisfied and for all* $x\in\Omega$ $(x\in\mathbb{R}^n)$ *and for any* $\boldsymbol{\zeta}\in\mathbb{R}^m$, $|\boldsymbol{\zeta}|=1$ *the inequality*

$$\sum_{i,j=1}^n b_{ij}(x)\Big[(\mathcal{A}_i(x)\boldsymbol{\zeta},\boldsymbol{\zeta})(\mathcal{A}_j(x)\boldsymbol{\zeta},\boldsymbol{\zeta})-(\mathcal{A}_i^*(x)\boldsymbol{\zeta},\mathcal{A}_j^*(x)\boldsymbol{\zeta})\Big]+4(\mathcal{A}_0(x)\boldsymbol{\zeta},\boldsymbol{\zeta})\ge 0$$

holds, where $((b_{ij}))$ *is the* $(n\times n)$*-matrix-valued function inverse of* $((a_{ij}))$ *and* $\mathcal{A}_j^*(x)$ *is the matrix transposed of* $\mathcal{A}_j(x)$.

REMARK 8.4. We give an example of the system whose principal part is not a scalar differential operator in the whole domain and for which the classical maximum modulus principle is valid in $\mathbb{R}^{n+1}_T$.

Consider the parabolic system

$$\frac{\partial \boldsymbol{v}}{\partial t}-\sum_{j,k=1}^n\mathcal{A}_{jk}(x,t)\frac{\partial^2\boldsymbol{v}}{\partial x_j\partial x_k}+\sum_{j=1}^n\mathcal{A}_j(x,t)\frac{\partial\boldsymbol{v}}{\partial x_j}+\mathcal{A}_0(x,t)\boldsymbol{v}+\lambda^2\boldsymbol{v}=\boldsymbol{0} \tag{8.3.28}$$

in $\mathbb{R}^{n+1}_T$, where

$$\mathcal{A}_{ij}\in[\mathrm{C}_{\mathrm{b}}^{2+\alpha,\alpha/2}(\overline{\mathbb{R}^{n+1}_T})]^{m\times m},\ \mathcal{A}_j\in[\mathrm{C}_{\mathrm{b}}^{1+\alpha,\alpha/2}(\overline{\mathbb{R}^{n+1}_T})]^{m\times m},\ \mathcal{A}_0\in[\mathrm{C}_{\mathrm{b}}^{\alpha,\alpha/2}(\overline{\mathbb{R}^{n+1}_T})]^{m\times m}$$

and λ=const. Suppose the coefficients of system (8.3.28) do not depend on t in the layer $\mathbb{R}^{n+1}_\delta$, $0<\delta<T$. Let the matrix-valued functions $\mathcal{A}_{jk}(x,0)$, $\mathcal{A}_j(x,0)$, $\mathcal{A}_0(x,0)$, denoted by $\mathcal{A}_{jk}(x)$, $\mathcal{A}_j(x)$, $\mathcal{A}_0(x)$, respectively, satisfy conditions (i), (ii) of Theorem 8.8. Suppose further that $|\boldsymbol{v}(x,0)|\le 1$. Then $|\boldsymbol{v}(x,t)|\le 1$ in $\mathbb{R}^{n+1}_\delta$. By M we denote the value $\mathcal{K}(\mathbb{R}^n,T)$ for system (8.1.2). Since

$$\boldsymbol{u}(x,t)=\boldsymbol{v}(x,t)\exp(\lambda^2 t)$$

satisfies (8.1.2), it follows that

$$\sup\Big\{|\boldsymbol{v}(x,t)| : (x,t)\in\overline{\mathbb{R}^{n+1}_T}\backslash\mathbb{R}^{n+1}_\delta\Big\}\le M\exp(-\lambda^2\delta).$$

Thus, the solution of the Cauchy problem for system (8.3.28) obeys the classical maximum modulus principle for sufficiently large values of λ.

8.3.2. The case of complex coefficients

We can extend the results of the first subsection to systems (8.1.2), (8.1.3) with complex coefficients and solutions $\boldsymbol{u} = \boldsymbol{v} + i\boldsymbol{w}$, where $\boldsymbol{v}$ and $\boldsymbol{w}$ are m-component vector-valued functions with real components. The results are obtained by application of corresponding assertions on the maximum modulus for the real case to systems obtained by separation of real and imaginary parts (see Subsection 8.2.2). Thus we can formulate analogous theorems and corollaries as in Subsection 8.3.1. By $\mathbb{C}^m$ we denote the complex linear m-dimensional space with elements $\boldsymbol{a} + i\boldsymbol{b}$, where $\boldsymbol{a}, \boldsymbol{b} \in \mathbb{R}^m$, with the scalar product $(\boldsymbol{c}, \boldsymbol{d}) = c_1\overline{d}_1 + \cdots + c_m\overline{d}_m, \boldsymbol{c} = (c_1, \ldots, c_m), \boldsymbol{d} = (d_1, \ldots, d_m)$. The length of the vector $\boldsymbol{d}$ in $\mathbb{C}^m$ is defined as $|\boldsymbol{d}| = (\boldsymbol{d}, \boldsymbol{d})^{1/2}$.

We retain the notations of Subsection 8.2.2 and use them putting $\ell = 1$. By analogy with definition (8.3.1) of the constant $\mathcal{K}(\Omega, T)$ let

$$\mathcal{K}'(\Omega, T) = \sup \frac{\|\boldsymbol{u}\|_{[\mathbf{C}(\overline{Q}_T)]^m}}{\|\boldsymbol{u}|_{\overline{\Gamma}_T}\|_{[\mathbf{C}(\overline{\Gamma}_T)]^m}},$$

where the supremum is taken over all vector-valued functions $\boldsymbol{u} = \boldsymbol{v} + i\boldsymbol{w}$ in the class $[\mathbf{C}^{(2,1)}(Q_T)]^m \cap [\mathbf{C}(\overline{Q}_T)]^m$ that satisfy system (8.1.2) with complex coefficients. Here $\boldsymbol{v}$ and $\boldsymbol{w}$ are m-component vector-valued functions with real components.

THEOREM 8.9. *The classical maximum modulus principle holds for solutions of the system with complex coefficients*

$$\frac{\partial \boldsymbol{u}}{\partial t} - \sum_{j,k=1}^{n} \mathcal{A}_{jk}(t) \frac{\partial^2 \boldsymbol{u}}{\partial x_j \partial x_k} + \sum_{j=1}^{n} \mathcal{A}_j(t) \frac{\partial \boldsymbol{u}}{\partial x_j} = \mathbf{0}$$

in $\mathbb{R}^{n+1}_T$ *if and only if*

$$\mathcal{A}_{jk}(t) = a_{jk}(t)I, \quad \mathcal{A}_j(t) = a_j(t)I, \quad 1 \leq j, k \leq n,$$

where $((a_{jk}))$ *is a real positive-definite* $(n \times n)$*-matrix-valued function and* a_j *are real scalar functions.*

THEOREM 8.10. *Let the classical maximum modulus principle be valid for system* (8.1.2) *with complex coefficients in* Q_T $(\mathbb{R}^{n+1}_T)$. *Then*

(i) *for all* $x \in \Omega$ $(x \in \mathbb{R}^n)$ *the equalities*

$$\mathcal{A}_{jk}(x, 0) = a_{jk}(x)I, \quad 1 \leq j, k \leq n,$$

hold, where $((a_{jk}))$ *is a real positive-definite* $(n \times n)$*-matrix-valued function;*

(ii) *for all* $x \in \Omega$ $(x \in \mathbb{R}^n)$ *and all* $\boldsymbol{\xi}_j$, $\boldsymbol{\zeta} \in \mathbb{C}^m$, $j = 1, \ldots, n$, *with* $\Re(\boldsymbol{\xi}_j, \boldsymbol{\zeta}) = 0$, *the inequality*

$$\Re\Big\{ \sum_{j,k=1}^{n} a_{jk}(x)(\boldsymbol{\xi}_j, \boldsymbol{\xi}_k) + \sum_{j=1}^{n} (\mathcal{A}_j(x, 0)\boldsymbol{\xi}_j, \boldsymbol{\zeta}) + (\mathcal{A}_0(x, 0)\boldsymbol{\zeta}, \boldsymbol{\zeta}) \Big\} \geq 0$$

holds.

THEOREM 8.11. *The classical maximum modulus principle is valid for solutions of system* (8.1.3) *with complex coefficients in* $Q_T(\mathbb{R}^{n+1}_T)$ *if and only if*

(i) *for all* $x \in \Omega$ $(x \in \mathbb{R}^n)$ *the equalities*

$$\mathcal{A}_{jk}(x) = a_{jk}(x)I, \quad 1 \leq j, k \leq n,$$

hold, where $((a_{jk}))$ *is a real positive-definite* $(n \times n)$*-matrix-valued function;*

(ii) *for all* $x \in \Omega$ $(x \in \mathbb{R}^n)$ *and for all* $\boldsymbol{\xi}_j,\ \boldsymbol{\zeta} \in \mathbb{C}^m$, $j = 1, \dots, n$, *with* $\Re(\boldsymbol{\xi}_j,\ \boldsymbol{\zeta}) = 0$, *the inequality*

$$\Re\Big\{ \sum_{j,k=1}^{n} a_{jk}(x)(\boldsymbol{\xi}_j, \boldsymbol{\xi}_k) + \sum_{j=1}^{n} (\mathcal{A}_j(x)\boldsymbol{\xi}_j, \boldsymbol{\zeta}) + (\mathcal{A}_0(x)\boldsymbol{\zeta}, \boldsymbol{\zeta}) \Big\} \geq 0$$

is satisfied.

COROLLARY 8.3. *The classical maximum modulus principle is valid for the system*

$$\frac{\partial \boldsymbol{u}}{\partial t} - \sum_{j,k=1}^{n} \mathcal{A}_{jk}(x) \frac{\partial^2 \boldsymbol{u}}{\partial x_j \partial x_k} + \sum_{j=1}^{n} \mathcal{A}_j(x) \frac{\partial \boldsymbol{u}}{\partial x_j} = \mathbf{0}$$

with complex coefficients in Q_T $(\mathbb{R}_T^{n+1})$ *if and only if for all* $x \in \Omega$ $(x \in \mathbb{R}^n)$ *the equalities*

$$\mathcal{A}_{jk}(x) = a_{jk}(x)I, \quad \mathcal{A}_j(x) = a_j(x)I, \quad 1 \leq j, k \leq n,$$

hold, where $((a_{jk}))$ *is a real positive-definite* $(n \times n)$*-matrix-valued function and* a_j *are real scalar functions.*

COROLLARY 8.4. *The classical maximum modulus principle is valid for solutions of system* (8.1.3) *with complex coefficients in* Q_T $(\mathbb{R}_T^{n+1})$ *if and only if the condition* (i) *of Theorem* 8.11 *is satisfied and for all* $x \in \Omega$ $(x \in \mathbb{R}^n)$ *an for any* $\boldsymbol{\zeta} \in \mathbb{C}^m$, $|\boldsymbol{\zeta}| = 1$ *the inequality*

$$\sum_{i,j=1}^{n} b_{ij}(x) \Re(\mathcal{A}_i(x)\boldsymbol{\zeta}, \boldsymbol{\zeta}) \Re(\mathcal{A}_j(x)\boldsymbol{\zeta}, \boldsymbol{\zeta})) - \sum_{i,j=1}^{n} b_{ij}(x) (\mathcal{A}_i^*(x)\boldsymbol{\zeta},\ \mathcal{A}_j^*(x)\boldsymbol{\zeta})$$
$$+ 4\Re(\mathcal{A}_0(x)\boldsymbol{\zeta}, \boldsymbol{\zeta}) \geq 0$$

holds. Here $((b_{jk}(x)))$ *is the* $(n \times n)$*-matrix inverse of* $((a_{jk}(x)))$ *and* $\mathcal{A}_j^*(x)$ *is the adjoint matrix of* $\mathcal{A}_j(x)$.

We remark that the second sum is real by symmetry of the matrix $((b_{ij}(x)))$.

In particular, the next assertion follows from Corollary 8.4 for the scalar parabolic equation with complex coefficients

$$\frac{\partial u}{\partial t} - \sum_{j,k=1}^{n} a_{jk}(x) \frac{\partial^2 u}{\partial x_j \partial x_k} + \sum_{j=1}^{n} a_j(x) \frac{\partial u}{\partial x_j} + a_0(x)u = 0. \tag{8.3.29}$$

COROLLARY 8.5. *The classical maximum modulus principle is valid for* (8.3.29) *in* Q_T $(\mathbb{R}_T^{n+1})$ *if and only if*

(i) *the* $(n \times n)$*-matrix-valued function* $((a_{jk}(x)))$ *is real and positive-definite*

(ii) *for all* $x \in \Omega$ $(x \in \mathbb{R}^n)$ *the inequality*

$$4\Re a_0(x) \geq \sum_{j,k=1}^{n} b_{jk}(x) \Im a_j(x)\ \Im a_k(x)$$

holds.

8.4. The parabolic Lamé system

Let $\omega > -1$. In the half-space $\mathbb{R}^{n+1}_+ = \{(x,t) \in \mathbb{R}^{n+1} : t > 0\}$, $n \geq 3$, we consider a strongly parabolic system (cf. Remark 7.4)

$$\frac{\partial \boldsymbol{u}}{\partial t} - \Delta\, \boldsymbol{u} - \omega \text{ grad div } \boldsymbol{u} = \boldsymbol{0}, \tag{8.4.1}$$

which involves the operator of elastostatics

$$\Delta\, + \omega \text{ grad div},$$

where $\boldsymbol{u} = (u_1, \dots, u_n)$.

According to Eidel'man [**Ei1**], there exists a unique solution bounded and continuous up to $\partial\mathbb{R}^{n+1}_+$, of the Cauchy problem

$$\frac{\partial \boldsymbol{u}}{\partial t} - \Delta\, \boldsymbol{u} - \omega \text{ grad div } \boldsymbol{u} = \boldsymbol{0} \text{ in } \mathbb{R}^{n+1}_+, \quad \boldsymbol{u} = \boldsymbol{\varphi} \text{ on } \partial\mathbb{R}^{n+1}_+, \tag{8.4.2}$$

where $\boldsymbol{\varphi}$ is a bounded and continuous vector-valued function on $\mathbb{R}^n$, and this solution can be represented in the form

$$\boldsymbol{u}(x,t) = \int_{\partial\mathbb{R}^{n+1}_+} G(t, x-y)\boldsymbol{\varphi}(y)\, dy. \tag{8.4.3}$$

LEMMA 8.6. *The elements of the fundamental matrix $G(t,x-y)$ of solutions of the Cauchy problem* (8.4.2) *are given by the equality*

$$G_{jk}(t,x) = \frac{1}{(2\sqrt{\pi})^n}\left[\left(t^{-n/2}e^{-|x|^2/4t} - \frac{1}{2}\int_t^{(\omega+1)t} \tau^{-(n+2)/2}e^{-|x|^2/4\tau}d\tau\right)\delta_{jk}\right.$$

$$\left. + \frac{x_j x_k}{4}\int_t^{(\omega+1)t} \tau^{-(n+4)/2}e^{-|x|^2/4\tau}d\tau\right]. \tag{8.4.4}$$

PROOF. For $\boldsymbol{u}$ defined on $\mathbb{R}^{n+1}_+$, consider the system which is equivalent to (8.4.1):

$$\frac{\partial \boldsymbol{u}}{\partial t} - \Delta \boldsymbol{u} + \text{grad}\, p = \boldsymbol{0}, \quad p = -\omega \text{ div } \boldsymbol{u}, \tag{8.4.5}$$

together with the initial condition

$$\boldsymbol{u}\big|_{t=0} = \boldsymbol{\varphi}. \tag{8.4.6}$$

We make the Fourier transform in (8.4.5) in x-variables and use the notations

$$F[\varphi] = \frac{1}{(2\pi)^{n/2}}\int_{\mathbb{R}^n} e^{-i(x,\xi)}\varphi(x)dx, \qquad F^{-1}[\psi] = \frac{1}{(2\pi)^{n/2}}\int_{\mathbb{R}^n} e^{i(x,\xi)}\psi(\xi)d\xi.$$

As usual, the summation over repeated indices is meant. Then

$$\frac{dF[u_j]}{dt} + |\xi|^2F[u_j] + i\xi_jF[p] = 0, \quad j = 1, \dots, n, \tag{8.4.7}$$

$$F[p] = -i\omega\xi_jF[u_j]. \tag{8.4.8}$$

Multiplying (8.4.7)by ξ_j and summing over j, we obtain

$$\frac{dF[u_j]\xi_j}{dt} + |\xi|^2F[u_j]\xi_j + i|\xi|^2F[p] = 0,$$

which together with (8.4.8) leads to the equation

$$\frac{dF[p]}{dt} + (\omega+1)|\xi|^2 F[p] = 0. \tag{8.4.9}$$

This equation has the solution

$$F[p] = F[p_0]e^{-(\omega+1)|\xi|^2 t}, \tag{8.4.10}$$

where $p_0(x) = p(x,0)$.

By (8.4.8), the equality (8.4.10) can be written as

$$F[p] = -i\omega\big(F[\varphi], \boldsymbol{\xi}\big)e^{-(\omega+1)|\xi|^2 t}\ , \tag{8.4.11}$$

which by (8.4.7) implies

$$\frac{dF[u_j]}{dt} + |\xi|^2 F[u_j] + \xi_j\omega\big(F[\varphi], \xi\big)e^{-(\omega+1)|\xi|^2 t} = 0\ . \tag{8.4.12}$$

The solution of this equation is

$$F[u_j] = C_j e^{-|\xi|^2 t} + \frac{\xi_j\big(F[\varphi], \xi\big)}{|\xi|^2}e^{-(\omega+1)|\xi|^2 t}.$$

Using (8.4.6), we find

$$F[u_j] = F[\boldsymbol{\varphi}_j]e^{-|\xi|^2 t} + \frac{\xi_j\big(F[\boldsymbol{\varphi}], \xi\big)}{|\xi|^2}\Big\{e^{-(\omega+1)|\xi|^2 t} - e^{-|\xi|^2 t}\Big\}. \tag{8.4.13}$$

This can be written as

$$F[\boldsymbol{u}] = \Psi(\xi, t)F[\boldsymbol{\varphi}], \tag{8.4.14}$$

where Ψ is the $(n \times n)$-matrix with elements

$$\Psi_{jk}(\xi, t) = e^{-|\xi|^2 t}\delta_{jk} + \frac{\xi_j\xi_k}{|\xi|^2}\Big\{e^{-(\omega+1)|\xi|^2 t} - e^{-|\xi|^2 t}\Big\}. \tag{8.4.15}$$

Hence

$$\boldsymbol{u}(x,t) = \frac{1}{(2\pi)^{n/2}}\int_{\mathbb{R}^n} e^{i(x,\xi)}\Psi(\xi,t)F[\boldsymbol{\varphi}](\xi)d\xi = \int_{\mathbb{R}^n} G(t, x-y)\boldsymbol{\varphi}(y)\, dy,$$

where

$$G(t,x) = \frac{1}{(2\pi)^n}\int_{\mathbb{R}^n} e^{i(x,\xi)}\Psi(\xi,t)d\xi.$$

Thus, by (8.4.15),

$$G_{jk}(t,x) = \frac{1}{(2\pi)^n}\int_{\mathbb{R}^n} e^{i(x,\xi)}\Big\{e^{-|\xi|^2 t}\delta_{jk} + \frac{\xi_j\xi_k}{|\xi|^2}\Big(e^{-(\omega+1)|\xi|^2 t} - e^{-|\xi|^2 t}\Big)\Big\}\, d\xi. \tag{8.4.16}$$

Since

$$\frac{1}{(2\pi)^n}\int_{\mathbb{R}^n} e^{-|\xi|^2 t + i(x,\xi)}d\xi = \frac{1}{(2\sqrt{\pi t})^n}e^{-|x|^2/4t}\ , \tag{8.4.17}$$

we get $G_{jk}(t,x)$ after evaluating the integral

$$\int_{\mathbb{R}^n} \frac{\xi_j\xi_k}{|\xi|^2}e^{-a|\xi|^2 t + i(x,\xi)}d\xi = -\frac{\partial^2}{\partial x_j \partial x_k}\int_{\mathbb{R}^n} \frac{1}{|\xi|^2}e^{-a|\xi|^2 t + i(x,\xi)}d\xi\ , \tag{8.4.18}$$

where a is a positive parameter.

To evaluate the integral on the right-hand side, we pass to spherical coordinates, directing the polar axis along the vector x:

$$\int_{\mathbb{R}^n} \frac{e^{-a|\xi|^2 t + i(x,\xi)}}{|\xi|^2} d\xi = \omega_{n-1} \int_0^\infty \left[\int_0^\pi e^{ir\rho\cos\theta} \sin^{n-2}\theta d\theta \right] e^{-a\rho^2 t} \rho^{n-3} d\rho, \tag{8.4.19}$$

$\rho = |\xi|$, $r = |x|$, θ is the angle between the vectors ξ and x. By (5.3.5), we find

$$\int_0^\pi e^{ir\rho\cos\theta} \sin^{n-2}\theta \, d\theta = \frac{2^{(n-2)/2}\sqrt{\pi}\,\Gamma\left(\frac{n-1}{2}\right)}{(r\rho)^{(n-2)/2}} J_{\frac{n-2}{2}}(r\rho) ,$$

which implies that (8.4.19) can be written as

$$\int_{\mathbb{R}^n} \frac{e^{-a|\xi|^2 t + i(x,\xi)}}{|\xi|^2} d\xi = \frac{(2\pi)^{n/2}}{r^{(n-2)/2}} \int_0^\infty e^{-a\rho^2 t} \rho^{(n-4)/2} J_{\frac{n-2}{2}}(r\rho) d\rho . \tag{8.4.20}$$

Hence by applying known formula for integral (see, e.g., Gradshtein and Ryzhik [**GR**], **6.631.5**),

$$\int_0^\infty e^{-\alpha x^2} x^{\nu-1} J_\nu(\beta x) dx = 2^{\nu-1}\beta^{-\nu} \int_0^{\beta^2/4\alpha} e^{-x} x^{\nu-1} dx ,$$

where J_ν is the Bessel function of the first kind, $\Re\, \alpha > 0$, $\beta > 0$, and $\nu > 0$, we write (8.4.20) in the form

$$\int_{\mathbb{R}^n} \frac{e^{-a|\xi|^2 t + i(x,\xi)}}{|\xi|^2} d\xi = \frac{2^{n-2}\pi^{n/2}}{r^{n-2}} \int_0^{\frac{r^2}{4at}} e^{-s} s^{(n-4)/2} ds .$$

Therefore, by (8.4.18)

$$\begin{aligned} &\frac{1}{(2\pi)^n} \int_{\mathbb{R}^n} \frac{\xi_j \xi_k}{|\xi|^2} \left(e^{-(\omega+1)|\xi|^2 t} - e^{-|\xi|^2 t} \right) e^{i(x,\xi)} d\xi \\ &= -\frac{1}{(2\pi)^n} \frac{\partial^2}{\partial x_j \partial x_k} \left\{ \frac{2^{n-2}\pi^{n/2}}{r^{n-2}} \int_{\frac{r^2}{4t}}^{\frac{r^2}{4(\omega+1)t}} e^{-s} s^{(n-4)/2} ds \right\} . \end{aligned} \tag{8.4.21}$$

Making the change of variable $\tau = r^2/(4s)$ in the integral on the right-hand side, we obtain

$$-\frac{1}{(2\pi)^n} \left\{ \frac{2^{n-2}\pi^{n/2}}{r^{n-2}} \int_{\frac{r^2}{4t}}^{\frac{r^2}{4(\omega+1)t}} e^{-s} s^{(n-4)/2} ds \right\} = \frac{1}{2^n \pi^{n/2}} \int_t^{(\omega+1)t} \tau^{-n/2} e^{-r^2/4\tau} d\tau ,$$

which together with (8.4.16), (8.4.17), and (8.4.21), results in the representation

$$G_{jk}(t,x) = \frac{1}{(2\sqrt{\pi})^n} \left\{ t^{-n/2} e^{-|x|^2/4t} \delta_{jk} + \frac{\partial^2}{\partial x_j \partial x_k} \int_t^{(\omega+1)t} \tau^{-n/2} e^{-|x|^2/4\tau} d\tau \right\},$$

equivalent to (8.4.4). □

REMARK 8.5. If $\operatorname{div} \boldsymbol{\varphi} = 0$, the vector-valued component $\boldsymbol{u} \in [\mathrm{C}^2(\mathbb{R}^{n+1}_+)]^n \cap [\mathrm{C_b}(\overline{\mathbb{R}^{n+1}_+})]^n$ of the solution $\{\boldsymbol{u}, p\}$ of the Cauchy problem (8.4.5), (8.4.6) satisfies

$$|\boldsymbol{u}(x,t)| \leq \sup_{\mathbb{R}^n} |\boldsymbol{\varphi}(x)|, \tag{8.4.22}$$

where $(x,t) \in \mathbb{R}^{n+1}_+$. In fact, if div $\boldsymbol{\varphi} = 0$, then, by (8.4.11), $p \equiv 0$. Hence, the first equation of system (8.4.5) becomes the heat equation with the solution satisfying (8.4.22) since

$$|(\boldsymbol{u}(x,t), \boldsymbol{z})| \le \sup_{\mathbb{R}^n} |(\boldsymbol{\varphi}(x), \boldsymbol{z})|$$

for any $\boldsymbol{z} \in \mathbb{R}^n$.

We note that solution (8.4.3) of the Cauchy problem (8.4.2) can be represented in the form

$$\boldsymbol{u}(x,t) = \int_{\partial\mathbb{R}^{n+1}_+} Z(t,x;\ y)\boldsymbol{\psi}(y)\ \lambda(t,x;dy), \tag{8.4.23}$$

where the components of the $(m \times m)$-matrix-valued function

$$Z(t,x;\ y) = t^{n/2} e^{\frac{|x-y|^2}{4\beta t}}\ G(t, x-y), \tag{8.4.24}$$

$\beta \ge \max\{1,\ \omega+1\}$, are bounded and continuous on $\mathbb{R}^n$ for any fixed $(x,t) \in \mathbb{R}^{n+1}_+$, and the set function

$$\lambda(t,x;\mathcal{B}) = \int_{\mathcal{B}} t^{-n/2} e^{-\frac{|x-y|^2}{4\beta t}}\ dy \tag{8.4.25}$$

is defined on Borel subsets $\mathcal{B}$ of $\partial\mathbb{R}^{n+1}_+ = \mathbb{R}^n$ for any fixed $(x,t) \in \mathbb{R}^{n+1}_+$.

THEOREM 8.12. *The exact constant* $\mathcal{K}(\mathbb{R}^{n+1}_+)$ *in the inequality*

$$|\boldsymbol{u}(x,t)| \le \mathcal{K} \sup\{\ |\boldsymbol{u}(y,0)| : y \in \mathbb{R}^n\}$$

for solutions $\boldsymbol{u} \in [\mathrm{C_b}(\overline{\mathbb{R}^{n+1}_+})]^n \bigcap [\mathrm{C}^{(2,1)}(\mathbb{R}^{n+1}_+)]^n$ *of system* (8.4.1), *where* $(x,t) \in \mathbb{R}^{n+1}_+$, *is equal to*

$$\frac{2}{\sqrt{\pi}\Gamma((n-1)/2)} \int_0^{\pi/2}\int_0^{\infty} \Big[A^2(\rho)\sin^2\vartheta + B^2(\rho)\cos^2\vartheta\Big]^{1/2} (\rho^{1/2}\sin\vartheta)^{n-2} d\rho\, d\vartheta.$$

Here

$$A(\rho) = e^{-\rho} - \int_\kappa^1 e^{-\rho t^2} t^{n-1} dt, \quad B(\rho) = \kappa^n e^{-\rho\kappa^2} + (n-1)\int_\kappa^1 e^{-\rho t^2} t^{n-1} dt,$$

where $\kappa = (\omega+1)^{-1/2}$.

PROOF. By Theorem 8.1 we have

$$\mathcal{K}(\mathbb{R}^{n+1}_+) = \sup_{(x,t)\in\mathbb{R}^{n+1}_+}\ \sup_{|\boldsymbol{z}|=1} \int_{\partial\mathbb{R}^{n+1}_+} |Z^*(t,x;\ y)\boldsymbol{z}|\ \lambda(t,x;dy).$$

From this, using (8.4.23) and (8.4.25), we obtain

$$\mathcal{K}(\mathbb{R}^{n+1}_+) = \sup_{(x,t)\in\mathbb{R}^{n+1}_+}\ \sup_{|\boldsymbol{z}|=1} \int_{\partial\mathbb{R}^{n+1}_+} |G^*(t, x-y)\boldsymbol{z}|\, dy. \tag{8.4.26}$$

Substituting $\tau = t/\xi$ into (8.4.4), we get

$$G_{ij}(t,x) = t^{-n/2} P_{ij}(x/t^{1/2}), \tag{8.4.27}$$

where $P_{ij}(x) = \Lambda(|x|)\delta_{ij} + x_i x_j \Phi(|x|)$. Here

$$\Lambda(r) = (2\sqrt{\pi})^{-n}\left(e^{-r^2/4} - \frac{1}{2}\int_{\kappa^2}^{1} \xi^{(n-2)/2} e^{-r^2\xi/4}\, d\xi\right),$$

$$4\Phi(r) = (2\sqrt{\pi})^{-n}\int_{\kappa^2}^{1} \xi^{n/2} e^{-r^2\xi/4}\, d\xi\,.$$

It follows from (8.4.26) and (8.4.27) that

$$\mathcal{K}(\mathbb{R}^{n+1}_+) = \sup_{(x,t)\in\mathbb{R}^{n+1}_+}\ \sup_{|\boldsymbol{z}|=1}\int_{\partial\mathbb{R}^{n+1}_+}\left|t^{-n/2}P^*\left(\frac{x-y}{t^{1/2}}\right)\boldsymbol{z}\right| dy.$$

Setting $\eta = (x-y)t^{-1/2}$ we get

(8.4.28)
$$\mathcal{K}(\mathbb{R}^{n+1}_+) = \sup_{|\boldsymbol{z}|=1}\int_{\partial\mathbb{R}^{n+1}_+} |P^*(\eta)\boldsymbol{z}|d\eta.$$

After simple transformations, we find that

$$|P^*(\eta)\boldsymbol{z}| = \Big[\Lambda^2(|\eta|)|\boldsymbol{z}|^2 + 2(\eta,\boldsymbol{z})^2\Lambda(|\eta|)\Phi(|\eta|) + |\eta|^2(\eta,\boldsymbol{z})^2\Phi^2(|\eta|)\Big]^{1/2}.$$

By (8.4.28) and spherical coordinates we arrive at

$$\begin{aligned}
\mathcal{K}(\mathbb{R}^{n+1}_+) &= \sup_{|\boldsymbol{z}|=1}\int_{\mathbb{R}^n}\Big[\Lambda^2(|\eta|) + 2(\eta,\boldsymbol{z})^2\Lambda(|\eta|)\Phi(|\eta|) + |\eta|^2(\eta,\boldsymbol{z})^2\Phi^2(|\eta|)\Big]^{1/2} d\eta\\
&= \sup_{|\boldsymbol{z}|=1}\int_{\mathbb{S}^{n-1}} d\sigma \int_0^\infty \{\Lambda^2(r)+(\boldsymbol{e}_\sigma,\boldsymbol{z})^2[2r^2\Lambda(r)\Phi(r)+r^4\Phi^2(r)]\}^{1/2} r^{n-1}dr\\
&= \sup_{|\boldsymbol{z}|=1}\int_{\mathbb{S}^{n-1}} d\sigma \int_0^\infty \{\Lambda^2(r)+(\boldsymbol{e}_\sigma,\boldsymbol{z})^2[(\Lambda(r)+r^2\Phi(r))^2-\Lambda^2(r)]\}^{1/2} r^{n-1}dr.
\end{aligned}$$

It is clear that the iterated integral in the last equality does not depend on $\boldsymbol{z}$, $|\boldsymbol{z}|=1$; therefore, the supremum over $\boldsymbol{z}$ is omitted in what follows. Using the change of variable $\rho = r^2/4$, we get the representation

$$\mathcal{K}(\mathbb{R}^{n+1}_+) = \frac{1}{2}\pi^{-n/2}\int_{\mathbb{S}^{n-1}} d\sigma \int_0^\infty \rho^{(n-2)/2}\{A^2(\rho) + (\boldsymbol{e}_\sigma,\boldsymbol{z})^2[B^2(\rho) - A^2(\rho)]\}^{1/2} d\rho,$$

where

$$A(\rho) = (2\sqrt{\pi})^n\Lambda(2\sqrt{\rho}),\quad B(\rho) = (2\sqrt{\pi})^n[\Lambda(2\sqrt{\rho}) + 4\rho\Phi(2\sqrt{\rho})].$$

Transforming the functions $A(\rho)$ and $B(\rho)$, we find that

$$A(\rho) = e^{-\rho} - \frac{1}{2}\int_{\kappa^2}^{1}\xi^{(n-2)/2}e^{-\rho\xi}d\xi = e^{-\rho} - \int_{\kappa}^{1} e^{-\rho t^2}t^{n-1}dt,$$

$$B(\rho) = \mathcal{A}(\rho) + \int_{\kappa^2}^{1}\xi^{n/2}e^{-\rho\xi}d\xi = \kappa^n e^{-\rho\kappa^2} + (n-1)\int_{\kappa}^{1} e^{-\rho t^2}t^{n-1}dt.$$

We introduce the angle ϑ between $\boldsymbol{e}_\sigma$ and $\boldsymbol{z}$. Then

$$
\begin{aligned}
\mathcal{K}(\mathbb{R}^{n+1}_+) &= \frac{1}{2}\pi^{-n/2}\int_0^\pi \dots \int_0^\pi \int_0^{2\pi}\int_0^\infty \{A^2(\rho)+\cos^2\vartheta[B^2(\rho)-A^2(\rho)]\}^{1/2} \\
&\times \ \rho^{(n-2)/2}\sin^{n-2}\vartheta\sin^{n-3}\vartheta_1\dots\sin\vartheta_{n-3}\,d\rho\,d\vartheta\,d\vartheta_1\cdots d\vartheta_{n-2} \\
&= \frac{1}{2}\pi^{-n/2}4\pi\Big(\prod_{k=1}^{n-3}2\int_0^{\pi/2}\sin^k\varphi\,d\varphi\Big) \\
&\times \int_0^{\pi/2}\int_0^\infty [A^2(\rho)\sin^2\vartheta+B^2(\rho)\sin^2\vartheta]^{1/2}(\rho^{1/2}\sin\vartheta)^{n-2}d\rho\,d\vartheta.
\end{aligned}
$$

□

8.5. Comments to Chapter 8

It is well known that solutions of parabolic second order equations with real coefficients in the cylinder

$$Q_T=\{(x,t): x\in\Omega,\ 0<t\le T\},\ \Omega\subset\mathbb{R}^n,$$

satisfy the maximum modulus principle. Namely, for any solution of the equation

$$\frac{\partial u}{\partial t}-\sum_{j,k=1}^n a_{jk}(x,t)\frac{\partial^2 u}{\partial x_j\partial x_k}+\sum_{j=1}^n a_j(x,t)\frac{\partial u}{\partial x_j}+a_0(x,t)u=0,$$

where $((a_{ij}))$ is a positive-definite $(n\times n)$-matrix-valued function and $a_0\ge 0$, the inequality

$$|u(x,t)|\le\sup\{|u(y,\tau)|:\ (y,\tau)\in\partial Q_T,\ \tau<T\}$$

holds. This classical fact was extended to parabolic second order systems with scalar coefficients in the first and second derivatives in Stys [**Sty2**], where a sufficient condition for validity of the classical maximum modulus principle was proved. We note that maximum principles for weakly coupled parabolic systems are discussed in the books by Protter and Weinberger [**PW**] and Walter [**Wal**] which also contain rich bibliographies in this subject.

The necessity of the condition $m=1$ for positivity of the fundamental solution of a parabolic equation of order $2m$ with constant coefficients was shown by Kimura and Otsuka [**KO**]. Note that the positivity of the fundamental solution of a parabolic equation is equivalent to the classical maximum modulus principle for its solutions (cf. the proof of Lemma (8.3)). Necessary and sufficient conditions for positivity of elements of the matrix of fundamental solutions of the Cauchy problem for the system (8.1.1) were found by Otsuka [**Ot**]. A componentwise maximum principle, described by Protter and Weinberger [**PW**], holds for systems with positive elements of the matrix of fundamental solutions. Conditions for a generalized componentwise maximum principle for parabolic strongly coupled systems are obtained by Kamynin and Khimchenko [**KK3**]. A sufficient condition for the maximum principle for $\alpha(x)|\mathbf{u}|$ with a certain function $\alpha(x)$ and the solution $\mathbf{u}$ of the parabolic system is due to Zhou [**Zh2**].

For $n=3$ Green's matrix for the parabolic Lamé system (8.4.1) was given in the paper [**Ei2**] by Eidel'man. Boundary value problems for the parabolic Lamé system were considered in the papers by Z. Shen [**Shen**] and Nyström [**Ny**].

CHAPTER 9

Maximum Modulus Principle for Parabolic Systems with Zero Boundary Data

9.1. Introduction

In the present chapter we consider the question of validity of the maximum modulus principle for strongly parabolic $(m \times m)$-system

$$\frac{\partial \boldsymbol{u}}{\partial t} - \sum_{j,k=1}^{n} \mathcal{A}_{jk}(x) \frac{\partial^2 \boldsymbol{u}}{\partial x_j \partial x_k} + \sum_{j=1}^{n} \mathcal{A}_j(x) \frac{\partial \boldsymbol{u}}{\partial x_j} + \mathcal{A}_0(x)\boldsymbol{u} = \boldsymbol{0} \tag{9.1.1}$$

with real or complex coefficients, $\mathcal{A}_{jk} = \mathcal{A}_{kj}$. Let us assume that (9.1.1) holds on the cylinder $Q_T = \Omega \times (0, T]$ and that solutions have zero Dirichlet data on $\partial\Omega \times [0, T]$. Here and elsewhere in this chapter Ω is a domain in $\mathbb{R}^n$ with compact closure and the boundary $\partial\Omega$ in the class $\mathrm{C}^{2,\alpha}$, $\alpha \in (0, 1]$.

We assume that $\mathcal{A}_{jk}, \mathcal{A}_j, \mathcal{A}_0$ are $(m \times m)$-matrix valued functions with elements in the spaces $\mathrm{C}^{2,\alpha}_{\mathrm{b}}(\mathbb{R}^n), \mathrm{C}^{1,\alpha}_{\mathrm{b}}(\mathbb{R}^n)$, $\mathrm{C}^{\alpha}_{\mathrm{b}}(\mathbb{R}^n)$ for real coefficients and in the spaces $\mathbf{C}^{2,\alpha}_{\mathbf{b}}(\mathbb{R}^n), \mathbf{C}^{1,\alpha}_{\mathbf{b}}(\mathbb{R}^n)$, $\mathbf{C}^{\alpha}_{\mathbf{b}}(\mathbb{R}^n)$ for complex coefficients, respectively.

In Section 9.2 we consider the case of real coefficients. Let F be a subset of $\mathbb{R}^n$. By $[\mathrm{C}_{\mathrm{b}}(F)]^m$ we mean the space of continuous and bounded m-component vector-valued functions on F supplied with the norm

$$\|\boldsymbol{f}\|_{[\mathrm{C}_{\mathrm{b}}(F)]^m} = \sup\{|\boldsymbol{f}(\zeta)| : \ \zeta \in F\}.$$

By $[\mathring{\mathrm{C}}(\overline{Q}_T)]^m$ we denote the space of m-component vector-valued functions from $[\mathrm{C}(\overline{Q}_T)]^m$ vanishing on the cylindrical surface $\overline{S}_T$, where $S_T = \partial\Omega \times (0, T]$. By $[\mathrm{C}^{(2,1)}(Q_T)]^m$ we mean the space of m-component vector-valued functions on Q_T whose derivatives with respect to x up to the second order and first derivative with respect to t are continuous.

Let $\mathring{\mathcal{K}}(\Omega, T)$ denote the best constant in the inequality

$$|\boldsymbol{u}(x, t)| \leq \mathcal{K} \sup\{|\boldsymbol{u}(y, 0)| : y \in \overline{\Omega}\},$$

where $(x, t) \in Q_T$ and $\boldsymbol{u}$ is a solution of system (9.1.1) in $[\mathrm{C}^{(2,1)}(Q_T)]^m \cap [\mathring{\mathrm{C}}(\overline{Q}_T)]^m$.

Let furthermore $\mathcal{K}(\Omega, T)$ be the best constant in the inequality

$$|\boldsymbol{u}(x, t)| \leq \mathcal{K} \sup\{|\boldsymbol{u}(y, \tau)| : (y, \tau) \in \partial Q_T, \ \tau < T\},$$

where $(x, t) \in Q_T$ and $\boldsymbol{u} \in [\mathrm{C}^{(2,1)}(Q_T)]^m \cap [\mathrm{C}(\overline{Q}_T)]^m$.

In the preceding chapter necessary and sufficient conditions for $\mathcal{K}(\Omega, T) = 1$ were obtained. Since $1 \leq \mathring{\mathcal{K}}(\Omega, T)$, these conditions are sufficient for $\mathring{\mathcal{K}}(\Omega, T) = 1$. Are they also necessary for $\mathring{\mathcal{K}}(\Omega, T) = 1$? This question is addressed in the present chapter, where it is shown that for the strongly elliptic system (9.1.1) the answer is positive.

The main result is:

THEOREM 1. *The maximum modulus principle is valid for solutions of system* (9.1.1) *in the class* $[\mathrm{C}^{(2,1)}(Q_T)]^m \cap [\mathring{\mathrm{C}}(\overline{Q}_T)]^m$ *if and only if:*

(i) *for all* $x \in \Omega$ *the equalities*

$$\mathcal{A}_{jk}(x) = a_{jk}(x)I, \quad 1 \leq j,k \leq n,$$

hold, where $((a_{jk}(x)))$ *is a positive-definite* $(n \times n)$*-matrix-valued function in* $\overline{\Omega}$*;*

(ii) *for all* $x \in \Omega$ *and for any* $\boldsymbol{\xi}_j, \boldsymbol{\xi} \in \mathbb{R}^m$, $j = 1, \dots, n$ *with* $(\boldsymbol{\xi}_j, \boldsymbol{\zeta}) = 0$ *the inequality*

$$\sum_{j,k=1}^{n} a_{jk}(x)(\boldsymbol{\xi}_j, \boldsymbol{\xi}_k) + \sum_{j=1}^{n} (\mathcal{A}_j(x)\boldsymbol{\xi}_j, \boldsymbol{\zeta}) + (A_0(x)\boldsymbol{\zeta}, \boldsymbol{\xi}) \geq 0$$

is valid.

This criterion admits the following equivalent formulation.

COROLLARY 1. *The maximum modulus principle holds for solutions of system* (9.1.1) *in the class* $[\mathrm{C}^{(2,1)}(Q_T)]^m \cap [\mathring{\mathrm{C}}(\overline{Q}_T)]^m$ *if and only if condition* (i) *of Theorem* 1 *is satisfied and for all* $x \in \Omega$ *and any* $\boldsymbol{\zeta} \in \mathbb{R}^m$, $|\boldsymbol{\zeta}| = 1$, *the inequality*

$$\sum_{i,j=1}^{n} b_{ij}(x)[(A_i(x)\boldsymbol{\zeta}, \boldsymbol{\zeta})(\mathcal{A}_j(x)\boldsymbol{\zeta}, \boldsymbol{\zeta}) - (A_i^*(x)\boldsymbol{\zeta}, \mathcal{A}_j^*(x)\boldsymbol{\zeta})] + 4(A_0(x)\boldsymbol{\zeta}, \boldsymbol{\zeta}) \geq 0$$

is valid. Here $((b_{ij}))$ *is the* $(n \times n)$*-matrix-valued function inverse of* $((a_{ij}))$ *and* $*$ *means the passage to the transposed matrix.*

Section 9.3 is devoted to similar results for the systems with complex coefficients which follow directly from the corresponding results obtained in Section 9.2 for the real coefficients case. As a particular case we formulate the following assertion for the scalar parabolic equation with complex coefficients.

COROLLARY 2. *For the scalar parabolic equation with complex coefficients*

$$\frac{\partial u}{\partial t} - \sum_{j,k=1}^{n} a_{jk}(x)\frac{\partial^2 u}{\partial x_j \partial x_k} + \sum_{j=1}^{n} a_j(x)\frac{\partial u}{\partial x_j} + a_0(x)u = 0, \tag{9.1.2}$$

the maximum modulus principle holds for solutions of (9.1.2) *in* $\mathbf{C}^{(2,1)}(Q_T) \cap \mathring{\mathbf{C}}(\overline{Q}_T)$ *if and only if:*

(i) *the* $(n \times n)$*-matrix-valued function* $((a_{jk}))$ *is real positive-definite in* $\overline{\Omega}$*;*

(ii) *for all* $x \in \Omega$ *the inequality*

$$4\Re\, a_0(x) \geq \sum_{j,k=1}^{n} b_{jk}(x)\, \Im\, a_j(x)\, \Im\, a_k(x)$$

holds.

9.2. The case of real coefficients

9.2.1. Notation

By $[\mathrm{C}_{\mathrm{b}}^{k,\alpha}(\mathbb{R}^n)]^m$ we denote the space of m-component real vector-valued functions with continuous and bounded in $\mathbb{R}^n$ derivatives up to order k which satisfy the uniform Hölder condition with exponent α. A similar space of $(m \times m)$-matrix-valued functions will be denoted by $[\mathrm{C}_{\mathrm{b}}^{k,\alpha}(\mathbb{R}^n)]^{m \times m}$.

We introduce the operator

$$\mathfrak{A}(x, D_x) = \sum_{j,k=1}^{n} \mathcal{A}_{jk}(x) \frac{\partial^2}{\partial x_j \partial x_k} - \sum_{j=1}^{n} \mathcal{A}_j(x) \frac{\partial}{\partial x_j} - \mathcal{A}_0(x),$$

where $\mathcal{A}_{jk} = \mathcal{A}_{kj}$, and

$$\mathcal{A}_{jk} \in [\mathrm{C}_{\mathrm{b}}^{2,\alpha}(\mathbb{R}^n)]^{m\times m}, \quad \mathcal{A}_j \in [\mathrm{C}_{\mathrm{b}}^{1,\alpha}(\mathbb{R}^n)]^{m\times m}, \quad \mathcal{A}_0 \in [\mathrm{C}_{\mathrm{b}}^{\alpha}(\mathbb{R}^n)]^{m\times m}.$$

We assume that $\mathfrak{A}(x, D_x)$ is uniformly strongly elliptic in $\mathbb{R}^n$ that is for all $x \in \mathbb{R}^n$, $\boldsymbol{\zeta} = (\zeta_1, \dots, \zeta_m) \in \mathbb{R}^m$, $\boldsymbol{\sigma} = (\sigma_1, \dots, \sigma_n) \in \mathbb{R}^n$ the inequality

$$\left(\sum_{j,k=1}^{n} \mathcal{A}_{jk}(x)\sigma_j\sigma_k \boldsymbol{\zeta},\ \boldsymbol{\zeta} \right) \geq \delta |\sigma|^2 |\boldsymbol{\zeta}|^2 \tag{9.2.1}$$

is valid, where $\delta = \mathrm{const} > 0$.

We shall use the constants $\mathcal{K}(\Omega, T)$ and $\mathring{\mathcal{K}}(\Omega, T)$ from Section 9.1 as well as the function

$$\mathbb{R}^n \ni y \to \mathcal{K}_0(y) = \sup \frac{\|\boldsymbol{u}\|_{[\mathrm{C}_{\mathrm{b}}(\overline{\mathbb{R}_T^{n+1}})]^m}}{\|\boldsymbol{u}\big|_{t=0}\|_{[\mathrm{C}_{\mathrm{b}}(\mathbb{R}^n)]^m}} ,$$

where $\mathbb{R}_T^{n+1} = \mathbb{R}^n \times (0, T]$ and the supremum is taken over all functions in the class $[\mathrm{C}^{(2,1)}(\mathbb{R}_T^{n+1})]^m \cap [\mathrm{C}_{\mathrm{b}}(\overline{\mathbb{R}_T^{n+1}})]^m$ satisfying the system

$$\frac{\partial \boldsymbol{u}}{\partial t} - \sum_{j,k=1}^{n} \mathcal{A}_{jk}(y) \frac{\partial^2 \boldsymbol{u}}{\partial x_j \partial x_k} = \boldsymbol{0}.$$

By $\|L\|$ we denote the norm of the $(m \times m)$-matrix L:

$$\max \{|L\boldsymbol{z}| : \boldsymbol{z} \in \mathbb{R}^m,\ |\boldsymbol{z}| = 1\}.$$

9.2.2. Necessary condition

According to Eidel'man [**Ei1**], there exists one and only one solution in the class

$$[\mathrm{C}^{(2,1)}\left(\mathbb{R}_T^{n+1}\right)]^m \cap [\mathrm{C}_{\mathrm{b}}\left(\overline{\mathbb{R}_T^{n+1}}\right)]^m$$

of the Cauchy problem

$$\frac{\partial \boldsymbol{u}}{\partial t} - \mathfrak{A}(x, D_x)\boldsymbol{u} = \boldsymbol{0} \quad \text{in} \quad \mathbb{R}_T^{n+1},\ \boldsymbol{u}\big|_{t=0} = \boldsymbol{\psi}, \tag{9.2.2}$$

with $\boldsymbol{\psi} \in [\mathrm{C}_{\mathrm{b}}(\mathbb{R}^n)]^m$. This solution can be represented in the form

$$\boldsymbol{u}(x, t) = \int_{\mathbb{R}^n} G(t, x, \eta) \boldsymbol{\psi}(\eta) d\eta,$$

where $G(t, x, \eta)$ is the fundamental matrix of solutions of the Cauchy problem (9.2.2).

The fundamental matrix $\mathcal{G}_0(t, x - \eta; y)$ of solutions of the Cauchy problem

$$\frac{\partial \boldsymbol{u}}{\partial t} - \sum_{j,k=1}^{n} \mathcal{A}_{jk}(y) \frac{\partial^2 \boldsymbol{u}}{\partial x_j \partial x_k} = \boldsymbol{0} \quad \text{in} \quad \mathbb{R}_T^{n+1}, \qquad \boldsymbol{u}\big|_{t=0} = \boldsymbol{\psi},$$

has the representation

$$\mathcal{G}_0(t, x - \eta; y) = (2\pi)^{-n} \int_{\mathbb{R}^n} \exp \left[- \sum_{j,k=1}^{n} \mathcal{A}_{jk}(y)\sigma_j\sigma_k t \right] e^{i(x-\eta,\sigma)} d\sigma,$$

where $\sigma = (\sigma_1, \dots, \sigma_n) \in \mathbb{R}^n$ (cf. Eidel'man [**Ei1**]). This implies

$$\mathcal{G}_0(t, x; y) = t^{-n/2} P\left(\frac{x}{\sqrt{t}}; y\right) \tag{9.2.3}$$

with

$$P(x; y) = (2\pi)^{-n} \int_{\mathbb{R}^n} e^{i(x,\sigma)} \exp\left[-\sum_{j,k=1}^{n} \mathcal{A}_{jk}(y)\sigma_j\sigma_k\right] d\sigma.$$

Henceforth all positive constants with non-significant values will be denoted by c and k with various indices.

PROPOSITION 9.1. *The inequality*

$$\mathring{\mathcal{K}}(\Omega, T) \geq \sup\{\mathcal{K}_0(y) : y \in \Omega\} \tag{9.2.4}$$

is valid.

PROOF. Let y be an arbitrary point of Ω and let the radius of the ball $\mathbb{B}_r(y)$ be so small that $\overline{\mathbb{B}_r(y)} \subset \Omega$. Further, let

$$\boldsymbol{\psi}_\varepsilon \in [\mathrm{C_b}(\mathbb{R}^n)]^m, \quad |\boldsymbol{\psi}_\varepsilon(x)| \leq 1, \ \operatorname{supp} \boldsymbol{\psi}_\varepsilon \subset \mathbb{B}_\varepsilon(y), \quad 0 < \varepsilon \leq r/2.$$

The vector-valued function

$$\boldsymbol{u}_\varepsilon(x, t) = \int_{\mathbb{R}^n} G(t, x, \eta)\boldsymbol{\psi}_\varepsilon(\eta) d\eta = \int_{\mathbb{B}_\varepsilon(y)} G(t, x, \eta)\boldsymbol{\psi}_\varepsilon(\eta) d\eta \tag{9.2.5}$$

is the solution of the Cauchy problem

$$\frac{\partial \boldsymbol{u}_\varepsilon}{\partial t} - \mathfrak{A}(x, D_x)\boldsymbol{u}_\varepsilon = \mathbf{0} \quad \text{in} \quad \mathbb{R}^{n+1}_T, \qquad \boldsymbol{u}_\varepsilon(x, 0) = \boldsymbol{\psi}_\varepsilon(x). \tag{9.2.6}$$

Since $\mathcal{A}_{jk}, \mathcal{A}_j, \mathcal{A}_0 \in [\mathrm{C}^\alpha_\mathrm{b}(\mathbb{R}^n)]^{m \times m}$, then, according to Eidel'man [**Ei1**], we have

$$\|G(t, x, \eta)\| \leq c_1 t^{-n/2} \exp\left(-c_2 \frac{|x - \eta|^2}{t}\right).$$

The last estimate, used for $(x, t) \in \partial\Omega \times (0, T]$, yields

$$\begin{aligned} |\boldsymbol{u}_\varepsilon(x, t)| &\leq \int_{\mathbb{B}_\varepsilon(y)} \|G(t, 0, x, \eta)\| d\eta \\ &\leq c_1 t^{-n/2} \int_{\mathbb{B}_\varepsilon(y)} \exp\left(-c_2 \frac{|x - \eta|^2}{t}\right) d\eta \\ &\leq c_3 \varepsilon^n t^{-n/2} \exp\left(-c_4 \frac{r^2}{t}\right) \leq c_4 \varepsilon^n r^{-n}. \end{aligned} \tag{9.2.7}$$

We introduce the boundary-value problem

$$\begin{cases} \partial \boldsymbol{v}_\varepsilon/\partial t - \mathfrak{A}(x, D_x)\boldsymbol{v}_\varepsilon = \mathbf{0} \quad \text{in} \quad Q_T, \\ \boldsymbol{v}_\varepsilon(x, 0) = \mathbf{0}, \quad x \in \Omega, \\ \boldsymbol{v}_\varepsilon\big|_{S_T} = \boldsymbol{\varphi}_\varepsilon, \end{cases} \tag{9.2.8}$$

where $\boldsymbol{\varphi}_\varepsilon = \boldsymbol{u}_\varepsilon\big|_{S_T}$.

The solution of this problem can be found as an integral of the heat potential type

$$\boldsymbol{v}_\varepsilon(x, t) = \int_0^t d\tau \int_{\partial\Omega} M(t, \tau, x, \xi)\boldsymbol{h}_\varepsilon(\xi, \tau) dS_\xi, \tag{9.2.9}$$

where $(x,t) \in Q_T$ (see Eidel'man [**Ei1**], Solonnikov [**So2**], and also Ladyzhenskaya, Solonnikov and Uraltseva [**LSU**]). Here $\boldsymbol{h}_\varepsilon(x,t)$ is a continuous vector-valued function on S_T which is to be defined, $M(t,\tau,x,\xi)$ is a special matrix of solutions to the boundary-value problem for system (9.1.1) with zero initial data, dS_ξ is the area element of $\partial\Omega$. The estimate holds

$$\|M(t,\tau,x,\xi)\| \le c_5(t-\tau)^{-\frac{n+1}{2}} \exp\left[-c_6\frac{|x-\xi|^2}{t-\tau}\right], \tag{9.2.10}$$

where $(x,t) \in Q_T$, $(\xi,\tau) \in S_T$.

It is shown in Eidel'man [**Ei1**] that the density $\boldsymbol{h}_\varepsilon(\xi,\tau)$ of potential (9.2.9), subject to the boundary condition of problem (9.2.8), is a solution of the integral equation

$$\boldsymbol{h}_\varepsilon(\eta,t) + \int_0^t d\tau \int_{\partial\Omega} M(t,\tau,\eta,\xi)\boldsymbol{h}_\varepsilon(\xi,\tau)dS_\xi = \boldsymbol{\varphi}_\varepsilon(\eta,t), \tag{9.2.11}$$

where $(\eta,t) \in S_T$. By the estimate

$$\|M(t,\tau,\eta,\xi)\| \le c_7(t-\tau)^{-\frac{n+1-\alpha}{2}} \exp\left[-c_8\frac{|\eta-\xi|^2}{t-\tau}\right], \tag{9.2.12}$$

obtained in Eidelman [**Ei1**], equation (9.2.11) has one and only one solution. When the solutions $\boldsymbol{u}_\varepsilon$ and $\boldsymbol{v}_\varepsilon$ of problems (9.2.6) and (9.2.8) have been found we introduce the function $\boldsymbol{w}_\varepsilon(x,t) = \boldsymbol{u}_\varepsilon(x,t) - \boldsymbol{v}_\varepsilon(x,t)$, which solves the boundary-value problem

$$\begin{cases} \partial\boldsymbol{w}_\varepsilon/\partial t - \mathfrak{A}(x,D_x)\boldsymbol{w}_\varepsilon = \boldsymbol{0} \quad \text{in} \quad Q_T, \\ \boldsymbol{w}_\varepsilon(x,0) = \boldsymbol{\psi}_\varepsilon(x), \quad x \in \Omega, \\ \boldsymbol{w}_\varepsilon\big|_{S_T} = 0. \end{cases} \tag{9.2.13}$$

The definition of the constant $\mathring{\mathcal{K}}(\Omega,T)$ implies

$$\begin{aligned} \mathring{\mathcal{K}}(\Omega,T) \;&\ge\; \sup_{y\in\Omega}\ \limsup_{\varepsilon\to 0}\ \limsup_{t\to+0} ||\boldsymbol{w}_\varepsilon(y,t)|| \\ &=\; \sup_{y\in\Omega}\ \limsup_{\varepsilon\to 0}\ \limsup_{t\to+0} ||\boldsymbol{u}_\varepsilon(y,t) - \boldsymbol{v}_\varepsilon(y,t)||, \end{aligned} \tag{9.2.14}$$

where

$$||\boldsymbol{w}_\varepsilon(y,t)|| = \sup\{|\boldsymbol{w}_\varepsilon(y,t)| : \ |\boldsymbol{\psi}_\varepsilon(x)| \le 1\}.$$

By (9.2.10) and taking into account that dist $(y,\partial\Omega) \ge r$, we obtain from (9.2.9) that

$$|\boldsymbol{v}_\varepsilon(y,t)| \le c_9(r)t\|\boldsymbol{h}_\varepsilon\|_{[\mathrm{C}(S_t)]^m}. \tag{9.2.15}$$

Estimate (9.2.12) implies that the norm of the integral operator

$$\mathfrak{M} : [\mathrm{C}(S_\delta)]^m \to [\mathrm{C}(S_\delta)]^m$$

in the left-hand side of (9.2.11) is less than 1 for sufficiently small δ. Hence, solutions of equation (9.2.11) satisfy

$$\|\boldsymbol{h}_\varepsilon\|_{[\mathrm{C}(S_\delta)]^m} \le c_{10}\|\boldsymbol{\varphi}\|_{[\mathrm{C}(S_\delta)]^m}, \tag{9.2.16}$$

where δ is sufficiently small. From (9.2.15) and (9.2.16) it follows that

$$|\boldsymbol{v}_\varepsilon(y,t)| \le c_{11}(r)t\|\boldsymbol{\varphi}\|_{[\mathrm{C}(S_\delta)]^m} \tag{9.2.17}$$

for $t \le \delta$. Taking into account the equality $\boldsymbol{\varphi}_\varepsilon = \boldsymbol{u}_\varepsilon\big|_{S_T}$, by (9.2.17) and (9.2.7) we get

$$|\boldsymbol{v}_\varepsilon(y,t)| \le c(r)\varepsilon^n t,$$

where $t \leq \delta$. Combining the last estimate with (9.2.14) gives

$$\mathring{\mathcal{K}}(\Omega, T) \geq \sup_{y\in\Omega} \limsup_{\varepsilon\to 0} \limsup_{t\to +0} ||\boldsymbol{u}_\varepsilon(y,t)||. \tag{9.2.18}$$

We find the norm $||\boldsymbol{u}_\varepsilon(y,t)||$ of the mapping

$$[\mathrm{C_b}(\mathbb{R}^n)]^m \ni \boldsymbol{\psi}_\varepsilon(x) \to \boldsymbol{u}_\varepsilon(y,t) \in \mathbb{R}^m,$$

where $\boldsymbol{u}_\varepsilon$ is defined by (9.2.5). Hence

$$\begin{aligned} ||\boldsymbol{u}_\varepsilon(y,t)|| &= \sup_{|\boldsymbol{\psi}_\varepsilon|\leq 1} \left| \int_{\mathbb{B}_\varepsilon(y)} G(t,y,\eta)\boldsymbol{\psi}_\varepsilon(\eta)d\eta \right| \\ &= \sup_{|\boldsymbol{\psi}_\varepsilon|\leq 1} \sup_{|\boldsymbol{z}|=1} \left(\boldsymbol{z}, \int_{\mathbb{B}_\varepsilon(y)} G(t,y,\eta)\boldsymbol{\psi}_\varepsilon(\eta)d\eta \right) \\ &= \sup_{|\boldsymbol{z}|=1} \sup_{|\boldsymbol{\psi}_\varepsilon|\leq 1} \int_{\mathbb{B}_\varepsilon(y)} (\boldsymbol{z}, G(t,y,\eta)\boldsymbol{\psi}_\varepsilon(\eta))\, d\eta \\ &= \sup_{|\boldsymbol{z}|=1} \sup_{|\boldsymbol{\psi}_\varepsilon|\leq 1} \int_{\mathbb{B}_\varepsilon(y)} (G^*(t,y,\eta)\boldsymbol{z}, \boldsymbol{\psi}_\varepsilon(\eta))\, d\eta \\ &= \sup_{|\boldsymbol{z}|=1} \int_{\mathbb{B}_\varepsilon(y)} |G^*(t,y,\eta)\boldsymbol{z}|\, d\eta. \end{aligned} \tag{9.2.19}$$

Since $\mathcal{A}_{jk}, \mathcal{A}_j, \mathcal{A}_0 \in [\mathrm{C}^\alpha_{\mathrm{b}}(\mathbb{R}^n)]^{m\times m}$, it follows from estimates given in Eidel'man [**Ei1**], that the matrix $G(t,y,\eta)$ admits the representation

$$\begin{aligned} G(t,y,\eta) &= \mathcal{G}_0(t, y-\eta; y) + [\mathcal{G}_0(t,y-\eta;\eta) - \mathcal{G}_0(t,y-\eta;y)] \\ &+ W(t,y,\eta), \end{aligned} \tag{9.2.20}$$

where

$$||\mathcal{G}_0(t,y-\eta;\eta) - \mathcal{G}_0(t,y-\eta;y)|| \leq k_1|y-\eta|^\alpha t^{-n/2} \exp\left(-k_2\frac{|y-\eta|^2}{t}\right) \tag{9.2.21}$$

and

$$||W(t,y,\eta)|| \leq k_3 t^{-(n-\alpha)/2} \exp\left(-k_4\frac{|y-\eta|^2}{t}\right). \tag{9.2.22}$$

Using (9.2.20)-(9.2.22) and representation (9.2.3), we obtain from (9.2.19) that

$$\begin{aligned} &\limsup_{\varepsilon\to 0} \limsup_{t\to +0} ||\boldsymbol{u}_\varepsilon(y,t)|| \\ =\ &\limsup_{\varepsilon\to 0} \limsup_{t\to +0} \sup_{|\boldsymbol{z}|=1} \int_{\mathbb{B}_\varepsilon(y)} |G^*(t,y,\eta)\boldsymbol{z}| d\eta \\ \geq\ &\sup_{|\boldsymbol{z}|=1} \limsup_{\varepsilon\to 0} \limsup_{t\to +0} \left\{ \int_{\mathbb{B}_\varepsilon(y)} |\mathcal{G}^*_0(t,y-\eta;y)\boldsymbol{z}| d\eta - k_5\varepsilon^\alpha - k_6 t^{\alpha/2} \right\} \\ =\ &\sup_{|\boldsymbol{z}|=1} \limsup_{\varepsilon\to 0} \limsup_{t\to +0} \int_{\mathbb{B}_\varepsilon(y)} t^{-n/2} \left| P^*\left(\frac{y-\eta}{\sqrt{t}}\,;y\right)\boldsymbol{z}\right| d\eta \\ =\ &\sup_{|\boldsymbol{z}|=1} \limsup_{\varepsilon\to 0} \limsup_{t\to +0} \int_{\mathbb{B}_\varepsilon/\sqrt{t}} |P^*(x;y)\boldsymbol{z}|\, dx = \\ =\ &\sup_{|\boldsymbol{z}|=1} \int_{\mathbb{R}^n} |P^*(x;y)\boldsymbol{z}|\, dx. \end{aligned}$$

Then by (9.2.18),

$$\mathring{\mathcal{K}}(\Omega, T) \geq \sup_{y\in\Omega} \sup_{|\boldsymbol{z}|=1} \int_{\mathbb{R}^n} |P^*(x; y)\boldsymbol{z}|dx,$$

which together with (8.2.8) gives the lower estimate (9.2.4) for $\mathring{\mathcal{K}}(\Omega, T)$ in the statement of Proposition. □

We need one more assertion to use in the proof of the necessary condition for validity of the maximum modulus principle for solutions of system (9.1.1) in the class $[\mathrm{C}^{(2,1)}(Q_T)]^m \cap [\mathring{\mathrm{C}}(\overline{Q}_T)]^m$. The assertion in question is given below. It is a direct corollary of Theorem 8.5.

PROPOSITION 9.2. *The classical maximum modulus principle is valid for solutions of the system with constant coefficients*

$$\frac{\partial \boldsymbol{u}}{\partial t} - \sum_{j,k=1}^{n} \mathcal{A}_{jk} \frac{\partial^2 \boldsymbol{u}}{\partial x_j \partial x_k} = \mathbf{0}$$

in $\mathbb{R}^{n+1}_T$ *if and only if* $\mathcal{A}_{jk} = a_{jk}I$, $1 \leq j, k \leq n$, *where* $((a_{jk}))$ *is a positive-definite* $(n \times n)$*-matrix.*

PROPOSITION 9.3. *Let the maximum modulus principle be valid for solutions of system* (9.1.1) *in the class* $[\mathrm{C}^{(2,1)}(Q_T)]^m \cap [\mathring{\mathrm{C}}(\overline{Q}_T)]^m$. *Then*

(i) *for all* $x \in \Omega$ *the equalities*

$$\mathcal{A}_{jk}(x) = a_{jk}(x)I, \ \ 1 \leq j, k \leq n,$$

hold, where $((a_{jk}))$ *is a positive-definite* $(n \times n)$*-matrix-valued function in* $\overline{\Omega}$*;*

(ii) *for all* $x \in \Omega$ *and for all* $\boldsymbol{\xi}_j, \boldsymbol{\zeta} \in \mathbb{R}^m$, $j = 1, 2, \dots, n$, *with* $(\boldsymbol{\xi}_j, \boldsymbol{\zeta}) = 0$ *the inequality*

$$\sum_{j,k=1}^{n} a_{jk}(x)(\boldsymbol{\xi}_j, \boldsymbol{\xi}_k) + \sum_{j=1}^{n} (A_j(x)\boldsymbol{\xi}_j, \boldsymbol{\zeta}) + (\mathcal{A}_0(x)\boldsymbol{\zeta}, \boldsymbol{\zeta}) \geq 0$$

holds.

PROOF. Let the solutions of system (9.1.1) satisfy the maximum modulus principle in the class $[\mathrm{C}^{(2,1)}(Q_T)]^m \cap [\mathring{\mathrm{C}}(\overline{Q}_T)]^m$. From Proposition 9.1 it follows that the equality $\mathring{\mathcal{K}}(\Omega, T) = 1$ is valid only if $\mathcal{K}_0(y) = 1$ for all $y \in \Omega$.

By Proposition 9.2 the equality $\mathcal{K}_0(y) = 1$ takes place if and only if $\mathcal{A}_{jk}(y) = a_{jk}(y)I$, $1 \leq j, k \leq n$. Hence (i) follows, and therefore it suffices to restrict consideration to parabolic systems with scalar principal parts.

Now we prove the necessity of (ii). Let y be an arbitrary fixed point of Ω and let the radius of the ball $\mathbb{B}_r(y)$ be so small that $\overline{\mathbb{B}_r(y)} \subset \Omega$. We introduce the vector-valued function

$$\boldsymbol{\psi}(x) = \left(\sum_{j=1}^{n} \boldsymbol{\xi}_j x_j + \boldsymbol{\zeta} \right) \left(\left| \sum_{j=1}^{n} \boldsymbol{\xi}_j x_j \right|^2 + |\boldsymbol{\zeta}|^2 \right)^{-1/2},$$

where $\boldsymbol{\xi}_j \in \mathbb{R}^m$, $\boldsymbol{\zeta} \in \mathbb{R}^m \backslash \{\mathbf{0}\}$, $(\boldsymbol{\xi}_j, \boldsymbol{\zeta}) = 0$, $j = 1, \dots, n$. Further, let $\chi_\varepsilon \in \mathrm{C}^\infty(\mathbb{R}^n)$, $\chi_\varepsilon(x) = 1$ for $|x| \leq \varepsilon/2$, $\chi_\varepsilon(x) = 0$ for $|x| \geq \varepsilon$ and $0 \leq \chi_\varepsilon(x) \leq 1$ for all $x \in \mathbb{R}^n$, where $0 < \varepsilon \leq r/2$.

We set $\boldsymbol{\psi}_\varepsilon(x) = \chi_\varepsilon(x-y)\boldsymbol{\psi}(x-y)$ and define the vector-valued function $\boldsymbol{w}_\varepsilon(x,t)$ as a solution of the boundary value problem

$$\frac{\partial \boldsymbol{w}_\varepsilon}{\partial t} - \sum_{j,k=1}^n a_{jk}(x)\frac{\partial^2 \boldsymbol{w}_\varepsilon}{\partial x_j \partial x_k} + \sum_{j=1}^n \mathcal{A}_j(x)\frac{\partial \boldsymbol{w}_\varepsilon}{\partial x_j} + \mathcal{A}_0(x)\boldsymbol{w}_\varepsilon = \boldsymbol{0} \text{ in } Q_T, \tag{9.2.23}$$

$$\boldsymbol{w}_\varepsilon(x,0) = \boldsymbol{\psi}_\varepsilon(x), \ x \in \Omega, \tag{9.2.24}$$

$$\boldsymbol{w}_\varepsilon\big|_{S_T} = \boldsymbol{0}. \tag{9.2.25}$$

Since $((a_{jk})) \in [\mathrm{C}^\alpha(\overline{\Omega})]^{n\times n}, \mathcal{A}_j, \mathcal{A}_0 \in [\mathrm{C}^\alpha(\overline{\Omega})]^{m\times m}$, $\boldsymbol{\psi}_\varepsilon \in [\mathrm{C}^{2+\alpha}(\overline{\Omega})]^m$ and the boundary of Ω is $\mathrm{C}^{2,\alpha}$-smooth, it follows by Solonnikov [**So2**], that the problem (9.2.23)-(9.2.25) has a solution $\boldsymbol{w}_\varepsilon$ in some subspace of the space $[\mathrm{C}^{(2,1)}(\overline{Q}_T)]^m$.

By taking the scalar product of system (9.2.23) and $\boldsymbol{w}_\varepsilon$ we transform the equality

$$\left(\frac{\partial \boldsymbol{w}_\varepsilon}{\partial t}, \boldsymbol{w}_\varepsilon\right) - \sum_{j,k=1}^n a_{jk}(x)\left(\frac{\partial^2 \boldsymbol{w}_\varepsilon}{\partial x_j \partial x_k}, \boldsymbol{w}_\varepsilon\right) + \sum_{j=1}^n \left(\mathcal{A}_j(x)\frac{\partial \boldsymbol{w}_\varepsilon}{\partial x_j}, \boldsymbol{w}_\varepsilon\right) + (\mathcal{A}_0(x)\boldsymbol{w}_\varepsilon, \boldsymbol{w}_\varepsilon) = 0.$$

As a result we find

$$\begin{aligned}\frac{1}{2}\frac{\partial |\boldsymbol{w}_\varepsilon|^2}{\partial t} &= \frac{1}{2}\sum_{j,k=1}^n a_{jk}(x)\frac{\partial^2 |\boldsymbol{w}_\varepsilon|^2}{\partial x_j \partial x_k} - \sum_{j,k=1}^n a_{jk}(x)\left(\frac{\partial \boldsymbol{w}_\varepsilon}{\partial x_j}, \frac{\partial \boldsymbol{w}_\varepsilon}{\partial x_k}\right) \\ &- \sum_{j=1}^n \left(\mathcal{A}_j(x)\frac{\partial \boldsymbol{w}_\varepsilon}{\partial x_j}, \boldsymbol{w}_\varepsilon\right) - (\mathcal{A}_0(x)\boldsymbol{w}_\varepsilon, \boldsymbol{w}_\varepsilon) = 0.\end{aligned}$$

Using the continuity of coefficients of system (9.2.23) and the inclusion $\boldsymbol{w}_\varepsilon \in [\mathrm{C}^{(2,1)}(\overline{Q}_T)]^m$ we can pass to the limit in the last equality as $(x,t) \to (y,+0)$. Then

$$\begin{aligned}\frac{\partial_+ |\boldsymbol{w}_\varepsilon(y,t)|^2}{\partial t}\bigg|_{t=0} &= 2\Big\{\frac{1}{2}\sum_{j,k=1}^n a_{jk}(y)\frac{\partial^2 |\boldsymbol{\psi}_\varepsilon|^2}{\partial x_j \partial x_k} - \sum_{j,k=1}^n a_{jk}(y)\left(\frac{\partial \boldsymbol{\psi}_\varepsilon}{\partial x_j}, \frac{\partial \boldsymbol{\psi}_\varepsilon}{\partial x_k}\right) \\ &- \sum_{j=1}^n \left(\mathcal{A}_j(y)\frac{\partial \boldsymbol{\psi}_\varepsilon}{\partial x_j}, \boldsymbol{\psi}_\varepsilon\right) - (\mathcal{A}_0(y)\boldsymbol{\psi}_\varepsilon, \boldsymbol{\psi}_\varepsilon)\Big\}\bigg|_{x=y},\end{aligned} \tag{9.2.26}$$

where

$$\frac{\partial_+ |\boldsymbol{w}_\varepsilon(y,t)|^2}{\partial t}\bigg|_{t=0}$$

is the right-hand derivative of $|\boldsymbol{w}_\varepsilon(y,t)|^2$ in t at $t=0$. Hence, using the equalities

$$\frac{\partial^2 |\boldsymbol{\psi}_\varepsilon|^2}{\partial x_j \partial x_k}\bigg|_{x=y} = 0, \quad \frac{\partial \boldsymbol{\psi}_\varepsilon}{\partial x_j}\bigg|_{x=y} = \frac{\boldsymbol{\xi}_j}{|\boldsymbol{\zeta}|}, \quad \boldsymbol{\psi}_\varepsilon(y) = \frac{\boldsymbol{\zeta}}{|\boldsymbol{\zeta}|},$$

where $j,k = 1,\dots,n$, we find from (9.2.26):

$$\begin{aligned}\frac{\partial_+ |\boldsymbol{w}_\varepsilon(y,t)|^2}{\partial t}\bigg|_{t=0} = -\frac{2}{|\boldsymbol{\zeta}|^2}\Big[\sum_{j,k=1}^n a_{jk}(y)(\boldsymbol{\xi}_j, \boldsymbol{\xi}_k) &+ \sum_{j=1}^n (\mathcal{A}_j(y)\boldsymbol{\xi}_j, \boldsymbol{\zeta}) \\ &+ (\mathcal{A}_0(y)\boldsymbol{\zeta}, \boldsymbol{\zeta})\Big].\end{aligned} \tag{9.2.27}$$

Since

$$|\boldsymbol{w}_\varepsilon(y,0)| = |\boldsymbol{\psi}_\varepsilon(y)| = |\boldsymbol{\psi}(0)| = 1$$

and since the maximum modulus principle holds for solutions of the boundary value problem (9.2.23)-(9.2.25), it follows that

$$\frac{\partial_+ |\boldsymbol{w}_\varepsilon(y,t)|^2}{\partial t}\Big|_{t=0} \leq 0.$$

Combining this with (9.2.27) and taking into account the arbitrariness of the point $y \in \Omega$, one gets the following necessary condition for validity of the maximum modulus principle for solutions of (9.2.23)-(9.2.25):

$$\sum_{j,k=1}^{n} a_{jk}(y)(\boldsymbol{\xi}_j, \boldsymbol{\xi}_k) + \sum_{j=1}^{n} (\mathcal{A}_j(y)\boldsymbol{\xi}_j, \boldsymbol{\zeta}) + (\mathcal{A}_0(y)\boldsymbol{\zeta}, \boldsymbol{\zeta}) \geq 0 \tag{9.2.28}$$

for all $y \in \Omega$ and for any $\boldsymbol{\xi}_j \in \mathbb{R}^m$, $\boldsymbol{\zeta} \in \mathbb{R}^m \backslash \{\boldsymbol{0}\}$, $j = 1, \dots, n$, with $(\boldsymbol{\xi}_j, \boldsymbol{\zeta}) = 0$.

We complete the proof of the proposition by noting that the condition $\boldsymbol{\zeta} \in \mathbb{R}^m \backslash \{\boldsymbol{0}\}$ in (9.2.28) can be omitted since the inequality

$$\sum_{j,k=1}^{n} a_{jk}(y)(\boldsymbol{\xi}_j, \boldsymbol{\xi}_k) \geq 0, \quad y \in \Omega,$$

holds for all $\boldsymbol{\xi}_j \in \mathbb{R}^m$, $j = 1, \dots, n$, by necessity of condition (i) for validity of the maximum modulus principle. □

9.2.3. Necessary and sufficient condition

Since $\mathcal{K}(\Omega, T) \geq \mathring{\mathcal{K}}(\Omega, T)$, a sufficient condition for validity of the maximum modulus principle for solutions of system (9.1.1) in the class

$$[\mathrm{C}^{(2,1)}(Q_T)]^m \cap [\mathrm{C}(\overline{Q}_T)]^m$$

is simultaneously a sufficient condition for the same property in the class

$$[\mathrm{C}^{(2,1)}(Q_T)]^m \cap [\mathring{\mathrm{C}}(\overline{Q}_T)]^m.$$

Therefore, we may use the sufficient condition in Theorem 8.7 adjusting its formulation to the problem under consideration.

PROPOSITION 9.4. *Let the coefficients of the parabolic system*

$$\frac{\partial \boldsymbol{u}}{\partial t} - \sum_{j,k=1}^{n} a_{jk}(x) \frac{\partial^2 \boldsymbol{u}}{\partial x_j \partial x_k} + \sum_{j=1}^{n} \mathcal{A}_j(x) \frac{\partial \boldsymbol{u}}{\partial x_j} + \mathcal{A}_0(x)\boldsymbol{u} = \boldsymbol{0}$$

satisfy the condition:

for all $x \in \Omega$ and for all vectors $\boldsymbol{\xi}_j, \boldsymbol{\zeta} \in \mathbb{R}^m$ with $(\boldsymbol{\xi}_j, \boldsymbol{\zeta}) = 0$, $j = 1, \dots, n$, the inequality

$$\sum_{j,k=1}^{n} a_{jk}(x)(\boldsymbol{\xi}_j, \boldsymbol{\xi}_k) + \sum_{j=1}^{n} (\mathcal{A}_j(x)\boldsymbol{\xi}_j, \boldsymbol{\zeta}) + (\mathcal{A}_0(x)\boldsymbol{\zeta}, \boldsymbol{\zeta}) \geq 0$$

holds. Then $\mathring{\mathcal{K}}(\Omega, T) = 1$.

Propositions 9.3 and 9.4 immediately imply the following assertion.

THEOREM 9.1. *The maximum modulus principle is valid for solutions of system (9.1.1) in the class $[\mathrm{C}^{(2,1)}(Q_T)]^m \cap [\mathring{\mathrm{C}}(\overline{Q}_T)]^m$ if and only if:*

(i) *for all $x \in \Omega$ the equalities*

$$\mathcal{A}_{jk}(x) = a_{jk}(x)I, \quad 1 \leq j, k \leq n,$$

hold, where $((a_{jk}))$ *is a positive-definite* $(n \times n)$*- matrix-valued function in* $\overline{\Omega}$*;*

(ii) *for all* $x \in \Omega$ *and all* $\boldsymbol{\xi}_j$, $\boldsymbol{\zeta} \in \mathbb{R}^m$, $j = 1, 2, \dots, n$, *with* $(\boldsymbol{\xi}_j, \boldsymbol{\zeta}) = 0$, *the inequality*

$$\sum_{j,k=1}^{n} a_{jk}(x)(\boldsymbol{\xi}_j, \boldsymbol{\xi}_k) + \sum_{j=1}^{n}(\mathcal{A}_{jk}(x)\boldsymbol{\xi}_j, \boldsymbol{\zeta}) + (A_0(x)\boldsymbol{\zeta}, \boldsymbol{\zeta}) \geq 0$$

is valid.

Thus, criteria for validity of the maximum modulus principle for solutions of system (9.1.1) in classes $[\mathrm{C}^{(2,1)}(Q_T)]^m \cap [\mathrm{C}(\overline{Q}_T)]^m$ and $[\mathrm{C}^{(2,1)}(Q_T)]^m \cap [\mathring{\mathrm{C}}(\overline{Q}_T)]^m$ coincide.

The following two corollaries of Theorem 9.1 are proved literally in the same way as Corollaries 8.1 and 8.2 in Chapter 8.

COROLLARY 9.1. *The maximum modulus principle is valid for solutions of the system*

$$\frac{\partial \boldsymbol{u}}{\partial t} - \sum_{j,k=1}^{n} \mathcal{A}_{jk}(x)\frac{\partial^2 \boldsymbol{u}}{\partial x_j \partial x_k} + \sum_{j=1}^{n} \mathcal{A}_j(x)\frac{\partial \boldsymbol{u}}{\partial x_j} = \mathbf{0}$$

in the class $[\mathrm{C}^{(2,1)}(Q_T)]^m \cap [\mathring{\mathrm{C}}(\overline{Q}_T)]^m$ *if and only if for all* $x \in \Omega$

$$\mathcal{A}_{jk}(x) = a_{jk}(x)I, \quad \mathcal{A}_j(x) = a_j(x)I,$$

where $((a_{jk}))$ *is a positive-definite* $(n \times n)$*-matrix-valued function in* $\overline{\Omega}$ *and* a_j *are scalar functions in* $\overline{\Omega}$.

COROLLARY 9.2. *The maximum modulus principle is valid for solutions of system* (9.1.1) *in the class* $[\mathrm{C}^{(2,1)}(Q_T)]^m \cap [\mathring{\mathrm{C}}(\overline{Q}_T)]^m$ *if and only if condition* (i) *of Theorem* 9.1 *is satisfied and for all* $x \in \Omega$ *and for any* $\boldsymbol{\zeta} \in \mathbb{R}^m$, $|\boldsymbol{\zeta}| = 1$, *the inequality*

$$\sum_{i,j=1}^{n} b_{ij}(x)\left[(\mathcal{A}_i(x)\boldsymbol{\zeta}, \boldsymbol{\zeta})(\mathcal{A}_j(x)\boldsymbol{\zeta}, \boldsymbol{\zeta}) - (\mathcal{A}_i^*(x)\boldsymbol{\zeta}, \mathcal{A}_j^*(x)\boldsymbol{\zeta})\right] + 4(A_0(x)\boldsymbol{\zeta}, \boldsymbol{\zeta}) \geq 0$$

holds, where $((b_{ij}))$ *is the* $(n \times n)$*-matrix-valued function inverse of* $((a_{ij}))$ *and* $\mathcal{A}_j^*$ *is the matrix transposed of* $\mathcal{A}_j$.

9.3. The case of complex coefficients

9.3.1. Notation

Let F be a subset of $\mathbb{R}^n$. The norm in the space $[\mathbf{C}(F)]^m$ of continuous and bounded on F vector-valued functions $\boldsymbol{u} = \boldsymbol{v} + i\boldsymbol{w}$ with m complex components is defined by the equality

$$\|\boldsymbol{u}\| = \sup\left\{ (|\boldsymbol{v}(\zeta)|^2 + |\boldsymbol{w}(\zeta)|^2)^{1/2} \ : \zeta \in F \right\}.$$

In general, for the spaces of vector-valued functions with complex components we retain the same notation as in the case of real components but use bold.

We introduce the operator

$$\mathfrak{C}(x, D_x) = \sum_{j,k=1}^{n} \mathcal{A}_{jk}(x)\frac{\partial^2}{\partial x_j \partial x_k} - \sum_{j=1}^{n} \mathcal{A}_j(x)\frac{\partial}{\partial x_j} - \mathcal{A}_0(x),$$

where $\mathcal{A}_{jk} = \mathcal{A}_{kj}$, and

$$\mathcal{A}_{jk} \in [\mathbf{C}_{\mathbf{b}}^{2,\alpha}(\mathbb{R}^n)]^{m\times m},\ \ \mathcal{A}_j \in [\mathbf{C}_{\mathbf{b}}^{1,\alpha}(\mathbb{R}^n)]^{m\times m},\ \ \mathcal{A}_0 \in [\mathbf{C}_{\mathbf{b}}^{\alpha}(\mathbb{R}^n)]^{m\times m}.$$

Suppose that the operator $\mathfrak{C}(x, D_x)$ is uniformly strongly elliptic in $\mathbb{R}^n$, that is for all $x \in \mathbb{R}^n$, $\boldsymbol{\zeta} = (\zeta_1, \dots, \zeta_m) \in \mathbb{C}^m$, $\boldsymbol{\sigma} = (\sigma_1, \dots, \sigma_n) \in \mathbb{R}^n$ the inequality

$$\Re\left(\sum_{j,k=1}^{n} \mathcal{C}_{jk}(x)\sigma_j\sigma_k\boldsymbol{\zeta},\ \boldsymbol{\zeta}\right) \geq \delta|\boldsymbol{\sigma}|^2|\boldsymbol{\zeta}|^2$$

holds, where $\delta = \text{const} > 0$.

Let $\mathcal{R}_{jk}$, $\mathcal{R}_j$, $\mathcal{R}_0$ and $\mathcal{H}_{jk}$, $\mathcal{H}_j$, $\mathcal{H}_0$ be real $(m \times m)$-matrix-valued functions defined on $\mathbb{R}^n$ such that

$$\mathcal{A}_{jk}(x) = \mathcal{R}_{jk}(x) + i\mathcal{H}_{jk}(x)\ ,\ \mathcal{A}_j(x) = \mathcal{R}_j(x) + i\mathcal{H}_j(x)\ ,\ \mathcal{A}_0(x) = \mathcal{R}_0(x) + i\mathcal{H}_0(x).$$

We use the following notation:

$$\mathfrak{R}(x, D_x) = \sum_{j,k=1}^{n} \mathcal{R}_{jk}(x)\frac{\partial^2}{\partial x_j \partial x_k} - \sum_{j=1}^{n} \mathcal{R}_j(x)\frac{\partial}{\partial x_j} - \mathcal{R}_0(x),$$

$$\mathfrak{H}(x, D_x) = \sum_{j,k=1}^{n} \mathcal{H}_{jk}(x)\frac{\partial^2}{\partial x_j \partial x_k} - \sum_{j=1}^{n} \mathcal{H}_j(x)\frac{\partial}{\partial x_j} - \mathcal{H}_0(x).$$

We define one more matrix differential operator:

$$\mathfrak{K}(x, D_x) = \begin{pmatrix} \mathfrak{R}(x, D_x) & -\mathfrak{H}(x, D_x) \\ \mathfrak{H}(x, D_x) & \mathfrak{R}(x, D_x) \end{pmatrix}.$$

9.3.2. Necessary and sufficient condition

Separating the real and imaginary parts of the system

$$\frac{\partial \boldsymbol{u}}{\partial t} - \mathfrak{C}(x, D_x)\boldsymbol{u} = \mathbf{0}, \tag{9.3.1}$$

we get the system with real coefficients

$$\left(\frac{\partial}{\partial t} - \mathfrak{K}(x, D_x)\right)\begin{pmatrix}\boldsymbol{v}\\ \boldsymbol{w}\end{pmatrix} = \mathbf{0}, \tag{9.3.2}$$

which like the original system (9.3.1) is strongly parabolic (see Subsection 2.3.2).

As in the definition of $\overset{\circ}{\mathcal{K}}(\Omega, T)$ we put

$$\overset{\circ}{\mathcal{K}}{}'(\Omega, T) = \sup \frac{\|\boldsymbol{u}\|_{[\mathbf{C}(\overline{Q}_T)]^m}}{\big\|\boldsymbol{u}\big|_{\Gamma_T}\big\|_{[\mathbf{C}(\overline{\Gamma}_T)]^m}},$$

where the supremum is taken over all functions $\boldsymbol{u} = \boldsymbol{v} + i\boldsymbol{w}$ in the class

$$[\mathbf{C}^{(2,1)}(Q_T)]^m \cap [\overset{\circ}{\mathbf{C}}(\overline{Q}_T)]^m$$

satisfying system (9.3.1).

Clearly, the constant $\overset{\circ}{\mathcal{K}}{}'(\Omega, T)$ for system (9.3.1) coincides with the constant $\overset{\circ}{\mathcal{K}}(\Omega, T)$ for system (9.3.2). Therefore, all the assertions concerning $\overset{\circ}{\mathcal{K}}{}'(\Omega, T)$ are immediate corollaries of analogous assertions about $\overset{\circ}{\mathcal{K}}(\Omega, T)$. Taking this into account, we obtain the assertions below from Theorem 9.1 and Corollaries 9.1, 9.2 of Subsection 9.2.3.

THEOREM 9.2. *The maximum modulus principle is valid for solutions of system* (9.1.1) *with complex coefficients in the class* $[\mathbf{C}^{(2,1)}(Q_T)]^m \cap [\mathring{\mathbf{C}}(\overline{Q}_T)]^m$ *if and only if:*

(i) *for all* $x \in \Omega$ *the equalities*

$$\mathcal{A}_{jk}(x) = a_{jk}(x)I, \quad 1 \le j, k \le n,$$

hold, where $((a_{jk}))$ *is a real positive-definite* $(n \times n)$*-matrix-valued function in* $\overline{\Omega}$;

(ii) *for all* $x \in \Omega$ *and for all* $\boldsymbol{\xi}_j, \boldsymbol{\zeta} \in \mathbb{C}^m$, $j = 1, \ldots, n$, *with* $\Re(\boldsymbol{\xi}_j, \boldsymbol{\zeta}) = 0$, *the inequality*

$$\Re\left\{\sum_{j,k=1}^{n} a_{jk}(x)(\boldsymbol{\xi}_j, \boldsymbol{\xi}_k) + \sum_{j=1}^{n}(\mathcal{A}_j(x)\boldsymbol{\xi}_j, \boldsymbol{\zeta}) + (\mathcal{A}_0(x)\boldsymbol{\zeta}, \boldsymbol{\zeta})\right\} \ge 0$$

is valid.

COROLLARY 9.3. *The maximum modulus principle is valid for solutions of the system*

$$\frac{\partial \boldsymbol{u}}{\partial t} - \sum_{j,k=1}^{n} \mathcal{A}_{jk}(x)\frac{\partial^2 \boldsymbol{u}}{\partial x_j \partial x_k} + \sum_{j=1}^{n} \mathcal{A}_j(x)\frac{\partial \boldsymbol{u}}{\partial x_j} = \mathbf{0}$$

with complex coefficients in the class $[\mathbf{C}^{(2,1)}(Q_T)]^m \cap [\mathring{\mathbf{C}}(Q_T)]^m$ *if and only if for all* $x \in \Omega$

$$\mathcal{A}_{jk}(x) = a_{jk}(x)I, \quad \mathcal{A}_j(x) = a_j(x)I, \quad 1 \le j, k \le n,$$

where $((a_{jk}))$ *is a real positive-definite* $(n \times n)$*-matrix-valued function in* $\overline{\Omega}$ *and* a_j *are real scalar functions.*

COROLLARY 9.4. *The maximum modulus principle is valid for solutions of system* (9.1.1) *with complex coefficients in the class* $[\mathbf{C}^{(2,1)}(Q_T)]^m \cap [\mathring{\mathbf{C}}(\overline{Q}_T)]^m$ *if and only if condition* (i) *of Theorem* 9.2 *is satisfied and for all* $x \in \Omega$ *and for any* $\boldsymbol{\zeta} \in \mathbb{C}^m$, $|\boldsymbol{\zeta}| = 1$ *the inequality*

$$\sum_{i,j=1}^{n} b_{ij}(x)\big[\Re\,(\mathcal{A}_i(x)\boldsymbol{\zeta}, \boldsymbol{\zeta})\,\Re\,(\mathcal{A}_j(x)\boldsymbol{\zeta}, \boldsymbol{\zeta}) - (\mathcal{A}_i^*(x)\boldsymbol{\zeta}, \mathcal{A}_j^*(x)\boldsymbol{\zeta})\big] + \Re\,(\mathcal{A}_0(x)\boldsymbol{\zeta}, \boldsymbol{\zeta}) \ge 0$$

holds. Here $((b_{jk}))$ *is the* $(n \times n)$*-matrix inverse of* $((a_{jk}))$ *and* $\mathcal{A}_j^*$ *is the adjoint matrix of* $\mathcal{A}_j$.

In particular, the next assertion follows from Corollary 9.4 for the scalar parabolic equation (9.1.2) with complex coefficients.

COROLLARY 9.5. *The maximum modulus principle is valid for solutions of equation* (9.1.2) *in the class* $\mathbf{C}^{(2,1)}(Q_T) \cap \mathring{\mathbf{C}}(\overline{Q}_T)$ *if and only if:*

(i) *the* $(n \times n)$*-matrix-valued function* $((a_{jk}))$ *is real and positive-definite in* $\overline{\Omega}$;

(ii) *for all* $x \in \Omega$ *the inequality*

$$4\,\Re\, a_0(x) \ge \sum_{j,k=1}^{n} b_{jk}(x)\,\Im\, a_j(x)\,\Im\, a_k(x)$$

holds.

9.4. Comments to Chapter 9

We note that the results stated in this chapter can be interpreted as criteria for the $[\mathrm{C}(\overline{\Omega})]^m$-contractivity of a semigroup generated by the operator

$$\sum_{j,k=1}^{n} \mathcal{A}_{jk}(x)\frac{\partial^2}{\partial x_j \partial x_k} - \sum_{j=1}^{n} \mathcal{A}_j(x)\frac{\partial}{\partial x_j} - \mathcal{A}_0(x)$$

defined on vector-valued functions $\boldsymbol{u}$ satisfying the boundary condition $\boldsymbol{u}\big|_{\partial\Omega} = \mathbf{0}$.

In the paper [**MS**] by Maz'ya and Sobolevskiĭ it was noted that scalar second order elliptic operators with real coefficients may generate contractive semigroups in L^p. Necessary and sufficient conditions for L^∞-contractivity for scalar elliptic operators with complex measurable and bounded coefficients were found by Auscher, Barthélemy, Bénilan and Ouhabaz [**ABBO**]. Dissipativity conditions for partial differential operators and corresponding conditions for the L^p-contractivity of semigroups generated by those operators are treated by Langer and Maz'ya [**LM**], Langer [**Lan**], Cialdea and Maz'ya [**CM1, CM2**], Cialdea [**Ci**]. The last paper contains an extensive bibliography.

CHAPTER 10

Maximum Norm Principle for Parabolic Systems without Lower Order Terms

10.1. Introduction

In the present chapter we study parabolic systems of the form

$$\frac{\partial \boldsymbol{u}}{\partial t} - \sum_{j,k=1}^{n} \mathcal{A}_{jk}(x,t)\frac{\partial^2 \boldsymbol{u}}{\partial x_j \partial x_k} = \boldsymbol{0}, \tag{10.1.1}$$

where $\boldsymbol{u}$ is a m-component vector-valued function and $\mathcal{A}_{jk} = \mathcal{A}_{kj}$ are real $(m \times m)$-matrix-valued functions.

We obtain criteria for validity of the so-called maximum norm principle for solutions of system (10.1.1) which is uniformly parabolic in the sense of Petrovskiĭ.

We shall consider system (10.1.1) in the layer $\mathbb{R}^{n+1}_T = \mathbb{R}^n \times (0,T]$ and in the cylinder $Q_T = \Omega \times (0,T]$, where Ω is a bounded subdomain of $\mathbb{R}^n$.

Throughout the chapter we make the following assumptions:

(A) The matrix-valued functions $\mathcal{A}_{jk}$ are defined in $\overline{\mathbb{R}^{n+1}_T}$ and have bounded derivatives in x up to the second order which satisfy the uniform Hölder condition on $\overline{\mathbb{R}^{n+1}_T}$ with exponent $\alpha, 0 < \alpha \leq 1$, with respect to the parabolic distance

$$d\left[(x,t),(x',t')\right] = \left(|x-x'|^2 + |t-t'|\right)^{1/2}$$

between points (x,t) and (x',t') in $\mathbb{R}^{n+1}$;

(B) For any point $(x,t) \in \overline{\mathbb{R}^{n+1}_T}$, the real parts of the λ-roots of the equation

$$\det\left(\sum_{j,k=1}^{n} \mathcal{A}_{jk}(x,t)\sigma_j\sigma_k + \lambda I\right) = 0$$

satisfy the inequality $\Re\lambda(x,t,\boldsymbol{\sigma}) \leq -\delta|\boldsymbol{\sigma}|^2$, where δ=const > 0, for any $\boldsymbol{\sigma} = (\sigma_1,\dots,\sigma_n) \in \mathbb{R}^n$, I is the identity matrix of order m, and $|\cdot|$ is the Euclidean length of a vector.

Before passing to the description of sections of the chapter we discuss the notions of the norm and the dual norm in $\mathbb{R}^m$ in the sense of Minkowski as well as relations between the norms and compact convex bodies in $\mathbb{R}^m$ (see Minkowski [**MI**]). While describing these notions and their properties we shall follow the book by Stoer and Witzgall [**SW**].

By definition, the norm in the Minkowski sense is a function $|\cdot| : \mathbb{R}^m \to \mathbb{R}$ with the following properties

$$\begin{aligned}
&(a) \quad |\zeta| \geq 0 \ \text{ for all } \zeta \in \mathbb{R}^m,\\
&(b) \quad |\zeta| = 0 \ \text{ if and only if } \ \zeta = 0,\\
&(c) \quad |\lambda\zeta| = \lambda|\zeta| \ \text{for all } \zeta \in \mathbb{R}^m \text{ and } \lambda \geq 0,\\
&(d) \quad |\zeta_1 + \zeta_2| \leq |\zeta_1| + |\zeta_2| \ \text{for all } \zeta_1, \zeta_2 \in \mathbb{R}^m.
\end{aligned}$$

The unit ball $\boldsymbol{B} = \{\zeta \in \mathbb{R}^m : |\zeta| \leq 1\}$ of an arbitrary norm in $\mathbb{R}^m$ is a compact convex set of dimension m which contains the origin in its interior. Conversely, for every compact convex body $B \subset \mathbb{R}^m$, which contains the origin in its interior, there is precisely one norm $|\cdot|$ such that $B = \boldsymbol{B}$, namely, $|\zeta| = \inf\{r^{-1}\zeta \in B, \ r > 0\}$. In other words, the compact convex bodies which contain the origin in their interior are in 1-1 correspondence with the norms in the Minkowski sense in $\mathbb{R}^m$.

We remark that the norm $|\cdot|$ is usually called the Minkowski functional of the convex compact body B and, sometimes, it is called the generalized norm (see Beckenbach and Bellman [**BB**], Ch. 1, § 25). Following Stoer and Witzgall [**SW**], we shall call it the norm.

Given the norm $|\cdot|$, one can introduce a new function $|\cdot|_* : \mathbb{R}^m \to \mathbb{R}$ by the equality

$$|\boldsymbol{z}|_* = \sup\{(\boldsymbol{z}, \zeta) : |\zeta| \leq 1\}. \tag{10.1.2}$$

The function $|\cdot|_*$, similarly to $|\cdot|$, satisfies (a)-(d), and is called the dual norm.

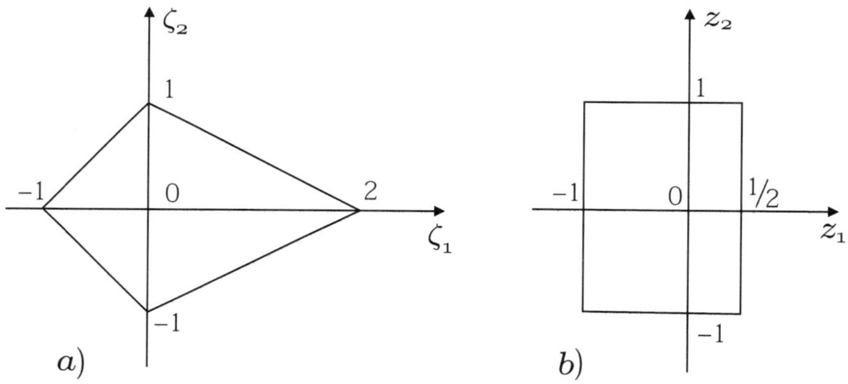

FIGURE 10.1. Example: ***a)*** The unit ball of the norm $|\zeta| = \alpha|\zeta_1| + |\zeta_2|$ with $\alpha = 1/2$ for $\zeta_1 \geq 0$, and $\alpha = 1$ for $\zeta_1 < 0$; ***b)*** The unit ball of the corresponding dual norm $|\boldsymbol{z}|_* = \max\{\beta|z_1|, |z_2|\}$, where $\beta = 2$ for $z_1 \geq 0$, and $\beta = 1$ for $z_1 < 0$.

The symbol (,) will be used as the standard notation of the inner product in $\mathbb{R}^m$. The original and dual norms satisfy the following duality relation

$$|\zeta| = \sup\{(\boldsymbol{z}, \zeta) : |\boldsymbol{z}|_* \leq 1\} \tag{10.1.3}$$

which is a corollary of the equality $|\cdot|_{**} = |\cdot|$. We note also that the Hölder inequality

$$|(\boldsymbol{z}, \zeta)| \leq |\boldsymbol{z}|_* |\zeta| \tag{10.1.4}$$

follows immediately from definition (10.1.2) of the dual norm.

Section 10.2 concerns notations used in this chapter.

In Section 10.3 we obtain an expression for the best constant $\mathcal{K}(\mathbb{R}^n, T)$ in the inequality

$$|\boldsymbol{u}(x,t)| \leq \mathcal{K}(\mathbb{R}^n, T) \sup\{|\boldsymbol{u}(y,0)| : y \in \mathbb{R}^n\},$$

where u is a solution of system (10.1.1) and $(x,t) \in \mathbb{R}^{n+1}_T$. Besides the constant $\mathcal{K}(\mathbb{R}^n, T)$, we study the best constant $\mathcal{K}(\Omega, T)$ in the inequality

$$|\boldsymbol{u}(x,t)| \leq \mathcal{K}(\Omega, T) \sup\{|\boldsymbol{u}(y,\tau)| : (y,\tau) \in \overline{\Gamma}_T\},$$

where $\boldsymbol{u}$ is a solution of system (10.1.1), $(x,t) \in Q_T$ and $\Gamma_T = \{(x,t) \in \partial Q_T : t < T\}$.

We give separate necessary (in Section 10.4) and sufficient (in Section 10.5) conditions for validity of the maximum norm principle (i.e., $\mathcal{K}(\Omega, T) = 1,\ \mathcal{K}(\mathbb{R}^n, T) = 1$) for solutions of system (10.1.1).

If the coefficients of system (10.1.1) do not depend on t, i.e. the system under consideration is

$$\frac{\partial \boldsymbol{u}}{\partial t} - \sum_{j,k=1}^{n} \mathcal{A}_{jk}(x) \frac{\partial^2 \boldsymbol{u}}{\partial x_j \partial x_k} = \mathbf{0}, \tag{10.1.5}$$

then the above mentioned necessary and sufficient conditions coincide. These criteria are addressed in Section 10.6. In order to formulate them, we introduce the following notation.

Let, for the sake of brevity, $\Pi_T = \mathcal{D} \times (0, T]$, where $\mathcal{D} = \Omega$ or $\mathcal{D} = \mathbb{R}^n$. Given the norm $|\cdot|$, by $\mathcal{S}^{m-1}_{\boldsymbol{\nu}}$ we denote the set of points of the unit sphere $\mathcal{S}^{m-1} = \{\zeta \in \mathbb{R}^m : |\zeta| = 1\}$ for which there exists the unit outward normal $\boldsymbol{\nu}(\zeta)$. By $\boldsymbol{\nu}_0(\zeta)$ we denote the vector obtained by parallel translation of $\boldsymbol{\nu}(\zeta)$ into the origin $0 \in \mathbb{R}^m$. Similar to $\boldsymbol{\nu}(\zeta)$, we call $\boldsymbol{\nu}_0(\zeta)$ the unit outward normal to $\mathcal{S}^{m-1}$ at the point ζ. Let further $\mathcal{N} = \{\boldsymbol{\nu}(\zeta) : \zeta \in \mathcal{S}^{m-1}_{\boldsymbol{\nu}}\}$. By $[\boldsymbol{v}_1, \dots, \boldsymbol{v}_m]$ we mean the $(m \times m)$-matrix whose columns are m-component vectors $\boldsymbol{v}_1, \dots, \boldsymbol{v}_m$. As before, the upper asterisk means passage to the transposed matrix and the following notation is used

$$\begin{aligned} A[\mathcal{D}] &= \{\mathcal{A}_{jk}(x) : 1 \leq j,k \leq n,\ \ x \in \mathcal{D}\}, \\ A^*[\mathcal{D}] &= \{\mathcal{A}^*_{jk}(x) : 1 \leq j,k \leq n,\ \ x \in \mathcal{D}\}. \end{aligned}$$

We formulate the main result of Section 10.6 and the present chapter restricting ourselves to three criteria out of a series of equivalent criteria for validity of the maximum norm principle for system (10.1.5).

THEOREM 1. *The maximum norm principle is valid for solutions of system (10.1.5) in $\Pi_T = \mathcal{D} \times (0, T]$ if and only if one of the following equivalent conditions holds:*

(i) *the family of eigenspaces $\mathcal{L}_1, \dots, \mathcal{L}_s$ of any matrix from $A^*[\mathcal{D}]$ satisfies*

$$\bigcup_{i=1}^{s} \mathcal{L}_i \supset \mathcal{N}; \tag{$\mathcal{L}$}$$

(ii) *the family of matrices $A^*[\mathcal{D}]$ is simultaneously diagonable by any matrix of the form $[\boldsymbol{\nu}_1, \dots, \boldsymbol{\nu}_m]$, where $\boldsymbol{\nu}_1, \dots, \boldsymbol{\nu}_m$ is a linear independent system of normals to $\mathcal{S}^{m-1}$;*

(iii) *any plane* $T_\zeta \mathcal{S}^{m-1}$, *tangent to* $\mathcal{S}^{m-1}$ *at* $\zeta \in \mathcal{S}^{m-1}$, *is a shift of an invariant subspace of any matrix* $\mathcal{A} \in A[\mathcal{D}]$.

This result implies, in particular, that the maximum norm principle holds for system (10.1.5) with at least one norm if and only if the system can be diagonalized (Theorem 10.4).

Let a certain norm and hence, the set $\mathcal{N}$ of normals to its unit sphere, be given. We denote by $\{L\}$ the family $\{L_1, \dots, L_s\}$ of subspaces $L_1, \dots, L_s \subset \mathbb{R}^m$, $1 \le s \le m$, satisfying

$$(L) \qquad \mathbb{R}^m = \bigoplus_{i=1}^{s} L_i, \quad \bigcup_{i=1}^{s} L_i \supset \mathcal{N}.$$

Further, let $s_{\max}$ be the maximal s for which (L) holds.

In Section 10.7 we describe some convex bodies for which condition (L) is satisfied with $s_{\max} = 1$, $s_{\max} = 2$ and $s_{\max} = m$. This enables us to obtain explicit necessary and sufficient conditions for validity of the maximum norm principle for system (10.1.5) for some particular cases. We study also the case $m = 2$ for which we characterize the relation between the structure of system (10.1.5) with the maximum norm principle and the geometry of the unit sphere. As corollaries of particular cases $m = 2$, $s_{\max} = 1$, and $s_{\max} = m$, we give criteria for validity of the maximum norm principle for systems of the form (10.1.5) and the p-norm in $\mathbb{R}^m$, $1 \le p \le \infty$:

$$\|\boldsymbol{\zeta}\|_p = \left(\sum_{j=1}^{m} |\zeta_j|^p \right)^{1/p}, \quad 1 \le p < \infty,$$

$$\|\boldsymbol{\zeta}\|_\infty = \max\{|\zeta_j| : 1 \le j \le m\}.$$

We state some criteria for validity of the maximum norm principle for the above mentioned particular cases. By $a_{jk}(x), a_{i,jk}(x)$ we denote some real-valued functions with the same smoothness as the matrix-valued functions $\mathcal{A}_{jk}(x)$.

COROLLARY 1. *Let* $m = 2$. *The maximum norm principle is valid for solutions of system* (10.1.5) *in* Π_T *if and only if:*

(i) $\mathcal{A}_{jk}(x) = a_{jk}(x) I$, $1 \le j,\ k \le n$,

for all $x \in \mathcal{D}$, *if the unit ball of the norm is different from a parallelogram, where* I *is the unit* (2×2)*-matrix;*

(ii) $\mathcal{A}_{jk}(x) = \left([\boldsymbol{\nu}_1, \boldsymbol{\nu}_2]^*\right)^{-1} \operatorname{diag}\{a_{1,jk}(x), a_{2,jk}(x)\}[\boldsymbol{\nu}_1, \boldsymbol{\nu}_2]^*$,

$1 \le j,\ k \le n$, *if the unit ball of the norm is a parallelogram. Here* $\boldsymbol{\nu}_1, \boldsymbol{\nu}_2$ *is a linear independent pair of normals to parallelogram's sides.*

In particular, for $\boldsymbol{\nu}_1 = \boldsymbol{e}_1$, $\boldsymbol{\nu}_2 = \boldsymbol{e}_2$ *one has*

$$\mathcal{A}_{jk}(x) = \operatorname{diag}\{a_{1,jk}(x),\ a_{2,jk}(x)\}.$$

REMARK. From the parabolicity in the sense of Petrovskiĭ of system (10.1.5) it follows immediately that the functions $a_{jk}(x)$, $a_{i,jk}(x)$ satisfy the inequalities

$$\sum_{j,k=1}^{n} a_{jk}(x)\sigma_j\sigma_k \ge \delta|\sigma|^2, \tag{10.1.6}$$

$$\sum_{j,k=1}^{n} a_{i,jk}(x)\sigma_j\sigma_k \geq \delta|\sigma|^2, \quad i=1,2. \tag{10.1.7}$$

Henceforth we assume all the systems under consideration to be parabolic and we shall omit inequalities similar to (10.1.6), (10.1.7) in all corresponding statements.

COROLLARY 2. *Let* $s_{\max} = 1$. *The maximum norm principle is valid for solutions of system* (10.1.5) *in* Π_T *if and only if*

$$\mathcal{A}_{jk}(x) = a_{jk}(x)I, \quad 1 \leq j,\ k \leq n,$$

for all $x \in \mathcal{D}$.

It is shown that the condition $s_{\max} = 1$ holds, for example, if the norm $|\!\cdot\!|$ is differentiable, that is at any point of the unit sphere $\mathcal{S}^{m-1}$ of the norm there exists a tangent hyperplane. Another sufficient condition for $s_{\max} = 1$ is that the ball $\boldsymbol{B}$ of the norm is a non-simple m-polytope (i.e. has a vertex common for more than m facets). For instance, among nonsymple m-polytopes for $m > 2$ there is the so-called m-bipyramid (in particular, m-crosspolytope). The m-bipyramid is a convex body composed of two m-pyramids with the common basement and the m-crosspolytope is a convex body of the form $\mathrm{Conv}\{\pm\boldsymbol{z}_1, \dots, \pm\boldsymbol{z}_m\}$, where $\boldsymbol{z}_1, \dots, \boldsymbol{z}_m$ is a system of linear independent vectors in $\mathbb{R}^m$ and Conv stands for the convex hull (see McMullen and Sheppard [**McMS**]). We also note that the condition $s_{\max} = 1$ holds if the ball $\boldsymbol{B}$ belongs to the class of m-pyramids (in particular, it is an m-simplex).

COROLLARY 3. *The maximum p-norm principle is valid for solutions of system* (10.1.5) *in* Π_T *if and only if:*

$$\text{(i)}\quad \mathcal{A}_{jk}(x) = \frac{1}{2}\begin{pmatrix} \alpha_{jk}(x) & \beta_{jk}(x) \\ \beta_{jk}(x) & \alpha_{jk}(x) \end{pmatrix}, \quad 1 \leq j,\ k \leq n,$$

for all $x \in D$, $p = 1$, $m = 2$, *where*

$$\alpha_{jk}(x) = a_{1,jk}(x) + a_{2,jk}(x),$$

$$\beta_{jk}(x) = a_{1,jk}(x) - a_{2,jk}(x).$$

(ii) $A_{jk}(x) = a_{jk}(x)I$, $1 \leq j,\ k \leq n$, *for either* $p = 1$, $m > 2$ *or* $1 < p < \infty$.

(iii) $A_{jk}(x) = \mathrm{diag}\,\{a_{1,jk}(x), \dots, a_{m,jk}(x)\}$, $1 \leq j,\ k \leq n$, *for* $p = \infty$.

10.2. Some notation

Let F be a set in $\mathbb{R}^n$. By $[\mathrm{C_b}(F)]^m$ we denote the space of continuous and bounded m-component vector-valued functions on F with the generalized norm

$$\|\boldsymbol{u}\|_{[\mathrm{C_b}(F)]^m} = \sup\,\{|\boldsymbol{u}(q)| : q \in F\}.$$

Let $\Pi_T = \mathcal{D} \times (0, T]$, where $\mathcal{D}$ is either a bounded domain in $\mathbb{R}^n$ or $\mathcal{D} = \mathbb{R}^n$ and $0 < T < \infty$. By $[\mathrm{C}^{(2,1)}(\Pi_T)]^m$ we mean the space of m-component vector-valued functions on Π_T whose derivatives with respect to x up to the second order and first derivative with respect to t are continuous.

We introduce the operator

$$\mathfrak{A}_0(x,t,D_x) = \sum_{j,k=1}^{n} \mathcal{A}_{jk}(x,t)\frac{\partial^2}{\partial x_j \partial x_k},$$

where $\mathcal{A}_{jk} = \mathcal{A}_{kj}$ are real $(m\times m)$-matrix-valued functions satisfying the conditions (A), (B) of Section 10.1.

Consider the system

$$\frac{\partial \boldsymbol{u}}{\partial t} - \mathfrak{A}_0(x,t,D_x)\boldsymbol{u} = \boldsymbol{0} \tag{10.2.1}$$

which is uniformly parabolic in the sense of Petrovskiĭ in the layer $\mathbb{R}^{n+1}_T = \mathbb{R}^n \times (0,T]$ and in the cylinder $Q_T = \Omega \times (0,T]$, where Ω is a bounded subdomain of $\mathbb{R}^n$.

We set

$$\mathcal{K}(\mathbb{R}^n, T) = \sup \frac{\|\boldsymbol{u}\|_{[\mathrm{C_b}(\overline{\mathbb{R}^{n+1}_T})]^m}}{\|\boldsymbol{u}\big|_{t=0}\|_{[\mathrm{C_b}(\mathbb{R}^n)]^m}}, \tag{10.2.2}$$

where the supremum is taken over the set of functions in the class $[\mathrm{C}^{(2,1)}(\mathbb{R}^{n+1}_T)]^m \cap [\mathrm{C_b}(\overline{\mathbb{R}^{n+1}_T})]^m$ satisfying system (10.2.1).

Let $\Gamma_T = \{(x,t) \in \partial Q_T : t < T\}$. We put

$$\mathcal{K}(\Omega, T) = \sup \frac{\|\boldsymbol{u}\|_{[\mathrm{C}(\overline{Q}_T)]^m}}{\|\boldsymbol{u}\big|_{\overline{\Gamma}_T}\|_{[\mathrm{C}(\overline{\Gamma}_T)]^m}}, \tag{10.2.3}$$

where the supremum is taken over the set of all solutions of system (10.2.1) in the class $[\mathrm{C}^{(2,1)}(Q_T)]^m \cap [\mathrm{C}(\overline{Q}_T)]^m$.

By $\mathcal{K}_0(y)$ we denote the constant $\mathcal{K}(\mathbb{R}^n, T)$ for the system

$$\frac{\partial \boldsymbol{u}}{\partial t} - \mathfrak{A}_0(y,0,D_x)\boldsymbol{u} = \boldsymbol{0}, \tag{10.2.4}$$

in which $y \in \mathbb{R}^n$ serves as a parameter.

When considering system (10.2.1) with constant coefficients we shall use the notation $\mathcal{K}$ instead of $\mathcal{K}(\mathbb{R}^n, T)$.

By $[\boldsymbol{v}_1, \dots, \boldsymbol{v}_m]$ we denote the $(m\times m)$-matrix whose columns are m-component vectors $\boldsymbol{v}_1, \dots, \boldsymbol{v}_m$. By $\mathrm{Span}\{\boldsymbol{v}_1, \dots, \boldsymbol{v}_k\}$ we mean the linear span of the vectors $\boldsymbol{v}_1, \dots, \boldsymbol{v}_k$.

10.3. Representation of the constant $\mathcal{K}(\mathbb{R}^n, T)$

According to Eidel'man [**Ei1**], there exists a function $\boldsymbol{u}(x,t)$ in the class

$$[\mathrm{C}^{(2,1)}(\mathbb{R}^{n+1}_T)]^m \cap [\mathrm{C_b}(\overline{\mathbb{R}^{n+1}_T})]^m$$

which is the only solution of the Cauchy problem

$$\frac{\partial \boldsymbol{u}}{\partial t} - \mathfrak{A}_0(x,t,D_x)\boldsymbol{u} = \boldsymbol{0} \ \text{ in } \ \mathbb{R}^{n+1}_T, \ \boldsymbol{u}\big|_{t=0} = \boldsymbol{\psi}, \tag{10.3.1}$$

where $\boldsymbol{\psi} \in [\mathrm{C_b}(\mathbb{R}^n)]^m$. This solution can be represented in the form

$$\boldsymbol{u}(x,t) = \int_{\mathbb{R}^n} G(t,0,x,\eta)\boldsymbol{\psi}(\eta)d\eta, \tag{10.3.2}$$

where $G(t, \tau, x, \eta)$ is the Green matrix (or the fundamental matrix of solutions of the Cauchy problem (10.3.1)). The Green matrix for system (10.2.4) will be denoted by $\mathcal{G}_0(t - \tau, x - \eta; y)$. It has the following representation

$$\mathcal{G}_0(t-\tau, x-\eta; y) = (2\pi)^{-n} \int_{\mathbb{R}^n} \exp\Big[- \sum_{j,k=1}^{n} \mathcal{A}_{jk}(y, 0)\sigma_j \sigma_k (t-\tau)\Big] e^{i(x-\eta, \sigma)} d\sigma.$$

This implies

$$\mathcal{G}_0(t-\tau, x-\eta; y) = (t-\tau)^{-n/2} P\left(\frac{x-\eta}{\sqrt{t-\tau}}\, ; y\right) \tag{10.3.3}$$

with

$$P(x; y) = (2\pi)^{-n} \int_{\mathbb{R}^n} e^{i(x,\sigma)} \exp\Big[- \sum_{j,k=1}^{n} \mathcal{A}_{jk}(y, 0)\sigma_j \sigma_k \Big] d\sigma. \tag{10.3.4}$$

When discussing system (10.2.1) with constant coefficients we shall use the notations $\mathcal{A}_{jk}$, $\mathfrak{A}_0(D_x)$, $\mathcal{G}_0(t-\tau, x-\eta)$, $P(x)$. In particular, for the system with constant coefficients formulas (10.3.3) and (10.3.4) take the form

$$\mathcal{G}_0(t-\tau, x-\eta) = (t-\tau)^{-n/2} P\left(\frac{x-\eta}{\sqrt{t-\tau}} \right), \tag{10.3.5}$$

$$P(x) = (2\pi)^{-n} \int_{\mathbb{R}^n} e^{i(x,\sigma)} \exp\left(- \sum_{j,k=1}^{n} \mathcal{A}_{jk}\sigma_j \sigma_k \right) d\sigma. \tag{10.3.6}$$

THEOREM 10.1. *The following formula holds*

$$\mathcal{K}(\mathbb{R}^n, T) = \sup_{x\in\mathbb{R}^n} \sup_{0<t\leq T} \sup_{|\boldsymbol{z}|_*=1} \int_{\mathbb{R}^n} |G^*(t, 0, x, \eta)\boldsymbol{z}|_* d\eta. \tag{10.3.7}$$

In particular,

$$\mathcal{K}_0(y) = \sup_{|\boldsymbol{z}|_*=1} \int_{\mathbb{R}^n} |P^*(\xi, y)\boldsymbol{z}|_* d\xi \tag{10.3.8}$$

and

$$\mathcal{K} = \sup_{|\boldsymbol{z}|_*=1} \int_{\mathbb{R}^n} |P^*(\xi)\boldsymbol{z}|_* d\xi. \tag{10.3.9}$$

PROOF. Let (x, t) be a fixed point in $\mathbb{R}^{n+1}_T$. We find the generalized norm $\|\boldsymbol{u}(x, t)\|$ of the mapping $[\mathrm{C_b}(\mathbb{R}^n)]^m \ni \boldsymbol{\psi} \to \boldsymbol{u}(x, t) \in \mathbb{R}^m$, where $\boldsymbol{u}$ is defined by (10.3.2). Using equalities (10.1.2), (10.1.3), properties of the inner product in $\mathbb{R}^m$ and the fact that the supremum operations commute, we obtain

$$\begin{aligned}
\|\boldsymbol{u}(x,t)\| &= \sup_{|\boldsymbol{\psi}|\le 1}\left|\int_{\mathbb{R}^n} G(t,0,x,\eta)\boldsymbol{\psi}(\eta)d\eta\right| \\
&= \sup_{|\boldsymbol{\psi}|\le 1}\ \sup_{|z|_*=1}\left(z, \int_{\mathbb{R}^n} G(t,0,x,\eta)\boldsymbol{\psi}(\eta)d\eta\right) \\
&= \sup_{|\boldsymbol{z}|_*=1}\ \sup_{|\boldsymbol{\psi}|\le 1}\int_{\mathbb{R}^n} (\boldsymbol{z}, G(t,0,x,\eta)\boldsymbol{\psi}(\eta))\, d\eta \\
&= \sup_{|\boldsymbol{z}|_*=1}\ \sup_{|\boldsymbol{\psi}|\le 1}\int_{\mathbb{R}^n} (G^*(t,0,x,\eta)\boldsymbol{z},\ \boldsymbol{\psi}(\eta))\, d\eta \\
&= \sup_{|\boldsymbol{z}|_*=1}\int_{\mathbb{R}^n} |G^*(t,0,x,\eta)\boldsymbol{z}|_* d\eta.
\end{aligned}$$

Using definition (10.2.2) of the constant $\mathcal{K}(\mathbb{R}^n, T)$, we obtain

$$\begin{aligned}
\mathcal{K}(\mathbb{R}^n,T) &= \sup\Big\{\|\boldsymbol{u}\|_{[\mathrm{C_b}(\overline{\mathbb{R}^{n+1}_T})]^m} : \|\boldsymbol{u}\big|_{t=0}\|_{[\mathrm{C_b}(\mathbb{R}^n)]^m} \le 1, \\
&\qquad \frac{\partial \boldsymbol{u}}{\partial t} - \mathfrak{A}(x,t,D_x)\boldsymbol{u} = \boldsymbol{0} \text{ in } \mathbb{R}^{n+1}_T\Big\} \\
&= \sup_{x\in\mathbb{R}^n}\ \sup_{0<t\le T}\ \sup\Big\{\|\boldsymbol{u}(x,t)\| : \|\boldsymbol{u}\big|_{t=0}\|_{[\mathrm{C_b}(\mathbb{R}^n)]^m} \le 1, \\
&\qquad \frac{\partial \boldsymbol{u}}{\partial t} - \mathfrak{A}(x,t,D_x)\boldsymbol{u} = \boldsymbol{0} \text{ in } \mathbb{R}^{n+1}_T\Big\} = \sup_{x\in\mathbb{R}^n}\ \sup_{0<t\le T}\|\boldsymbol{u}(x,t)\| \\
&= \sup_{x\in\mathbb{R}^n}\ \sup_{0<t\le T}\ \sup_{|\boldsymbol{z}|_*=1}\int_{\mathbb{R}^n} |G^*(t,0,x,\eta)\boldsymbol{z}|_* d\eta,
\end{aligned}$$

which gives (10.3.7).

Substituting $\mathcal{G}_0(t, x-\eta; y)$ from (10.3.3) into (10.3.7) in place of $G(t, 0, x, \eta)$ we arrive at representation (10.3.8).

Similarly, substituting $\mathcal{G}_0(t, x-\eta)$ from (10.3.5) into (10.3.7) in place of $G(t, 0, x, \eta)$ we obtain the representation of the best constant $\mathcal{K}$ for the system with constant coefficients in the form (10.3.9). □

REMARK 10.1. Theorem 10.1 for the system

$$\frac{\partial \boldsymbol{u}}{\partial t} - \mathfrak{A}_0(D_x)\boldsymbol{u} = \boldsymbol{0}$$

with constant coefficients implies (10.3.9).

When willing to underline that the constant $\mathcal{K}$ is sought for the generalized norm $|\cdot|$ we shall use the notation $\mathcal{K}(|\cdot|)$. Now we give a two-sided estimate for $\mathcal{K}(|\cdot|)$.

Let $\boldsymbol{z}$ be some fixed vector in $\mathbb{R}^m$. Clearly, for any parallel vectors $P^*(\xi)\boldsymbol{z}$ and $\boldsymbol{\mathcal{Z}}(\xi)$, $|\boldsymbol{\mathcal{Z}}(\xi)|_* = 1$, directed in the same way, one has

$$|P^*(\xi)\boldsymbol{z}|_* = \frac{|P^*(\xi)\boldsymbol{z}|}{|\boldsymbol{\mathcal{Z}}(\xi)|}, \tag{10.3.10}$$

where $|\cdot|$ is the Euclidean length of an m-component vector. Let $\boldsymbol{z}_{max}$ denote an m-component vector such that

$$|\boldsymbol{z}_{max}| = \max\{|\boldsymbol{z}| : |\boldsymbol{z}|_* = 1\}.$$

Similarly, let $\boldsymbol{z}_{min}$ be a vector such that

$$|\boldsymbol{z}_{min}| = \min\{|\boldsymbol{z}| : \| \boldsymbol{z} \|_* = 1\}.$$

The equality (10.3.10) implies

$$\| P^*(\xi)\boldsymbol{z}_{max} \|_* \geq \frac{|P^*(\xi)\boldsymbol{z}_{max}|}{|\boldsymbol{z}_{max}|},$$

$$\| P^*(\xi)\boldsymbol{z} \|_* \leq \frac{|P^*(\xi)\boldsymbol{z}|}{|\boldsymbol{z}|} \cdot \frac{|\boldsymbol{z}|}{|\boldsymbol{z}_{min}|} \leq \frac{|P^*(\xi)\boldsymbol{z}|}{|\boldsymbol{z}|} \cdot \frac{|\boldsymbol{z}_{max}|}{|\boldsymbol{z}_{min}|}.$$

From this and (10.3.9) one obtains

$$\frac{1}{|\boldsymbol{z}_{max}|} \int_{\mathbb{R}^n} |P^*(\xi)\boldsymbol{z}_{max}| d\xi \leq \mathcal{K}(\| \cdot \|) \leq \frac{|\boldsymbol{z}_{max}|}{|\boldsymbol{z}_{min}|} \sup_{|\boldsymbol{z}|=1} \int_{\mathbb{R}^n} |P^*(\xi)\boldsymbol{z}| d\xi. \tag{10.3.11}$$

Suppose the integral

$$\int_{\mathbb{R}^n} |P^*(\xi)\boldsymbol{z}| d\xi, \quad |\boldsymbol{z}| = 1, \tag{10.3.12}$$

does not depend on the choice of $\boldsymbol{z}$. This is the case, for example, if

$$P_{ij}(\xi) = \lambda(|\xi|)\delta_{ij} + \xi_i \xi_j \mu(|\xi|),$$

where λ and μ are scalar functions. Using the notation $\gamma(\| \cdot \|) = |\boldsymbol{z}_{max}|/|\boldsymbol{z}_{min}|$, we find by (10.3.11)

$$\mathcal{K}(| \cdot |) \leq \mathcal{K}(\| \cdot \|) \leq \gamma(\| \cdot \|)\mathcal{K}(| \cdot |). \tag{10.3.13}$$

Hence it follows, that if integral (10.3.12) does not depend on the choice of $\boldsymbol{z}$, then

$$\mathcal{K}(| \cdot |) = \inf_{\| \cdot \|} \mathcal{K}(\| \cdot \|), \tag{10.3.14}$$

where the infimum is taken over the set of all generalized norms in $\mathbb{R}^m$.

The next assertion is established in the same way as Lemmas 8.1 and 8.5 in Chapter 8, where conditions for validity of the maximum modulus principle are considered.

LEMMA 10.1. *The inequality*

$$\mathcal{K}(\mathcal{D}, T) \geq \sup\{\mathcal{K}_0(y) : \ y \in \mathcal{D}\}$$

is valid, where $\mathcal{D}$ is either a bounded domain $\Omega \subset \mathbb{R}^n$, or $\mathcal{D} = \mathbb{R}^n$.

10.4. Necessary condition for validity of the maximum norm principle for the system $\partial \boldsymbol{u}/\partial t - \mathfrak{A}_0(x, t, D_x)\boldsymbol{u} = \boldsymbol{0}$

Let $\mathcal{S}^{m-1}_{\boldsymbol{\nu}}$ denote a subset of points ζ of the unit sphere $\mathcal{S}^{m-1}$ for which there exists a unit outward normal $\boldsymbol{\nu}(\zeta)$ to $\mathcal{S}^{m-1}$.

The next assertion, crucial in what follows, contains a necessary condition for validity of the maximum norm principle for parabolic systems with constant coefficients in $\mathbb{R}^{n+1}_T$.

PROPOSITION 10.1. *Let the maximum norm principle hold for solutions of the system*

$$\frac{\partial \boldsymbol{u}}{\partial t} - \sum_{j,k=1}^{n} \mathcal{A}_{jk} \frac{\partial^2 \boldsymbol{u}}{\partial x_j \partial x_k} = \boldsymbol{0} \tag{10.4.1}$$

with constant coefficients in $\mathbb{R}^{n+1}_T$. *Then there exist functions*

$$\mu_{jk} : \mathcal{S}^{m-1}_{\boldsymbol{\nu}} \to \mathbb{R},\ 1 \le j,k \le n,$$

such that

$$\mathcal{A}^*_{jk}\nabla|\boldsymbol{\zeta}| = \mu_{jk}(\zeta)\nabla|\boldsymbol{\zeta}|,\ \ 1 \le j,k \le n, \tag{10.4.2}$$

for any $\zeta \in \mathcal{S}^{m-1}_{\boldsymbol{\nu}}$.

PROOF. *1. Structure of the matrix* $\mathcal{G}_0(t-\tau, x-\eta)$. It is well-known that a convex function has the complete differential almost everywhere (cf. for instance, Busemann [**Bus**], Ch. 1). Therefore, the set $\mathcal{S}^{m-1}_{\boldsymbol{\nu}}$ of points, where the tangent plane to $\mathcal{S}^{m-1}$ exists, is everywhere dense on $\mathcal{S}^{m-1}$. Fix a point $\zeta \in \mathcal{S}^{m-1}_{\boldsymbol{\nu}}$. Then there exists the only vector $\boldsymbol{z} = \nabla|\boldsymbol{\zeta}|$ such that $|\boldsymbol{z}|_* = 1$ and $(\boldsymbol{z}, \boldsymbol{\zeta}) = 1$ (cf. for example, Phelps [**Ph**]). Let $V(\boldsymbol{z}) = \{\xi \in \mathbb{R}^n : P^*(\xi)\boldsymbol{z} = \mathbf{0}\}$, where the matrix $P(\xi)$ is related with the Green matrix $\mathcal{G}_0(t-\tau, x-\eta)$ of system (10.4.1) by equality (10.3.5).

The constant $\mathcal{K}$ for system (10.4.1) is defined by (10.3.9). Suppose $\mathcal{K} = 1$ and let there exist a set $\mathcal{M} \subset \mathbb{R}^n$, $\text{mes}_n\mathcal{M} > 0$ such that for all $\zeta \in \mathcal{M}$ the inequality

$$|P^*(\xi)\boldsymbol{z}|_* > (P^*(\xi)\boldsymbol{z}, \boldsymbol{\zeta}) \tag{10.4.3}$$

holds and for all $\xi \in \mathbb{R}^n \backslash \mathcal{M}$ the equality

$$|P^*(\xi)\boldsymbol{z}|_* = (P^*(\xi)\boldsymbol{z}, \boldsymbol{\zeta}) \tag{10.4.4}$$

is valid.

Since by (10.3.5)

$$\int_{\mathbb{R}^n} \mathcal{G}_0(t, x-\eta)d\eta = \int_{\mathbb{R}^n} t^{-n/2}P\left(\frac{x-\eta}{\sqrt{t}}\right)d\eta = \int_{\mathbb{R}^n} P(\xi)d\xi,$$

using the equality

$$\int_{\mathbb{R}^n} \mathcal{G}_0(t, x-\eta)d\eta = I,$$

we get from (10.3.9)

$$\begin{aligned} 1 &= \mathcal{K} \ge \int_{\mathbb{R}^n} |P^*(\xi)\boldsymbol{z}|_* d\xi \\ &= \int_{\mathbb{R}^n\backslash\mathcal{M}} (P^*(\xi)\boldsymbol{z}, \boldsymbol{\zeta})d\xi + \int_{\mathcal{M}} |P^*(\xi)\boldsymbol{z}|_* d\xi \\ &> \int_{\mathbb{R}^n\backslash\mathcal{M}} (P^*(\xi)\boldsymbol{z}, \boldsymbol{\zeta})d\xi + \int_{\mathcal{M}} (P^*(\xi)\boldsymbol{z}, \boldsymbol{\zeta})d\xi \\ &= \int_{\mathbb{R}^n} (P^*(\xi)\boldsymbol{z}, \boldsymbol{\zeta})d\xi = (\boldsymbol{z}, \boldsymbol{\zeta}) = 1. \end{aligned}$$

This shows that our assumption of validity of inequality (10.4.3) on the set $\mathcal{M}$ of positive measure contradicts the maximum norm principle.

Therefore, if $\mathcal{K} = 1$, then for any vectors $\boldsymbol{\zeta}$ and $\boldsymbol{z}$ such that $|\boldsymbol{\zeta}| = 1$, $|\boldsymbol{z}|_* = 1$, $(\boldsymbol{z}, \boldsymbol{\zeta}) = 1$ and for almost all $\xi \in \mathbb{R}^n$ (10.4.4) holds. By continuity of the matrix P, (10.4.4) is valid for all $\xi \in \mathbb{R}^n$. In other words, if $\mathcal{K} = 1$, then for any pair of vectors $\boldsymbol{\zeta}, \boldsymbol{z}$ such that $|\boldsymbol{\zeta}| = 1$, $|\boldsymbol{z}|_* = 1$, $(\boldsymbol{z}, \boldsymbol{\zeta}) = 1$ and for all $\xi \in \mathbb{R}^n \backslash V(z)$ one has

$$\left(\frac{P^*(\xi)\boldsymbol{z}}{|P^*(\xi)\boldsymbol{z}|_*}, \boldsymbol{\zeta}\right) = 1.$$

Hence, taking into account that to any $\zeta \in \mathcal{S}_\nu^{m-1}$ there corresponds only one vector $\boldsymbol{z}$ with $|\boldsymbol{z}|_* = 1$ such that $(\boldsymbol{z}, \boldsymbol{\zeta}) = 1$, we get

$$\frac{P^*(\xi)\boldsymbol{z}}{|P^*(\xi)\boldsymbol{z}|_*} = 1, \tag{10.4.5}$$

with $\xi \in \mathbb{R}^n \backslash V(\boldsymbol{z})$. Since $\boldsymbol{z} = \nabla|\boldsymbol{\zeta}|$, it follows that (10.4.5) implies

$$P^*(\xi)\nabla|\boldsymbol{\zeta}| = |P^*(\xi)\nabla|\boldsymbol{\zeta}||_* \nabla|\boldsymbol{\zeta}|$$

for all $\xi \in \mathbb{R}^n$. This and (10.3.5) show that

$$\mathcal{G}_0^*(t-\tau, x-\eta)\nabla|\boldsymbol{\zeta}| = g(t-\tau, x-\eta; \zeta)\nabla|\boldsymbol{\zeta}|, \tag{10.4.6}$$

where

$$g(t-\tau, x-\eta; \zeta) = |\mathcal{G}_0^*(t-\tau, x-\eta)\nabla|\boldsymbol{\zeta}||_*.$$

2. The structure of the operator $\mathfrak{A}_0(D_x)$. We fix $\zeta \in \mathcal{S}_\nu^{m-1}$ and denote by $T_\zeta \mathcal{S}^{m-1}$ the tangent plane to the sphere $\mathcal{S}^{m-1}$ at the point $\zeta \in \mathcal{S}_\nu^{m-1}$. Let $\boldsymbol{\chi} \in T_\zeta \mathcal{S}^{m-1}$. Consider the vector-valued function

$$\boldsymbol{v}(x,t) = \int_{\mathbb{R}^n} \mathcal{G}_0(t, x-\eta)\psi(\eta)\boldsymbol{\chi} d\eta, \tag{10.4.7}$$

where $\psi(\eta)$ is a continuous scalar function in $\mathbb{R}^n$ with compact support.

The vector-valued function $\boldsymbol{v}$ satisfies the Cauchy problem

$$\frac{\partial \boldsymbol{v}}{\partial t} - \sum_{j,k=1}^n \mathcal{A}_{jk} \frac{\partial^2 \boldsymbol{v}}{\partial x_j \partial x_k} = \boldsymbol{0} \text{ in } \mathbb{R}_T^{n+1}, \tag{10.4.8}$$

$$\boldsymbol{v}(x,0) = \psi(x)\boldsymbol{\chi}. \tag{10.4.9}$$

Applying the Fourier transform with respect to the variables $x_1, \dots, x_n$ for a fixed $t \in (0, T]$ to (10.4.7) and (10.4.8), we obtain

$$F[\boldsymbol{v}] = F[\mathcal{G}_0]F[\psi]\boldsymbol{\chi}, \tag{10.4.10}$$

$$\frac{dF[\boldsymbol{v}]}{dt} + \sum_{j,k=1}^n \mathcal{A}_{jk} F[\boldsymbol{v}]\sigma_j\sigma_k = 0. \tag{10.4.11}$$

Substituting (10.4.10) into (10.4.11), we find

$$F[\psi]\frac{dF[\mathcal{G}_0]}{dt}\chi + F[\psi]\sum_{j,k=1}^n \mathcal{A}_{jk} F[\mathcal{G}_0]\chi\sigma_j\sigma_k = 0,$$

which, by virtue of arbitrariness of ψ in (10.4.7), implies

$$\frac{dF[\mathcal{G}_0]}{dt}\boldsymbol{\chi} + \sum_{j,k=1}^n \mathcal{A}_{jk} F[\mathcal{G}_0]\boldsymbol{\chi}\sigma_j\sigma_k = 0.$$

Taking the scalar product of this equality with $\nabla|\boldsymbol{\zeta}|$ and using the relation $(F[\mathcal{G}_0])^* = F[\mathcal{G}_0^*]$, we get

$$\frac{d}{dt}(\boldsymbol{\chi}, F[\mathcal{G}_0^*]\nabla|\boldsymbol{\zeta}|) + \sum_{j,k=1}^n \left(F[\mathcal{G}_0]\boldsymbol{\chi}, \mathcal{A}_{jk}^*\nabla|\boldsymbol{\zeta}|\right)\sigma_j\sigma_k = 0. \tag{10.4.12}$$

According to (10.4.6),

$$F[\mathcal{G}_0^*]\nabla|\boldsymbol{\zeta}| = F[g]\nabla|\boldsymbol{\zeta}|.$$

Hence, due to orthogonality of vectors $\nabla|\zeta|$ and $\chi \in T_\zeta \mathcal{S}^{m-1}$, equality (10.4.12) implies

$$\sum_{j,k=1}^{n} \left(F[\mathcal{G}_0]\chi, \mathcal{A}_{jk}^* \nabla|\zeta|\right) \sigma_j \sigma_k = 0.$$

Taking into account that $F[\mathcal{G}_0]$ tends to the unit matrix as $t \to 0$, we obtain

$$\sum_{j,k=1}^{n} \left(\chi, \mathcal{A}_{jk}^* \nabla|\zeta|\right) \sigma_j \sigma_k = 0.$$

Due to the arbitrariness of $\chi \in T_\zeta \mathcal{S}^{m-1}$ and $\sigma \in \mathbb{R}^n$ the last equality implies the linear dependence of the vectors $\mathcal{A}_{jk}^* \nabla|\zeta|$ and $\nabla|\zeta|$. Consequently, there exist functions $\mu_{jk} : \mathcal{S}_\nu^{m-1} \to \mathbb{R}$ such that for all $\zeta \in \mathcal{S}_\nu^{m-1}$

$$\mathcal{A}_{jk}^* \nabla|\zeta| = \mu_{jk}(\zeta)|\nabla\zeta|, \quad 1 \le j,k \le n.$$

The proof is complete. □

We remind that $\mathcal{D}$ is either a bounded domain $\Omega \subset \mathbb{R}^n$ or $\mathcal{D} = \mathbb{R}^n$, so that $\Pi_T = \mathcal{D} \times (0,T]$ coincides either with Q_T or with $\mathbb{R}_T^{n+1}$.

The following assertion follows from Proposition 10.1 and Lemma 10.1.

PROPOSITION 10.2. *Let the maximum norm principle be valid for solutions of the system*

$$\frac{\partial \boldsymbol{u}}{\partial t} - \sum_{j,k=1}^{n} \mathcal{A}_{jk}(x,t) \frac{\partial^2 \boldsymbol{u}}{\partial x_j \partial x_k} = \boldsymbol{0}$$

in Π_T. Then there exist functions $\mu_{jk} : \mathcal{D} \times \mathcal{S}_\nu^{m-1} \to \mathbb{R}$, $1 \le j,k \le n$, such that for any $\zeta \in \mathcal{S}_\nu^{m-1}$ and for all $x \in \mathcal{D}$

$$\mathcal{A}_{jk}^*(x,0)\nabla|\zeta| = \mu_{jk}(x,\zeta)\nabla|\zeta|, \qquad 1 \le j,k \le n. \tag{10.4.13}$$

10.5. Sufficient condition for validity of the maximum norm principle for the system $\partial \boldsymbol{u}/\partial t - \mathfrak{A}_0(x,t,D_x)\boldsymbol{u} = \boldsymbol{0}$

In this section we prove the following assertion.

PROPOSITION 10.3. *Let the equalities*

$$\mathcal{A}_{jk}^*(x,t)\nabla|\zeta| = \mu_{jk}(x,t,\zeta)\nabla|\zeta|, \ 1 \le j,k \le n, \tag{10.5.1}$$

be valid for any $\zeta \in \mathcal{S}_\nu^{m-1}$ and for all $(x,t) \in \Pi_T$ with

$$\mu_{jk} : \Pi_T \times \mathcal{S}_\nu^{m-1} \to \mathbb{R}, \quad 1 \le j,k \le n.$$

Then the maximum norm principle holds for solutions of the system

$$\frac{\partial \boldsymbol{u}}{\partial t} - \sum_{j,k=1}^{n} \mathcal{A}_{jk}(x,t) \frac{\partial^2 \boldsymbol{u}}{\partial x_j \partial x_k} = \boldsymbol{0} \tag{10.5.2}$$

in Π_T.

PROOF. Let $\Pi_T = Q_T$. The case $\Pi_T = \mathbb{R}^{n+1}_T$ can be considered in the same way as in Subsection 8.3.1.2, where $|\cdot|$ should be replaced by $\boldsymbol{|}\cdot\boldsymbol{|}$.

Let $\boldsymbol{u} \in [\mathrm{C_b}(\overline{Q}_T)]^m \cap [\mathrm{C}^{(2,1)}(Q_T)]^m$ be a solution of the initial-value problem

$$\frac{\partial \boldsymbol{u}}{\partial t} - \sum_{j,k=1}^{n} \mathcal{A}_{jk}(x,t)\frac{\partial^2 \boldsymbol{u}}{\partial x_j \partial x_k} = \boldsymbol{0} \text{ in } Q_T, \qquad \boldsymbol{u}\Big|_{\overline{\Gamma}_T} = \boldsymbol{\psi}.$$

Suppose the function $\boldsymbol{|u}(x,t)\boldsymbol{|}$ attains its global maximum at a point $(x_0,t_0) \in \overline{Q}_T$. We introduce the notation

$$\boldsymbol{\zeta} = \boldsymbol{u}(x_0,t_0)/\boldsymbol{|u}(x_0,t_0)\boldsymbol{|}. \tag{10.5.3}$$

We consider two cases.

(i) Let $\zeta \in \mathcal{S}^{m-1}_{\boldsymbol{\nu}}$. Taking the scalar product of both sides of (10.5.2) with $\nabla\boldsymbol{|}\zeta\boldsymbol{|}$ and using (10.5.1), one obtains

$$\begin{aligned}
0 &= \frac{\partial}{\partial t}(\boldsymbol{u}, \nabla\boldsymbol{|\zeta|}) - \sum_{j,k=1}^{n}\left(\mathcal{A}_{jk}(x,t)\frac{\partial^2 \boldsymbol{u}}{\partial x_j \partial x_k},\ \nabla\boldsymbol{|\zeta|}\right)\\
&= \frac{\partial}{\partial t}(\boldsymbol{u}, \nabla\boldsymbol{|\zeta|}) - \sum_{j,k=1}^{n}\left(\frac{\partial^2 \boldsymbol{u}}{\partial x_j \partial x_k},\ \mathcal{A}^*_{jk}(x,t)\nabla\boldsymbol{|\zeta|}\right)\\
&= \frac{\partial}{\partial t}(\boldsymbol{u}, \nabla\boldsymbol{|\zeta|}) - \sum_{j,k=1}^{n}\left(\frac{\partial^2 \boldsymbol{u}}{\partial x_j \partial x_k},\ a_{jk}(x,t,\zeta)\nabla\boldsymbol{|\zeta|}\right)\\
&= \frac{\partial}{\partial t}(\boldsymbol{u}, \nabla\boldsymbol{|\zeta|}) - \sum_{j,k=1}^{n} a_{jk}(x,t,\zeta)\frac{\partial^2}{\partial x_j \partial x_k}(\boldsymbol{u}, \nabla\boldsymbol{|\zeta|}).
\end{aligned}$$

Thus the scalar function $v = (\boldsymbol{u}, \nabla\boldsymbol{|\zeta|})$ satisfies the problem

$$\frac{\partial v}{\partial t} - \sum_{j,k=1}^{n} a_{jk}(x,t,\zeta)\frac{\partial^2 v}{\partial x_j \partial x_k} = 0 \text{ in } Q_T,\ v\Big|_{\overline{\Gamma}_T} = (\boldsymbol{\psi}, \nabla\boldsymbol{|\zeta|}).$$

Hence, by the maximum modulus principle for the scalar parabolic equation, one has

$$\max_{\overline{Q}_T}|v| = \max_{\overline{\Gamma}_T}|v|$$

which can be written as

$$\max_{\overline{Q}_T}|(\boldsymbol{u}, \nabla\boldsymbol{|\zeta|})| = \max_{\overline{\Gamma}_T}|(\boldsymbol{u}, \nabla\boldsymbol{|\zeta|})|. \tag{10.5.4}$$

Since $\boldsymbol{|}\nabla\boldsymbol{|\zeta||}_* = 1$, by (10.1.3) we find

$$\max_{\overline{\Gamma}_T}|(\boldsymbol{u}, \nabla\boldsymbol{|\zeta|}) \le \max_{\overline{\Gamma}_T}\boldsymbol{|u|}. \tag{10.5.5}$$

Then, using (10.5.3) and the equality $(\boldsymbol{\zeta}, \nabla\boldsymbol{|\zeta|}) = 1$, we obtain

$$\begin{aligned}
\max_{\overline{Q}_T}|(\boldsymbol{u}, \nabla\boldsymbol{|\zeta|})| &\ge |(\boldsymbol{|u}(x_0,t_0)\boldsymbol{|\zeta}, \nabla\boldsymbol{|\zeta|})|\\
&= \boldsymbol{|u}(x_0,t_0)\boldsymbol{|}\ |(\boldsymbol{\zeta}, \nabla\boldsymbol{|\zeta|})| = \max_{\overline{Q}_T}\boldsymbol{|u|}.
\end{aligned} \tag{10.5.6}$$

From (10.5.4)-(10.5.6) it follows

$$\max_{\overline{Q}_T}\boldsymbol{|u|} \le \max_{\overline{\Gamma}_T}\boldsymbol{|u|},$$

consequently, the maximum norm principle

$$\max_{\overline{Q}_T} |\boldsymbol{u}| = \max_{\overline{\Gamma}_T} |\boldsymbol{u}| \tag{10.5.7}$$

is valid.

(ii) Let $\zeta \in \mathcal{S}^{m-1} \backslash \mathcal{S}_{\nu}^{m-1}$. We use the fact that the set $\mathcal{S}_{\nu}^{m-1}$ is everywhere dense on $\mathcal{S}^{m-1}$. Given an arbitrary $\varepsilon > 0$ one can find $\zeta_\varepsilon \in \mathcal{S}_{\nu}^{m-1}$ such that $|\boldsymbol{\zeta} - \boldsymbol{\zeta}_\varepsilon| < \varepsilon$. Using the same arguments as in getting (10.5.4), one obtains

$$\max_{\overline{Q}_T} |(\boldsymbol{u}, \nabla|\boldsymbol{\zeta}_\varepsilon|)| = \max_{\overline{\Gamma}_T} |(\boldsymbol{u}, \nabla|\boldsymbol{\zeta}_\varepsilon|)|. \tag{10.5.8}$$

Since $|\nabla|\boldsymbol{\zeta}_\varepsilon||_* = 1$, by (10.1.3) we find

$$\max_{\overline{\Gamma}_T} |(\boldsymbol{u}, \nabla|\boldsymbol{\zeta}_\varepsilon|)| \leq \max_{\overline{\Gamma}_T} |\boldsymbol{u}|. \tag{10.5.9}$$

Then, using

$$|(\boldsymbol{\zeta} - \boldsymbol{\zeta}_\varepsilon, \nabla|\boldsymbol{\zeta}_\varepsilon|)| \leq |\boldsymbol{\zeta} - \boldsymbol{\zeta}_\varepsilon| \leq \varepsilon$$

the equality $(\boldsymbol{\zeta}_\varepsilon, \nabla|\boldsymbol{\zeta}_\varepsilon|) = 1$ and (10.5.3) we obtain

$$\begin{aligned} &\max_{\overline{Q}_T} |(\boldsymbol{u}, \nabla|\boldsymbol{\zeta}_\varepsilon|)| \geq (\boldsymbol{u}(x_0, t_0), \nabla|\boldsymbol{\zeta}_\varepsilon|) \\ = \;& |\boldsymbol{u}(x_0, t_0)|(\boldsymbol{\zeta}, \nabla|\boldsymbol{\zeta}_\varepsilon|) = |\boldsymbol{u}(x_0, t_0)|[(\boldsymbol{\zeta}_\varepsilon, \nabla|\boldsymbol{\zeta}_\varepsilon|) \\ + \;& (\boldsymbol{\zeta} - \boldsymbol{\zeta}_\varepsilon, \nabla|\boldsymbol{\zeta}_\varepsilon|)] \geq |\boldsymbol{u}(x_0, t_0)|(1-\varepsilon) = (1-\varepsilon) \max_{\overline{Q}_T} |\boldsymbol{u}|. \end{aligned}$$

This together with (10.5.8), (10.5.9) implies

$$(1-\varepsilon) \max_{\overline{Q}_T} |\boldsymbol{u}| \leq \max_{\overline{\Gamma}_T} |\boldsymbol{u}|.$$

Using the arbitrariness of ε, we arrive at (10.5.7) which completes the proof. □

10.6. Necessary and sufficient condition for validity of the maximum norm principle for the system $\partial \boldsymbol{u}/\partial t - \mathfrak{A}_0(x, D_x)\boldsymbol{u} = \boldsymbol{0}$

Before stating the main theorem about a series of equivalent criteria for validity of the maximum norm principle for solutions of systems with coefficients which depend only on x we give an auxiliary assertion of geometric character. Henceforth, when mentioning polytopes we mean convex m-polytopes.

LEMMA 10.2. *One can always choose m points $p_1, \dots, p_m$ on the boundary of a compact convex body $B \subset \mathbb{R}^m$ so that the external normals $\boldsymbol{\nu}(p_1), \dots, \boldsymbol{\nu}(p_m)$ are linear independent. Moreover, one of the points $p_1, \dots, p_m$ can be chosen arbitrarily in the set $\partial B_{\boldsymbol{\nu}} \subset \partial B$ such that the tangent plane to ∂B exists at each point of $\partial B_{\boldsymbol{\nu}}$.*

PROOF. **1.** Suppose that B is a polytope. Let F_1 be one of its facets, and let T_1 denote the supporting plane of this facet. By q we mean one of polytope's vertices in T_1. The point q is the intersection point of at least m hyperplanes $T_1, \dots, T_m$ containing correspondent facets $F_1, \dots, F_m$. Take an arbitrary point $O \in$ int B and draw the rays $OL_1, \dots, OL_m$ orthogonal to $T_1, \dots, T_m$. For $q_i = T_i \cap OL_i$ let $\boldsymbol{\nu}(q_i)$ denote the unit outward normal to T_i at q_i. Further, let $d_i = |q_i - O|$, $i = 1, \dots, m$. We place the origin of the coordinate system with orthonormal basis $\boldsymbol{e}_1, \dots, \boldsymbol{e}_m$ at the point O and use the notation $x = Oq$. Since

$$q = \bigcap_{i=1}^{m} T_i,$$

it follows that $x = (x_1, \dots, x_m)$ is the only solution of the system

$$(\boldsymbol{\nu}(q_i), x) = d_i, \quad i = 1, 2, \dots, m,$$

or, which is the same,

$$\sum_{j=1}^{m} (\boldsymbol{\nu}(q_i), \boldsymbol{e}_j) x_j = d_i, \quad i = 1, 2, \dots, m.$$

The matrix of this system is $[\boldsymbol{\nu}(q_i), \dots, \boldsymbol{\nu}(q_m)]^*$. Consequently,

$$\det[\boldsymbol{\nu}(q_1), \dots, \boldsymbol{\nu}(q_m)]^* \neq 0.$$

This implies linear independence of the system $\boldsymbol{\nu}(q_1), \dots, \boldsymbol{\nu}(q_m)$.

2. Consider now the case of an arbitrary compact convex body B in $\mathbb{R}^m$. We use Theorem 18.8 from Rockafellar [**Ro**] which states that any m-dimensional closed convex set $\mathcal{M} \subset \mathbb{R}^m$ is the intersection of closed half-spaces, tangent to this set. This implies the existence of a polytope $\mathcal{P}$ circumscribing B with facets whose supporting planes are tangent to ∂B. Take an arbitrary $p_1 \in \partial B_{\boldsymbol{\nu}}$ and let $\mathbb{R}^n_+(p_1)$ be that of two closed half-spaces tangent to ∂B at p_1 which contains B. Then the body $\mathcal{P}' = \mathcal{P} \cap \mathbb{R}^n_+(p_1)$ is also a polytope. Now apply the proof of part 1 of the present lemma to the polytope $\mathcal{P}'$ putting $T_1 = \partial \mathbb{R}^n_+(p_1)$ and choosing as q one of polytope's $\mathcal{P}'$ vertices in T_1.

As a result we construct the linear independent system of normals $\boldsymbol{\nu}(q_1), \dots, \boldsymbol{\nu}(q_m)$ to m facets of the polytope $\mathcal{P}'$ with supporting planes $T_1, \dots, T_m$ intersecting at q. Each of planes $T_1, \dots, T_m$ is tangent to ∂B. Choose the points $p_2, \dots, p_m$ arbitrarily in the sets $\partial B \cap T_2, \dots, \partial B \cap T_m$. Taking into account linear dependence of the vectors $\boldsymbol{\nu}(q_i)$, $\boldsymbol{\nu}(p_i)$, $i = 1, \dots, m$, we complete the proof. □

Further we need the following notation:

$$\mathcal{N} = \{\boldsymbol{\nu}(\zeta) : \zeta \in \mathcal{S}^{m-1}_{\boldsymbol{\nu}}\}, \ A[\mathcal{D}] = \{\mathcal{A}_{jk}(x) : x \in \mathcal{D};\ 1 \le j,k \le n\},$$
$$A^*[\mathcal{D}] = \{\mathcal{A}^*_{jk}(x) : x \in \mathcal{D},\ 1 \le j,k \le n\}.$$

Now we are in a position to give the principal assertion of this section.

THEOREM 10.2. *The maximum norm principle is valid for solutions of the system*

$$\frac{\partial \boldsymbol{u}}{\partial t} - \sum_{j,k=1}^{n} \mathcal{A}_{jk}(x) \frac{\partial^2 \boldsymbol{u}}{\partial x_j \partial x_k} = \boldsymbol{0} \tag{10.6.1}$$

in $\Pi_T = \mathcal{D} \times (0, T]$ *if and only if one of the following equivalent conditions holds:*

(i) *for any matrix* $\mathcal{A}^* \in A^*[\mathcal{D}]$ *there exists a function* $\mu : \mathcal{S}^{m-1}_{\boldsymbol{\nu}} \to \mathbb{R}$ *such that*

$$\mathcal{A}^* \boldsymbol{\nu}(\zeta) = \mu(\zeta) \boldsymbol{\nu}(\zeta)$$

for all $\zeta \in \mathcal{S}^{m-1}_{\boldsymbol{\nu}}$,

(ii) *the family of eigenspaces* $\mathcal{L}_1, \dots, \mathcal{L}_s$ *of each matrix* $\mathcal{A}^* \in A^*[\mathcal{D}]$ *is such that*

$$\bigcup_{i=i}^{s} \mathcal{L}_i \supset \mathcal{N}, \tag{$\mathcal{L}$}$$

(iii) *the family of matrices* $A^*[\mathcal{D}]$ *is simultaneously diagonable by any matrix of the form* $[\boldsymbol{\nu}_1, \dots, \boldsymbol{\nu}_m]$, *where* $\boldsymbol{\nu}_1, \dots, \boldsymbol{\nu}_m$ *is a linear independent system of normals to* $\mathcal{S}^{m-1}$,

(iv) *any plane* $T'_\zeta \mathcal{S}^{m-1}$, *passing through the origin of* $\mathbb{R}^m$ *and parallel to the plane* $T_\zeta \mathcal{S}^{m-1}$, *tangent to the sphere* $\mathcal{S}^{m-1}$ *at* $\zeta \in \mathcal{S}_{\boldsymbol{\nu}}^{m-1}$, *is an invariant subspace of any matrix* $\mathcal{A} \in A[\mathcal{D}]$.

PROOF. (i) The necessity and sufficiency of this condition for validity of the maximum norm principle for solutions of system (10.6.1) in Π_T follows from Propositions 10.2, 10.3 and the linear dependence of vectors $\nabla|\zeta|$, $\boldsymbol{\nu}(\zeta)$ for $\zeta \in \mathcal{S}_{\boldsymbol{\nu}}^{m-1}$.

(ii) *Necessity.* Suppose that the maximum norm principle is valid for solutions of system (10.6.1) in Π_T. Then, by (i), each of vectors $\boldsymbol{\nu} \in \mathcal{N}$ is an eigenvector of any matrix $\mathcal{A}^* \in A^*[\mathcal{D}]$. Hence, the set $\mathcal{N}$ belongs to the union of eigenspaces of $\mathcal{A}^*$.

Sufficiency. Let (ii) be valid. Then for a fixed matrix $\mathcal{A}^* \in A^*[\mathcal{D}]$ and for $\zeta \in \mathcal{S}_{\boldsymbol{\nu}}^{m-1}$ there exists a real number $\mu(\zeta)$ such that

$$\mathcal{A}^* \boldsymbol{\nu}(\zeta) = \mu(\zeta)\boldsymbol{\nu}(\zeta),$$

i.e. (i) holds and, consequently, the maximum norm principle is valid.

(iii) *Necessity.* Let the maximum norm principle be valid for solutions of the system (10.6.1). We choose m points $\zeta_1, \dots, \zeta_m \in \mathcal{S}_{\boldsymbol{\nu}}^{m-1}$ so that the normals $\boldsymbol{\nu}(\zeta_1), \dots, \boldsymbol{\nu}(\zeta_m)$ are linear independent, which is possible by Lemma 10.2. Then by (i) one obtains

$$\mathcal{A}^* \boldsymbol{\nu}(\zeta_i) = \mu(\zeta_i)\boldsymbol{\nu}(\zeta_i),$$

where $1 \le i \le m$. Consequently,

$$[\boldsymbol{\nu}(\zeta_1), \dots, \boldsymbol{\nu}(\zeta_m)]^{-1} \mathcal{A}^* [\boldsymbol{\nu}(\zeta_1), \dots, \boldsymbol{\nu}(\zeta_m)] = \text{diag}\,\{\mu(\zeta_1), \dots, \mu(\zeta_m)\}, \tag{10.6.2}$$

that is any matrix $\mathcal{A}^* \in A^*[\mathcal{D}]$ can be diagonalized by a fixed matrix of the form $[\boldsymbol{\nu}, \dots, \boldsymbol{\nu}_m]$, where $\boldsymbol{\nu}_1, \dots, \boldsymbol{\nu}_m$ is a linear independent system of normals to $\mathcal{S}^{m-1}$.

Sufficiency. Let $\boldsymbol{\nu}(\zeta_1), \dots, \boldsymbol{\nu}(\zeta_m)$ denote a linear independent system of normals to $\mathcal{S}^{m-1}$ at the points $\zeta_1, \dots, \zeta_m \in \mathcal{S}_{\boldsymbol{\nu}}^{m-1}$. Suppose that condition (iii) holds. Then to any matrix $\mathcal{A}^* \in A^*[\mathcal{D}]$ there corresponds a matrix $M = \text{diag}\,\{\mu_1, \dots, \mu_m\}$ such that

$$[\boldsymbol{\nu}(\zeta_1), \dots, \boldsymbol{\nu}(\zeta_m)]^{-1} \mathcal{A}^* [\boldsymbol{\nu}(\zeta_1), \dots, \boldsymbol{\nu}(\zeta_m)] = M,$$

or, equivalently,

$$\mathcal{A}^* [\boldsymbol{\nu}(\zeta_1), \dots, \boldsymbol{\nu}(\zeta_m)] = [\boldsymbol{\nu}(\zeta_1), \dots, \boldsymbol{\nu}(\zeta_m)] M. \tag{10.6.3}$$

Here we may assume that the point $\zeta_1 \in \mathbf{S}_{\nu}^{m-1}$ is chosen arbitrarily and that the points $\zeta_2, \dots, \zeta_m \in \mathcal{S}_{\boldsymbol{\nu}}^{m-1}$ are chosen to satisfy the condition

$$\det\,[\boldsymbol{\nu}(\zeta_1), \dots, \boldsymbol{\nu}(\zeta_m)] \neq 0,$$

which is possible by Lemma 10.2. Then, by (10.6.2),

$$\mathcal{A}^* \boldsymbol{\nu}(\zeta_1) = \mu_1 \boldsymbol{\nu}(\zeta_1).$$

With ζ_1 replaced by ζ and μ_1 replaced by $\mu(\zeta)$ one can rewrite the last equality as

$$\mathcal{A}^* \boldsymbol{\nu}(\zeta) = \mu(\zeta)\boldsymbol{\nu}(\zeta),$$

where ζ is any point of $\mathcal{S}_{\boldsymbol{\nu}}^{m-1}$. Thus condition (i) of the present theorem is fulfilled and so the maximum norm principle holds for solutions of system (10.6.1).

(iv) *Necessity.* Let the maximum norm principle be valid for solutions of (10.6.1). By $T'_\zeta \mathcal{S}^{m-1}$ we denote a plane, passing through the origin $O \in \mathbb{R}^m$ and parallel to the plane $T_\zeta \mathcal{S}^{m-1}$, tangent to $\mathcal{S}^{m-1}$ at $\zeta \in \mathcal{S}_{\boldsymbol{\nu}}^{m-1}$. Then (i) holds and, consequently, for any $\boldsymbol{\chi} \in T'_\zeta \mathcal{S}^{m-1}$, $\zeta \in \mathcal{S}_{\boldsymbol{\nu}}^{m-1}$ one obtains

$$0 = (\mu(\zeta)\boldsymbol{\nu}(\zeta), \boldsymbol{\chi}) = (\mathcal{A}^* \boldsymbol{\nu}(\zeta), \boldsymbol{\chi}) = (\boldsymbol{\nu}(\zeta), \mathcal{A}\boldsymbol{\chi}) .$$

Thus,

$$\mathcal{A}\left(T'_\zeta \mathcal{S}^{m-1}\right) \subset T'_\zeta \mathcal{S}^{m-1}.$$

Sufficiency. Suppose (iv) holds. Then, for all $\boldsymbol{\chi} \in T'_\zeta \mathcal{S}^{m-1}$,

$$0 = (\mathcal{A}\boldsymbol{\chi}, \boldsymbol{\nu}(\zeta)) = (\boldsymbol{\chi}, \mathcal{A}^* \boldsymbol{\nu}(\zeta)),$$

that is there exists $\mu : \mathcal{S}_{\boldsymbol{\nu}}^{m-1} \to \mathbb{R}$ such that

$$\mathcal{A}^* \boldsymbol{\nu}(\zeta) = \mu(\zeta)\boldsymbol{\nu}(\zeta).$$

Thus, (i) holds and the maximum norm principle is valid for solutions of (10.6.1). □

In the following theorem, the spaces $L_1, \dots, L_s \subset \mathbb{R}^m$, $1 \leq s \leq m$, form a family $\{L_i\}$ such that

$$(L) \qquad \mathbb{R}^m = \bigoplus_{i=1}^{s} L_i, \quad \bigcup_{i=1}^{s} L_i \supset \mathcal{N}.$$

Let $s_{\max}$ denote the maximum value of s for which (L) holds. Further, let $\boldsymbol{\nu}_1, \dots, \boldsymbol{\nu}_m$ be a fixed linear independent system of normals to $\mathcal{S}^{m-1}$.

The next assertion is also a criterion of validity of the maximum norm principle for solutions of system (10.6.1). However, we formulate it separately as a theorem on a representation of the matrix coefficients $\mathcal{A}_{jk}(x)$.

THEOREM 10.3. *The maximum norm principle is valid for solutions of the system* (10.6.1) *in* $\Pi_T = \mathcal{D} \times (0, T]$ *if and only if*

(*a*) *for all* $x \in \mathcal{D}$,

$$(10.6.4) \quad \mathcal{A}_{jk}(x) = ([\boldsymbol{\nu}_1, \dots, \boldsymbol{\nu}_m]^*)^{-1} \operatorname{diag}\{a_{1,jk}(x), \dots, a_{m,jk}(x)\}[\boldsymbol{\nu}_1, \dots, \boldsymbol{\nu}_m]^*,$$

with $1 \leq j, k \leq n$;

(*b*) *to any matrix* $\mathcal{A}_{jk}(x)$ *there corresponds a partition of the set* $\{\boldsymbol{\nu}_1, \dots, \boldsymbol{\nu}_m\}$ *into subsets* $\mathcal{N}_1, \dots, \mathcal{N}_s$, $1 \leq s \leq m$, *such that*

1) *subspaces* $L_1 = \operatorname{Span}\{\mathcal{N}_1\}, \dots, L_s = \operatorname{Span}\{\mathcal{N}_s\}$ *satisfy*

$$\bigcup_{i=1}^{s} L_i \supset \mathcal{N}$$

and

2) *if the vectors* $\boldsymbol{\nu}_p$ *and* $\boldsymbol{\nu}_q$, $1 \leq p, q \leq m$, $p \neq q$, *belong to the same set* $\mathcal{N}_i$, $1 \leq i \leq s$, *then the eigenvalues* $a_{p,jk}(x)$ *and* $a_{q,jk}(x)$ *of the matrix* $\mathcal{A}_{jk}(x)$ *are equal.*

PROOF. *Necessity.* Suppose that the maximum norm principle holds for (10.6.1) in Π_T. We fix a point $x \in \mathcal{D}$ and a pair of indices $j, k \in \{1, \dots, n\}$. Let $\lambda_{1,jk}(x), \dots, \lambda_{s,jk}(x)$ be eigenvalues of the matrix $\mathcal{A}^*_{jk}(x)$ and let $\mathcal{L}_1, \dots, \mathcal{L}_s$ be corresponding eigenspaces. We choose a linear independent system of normals $\boldsymbol{\nu}_1, \dots, \boldsymbol{\nu}_m$ to $\mathcal{S}^{m-1}$ which is possible by Lemma 10.2.

Taking into account criterion (iii) of Theorem 10.2, we arrive at the representation

$$\mathcal{A}_{ik}(x) = ([\boldsymbol{\nu}_1, \dots, \boldsymbol{\nu}_m]^*)^{-1} \{a_{1,jk}(x), \dots, a_{m,jk}(x)\}[\boldsymbol{\nu}_1, \dots, \boldsymbol{\nu}_m]^*.$$

Since the eigenvectors $\boldsymbol{\nu}_1, \dots, \boldsymbol{\nu}_m$ of the $(m \times m)$-matrix $\mathcal{A}^*_{jk}(x)$ are linear independent, it follows that

$$\mathbb{R}^m = \bigoplus_{i=1}^{s} \mathcal{L}_i. \tag{10.6.5}$$

Clearly, each vector of the set $\{\boldsymbol{\nu}_1, \dots, \boldsymbol{\nu}_m\}$ belongs to one of subspaces $\mathcal{L}_1, \dots, \mathcal{L}_s$. Therefore the subsets $\mathcal{N}_i = \mathcal{L}_i \cap \{\boldsymbol{\nu}_1, \dots, \boldsymbol{\nu}_m\}$, $i = 1, 2, \dots, s$, form a partition of $\{\boldsymbol{\nu}_1, \dots, \boldsymbol{\nu}_m\}$. Let $L_i = \text{Span}\{\mathcal{N}_i\}, i = 1, 2, \dots, s$. Then

$$\sum_{i=1}^{n} \dim L_i = m. \tag{10.6.6}$$

On the other hand, (10.6.5) implies

$$\sum_{i=1}^{s} \dim \mathcal{L}_i = m. \tag{10.6.7}$$

Since $L_i \subset \mathcal{L}_i$, it follows that $\dim L_i \leq \dim \mathcal{L}_i$, $i = 1, 2, \dots, s$. Suppose that the strict inequality $\dim L_i < \dim \mathcal{L}_i$ holds for some i, $1 \leq i \leq s$. Then, by (10.6.7)

$$\sum_{i=1}^{s} \dim L_i < \sum_{i=1}^{s} \dim \mathcal{L}_i = m,$$

which contradicts (10.6.6). Consequently, $\dim L_i = \dim \mathcal{L}_i$, $i = 1, \dots, s$, and since $L_i \subset \mathcal{L}_i$, we have $L_i = \mathcal{L}_i$.

By criterion (ii) of Theorem 10.2,

$$\bigcup_{i=1}^{s} \mathcal{L}_i \supset \mathcal{N}. \tag{10.6.8}$$

Since $L_i = \mathcal{L}_i$, (10.6.8) implies

$$\bigcup_{i=1}^{s} L_i \supset \mathcal{N},$$

that is 1) in condition (b) holds. Further, since $\mathcal{N}_i \subset \mathcal{L}_i$, then $a_{p,jk}(x) = \lambda_{i,jk}(x)$ if $\nu_p \in \mathcal{N}_i$ $(1 \leq p \leq m, 1 \leq i \leq s)$. Thus if vectors $\boldsymbol{\nu}_p$ and $\boldsymbol{\nu}_q$ $(1 \leq p, q \leq m, p \neq q)$ belong to the same set $\mathcal{N}_i (1 \leq i \leq s)$, then the eigenvalues $a_{p,jk}(x)$, $a_{q,jk}(x)$ of the matrix $\mathcal{A}_{jk}(x)$ are equal. This means that 2) in condition (b) is true.

Sufficiency. Let the conditions of the present theorem be valid. Then the spaces $L_1, \dots, L_s$ form complete set of eigenspaces of the matrix $\mathcal{A}^*_{jk}(x)$. Moreover,

$$\bigcup_{i=1}^{s} L_i \supset \mathcal{N}.$$

By condition (ii) of Theorem 10.2, the maximum norm principle is valid for system (10.6.1). □

REMARK 10.2. We observe that, for $s = 1$ condition (L) holds on the set $\mathcal{N}$ of normals to ∂B, where B is an arbitrary convex compact body in $\mathbb{R}^m$. Moreover, $s = 1$ can be the only value of s for which (L) holds. In what follows we shall show that (L) is valid for a smooth surface ∂B only for $s = 1$. In other cases (L) can hold for several values of s. For instance, if ∂B is a cylindrical surface with a smooth directrix, then (L) holds for $s = 1$ and for $s = 2$. If ∂B is the surface of an m-dimensional parallelepiped, then (L) holds for $s = 1, \dots, m$. Here, it is easily seen, that the choice of subspaces $L_1, \dots, L_s$ for a fixed s can be done in different ways (for instance, this is the case with a parallelepiped).

REMARK 10.3. Let B be a convex compact body in $\mathbb{R}^m$, int $B \overline{\ni} O$. We are looking for the answer to the question whether B is an invariant set of the system (10.1.1).

Let $\boldsymbol{u}_0$ denote an m-component vector directed from O to $u_0 \in$ int B. If a vector-valued function $\boldsymbol{u}$ is a solution of system (10.1.1), then $\boldsymbol{v} = \boldsymbol{u} - \boldsymbol{u}_0$ is also a solution. Let $B' = \{v : v = u - u_0,\ u \in B\}$. Since $u \in B \Leftrightarrow v \in B'$ and $B' \ni O$, it follows that the conditions of invariance of the set B for system (10.1.1) in the case int $B \overline{\ni} O$ are the same as in the case int $B \ni O$ (cf. Theorem 10.2).

THEOREM 10.4. *The maximum norm principle is valid for solutions of system* (10.6.1) *with at least one norm if and only if for all $x \in \mathcal{D}, 1 \le j, k \le n$, there exists a constant nondegenerate $(m \times m)$-matrix $\mathcal{Q}$ such that*

$$\mathcal{A}_{jk}(x) = \mathcal{Q}^{-1}\text{diag}\,\{a_{1,jk}(x), \dots, a_{m,jk(x)}\}\mathcal{Q}. \tag{10.6.9}$$

PROOF. *Necessity.* Let there exist a norm $|\!\cdot\!|$ such that the maximum norm principle is valid for (10.6.1). By Theorem 10.3 representation (10.6.4) holds. Denote by $\nu_{j1}, \dots, \nu_{jm}$ the components of the unit vector $\boldsymbol{\nu}_j$ normal to the sphere $\mathcal{S}^{m-1}$ and consider the matrix $\mathcal{Q}$ with the elements ν_{jk}. Since the vectors $\boldsymbol{\nu}_1, \dots, \boldsymbol{\nu}_m$ in the statement of Theorem 10.3 are linear independent, $\mathcal{Q}$ is nondegenerate. Hence (10.6.4) can be written as (10.6.9).

Sufficiency. Let (10.6.9) hold with some $\mathcal{Q}$ and set $\boldsymbol{\nu}_j = c_j\{\nu_{j1}, \dots, \nu_{jm}\}$ where the constants c_j are chosen to satisfy $|\boldsymbol{\nu}_j| = 1$. Denote $C = \text{diag}\,\{c_1, \dots, c_m\}$ and write (10.6.9) as

$$\mathcal{A}_{jk}(x) = (C\mathcal{Q})^{-1}\text{diag}\,\{a_{1,jk}(x), \dots, a_{m,jk}(x)\}C\mathcal{Q}. \tag{10.6.10}$$

Since $\mathcal{Q}$ is nondegenerate, the vectors $\boldsymbol{\nu}_j$ are linear independent and we write (10.6.10) as (10.6.4). The vectors $\boldsymbol{\nu}_1, -\boldsymbol{\nu}_1, \boldsymbol{\nu}_2, -\boldsymbol{\nu}_2, \dots, \boldsymbol{\nu}_m, -\boldsymbol{\nu}_m$ can be interpreted as normals to the faces of a parallelepiped $\mathcal{P} \subset \mathbb{R}^m$ which contains the origin O. We put $L_i = \text{Span}\{\boldsymbol{\nu}_i\}$, and denote the set of normals to the faces of $\mathcal{P}$ by $\mathcal{N}$. Then all the conditions of Theorem 10.3 are satisfied and $\mathcal{P}$ can serve as the unit ball of a certain norm $|\!\cdot\!|$ for which the solutions of (10.6.1) in Π_T satisfy the maximum norm principle. □

10.7. Certain particular cases and examples

In this section we consider particular cases of Theorem 10.3 on necessary and sufficient conditions for validity of the maximum norm principle for solutions of system (10.6.1). We give a geometric description of certain norms for which $s_{\max} = 1$, $s_{\max} = 2$, $s_{\max} = m$. Special attention is paid to the case $m = 2$. At the end of the section we give a criterion for validity of the maximum p-norm principle, $1 \le p \le \infty$, in $\mathbb{R}^m$ for solutions of system (10.6.1).

10.7.1. The case $m = 2$

The following assertion contains a complete description of the relation between the structure of system (10.6.1) with two unknown functions and the geometry of the unit ball of the corresponding norm which follows from the maximum norm principle.

COROLLARY 10.1. *Let $m = 2$. The maximum norm principle is valid for solutions of system (10.6.1) in Π_T if and only if, for all $x \in \mathcal{D}$*

$$\text{(10.7.1) (i)} \quad \mathcal{A}_{jk}(x) = a_{jk}(x)I, \quad 1 \le j,k \le n,$$

in the case when the unit ball of the norm is not a parallelogram. Here I is a unit (2×2)-matrix.

$$\text{(10.7.2) (ii)} \quad \mathcal{A}_{jk}(x) = ([\boldsymbol{\nu}_1, \boldsymbol{\nu}_2]^*)^{-1} \operatorname{diag}\{a_{1,jk}(x) a_{2,jk}(x)\}[\boldsymbol{\nu}_1, \boldsymbol{\nu}_2]^*,$$

$1 \le j,k \le n$, in the case when the unit ball of the norm is a parallelogram. Here $\boldsymbol{\nu}_1, \boldsymbol{\nu}_2$ are linear independent normals to the sides of the parallelogram and $a_{1,jk}(x)$, $a_{2,jk}(x)$ are real-valued functions.

In particular, if $\boldsymbol{\nu}_1 = \boldsymbol{e}_1$, $\boldsymbol{\nu}_2 = \boldsymbol{e}_2$, then (10.7.2) takes the form

$$\mathcal{A}_{jk}(x) = \operatorname{diag}\{a_{1,jk}(x),\ a_{2,jk}(x)\}.$$

PROOF. For $m = 2$ the condition (L) holds either for $s_{\max} = 1$, or for $s_{\max} = 2$. Obviously, condition (L) with $s_{\max} = 2$ holds if and only if B is a parallelogram. For all other norms (L) holds with $s_{\max} = 1$.

By Theorem 10.3,

$$\mathcal{A}_{jk}(x) = ([\boldsymbol{\nu}_1, \boldsymbol{\nu}_2]^*)^{-1} \operatorname{diag}\{a_{1,jk}(x) a_{2,jk}(x)\}[\boldsymbol{\nu}_1, \boldsymbol{\nu}_2]^*, \tag{10.7.3}$$

for $1 \le j,k \le n$ and all $x \in \mathcal{D}$.

(i) Let the unit ball of the norm be not a parallelogram. Since $s_{\max} = 1$ it follows from Theorem 10.3, that $\mathcal{N}_1 = \{\boldsymbol{\nu}_1, \boldsymbol{\nu}_2\}$, $L_1 = \operatorname{Span}\{\mathcal{N}_1\} = \mathbb{R}^2$, and $\mathcal{N} \subset L_1$. Consequently, $a_{1,jk}(x) = a_{2,jk}(x)$. Denoting the common value of the functions in the last equality by $a_{jk}(x)$, we find by (10.7.3) that

$$\mathcal{A}_{jk}(x) = a_{jk}(x)I, \quad 1 \le j,k \le n, \tag{10.7.4}$$

i.e. we arrive at (10.7.1).

(ii) Let the unit ball of the norm be a parallelogram. Then $s_{\max} = 2$. Let $\boldsymbol{\nu}_1, \boldsymbol{\nu}_2$ be outward unit normals to the sides of a parallelogram with common vertex. We put $L_1 = \operatorname{Span}\{\boldsymbol{\nu}_1\}$, $L_2 = \operatorname{Span}\{\boldsymbol{\nu}_2\}$. Clearly,

$$L_1 \cup L_2 \supset \mathcal{N}$$

and hence the conditions of Theorem 10.3 hold. In the case $s = 1$ one arrives at (10.7.4) in the same way as in (i). Thus, if the unit ball of the norm is a parallelogram, then a necessary and sufficient condition for validity of the maximum norm principle for system (10.6.1) is the representability of coefficients of the system in the form (10.7.2). Here the equality $a_{1,jk}(x) = a_{2,jk}(x)$ is not excluded, which leads to (10.7.1). □

10.7.2. The case $m > 2$, $s_{\max} = 1$

The following assertion is proved in the same way as the first statement of Corollary 10.1. Throughout this section we assume that $m > 2$.

COROLLARY 10.2. *Let condition* (L) *hold for the given norm for* $s_{\max} = 1$. *Then the maximum norm principle is valid for solutions of system* (10.6.1) *in* Π_T *if and only if, for all* $x \in \mathcal{D}$, *the equalities*

$$\mathcal{A}_{jk}(x) = a_{jk}(x)I, \quad 1 \le j, k \le n,$$

hold.

Each of the statements below describes a certain class of convex compact bodies in $\mathbb{R}^m$ for which (L) holds with $s_{\max} = 1$. According to Corollary 10.2, for corresponding norms, the maximum norm principle is valid only for scalar systems. Of all these statements the following one is principal.

LEMMA 10.3. *Suppose at points* $\zeta_1, \dots, \zeta_m, \zeta_{m+1}$ *of the boundary* ∂B *of the compact convex body* $B \subset \mathbb{R}^m$ *there exist a family of normals*

$$\boldsymbol{\nu}_1 = \boldsymbol{\nu}(\zeta_1), \dots, \boldsymbol{\nu}_m = \boldsymbol{\nu}(\zeta_m), \boldsymbol{\nu}_{m+1} = \boldsymbol{\nu}(\zeta_{m+1})$$

such that any proper subspace of this family forms a linear independent system. Then condition (L) *holds for* B *with* $s_{\max} = 1$.

PROOF. We introduce the notation $\mathcal{N}_k = \{\boldsymbol{\nu}_1, \dots, \boldsymbol{\nu}_k\}$, $1 \le k \le m+1$. Suppose that (L) holds with $s_{\max} > 1$. Then there exist subspaces $L_1, \dots, L_s \subset \mathbb{R}^m$, $1 < s \le s_{\max}$ such that

$$\mathcal{N}_{m+1} \subset \bigcup_{i=1}^{s} L_i, \quad \mathbb{R}^m = \bigoplus_{i=1}^{s} L_i.$$

Since the elements of $\mathcal{N}_m$ are linear independent and $L_i \cap L_j = \{0\}$ for $i \ne j$, it follows that $\mathcal{N}_m$ can be decomposed into subsets $\{\boldsymbol{\nu}_{i1}, \dots, \boldsymbol{\nu}_{im_i}\}$, $i = 1, \dots, s$, such that $\{\boldsymbol{\nu}_{i1}, \dots, \boldsymbol{\nu}_{im_i}\} \subset L_i$ and $m_i = \dim L_i$. The vector $\boldsymbol{\nu}_{m+1}$ belongs to the union of subspaces $L_1, \dots, L_s$, therefore there exists a number k, $1 \le k \le s$, such that $\boldsymbol{\nu}_{m+1} \in L_k$. Taking into account that $s > 1$ and hence $m_k = \dim L_k < m$, we conclude that the system $\boldsymbol{\nu}_{k1}, \dots, \boldsymbol{\nu}_{km_k}, \boldsymbol{\nu}_{m+1}$ is linear dependent. This contradicts our assumption and thus (L) holds with $s_{\max} = 1$. □

The next assertion shows that condition (L) holds for differentiable norms with $s_{\max} = 1$.

LEMMA 10.4. *Let at any point of the boundary* ∂B *of a compact convex body* $B \subset \mathbb{R}^m$ *there exists a tangent plane. Then condition* (L) *holds for* B *with* $s_{\max} = 1$.

PROOF. The set of normals to the smooth boundary of a compact convex body $B \subset \mathbb{R}^m$ coincides with the set of unit vectors in $\mathbb{R}^m$. Therefore, one can choose points $\zeta_1, \dots, \zeta_m, \zeta_{m+1}$ on ∂B in such a way that

$$\boldsymbol{\nu}(\zeta_1) = \boldsymbol{e}_1, \dots, \ \boldsymbol{\nu}(\zeta_m) = \boldsymbol{e}_m, \ \ \boldsymbol{\nu}(\zeta_{m+1}) = m^{-1/2}(\boldsymbol{e}_1 + \dots + \boldsymbol{e}_m),$$

where $\boldsymbol{e}_1, \dots, \boldsymbol{e}_m$ is an orthonormal basis of the Cartesian coordinate system in $\mathbb{R}^m$. After applying Lemma 10.3 one obtains $s_{\max} = 1$. □

All statements below in this section concern polytopes. In the first statement we consider non-simple polytopes. By definition (cf., for instance, McMullen and Sheppard [**McMS**]), m-polytope is called simple if its vertices belong to exactly m facets. We observe that for $m = 2$ any polytope is simple.

LEMMA 10.5. *Condition* (L) *holds for a non-simple polytope with* $s_{\max} = 1$.

PROOF. Take a vertex of a non-simple polytope B which is common for k facets with $k \geq m+1$. We choose arbitrary $m+1$ facets for which this vertex is common. Let $\boldsymbol{\nu}_1, \dots, \boldsymbol{\nu}_m, \boldsymbol{\nu}_{m+1}$ denote normals to these facets. While proving Lemma 10.2 we have shown that any m normals from the set $\{\boldsymbol{\nu}_1, \dots, \boldsymbol{\nu}_m, \boldsymbol{\nu}_{m+1}\}$ are linear independent. Consequently, any proper subset of the set $\{\boldsymbol{\nu}_1, \dots, \boldsymbol{\nu}_m, \boldsymbol{\nu}_{m+1}\}$ is also linear independent. Hence by Lemma 10.3 we conclude that the set of normals to ∂B satisfies condition (L) with $s_{\max} = 1$. □

The class of polytopes for which $s_{\max} = 1$ is not exhausted by non-simple polytopes. For example, $s_{\max} = 1$ for m-simplexes.

LEMMA 10.6. *Condition* (L) *holds with* $s_{\max} = 1$ *for* m*-pyramids (in particular, for* m*-simplexes).*

PROOF. **1**. Let an m-pyramid be a non-simple polytope. Then, by Lemma 10.5, $s_{\max} = 1$.

2. Let an m-pyramid be a simple polytope, that is an m-simplex. Let $\boldsymbol{\nu}_1, \dots, \boldsymbol{\nu}_m, \boldsymbol{\nu}_{m+1}$ denote normals to the sides of the m-simplex. The set of normals to the sides which meet at a vertex of the simplex is one of $m+1$ combinations of m elements chosen from the set $\{\boldsymbol{\nu}_1, \dots, \boldsymbol{\nu}_m, \boldsymbol{\nu}_{m+1}\}$. While proving Lemma 10.2 it was shown that each of such combinations is linear independent. Applying Lemma 10.3, we find that the set of normals to the surface of the simplex satisfies condition (L) with $s_{\max} = 1$. □

We conclude this section with a lemma concerning the m-bipyramids defined on page 255.

LEMMA 10.7. *Condition* (L) *holds with* $s_{\max} = 1$ *for* m*-bipyramids.*

PROOF. According to the definition of the m-bipyramid, it has a vertex which is common for at least $2m - 2$ facets. Since $2m - 2 > m$ for $m > 2$, then the m-bipyramid is a non-simple polytope. Hence, by Lemma 10.5, $s_{\max} = 1$. □

10.7.3. The case $m > 2$, $s_{\max} = 2$

In this section we give a criterion for validity of the maximum norm principle for solutions of system (10.6.1) under the assumption that the unit ball of the norm is a cylindric body for which condition (L) holds with $s_{\max} = 2$.

COROLLARY 10.3. *Let the unit ball of the norm be a cylindric body. Suppose there exists a family of normals* $\boldsymbol{\nu}_0, \boldsymbol{\nu}_1, \dots, \boldsymbol{\nu}_{m-1}$ *to its lateral surface such that each proper subset of the family forms a linear independent system. Further, let* $\boldsymbol{\nu}_m$ *denote the normal to one of cylinder's basements.*

The maximum norm principle for solutions of system (10.6.1) *is valid in* Π_T *if and only if, for all* $x \in \mathcal{D}$,

$$\mathcal{A}_{jk}(x) = ([\boldsymbol{\nu}_1, \dots, \boldsymbol{\nu}_m]^*)^{-1} \Lambda_{jk}(x) [\boldsymbol{\nu}_1, \dots, \boldsymbol{\nu}_m]^*, \quad 1 \leq j, k \leq n, \tag{10.7.5}$$

where the $(m \times m)$*-matrix-valued function* $\Lambda_{jk}(x)$ *has the form*

$$\Lambda_{jk}(x) = \operatorname{diag}\{a_{1,jk}(x), \ldots, a_{1,jk}(x),\ \ a_{2,jk}(x)\}. \tag{10.7.6}$$

In particular, if the body is a right cylinder and $\boldsymbol{\nu}_m = \boldsymbol{e}_m$*, then*

$$\mathcal{A}_{jk}(x) = \operatorname{diag}\{a_{1,jk}(x), \ldots, a_{1,jk}(x),\ \ a_{2,jk}(x)\}. \tag{10.7.7}$$

PROOF. Let Q be the projection of the cylindric body U onto the hyperplane orthogonal to the generatrix. The assumptions of this corollary ensure the existence of a family of normals $\boldsymbol{\nu}_0, \boldsymbol{\nu}_1, \ldots, \boldsymbol{\nu}_{m-1}$ to the boundary ∂Q of the $(m-1)$-dimensional compact convex body Q such that any proper subset of the family $\{\boldsymbol{\nu}_0, \boldsymbol{\nu}_1, \ldots, \boldsymbol{\nu}_{m-1}\}$ is linear independent. We put

$$L_1 = \operatorname{Span}\{\boldsymbol{\nu}_1, \ldots, \boldsymbol{\nu}_{m-1}\}, \quad L_2 = \operatorname{Span}\{\boldsymbol{\nu}_m\}.$$

Since the set $\mathcal{N}_Q$ of normals to ∂Q (i.e. to the lateral surface of the cylinder U) is a subset of L_1, then

$$\mathcal{N} \subset L_1 \cup L_2, \quad \mathbb{R}^m = L_1 \oplus L_2.$$

We claim that $s_{\max} = 2$. In fact, by Lemma 10.3, the set $\mathcal{N}_Q$ satisfies condition (L)

$$\mathcal{N}_Q \subset L_1, \quad L_1 = \mathbb{R}^{m-1}$$

with $s_{\max} = 1$. Consequently, condition (L) for the set $\mathcal{N}$ holds with $s_{\max} = 2$.

Applying Theorem 10.3, we arrive at representation (10.7.5). In particular, if the cylinder U is right and $\boldsymbol{\nu}_m = \boldsymbol{e}_m$ then the matrix $[\boldsymbol{\nu}_1, \ldots, \boldsymbol{\nu}_m]$ is block diagonal which together with (10.7.5), (10.7.6) implies (10.7.7). □

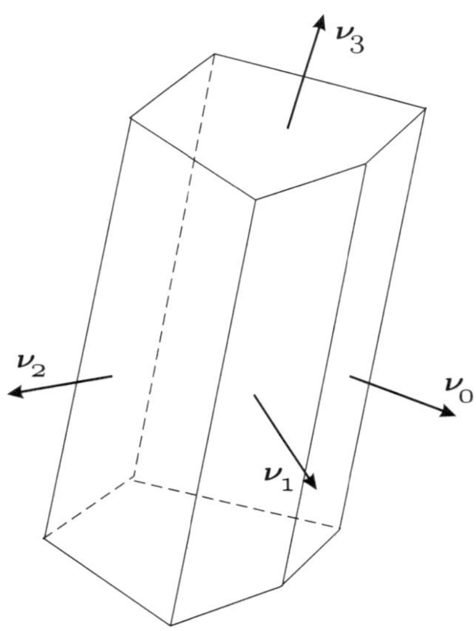

FIGURE 10.2. The unit ball of a norm in Corollary 10.3 ($s_{\max} = 2$)

Using Lemmas 10.3-10.7, one can give geometric descriptions of some norms for which the conditions of the last corollary are satisfied. For example, these conditions hold if ∂Q is a smooth $(m-2)$-dimensional manifold in $\mathbb{R}^m$ (Lemma 10.4) or if Q is an $(m-1)$-simplex (Lemma 10.6).

10.7.4. The case $m > 2$, $s_{\max} = m$

Obviously, condition (L) holds for a compact convex body $B \subset \mathbb{R}^m$ with $s_{\max} = m$ if and only if B is a parallelepiped. The next assertion follows immediately from Theorem 10.3.

COROLLARY 10.4. *Let the unit ball of the norm be a parallelepiped and let* $\boldsymbol{\nu}_1, \dots, \boldsymbol{\nu}_m$ *denote the normals to its facets which have a common vertex.*

Then the maximum norm principle is valid for solutions of system (10.6.1) *in* Π_T *if and only if, for all* $x \in \mathcal{D}$,

$$\mathcal{A}_{jk}(x) = ([\boldsymbol{\nu}_1, \dots, \boldsymbol{\nu}_m]^*)^{-1} \Lambda_{jk}(x) [\boldsymbol{\nu}_1, \dots, \boldsymbol{\nu}_m]^{-1}, \tag{10.7.8}$$

$1 \leq j, k \leq n$, *where*

$$\Lambda_{jk}(x) = \text{diag}\,\{a_{1,jk}(x), \dots, a_{m,jk}(x)\}.$$

In particular, if $\boldsymbol{\nu}_1 = \boldsymbol{e}_1, \dots, \boldsymbol{\nu}_m = \boldsymbol{e}_m$, *then* (10.7.8) *takes the form*

$$\mathcal{A}_{jk}(x) = \text{diag}\,\{a_{1,jk}(x), \dots, a_{m,jk}(x)\}.$$

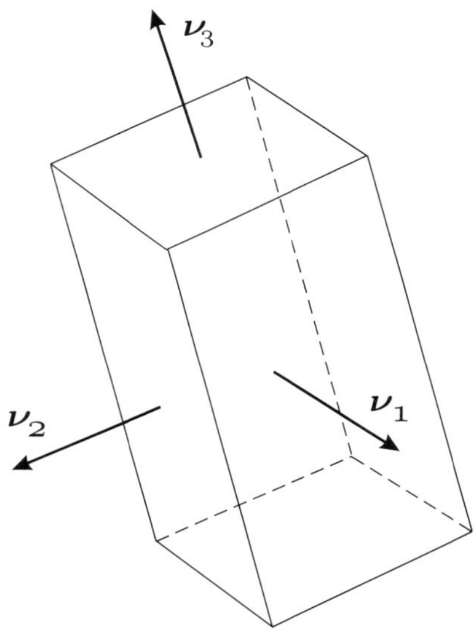

FIGURE 10.3. The unit ball of a norm subject to the condition $s_{\max} = m$

10.7.5. Example: criterion for validity of the maximum p-norm principle, $1 \leq p \leq \infty$

In this section we give applications of above results to p-norms in $\mathbb{R}^m$, that is to the norms

$$\|\boldsymbol{\zeta}\|_p = \left(\sum_{j=1}^{m} |\zeta_j|^p \right)^{1/p}, \quad 1 \leq p < \infty,$$

and

$$\|\boldsymbol{\zeta}\|_\infty = \max\{|\zeta_j| : 1 \leq j \leq m\},$$

where $\boldsymbol{\zeta} = (\zeta_1, \dots, \zeta_m) \in \mathbb{R}^m$.

COROLLARY 10.5. *The maximum p-norm principle is valid for solutions of system* (10.6.1) *in* Π_T *if and only if, for all* $x \in \mathcal{D}$,

$$\text{(10.7.9)} \quad \text{(i)}\ \mathcal{A}_{jk}(x) = \frac{1}{2}\begin{pmatrix} \alpha_{jk}(x) & \beta_{jk}(x) \\ \beta_{jk}(x) & \alpha_{jk}(x) \end{pmatrix}, \quad 1 \le j,k \le n,$$

for $p = 1,\ m = 2$, *where*

$$\text{(10.7.10)} \qquad \alpha_{jk}(x) = a_{1,jk}(x) + a_{2,jk}(x), \quad \beta_{jk}(x) = a_{1,jk}(x) - a_{2,jk}(x);$$

$$\text{(ii)}\ \mathcal{A}_{jk}(x) = a_{jk}(x)I,$$

for $p = 1,\ m > 2$ *and for* $1 < p < \infty$,

$$\text{(iii)}\ \mathcal{A}_{jk}(x) = \operatorname{diag}\{a_{1,jk}(x), \dots, a_{m,jk}(x)\}$$

for $p = \infty$.

PROOF. (i) Let $m = 2,\ p = 1$. Then the unit ball of the p-norm is the square $\operatorname{Conv}\{\pm \boldsymbol{e}_1, \pm \boldsymbol{e}_2\}$. As the pair of linear independent normals to its sides one can take

$$\boldsymbol{\nu}_1 = \frac{1}{\sqrt{2}}(1,1), \quad \boldsymbol{\nu}_2 = \frac{1}{\sqrt{2}}(1,-1).$$

Applying Corollary 10.1, we find

$$\mathcal{A}_{jk}(x) = \frac{1}{2}\begin{pmatrix} a_{1,jk}(x) + a_{2,jk}(x) & a_{1,jk}(x) - a_{2,jk}(x) \\ a_{1,jk}(x) - a_{2,jk}(x) & a_{1,jk}(x) + a_{2,jk}(x) \end{pmatrix}$$

Using the notation (10.7.10) we arrive at (10.7.9).

(ii) Let $p = 1,\ m > 2$. Then the unit ball $\boldsymbol{B}$ of the p-norm can be represented as $\operatorname{Conv}\{\pm \boldsymbol{e}_1, \dots, \pm \boldsymbol{e}_m\}$, that is $\boldsymbol{B}$ is the m-crosspolytope, a representative of the class of m-bipyramids. In this case, by Lemma 10.7, $s_{\max} = 1$. If $1 < p < \infty$, then the p-norm is differentiable and hence, by Lemma 10.4, again $s_{\max} = 1$. Applying Corollary 10.2, we complete the proof.

(iii) If $p = \infty$, then the unit ball of the p-norm is the cube $\boldsymbol{B} = \{\zeta = (\zeta_1, \dots, \zeta_m) \in \mathbb{R}^m : -1 \le \zeta_i \le 1,\quad i = 1, \dots, m\}$. In this case one can put $\boldsymbol{\nu}_1 = \boldsymbol{e}_1, \dots, \boldsymbol{\nu}_m = \boldsymbol{e}_m$. Now, application of Corollary 10.4 completes the proof. □

10.8. Comments to Chapter 10

The case $m = 1$ in (10.1.1) is classical. It is well-known that solutions of the second order uniformly parabolic equation

$$\text{(10.8.1)} \qquad \frac{\partial u}{\partial t} - \sum_{j,k=1}^{n} a_{jk}(x,t)\frac{\partial^2 u}{\partial x_j \partial x_k} = 0$$

with real coefficients in the cylinder

$$Q_T = \{(x,t) : x \in \Omega,\ 0 < t \le T\}, \quad \Omega \subset \mathbb{R}^n,$$

satisfy the maximum principle (see Friedman [**Fri**], Protter and Weinberger [**PW**]). In other words, for any solution of (10.8.1) which is continuous in $\overline{Q}_T$ together with the first derivative with respect to t and derivatives with respect to x up to order two in Q_T, the inequality

$$\min\{u(y,\tau) : (y,\tau) \in \overline{\Gamma}_T\} \le u(x,t) \le \max\{u(y,\tau) : (y,\tau) \in \overline{\Gamma}_T\}$$

is valid. Here $(x,t) \in \overline{Q}_T$ and $\Gamma_T = \{(y,\tau) \in \partial Q_T : 0 \le \tau < T\}$. This means that if the initial-boundary conditions belong to the interval $[\alpha, \beta]$ then the same is valid for the solution of equation (10.8.1), i.e. any closed interval is an invariant set of the equation.

We note that maximum principles for weakly coupled parabolic systems are discussed in the books Protter and Weinberger [**PW**], Walter [**Wal**] which also contain rich bibliographies on this subject. The notion of invariant sets for parabolic and elliptic systems and the first results concerning these sets appeared in the paper by Weinberger [**We1**]. Nowadays, there exists a large literature on invariant sets for nonlinear parabolic and elliptic systems with principal part subjected to various structural conditions such as scalarity, diagonality and others (see, for example, Alikakos [**Ali1, Ali2**], Amann [**Am**], Bates [**Bat**], Bebernes and Schmitt [**BeSch**], Bebernes, Chueh and Fulks [**BCM**], Chueh, Conley and Smoller [**CCS**], Conway, Hoff and Smoller [**CHS**], Cosner and Schaefer [**CS**], Kuiper [**Ku**], Lemmert [**Lem**], Redheffer and Walter [**RW1, RW2**], Schaefer [**Shae**], Smoller [**Sm**], Weinberger [**We2**] and references there).

In particular, we note that a condition similar to (i) of Theorem 10.2 was obtained earlier in Chueh, Conley and Smoller [**CCS**] as a criterion of positively invariant regions for systems of nonlinear diffusion equations

$$\frac{\partial \boldsymbol{u}}{\partial t} = A(x, \boldsymbol{u}) \sum_{j,k=1}^{n} a_{jk}(x) \frac{\partial^2 \boldsymbol{u}}{\partial x_j \partial x_k} + \boldsymbol{F}(\boldsymbol{u}).$$

The geometric aspect of the criteria obtained in this chapter consists of establishing the correspondence between a compact convex body $B \subset \mathbb{R}^m$ and a subclass of systems (10.1.1) for which B is invariant.

The present chapter is related to Chapter 8, where criteria for validity of the maximum modulus principle for system (10.1.5) were obtained (cf. Corollary 10.5 for $p = 2$).

CHAPTER 11

Maximum Norm Principle with Respect to Smooth Norms for Parabolic Systems

11.1. Introduction

In the present chapter we study parabolic systems

$$\frac{\partial \boldsymbol{u}}{\partial t} - \sum_{j,k=1}^{n} \mathcal{A}_{jk}(x,t)\frac{\partial^2 \boldsymbol{u}}{\partial x_j \partial x_k} + \sum_{j=1}^{n} \mathcal{A}_j(x,t)\frac{\partial \boldsymbol{u}}{\partial x_j} + \mathcal{A}_0(x,t)\boldsymbol{u} = \boldsymbol{0}, \tag{11.1.1}$$

where $\boldsymbol{u}$ is an m-component vector-valued function, $\mathcal{A}_{jk} = \mathcal{A}_{kj}$ and $\mathcal{A}_{jk}, \mathcal{A}_j, \mathcal{A}_0$ are real $(m \times m)$-matrix-valued functions.

We obtain criteria for validity of the maximum norm principle (see Section 10.1) for solutions of system (11.1.1) which is uniformly parabolic in the sense of Petrovskiĭ. As in Chapter 10, the norm $\lvert \cdot \rvert$ is understood in a generalized sense, namely, as the Minkowski functional of a compact convex body in $\mathbb{R}^m$ containing the origin in its interior (see Section 10.1). Unlike Chapter 10, we assume that the norm $\lvert \cdot \rvert$ is twice continuously differentiable, that is the function $\lvert \cdot \rvert : \mathbb{R}^m \to \mathbb{R}$ belongs to $\mathrm{C}^2(\mathbb{R}^m \backslash \{\boldsymbol{0}\})$. The obtained criteria establish a correspondence between a compact convex body $B \subset \mathbb{R}^m$ containing the origin in its interior and a subclass of systems (11.1.1) for which B is invariant.

We shall consider system (11.1.1) in the layer $\mathbb{R}^{n+1}_T = \mathbb{R}^n \times (0,T]$ and in the cylinder $Q_T = \Omega \times (0,T]$, where Ω is a bounded subdomain of $\mathbb{R}^n$.

We recall some notations introduced in Subsection 8.2.1.

By $[\mathrm{C}^{(2,1)}(\Pi_T)]^m$ we mean the space of m-component vector-valued functions on Π_T whose derivatives with respect to x up to the second order and the first derivative with respect to t are continuous. Here $\Pi_T = \mathcal{D} \times (0,T]$, where $\mathcal{D}$ is either a bounded domain Ω in $\mathbb{R}^n$ or $\mathcal{D} = \mathbb{R}^n$ and $0 < T < \infty$.

By $[\mathrm{C}_{\mathrm{b}}^{k+\alpha,\alpha/2}(\overline{\mathbb{R}^{n+1}_T})]^m$ we denote the space of m-component vector-valued functions defined on $\overline{\mathbb{R}^{n+1}_T}$ with bounded derivatives in x up to order k which satisfy the uniform Hölder condition on $\overline{\mathbb{R}^{n+1}_T}$ with exponent $\alpha, 0 < \alpha \le 1$, with respect to the parabolic distance $(|x-x'|^2 + |t-t'|)^{1/2}$ between points (x,t) and (x',t') in $\overline{\mathbb{R}^{n+1}_T}$. A similar space of $(m \times m)$-matrix-valued functions will be denoted by $[\mathrm{C}_{\mathrm{b}}^{k+\alpha,\alpha/2}(\overline{\mathbb{R}^{n+1}_T})]^{m \times m}$.

As in the preceding chapter, by $[\mathrm{C}_{\mathrm{b}}(F)]^m$ we denote the space of continuous and bounded m-component vector-valued functions on a set $F \subset \mathbb{R}^n$ endowed with the generalized norm

$$\lVert \boldsymbol{u} \rVert_{[\mathrm{C}_{\mathrm{b}}(F)]^m} = \sup \{ \lvert \boldsymbol{u}(q) \rvert : \ q \in F \}.$$

Throughout the chapter we make the following assumptions concerning system (11.1.1):

(A) $\mathcal{A}_{jk} \in \left[\mathrm{C}_{\mathrm{b}}^{2+\alpha,\alpha/2}(\overline{\mathbb{R}_T^{n+1}})\right]^{m\times m}$, $\mathcal{A}_j \in \left[\mathrm{C}_{\mathrm{b}}^{1+\alpha,\alpha/2}(\overline{\mathbb{R}_T^{n+1}})\right]^{m\times m}$, and $\mathcal{A}_0 \in \left[\mathrm{C}_{\mathrm{b}}^{\alpha,\alpha/2}(\overline{\mathbb{R}_T^{n+1}})\right]^{m\times m}$, where $1 \le j,k \le n$;

(B) For any point $(x,t) \in \overline{\mathbb{R}_T^{n+1}}$, the real parts of the λ-roots of the equation

$$\det\left(\sum_{j,k=1}^{n} \mathcal{A}_{jk}(x,t)\sigma_j\sigma_k + \lambda I\right) = 0$$

satisfy the inequality $\Re\lambda(x,t,\boldsymbol{\sigma}) \le -\delta|\boldsymbol{\sigma}|^2$, where δ=const > 0, for any $\boldsymbol{\sigma} = (\sigma_1,\dots,\sigma_n) \in \mathbb{R}^n$, I is the identity matrix of order m, and $|\cdot|$ is the Euclidean length of a vector.

In Section 11.2 we obtain an expression for the best constant $\mathcal{K}(\mathbb{R}^n,T)$ in the inequality

$$|\boldsymbol{u}(x,t)| \le \mathcal{K}\sup\{|\boldsymbol{u}(y,0)| : y \in \mathbb{R}^n\},$$

where $(x,t) \in \mathbb{R}_T^{n+1}$ and $\boldsymbol{u}$ is a solution of system (11.1.1) in $[\mathrm{C}^{(2,1)}(\mathbb{R}_T^{n+1})]^m \cap [\mathrm{C}_{\mathrm{b}}(\overline{\mathbb{R}_T^{n+1}})]^m$. Besides the constant $\mathcal{K}(\mathbb{R}^n,T)$, we study the best constant $\mathcal{K}(\Omega,T)$ in the inequality

$$|\boldsymbol{u}(x,t)| \le \mathcal{K}\sup\{|\boldsymbol{u}(y,\tau)| : (y,\tau) \in \overline{\Gamma}_T\},$$

where $(x,t) \in Q_T$ and $\boldsymbol{u}$ is a solution of system (11.1.1) in $[\mathrm{C}^{(2,1)}(Q_T)]^m \cap [\mathrm{C}(\overline{Q_T})]^m$.

Assuming that the norm $|\cdot|$ is twice continuously differentiable, we give separate necessary (Section 11.3) and sufficient (Section 11.4) conditions for validity of the maximum norm principle (i.e. $\mathcal{K}(\Omega,T)=1$, $\mathcal{K}(\mathbb{R}^n,T)=1$) for solutions of system (11.1.1).

If the coefficients of system (11.1.1) do not depend on t, i.e. the system under consideration is

$$\frac{\partial \boldsymbol{u}}{\partial t} - \sum_{j,k=1}^{n} \mathcal{A}_{jk}(x)\frac{\partial^2 \boldsymbol{u}}{\partial x_j \partial x_k} + \sum_{j=1}^{n} \mathcal{A}_j(x)\frac{\partial \boldsymbol{u}}{\partial x_j} + \mathcal{A}_0(x)\boldsymbol{u} = \boldsymbol{0}, \tag{11.1.2}$$

then the above mentioned necessary and sufficient conditions coincide and thus we arrive (Section 11.5) at criteria of the validity of the maximum norm principle for system (11.1.2). These criteria are stated in the next assertion, where by $\mathcal{H}(|\boldsymbol{\zeta}|)$ we denote the Hessian $((\partial^2|\boldsymbol{\zeta}|/\partial\zeta_j\partial\zeta_k))$ of the norm $|\cdot|$.

THEOREM 1. *The maximum norm principle is valid for solutions of system (11.1.2) in the class* $[\mathrm{C}^{(2,1)}(\Pi_T)]^m \cap [\mathrm{C}_{\mathrm{b}}(\overline{\Pi}_T)]^m$ *if and only if:*

(i) *for all* $x \in \mathcal{D}$ *the equalities*

$$\mathcal{A}_{jk}(x) = a_{jk}(x)I, \quad 1 \le j,k \le n,$$

hold, where $((a_{jk}(x)))$ *is a positive-definite* $(n\times n)$*-matrix-valued function in* $\overline{\mathcal{D}}$*;*

(ii) *for all* $x \in \mathcal{D}$ *and for all vectors* $\boldsymbol{\xi}_j \in \mathbb{R}^m, \boldsymbol{\zeta} \in \mathbb{R}^m\backslash\{\boldsymbol{0}\}$, *with* $(\boldsymbol{\xi}_j, \nabla|\boldsymbol{\zeta}|) = 0$, $j = 1,\dots,n$, *the inequality*

$$\sum_{j,k=1}^{n} a_{jk}(x)(\mathcal{H}(|\boldsymbol{\zeta}|)\boldsymbol{\xi}_j,\boldsymbol{\xi}_k) + \sum_{j=1}^{n}(\mathcal{A}_j(x)\boldsymbol{\xi}_j,\nabla|\boldsymbol{\zeta}|) + (\mathcal{A}_0(x)\boldsymbol{\zeta},\nabla|\boldsymbol{\zeta}|) \ge 0$$

is valid.

The following assertion is an application of Theorem 1 to the particular case $\mathcal{A}_0 = 0$.

COROLLARY 1. *The maximum norm principle is valid for solutions of the system*

$$\frac{\partial \boldsymbol{u}}{\partial t} - \sum_{j,k=1}^{n} \mathcal{A}_{jk}(x)\frac{\partial^2 \boldsymbol{u}}{\partial x_j \partial x_k} + \sum_{j=1}^{n} \mathcal{A}_j(x)\frac{\partial \boldsymbol{u}}{\partial x_j} = \mathbf{0}$$

in the class $[\mathrm{C}^{(2,1)}(\Pi_T)]^m \cap [\mathrm{C_b}(\overline{\Pi}_T)]^m$ *if and only if for all* $x \in \mathcal{D}$

$$\mathcal{A}_{jk}(x) = a_{jk}(x)I, \ \ \mathcal{A}_j(x) = a_j(x)I, \ \ 1 \le j,k \le n,$$

where $((a_{jk}(x)))$ *is a positive-definite* $(n \times n)$*-matrix-valued function and* $a_1(x), \dots, a_n(x)$ *are scalar functions in* $\overline{\mathcal{D}}$.

To formulate corollaries of Theorem 1 we introduce the following notation.

Let $\mathcal{T}_\zeta = \{\boldsymbol{\xi} \in \mathbb{R}^m : (\nabla|\boldsymbol{\zeta}|, \ \boldsymbol{\xi}) = 0\}$ be the space of vectors lying in the tangent hyperplane to the sphere of the norm at the point $\zeta \in \mathbb{R}^m\backslash\{0\}$. Let further $\mathcal{Z} = \{\boldsymbol{\zeta} \in \mathbb{R}^m \ : \ \mathcal{H}(|\boldsymbol{\zeta}|)\mathcal{T}_\zeta = 0 \ \}$. By $\boldsymbol{\nu}(\zeta)$ we denote the unit outward normal to the sphere $\mathcal{S}^{m-1} = \{\zeta \in \mathbb{R}^m : |\boldsymbol{\zeta}| = 1\}$ at the point $\zeta \in \mathcal{S}^{m-1}$. Let $\mathcal{N} = \{\boldsymbol{\nu}(\zeta) : \zeta \in \mathcal{S}^{m-1} \bigcap \mathcal{Z}\}$. By $[\boldsymbol{v}_1, \dots, \boldsymbol{v}_m]$ we mean the $(m \times m)$-matrix whose columns are m-component vectors $\boldsymbol{v}_1, \dots, \boldsymbol{v}_m$.

The following result is a direct consequence of Theorem 1.

COROLLARY 2. *Let* $\mathcal{N} \supset \{\boldsymbol{\nu}_1, \dots, \boldsymbol{\nu}_m\}$ *and suppose that the set* $\{\boldsymbol{\nu}_1, \dots, \boldsymbol{\nu}_m\}$ *forms a linear independent system. Then the maximum norm principle is valid for solutions of system* (11.1.2) *in the class* $[\mathrm{C}^{(2,1)}(\Pi_T)]^m \cap [\mathrm{C_b}(\overline{\Pi}_T)]^m$ *if and only if:*

(i) *for all* $x \in \mathcal{D}$ *the equalities*

$$\mathcal{A}_{jk}(x) = a_{jk}(x)I,$$

$$\mathcal{A}_j(x) = \left([\boldsymbol{\nu}_1, \dots, \boldsymbol{\nu}_m]^*\right)^{-1} \mathrm{diag}\,\{a_{1,j}(x), \dots, a_{m,j}(x)\}[\boldsymbol{\nu}_1, \dots, \boldsymbol{\nu}_m]^*,$$

$1 \le j,k \le n$, *hold, where* $((a_{jk}(x)))$ *is a positive-definite* $(n \times n)$*-matrix-valued function and* $a_{1,j}(x), \dots, a_{m,j}(x)$ *are scalar functions in* $\overline{\mathcal{D}}$;

(ii) *for all* $x \in \mathcal{D}$ *and for all vectors* $\boldsymbol{\xi}_j \in \mathbb{R}^m, \boldsymbol{\zeta} \in \mathbb{R}^m\backslash\{\mathbf{0}\}$, *with* $(\boldsymbol{\xi}_j, \nabla|\boldsymbol{\zeta}|) = 0$, $j = 1, \dots, n$, *the inequality*

$$\sum_{j,k=1}^{n} a_{jk}(x)(\mathcal{H}(|\boldsymbol{\zeta}|)\boldsymbol{\xi}_j, \boldsymbol{\xi}_k) + \sum_{j=1}^{n}(\mathcal{A}_j(x)\boldsymbol{\xi}_j, \nabla|\boldsymbol{\zeta}|) + (\mathcal{A}_0(x)\boldsymbol{\zeta}, \nabla|\boldsymbol{\zeta}|) \ge 0$$

holds.

The next assertion results from Corollary 2.

COROLLARY 3. *Let* $\mathcal{N} \supset \{\boldsymbol{\nu}_1, \dots, \boldsymbol{\nu}_{m+1}\}$ *and suppose that any proper subset of* $\{\boldsymbol{\nu}_1, \dots, \boldsymbol{\nu}_{m+1}\}$ *forms a linear independent system. Then the maximum norm principle is valid for solutions of system* (11.1.2) *in the class* $[\mathrm{C}^{(2,1)}(\Pi_T)]^m \cap [\mathrm{C_b}(\overline{\Pi}_T)]^m$ *if and only if:*

(i) *for all* $x \in \mathcal{D}$ *the equalities*

$$\mathcal{A}_{jk}(x) = a_{jk}(x)I, \ \ \mathcal{A}_j(x) = a_j(x)I, \ \ 1 \le j,k \le n,$$

hold, where $((a_{jk}(x)))$ *is a positive-definite* $(n \times n)$*-matrix-valued function and* $a_1(x), \dots, a_m(x)$ *are scalar functions in* $\overline{\mathcal{D}}$;

(ii) *for all* $x \in \mathcal{D}$ *and for all vectors* $\boldsymbol{\zeta} \in \mathbb{R}^m\backslash\{\mathbf{0}\}$ *the inequality*

$$(\mathcal{A}_0(x)\boldsymbol{\zeta}, \nabla|\boldsymbol{\zeta}|) \ge 0$$

holds.

In Section 11.6 we give an application of Corollary 2 to the p-norm in $\mathbb{R}^m$, that is to the norm

$$\|\boldsymbol{\zeta}\|_p = \Big(\sum_{j=1}^{m} |\zeta_j|^p \Big)^{1/p},$$

where $\boldsymbol{\zeta} = (\zeta_1, \dots, \zeta_m) \in \mathbb{R}^m$. Namely, we prove the following assertion.

COROLLARY 4. *The maximum p-norm principle with $p \in (2, \infty)$ is valid for solutions of system* (11.1.2) *in the class* $[\mathrm{C}^{(2,1)}(\Pi_T)]^m \cap [\mathrm{C_b}(\overline{\Pi}_T)]^m$ *if and only if:*

(i) *for all $x \in \mathcal{D}$*

$$\mathcal{A}_{jk}(x) = a_{jk}(x)I, \quad \mathcal{A}_j(x) = \mathrm{diag}\,\{a_{1,j}(x), \dots, a_{m,j}(x)\}, \quad 1 \le j, k \le n,$$

where $((a_{jk}(x)))$ is a positive-definite $(n \times n)$-matrix-valued function and $a_{1,j}(x)$, $\dots, a_{m,j}(x)$ are scalar functions in $\overline{\mathcal{D}}$;

(ii) *for all $x \in \mathcal{D}$ and for all vectors $\boldsymbol{\xi}_j$, $\boldsymbol{\zeta} \in \mathbb{R}^m$, with $(\boldsymbol{\xi}_j, \Lambda_p \boldsymbol{\zeta}) = 0$, $j = 1, \dots, n$,*

$$(p-1) \sum_{j,k=1}^{n} a_{jk}(x) \left(\Lambda_p \boldsymbol{\xi}_j, \boldsymbol{\xi}_k\right) + \sum_{j=1}^{n} \left(\mathcal{A}_j(x) \boldsymbol{\xi}_j, \Lambda_p \boldsymbol{\zeta}\right) + (\mathcal{A}_0(x)\boldsymbol{\zeta}, \Lambda_p \boldsymbol{\zeta}) \ge 0,$$

where $\Lambda_p = \mathrm{diag}\,\{\, |\zeta_1|^{p-2}, \dots, |\zeta_m|^{p-2} \,\}$.

11.2. Representation for the constant $\mathcal{K}(\mathbb{R}^n, T)$

We introduce the operators

$$\mathfrak{A}(x, t, D_x) = \sum_{j,k=1}^{n} \mathcal{A}_{jk}(x,t) \frac{\partial^2}{\partial x_j \partial x_k} - \sum_{j=1}^{n} \mathcal{A}_j(x,t) \frac{\partial}{\partial x_j} - \mathcal{A}_0(x,t),$$

$$\mathfrak{A}_0(x, t, D_x) = \sum_{j,k=1}^{n} \mathcal{A}_{jk}(x,t) \frac{\partial^2}{\partial x_j \partial x_k},$$

where $\mathcal{A}_{jk} = \mathcal{A}_{kj}$ and $\mathcal{A}_{jk}, \mathcal{A}_j, \mathcal{A}_0$ are real $(m \times m)$-matrix-valued functions subject to conditions (A) and (B), formulated in Section 11.1.

We shall use the constants $\mathcal{K}(\Omega, T)$ and $\mathcal{K}(R^n, T)$ from Section 11.1 as well as the function

$$\mathbb{R}^n \ni y \to \mathcal{K}_0(y) = \sup \frac{\|\boldsymbol{u}\|_{[\mathrm{C_b}(\overline{\mathbb{R}^{n+1}_T})]^m}}{\|\boldsymbol{u}\big|_{t=0}\|_{[\mathrm{C_b}(\mathbb{R}^n)]^m}},$$

where the supremum is taken over all vector-valued functions $\boldsymbol{u}$ in the class

$$[\mathrm{C}^{(2,1)}(\mathbb{R}^{n+1}_T)]^m \cap [\mathrm{C_b}(\overline{\mathbb{R}^{n+1}_T})]^m$$

satisfying the system

$$\frac{\partial \boldsymbol{u}}{\partial t} - \mathfrak{A}_0(y, 0, D_x)\boldsymbol{u} = \mathbf{0}.$$

According to Eidel'man [**Ei1**], there exists a unique function

$$\boldsymbol{u} \in [\mathrm{C}^{(2,1)}(\mathbb{R}^{n+1}_T)]^m \cap [\mathrm{C_b}(\overline{\mathbb{R}^{n+1}_T})]^m$$

solving the Cauchy problem

$$\frac{\partial \boldsymbol{u}}{\partial t} - \mathfrak{A}(x, t, D_x)\boldsymbol{u} = \mathbf{0} \quad \text{in } \mathbb{R}^{n+1}_T, \ \boldsymbol{u}\big|_{t=0} = \boldsymbol{\psi}, \tag{11.2.1}$$

where $\boldsymbol{\psi} \in [\mathrm{C_b}(\mathbb{R}^n)]^m$. This solution can be represented in the form

$$\boldsymbol{u}(x,t) = \int_{\mathbb{R}^n} G(t,0,x,\eta)\boldsymbol{\psi}(\eta)d\eta. \tag{11.2.2}$$

Here $G(t,\tau,x,\eta)$ is the fundamental matrix of solutions for the system

$$\frac{\partial \boldsymbol{u}}{\partial t} - \mathfrak{A}(x,t,D_x)\boldsymbol{u} = \mathbf{0}.$$

The fundamental matrix of solutions for the system

$$\frac{\partial \boldsymbol{u}}{\partial t} - \mathfrak{A}_0(y,0,D_x)\boldsymbol{u} = \mathbf{0}$$

will be denoted by $\mathcal{G}_0(t-\tau, x-\eta; y)$. It is represented as

$$\mathcal{G}_0(t-\tau,x-\eta;y) = (2\pi)^{-n}\int_{\mathbb{R}^n} \exp\Big[-\sum_{j,k=1}^{n} \mathcal{A}_{jk}(y,0)\sigma_j\sigma_k(t-\tau)\Big]e^{i(x-\eta,\sigma)}d\sigma,$$

where $\boldsymbol{\sigma} = (\sigma_1,\dots,\sigma_n) \in \mathbb{R}^n$. This implies

$$\mathcal{G}_0(t-\tau,x-\eta;y) = (t-\tau)^{-n/2}P\left(\frac{x-\eta}{\sqrt{t-\tau}};y\right) \tag{11.2.3}$$

with

$$P(x;y) = (2\pi)^{-n}\int_{\mathbb{R}^n} e^{i(x,\sigma)}\exp\Big[-\sum_{j,k=1}^{n}\mathcal{A}_{jk}(y,0)\sigma_j\sigma_k\Big]d\sigma. \tag{11.2.4}$$

THEOREM 11.1. *The following formula holds*

$$\mathcal{K}(\mathbb{R}^n,T) = \sup_{x\in\mathbb{R}^n}\ \sup_{0<t\le T}\ \sup_{|\boldsymbol{z}|_*=1}\int_{\mathbb{R}^n}|G^*(t,0,x,\eta)\boldsymbol{z}|_* d\eta. \tag{11.2.5}$$

In particular,

$$\mathcal{K}_0(y) = \sup_{|\boldsymbol{z}|_*=1}\int_{\mathbb{R}^n}|P^*(\eta;y)\boldsymbol{z}|_* d\eta. \tag{11.2.6}$$

PROOF. Let (x,t) be a fixed point in $\mathbb{R}^{n+1}_T$. We find the generalized norm $\|\boldsymbol{u}(x,t)\|$ of the mapping $[\mathrm{C_b}(\mathbb{R}^n)]^m \ni \boldsymbol{\psi} \to \boldsymbol{u}(x,t) \in \mathbb{R}^m$, where $\boldsymbol{u}$ is defined by (11.2.2). Using equalities (10.1.2), (10.1.3), properties of the inner product in $\mathbb{R}^m$, and the fact that the supremum operations commute, we obtain

$$\begin{aligned}
\|\boldsymbol{u}(x,t)\| &= \sup_{|\boldsymbol{\psi}|\le 1}\left|\int_{\mathbb{R}^n} G(t,0,x,\eta)\boldsymbol{\psi}(\eta)d\eta\right| \\
&= \sup_{|\boldsymbol{\psi}|\le 1}\ \sup_{|\boldsymbol{z}|_*=1}\left(\boldsymbol{z}, \int_{\mathbb{R}^n} G(t,0,x,\eta)\boldsymbol{\psi}(\eta)d\eta\right) \\
&= \sup_{|\boldsymbol{z}|_*=1}\ \sup_{|\boldsymbol{\psi}|\le 1}\int_{\mathbb{R}^n}\left(\boldsymbol{z}, G(t,0,x,\eta)\boldsymbol{\psi}(\eta)\right)d\eta \\
&= \sup_{|\boldsymbol{z}|_*=1}\ \sup_{|\boldsymbol{\psi}|\le 1}\int_{\mathbb{R}^n}\left(G^*(t,0,x,\eta)\boldsymbol{z},\ \boldsymbol{\psi}(\eta)\right)d\eta \\
&= \sup_{|\boldsymbol{z}|_*=1}\int_{\mathbb{R}^n}|G^*(t,0,x,\eta)\boldsymbol{z}|_* d\eta. \qquad (11.2.7)
\end{aligned}$$

Using the definition of $\mathcal{K}(\mathbb{R}^n, T)$, along (11.2.2) and (11.2.7), we obtain

$$\mathcal{K}(\mathbb{R}^n, T) = \sup \left\{ \frac{\|\boldsymbol{u}\|_{[\mathrm{C_b}(\overline{\mathbb{R}^{n+1}_T})]^m}}{\|\boldsymbol{u}|_{t=0}\|_{[\mathrm{C_b}(\mathbb{R}^n)]^m}} : \partial \boldsymbol{u}/\partial t - \mathfrak{A}(x,t;D_x)\boldsymbol{u} = \boldsymbol{0} \text{ in } \mathbb{R}^{n+1}_T \right\}$$
$$= \sup_{\|\boldsymbol{\psi}\|\le 1} \sup_{x\in\mathbb{R}^n} \sup_{0<t\le T} \left\| \int_{\mathbb{R}^n} G(t,0,x,\eta)\boldsymbol{\psi}(\eta)d\eta \right\|$$
$$= \sup_{x\in\mathbb{R}^n} \sup_{0<t\le T} \sup_{\|\boldsymbol{\psi}\|\le 1} \left\| \int_{\mathbb{R}^n} G(t,0,x,\eta)\boldsymbol{\psi}(\eta)d\eta \right\|$$
$$= \sup_{x\in\mathbb{R}^n} \sup_{0<t\le T} \|\boldsymbol{u}(x,t)\| = \sup_{x\in\mathbb{R}^n} \sup_{0<t\le T} \sup_{|\boldsymbol{z}|_*=1} \int_{\mathbb{R}^n} |G^*(t,0,x,\eta)\boldsymbol{z}|_* d\eta,$$

which gives representation (11.2.5).

Substituting $\mathcal{G}_0(t, x-\eta; y)$ into (11.2.5) in place of $G(t,0,x,\eta)$ we arrive at the representation of $\mathcal{K}_0(y)$ in the form (11.2.6). □

Henceforth all positive constants with non-significant values will be denoted by c or k with various indices.

LEMMA 11.1. *The inequality*

$$\mathcal{K}(\mathcal{D}, T) \ge \sup\{\mathcal{K}_0(y) : y \in \mathcal{D}\},$$

is valid, where $\mathcal{D}$ is either a bounded domain $\Omega \subset \mathbb{R}^n$, or $\mathcal{D} = \mathbb{R}^n$.

PROOF. *1. The case $\mathcal{D} = \mathbb{R}^n$.* By (11.2.5),

$$\mathcal{K}(\mathbb{R}^n, T) \ge \lim_{r\to 0} \sup_{y\in\mathbb{R}^n} \sup_{0<t\le T} \sup_{|\boldsymbol{z}|_*=1} \int_{\mathbb{B}_r(y)} |G^*(t,0,y,\eta)\boldsymbol{z}|_* d\eta, \tag{11.2.8}$$

where $\mathbb{B}_r(y) = \{x \in \mathbb{R}^n : |x-y| < r\}$.

We denote by $\|L\|_*$ the generalized norm $\sup\{\, |L\boldsymbol{z}|_* : \boldsymbol{z} \in \mathbb{R}^m, |\boldsymbol{z}|_* = 1\}$ of the $(m \times m)$-matrix L.

Since $\mathcal{A}_{jk}, \mathcal{A}_j, \mathcal{A}_0 \in [\mathrm{C}_{\mathrm{b}}^{\alpha,\alpha/2}(\overline{\mathbb{R}^{n+1}_T})]^{m\times m}$, it follows from estimates given in Eidel'man [**Ei1**] that the matrix $G(t,0,y,\eta)$ admits the representation

$$G(t,0,y,\eta) = \mathcal{G}_0(t, y-\eta; y) + [\mathcal{G}_0(t, y-\eta; \eta) - \mathcal{G}_0(t, y-\eta; y)] + W(t,y,\eta), \tag{11.2.9}$$

where

$$\|[\mathcal{G}_0(t, y-\eta; \eta) - \mathcal{G}_0(t, y-\eta; y)]^*\|_* \le k_1 |y-\eta|^\alpha t^{-n/2} \exp\left(-k_2 \frac{|y-\eta|^2}{t}\right), \tag{11.2.10}$$

$$\|W^*(t,y,\eta)\|_* \le k_3 t^{-(n-\alpha)/2} \exp\left(-k_4 \frac{|y-\eta|^2}{t}\right). \tag{11.2.11}$$

Using (11.2.9)-(11.2.11) and the representation (11.2.3), from (11.2.8) we obtain

$$\mathcal{K}(\mathbb{R}^n, T) \geq \sup_{y\in\mathbb{R}^n} \sup_{|\boldsymbol{z}|_*=1} \lim_{r\to 0} \limsup_{t\to+0} \int_{\mathbb{B}_r(y)} |G^*(t,0,y,\eta)\boldsymbol{z}|_* d\eta$$

$$\geq \sup_{y\in\mathbb{R}^n} \sup_{|\boldsymbol{z}|_*=1} \limsup_{r\to 0} \limsup_{t\to+0} \left\{ \int_{\mathbb{B}_r(y)} |\mathcal{G}_0^*(t, y-\eta; y)\boldsymbol{z}|_* d\eta - k_5 r^\alpha - k_6 t^{\alpha/2} \right\}$$

$$= \sup_{y\in\mathbb{R}^n} \sup_{|\boldsymbol{z}|_*=1} \limsup_{r\to 0} \limsup_{t\to+0} \int_{\mathbb{B}_r(y)} t^{-n/2} \left| P^* \left(\frac{y-\eta}{\sqrt{t}}\, ; y \right) \boldsymbol{z} \right|_* d\eta$$

$$= \sup_{y\in\mathbb{R}^n} \sup_{|\boldsymbol{z}|_*=1} \limsup_{r\to 0} \limsup_{t\to+0} \int_{\mathbb{B}_{r/\sqrt{t}}} |P^*(x;y)\boldsymbol{z}|_* dx$$

$$= \sup_{y\in R^n} \sup_{|\boldsymbol{z}|_*=1} \int_{\mathbb{R}^n} |P^*(x;y)\boldsymbol{z}|_* dx,$$

where $\mathbb{B}_r = \{x \in \mathbb{R}^n : |x| < r\}$.

This and (11.2.6) yield the inequality for $\mathcal{K}(\mathbb{R}^n, T)$ given in the statement of the lemma.

2. The case $\mathcal{D} = \Omega$. Let y be an arbitrary point of Ω and let the radius of the ball $\mathbb{B}_r(y)$ be so small that $\overline{\mathbb{B}_r(y)} \subset \Omega$. Further, let

$$\boldsymbol{\psi}_\varepsilon \in [\mathrm{C_b}(\mathbb{R}^n)]^m, \ \ |\boldsymbol{\psi}_\varepsilon(x)| \leq 1, \ \ \mathrm{supp}\, \boldsymbol{\psi}_\varepsilon \subset \mathbb{B}_\varepsilon(y), \ \ 0 < \varepsilon \leq r/2.$$

The vector-valued function

$$\boldsymbol{u}_\varepsilon(x,t) = \int_{\mathbb{R}^n} G(t,0,x,\eta)\boldsymbol{\psi}_\varepsilon(\eta) d\eta = \int_{\mathbb{B}_\varepsilon(y)} G(t,0,x,\eta)\boldsymbol{\psi}_\varepsilon(\eta) d\eta \tag{11.2.12}$$

is a solution of the Cauchy problem

$$\frac{\partial \boldsymbol{u}_\varepsilon}{\partial t} - \mathfrak{A}(x,t;D_x)\boldsymbol{u}_\varepsilon = \boldsymbol{0} \quad \text{in} \quad \mathbb{R}^{n+1}_T, \ \ \boldsymbol{u}_\varepsilon(x,0) = \boldsymbol{\psi}_\varepsilon(x). \tag{11.2.13}$$

Since $\mathcal{A}_{jk}, \mathcal{A}_j, \mathcal{A}_0 \in [\mathrm{C}_\mathrm{b}^{\alpha,\alpha/2}(\overline{\mathbb{R}^{n+1}_T})]^{m\times m}$, one has

$$\|G^*(t,0,x,\eta)\|_* \leq c_1 t^{-n/2} \exp\left(-c_2 \frac{|x-\eta|^2}{t}\right). \tag{11.2.14}$$

(see Eidel'man [**Ei1**]). The last estimate used for $(x,t) \in \partial\Omega \times (0,T]$ yields

$$\|\boldsymbol{u}_\varepsilon(x,t)\| \leq \int_{\mathbb{B}_\varepsilon(y)} \|G^*(t,0,x,\eta)\|_* d\eta$$

$$\leq c_1 t^{-n/2} \int_{\mathbb{B}_\varepsilon(y)} \exp\left(-c_2 \frac{|x-\eta|^2}{t}\right) d\eta$$

$$\leq c_3 \varepsilon^n t^{-n/2} \exp\left(-c_4 \frac{r^2}{t}\right) \leq c_4 \varepsilon^n r^{-n},$$

where $\|\boldsymbol{u}_\varepsilon(x,t)\|$ is the norm of the mapping $[\mathrm{C_b}(\mathbb{R}^n)]^m \ni \boldsymbol{\psi}_\varepsilon \to \boldsymbol{u}_\varepsilon(x,t) \in \mathbb{R}^m$. Hence, for sufficiently small ε

$$\sup\{\|\boldsymbol{u}_\varepsilon(x,t)\| : (x,t) \in \partial\Omega \times [0,T]\} \leq 1. \tag{11.2.15}$$

From this and the definition of the constant $\mathcal{K}(\Omega, T)$ we get

$$\mathcal{K}(\Omega, T) \geq \sup_{y\in\Omega} \limsup_{\varepsilon\to 0} \limsup_{t\to+0} \|\boldsymbol{u}_\varepsilon(y,t)\|. \tag{11.2.16}$$

Using (11.2.12) and (11.2.14), in the same way as in the proof of Theorem 10.1, we find

$$\|\boldsymbol{u}_\varepsilon(y,t)\| = \sup_{|\boldsymbol{z}|_*=1} \int_{\mathbb{B}_\varepsilon(y)} |G^*(t,0,y,\eta)\boldsymbol{z}|_* d\eta.$$

The last equality and (11.2.16) yield

$$\mathcal{K}(\Omega,T) \geq \sup_{y\in\Omega} \lim_{\varepsilon\to 0} \limsup_{t\to+0} \sup_{|\boldsymbol{z}|_*=1} \int_{\mathbb{B}_\varepsilon(y)} |G^*(t,0,y,\eta)\boldsymbol{z}|_* d\eta$$

$$(11.2.17) \qquad \geq \sup_{y\in\Omega} \sup_{|\boldsymbol{z}|_*=1} \lim_{\varepsilon\to 0} \limsup_{t\to+0} \int_{\mathbb{B}_\varepsilon(y)} |G^*(t,0,y,\eta)\boldsymbol{z}|_* d\eta.$$

It was shown in the first part of the lemma that

$$\sup_{|\boldsymbol{z}|_*=1} \lim_{\varepsilon\to 0} \limsup_{t\to+0} \int_{\mathbb{B}_\varepsilon(y)} |G^*(t,0,y,\eta)\boldsymbol{z}|_* d\eta$$

$$\geq \sup_{|\boldsymbol{z}|_*=1} \int_{\mathbb{R}^n} |P^*(x;y)\boldsymbol{z}|_* dx = \mathcal{K}_0(y)$$

which together with (11.2.17) gives the lower estimate for $\mathcal{K}(\Omega,T)$ in the statement of the lemma. □

11.3. Necessary condition for validity of the maximum norm principle for the system $\partial \boldsymbol{u}/\partial t - \mathfrak{A}(x,t,D_x)\boldsymbol{u} = \boldsymbol{0}$

If at any point of the unit sphere $\mathcal{S}^{m-1}$ of the norm there exists a tangent hyperplane, then we obtain the following corollary of Proposition 10.2.

PROPOSITION 11.1. *Suppose the norm $|\cdot|$ is differentiable. Let the maximum norm principle be valid for solutions of system*

$$(11.3.1) \qquad \frac{\partial \boldsymbol{u}}{\partial t} - \sum_{j,k=1}^{n} \mathcal{A}_{jk}(x,t)\frac{\partial^2 \boldsymbol{u}}{\partial x_j \partial x_k} = \boldsymbol{0}$$

in the class $[\mathrm{C}^{(2,1)}(\Pi_T)]^m \cap [\mathrm{C_b}(\overline{\Pi}_T)]^m$. Then

$$\mathcal{A}_{jk}(x,0) = a_{jk}(x)I, \quad 1 \leq j,k \leq n,$$

where $((a_{jk}))$ is a positive-definite $(n\times n)$-matrix-valued function in $\overline{\mathcal{D}}$.

PROOF. Let $\zeta \in \mathcal{S}^{m-1}$. Since the vectors $\nabla|\zeta|$ and $\boldsymbol{\nu}(\zeta)$ are linear dependent, condition (10.4.13) can be represented in the form

$$(11.3.2) \qquad \mathcal{A}^*_{jk}(x,0)\boldsymbol{\nu}(\zeta) = \mu_{jk}(x,\zeta)\boldsymbol{\nu}(\zeta).$$

The set of normals to $\mathcal{S}^{m-1}$ coincides with the set of unit vectors in $\mathbb{R}^m$. Therefore, one can choose the points $\zeta_1,\dots,\zeta_m,\zeta_{m+1}$ on $\mathcal{S}^{m-1}$ in such a way that

$$\boldsymbol{\nu}(\zeta_1) = \boldsymbol{e}_1,\dots,\boldsymbol{\nu}(\zeta_m) = \boldsymbol{e}_m, \quad \boldsymbol{\nu}(\zeta_{m+1}) = m^{-1/2}(\boldsymbol{e}_1 + \dots + \boldsymbol{e}_m),$$

where $\boldsymbol{e}_1,\dots,\boldsymbol{e}_m$ is an orthonormal basis of the Cartesian coordinate system in $\mathbb{R}^m$.

Using equality (11.3.2), we obtain

$$\mathcal{A}^*_{jk}(x,0)\boldsymbol{e}_1 = \mu_{jk}(x,\zeta_1)\boldsymbol{e}_1,\dots,\mathcal{A}^*_{jk}(x,0)\boldsymbol{e}_m = \mu_{jk}(x,\zeta_m)\boldsymbol{e}_m.$$

From this we get

$$\mathcal{A}_{jk}(x,0) = \mathrm{diag}\,\{\mu_{jk}(x,\zeta_1),\dots,\mu_{jk}(x,\zeta_m)\}.$$

The last representation and the equality

$$\mathcal{A}^*_{jk}(x,0)\boldsymbol{\nu}(\zeta_{m+1}) = \mu_{jk}(x,\zeta_{m+1})\boldsymbol{\nu}(\zeta_{m+1})$$

yield

$$\mu_{jk}(x,\zeta_1) = \cdots = \mu_{jk}(x,\zeta_m) = \mu_{jk}(x,\zeta_{m+1}).$$

Let $a_{jk}(x) = \mu_{jk}(x,\zeta_i)$, $1 \le i \le m+1$. Taking into account condition (B) from Section 11.1, we complete the proof. □

Now we suppose that the norm $|\cdot|$ is twice continuously differentiable.

The next assertion, crucial in what follows, contains a necessary condition for validity of the maximum norm principle for system (11.1.1).

PROPOSITION 11.2. *Let the maximum norm principle holds for solutions of system* (11.1.1) *in the class* $[\mathrm{C}^{(2,1)}(\Pi_T)]^m \cap [\mathrm{C_b}(\overline{\Pi}_T)]^m$. *Then*

(i) *for all* $x \in \mathcal{D}$

$$\mathcal{A}_{jk}(x,0) = a_{jk}(x)I, \quad 1 \le j,k \le n,$$

where $((a_{jk}(x)))$ *is a positive-definite* $(n\times n)$*-matrix-valued function in* $\overline{\mathcal{D}}$;

(ii) *for all* $x \in \mathcal{D}$ *and for all vectors* $\boldsymbol{\xi}_j \in \mathbb{R}^m, \boldsymbol{\zeta} \in \mathbb{R}^m\backslash\{\mathbf{0}\}$, *with* $(\boldsymbol{\xi}_j, \nabla|\boldsymbol{\zeta}|) = 0$, $j = 1,\dots,n$, *the inequality*

$$\sum_{j,k=1}^n a_{jk}(x)(\mathcal{H}(|\boldsymbol{\zeta}|)\boldsymbol{\xi}_j,\boldsymbol{\xi}_k) + \sum_{j=1}^n(\mathcal{A}_j(x,0)\boldsymbol{\xi}_j, \nabla|\boldsymbol{\zeta}|) + (\mathcal{A}_0(x,0)\boldsymbol{\zeta},\nabla|\boldsymbol{\zeta}|) \ge 0$$

is valid.

PROOF. Let the solutions of system (11.1.1) satisfy the maximum norm principle in the class $[\mathrm{C}^{(2,1)}(\Pi_T)]^m \cap [\mathrm{C_b}(\overline{\Pi}_T)]^m$. From Lemma 11.1 it follows that equality $\mathcal{K}(\mathcal{D},T) = 1$ implies $\mathcal{K}_0(y) = 1$ for all $y \in \mathcal{D}$, where $\mathcal{K}_0(y)$ is the constant $\mathcal{K}(\mathbb{R}^n,T)$ defined for the system

$$\frac{\partial \boldsymbol{u}}{\partial t} - \sum_{j,k=1}^n \mathcal{A}_{jk}(y,0)\frac{\partial^2 \boldsymbol{u}}{\partial x_j \partial x_k} = \mathbf{0}$$

and $y \in \mathcal{D}$ plays the role of parameter.

By Proposition 11.1, the equality $\mathcal{K}_0(y) = 1$ takes place for all $y \in \mathcal{D}$ only if the equalities $\mathcal{A}_{jk}(y,0) = a_{jk}(y)I$, $1 \le j,k \le n$, hold for all $y \in \mathcal{D}$. Therefore, condition (i) is necessary for validity of the maximum norm principle for solutions of system (11.1.1). In what follows in this proof we assume that condition (i) is satisfied.

Now we prove the necessity of (ii). Suppose there exist a point $y \in \mathcal{D}$ and vectors $\boldsymbol{\zeta} \in \mathbb{R}^m\backslash\{\mathbf{0}\}$, $\boldsymbol{\xi}_j \in \mathbb{R}^m$, with $(\boldsymbol{\xi}_j, \nabla|\boldsymbol{\zeta}|) = 0$, $j = 1,\dots,n$, for which

(11.3.3)

$$\sum_{j,k=1}^n a_{jk}(y)(\mathcal{H}(|\boldsymbol{\zeta}|)\boldsymbol{\xi}_j,\boldsymbol{\xi}_k) + \sum_{j=1}^n(\mathcal{A}_j(y,0)\boldsymbol{\xi}_j, \nabla|\boldsymbol{\zeta}|) + (\mathcal{A}_0(y)\boldsymbol{\zeta},\nabla|\boldsymbol{\zeta}|) = \gamma < 0.$$

We choose a constant $\vartheta > 0$ in such way that the inequality

$$\vartheta\sum_{j=1}^n a_{jj}(y) < -\gamma \tag{11.3.4}$$

is satisfied. Further, we introduce the vector-valued function

$$(11.3.5)\quad \boldsymbol{\psi}_\vartheta(x) = \frac{\boldsymbol{\zeta}}{|\boldsymbol{\zeta}|} + \frac{1}{|\boldsymbol{\zeta}|}\sum_{j=1}^{n} \boldsymbol{\xi}_j x_j - \frac{\boldsymbol{\zeta}}{2|\boldsymbol{\zeta}|^2}\sum_{j,k=1}^{n} (\mathcal{H}(|\boldsymbol{\zeta}|)\boldsymbol{\xi}_j, \boldsymbol{\xi}_k) x_j x_k - \frac{\vartheta\boldsymbol{\zeta}}{2|\boldsymbol{\zeta}|^2}\sum_{j=1}^{n} x_j^2,$$

where $\boldsymbol{\zeta} \in \mathbb{R}^m \backslash \{\mathbf{0}\}$, $\boldsymbol{\xi}_j \in \mathbb{R}^m$, $(\boldsymbol{\xi}_j, \nabla|\boldsymbol{\zeta}|) = 0$, $j = 1, \dots, n$.

The Hessian $\mathcal{H}(|\boldsymbol{\zeta}|)$ of the twice continuously differentiable norm $|\cdot|$ is a symmetric positive-semidefinite matrix (see, for example Elster, Reinhardt, Schäuble and Donath [**ERSD**]). For the function $|\boldsymbol{\psi}_\vartheta(x)|$ we have the expansion

$$|\boldsymbol{\psi}_\vartheta(x)| = |\boldsymbol{\psi}_\vartheta(0)| + \sum_{j=1}^{n} \frac{\partial|\boldsymbol{\psi}_\vartheta(x)|}{\partial x_j}\Big|_{x=0} x_j + \frac{1}{2}\sum_{j,k=1}^{n} \frac{\partial^2|\boldsymbol{\psi}_\vartheta(x)|}{\partial x_j \partial x_k}\Big|_{x=0} x_j x_k + |x|^2\eta(x),$$

where $\eta(x) \to 0$ as $x \to 0$. By direct calculation we find

$$|\boldsymbol{\psi}_\vartheta(0)| = 1,$$
$$\frac{\partial|\boldsymbol{\psi}_\vartheta(x)|}{\partial x_j}\Big|_{x=0} = \frac{(\nabla|\boldsymbol{\zeta}|, \boldsymbol{\xi}_j)}{|\boldsymbol{\zeta}|} = 0,$$
$$\frac{\partial^2|\boldsymbol{\psi}_\vartheta(x)|}{\partial x_j \partial x_k}\Big|_{x=0} = -\frac{\vartheta}{|\boldsymbol{\zeta}|}\,\delta_{jk}.$$

Hence,

$$|\boldsymbol{\psi}_\vartheta(x)| = 1 - \frac{\vartheta}{2|\boldsymbol{\zeta}|}|x|^2 + |x|^2\eta(x).$$

In the case $\mathcal{D} = \Omega$ let r be so small that $\overline{\mathbb{B}_r(y)} \subset \mathcal{D}$. The last equality implies the existence of a constant $\varepsilon_0 = \varepsilon_0(\vartheta)$, $0 < \varepsilon_0 \le r/2$, such that

$$(11.3.6)\qquad |\boldsymbol{\psi}_\vartheta(x)| \le 1$$

for $|x| \le \varepsilon_0$. Let $\chi_\varepsilon \in \mathrm{C}^\infty(\mathbb{R}^n)$, $\chi_\varepsilon(x) = 1$ for $|x| \le \varepsilon/2$, $\chi_\varepsilon(x) = 0$ for $|x| \ge \varepsilon$ and $0 \le \chi_\varepsilon(x) \le 1$ for all $x \in \mathbb{R}^n$. Further, let $\boldsymbol{v}_\varepsilon(x) = \chi_\varepsilon(x-y)\boldsymbol{\psi}_\vartheta(x-y)$, where $\varepsilon \le \varepsilon_0$. Then, by (11.3.6) we have

$$(11.3.7)\qquad |\boldsymbol{v}_\varepsilon(x)| \le 1$$

for all $x \in \mathbb{R}^n$.

The vector-valued function

$$\boldsymbol{u}_\varepsilon(x,t) = \int_{\mathbb{R}^n} G(t,0,x,\eta)\boldsymbol{v}_\varepsilon(\eta)d\eta,$$

(where $G(t,\tau,x,\eta)$ is the fundamental matrix of solutions for system (11.1.1)) is the solution of the Cauchy problem

$$\frac{\partial \boldsymbol{u}_\varepsilon}{\partial t} - \mathfrak{A}(x,t,D_x)\boldsymbol{u}_\varepsilon = 0 \quad \text{in} \quad \mathbb{R}^{n+1}_T, \qquad \boldsymbol{u}_\varepsilon(x,0) = \boldsymbol{v}_\varepsilon(x).$$

If $\mathcal{D}$ is a bounded subdomain of $\mathbb{R}^n$, then by (11.3.7) and (11.2.15) the value of ε can be chosen so small that

$$\sup\{|\boldsymbol{u}_\varepsilon(x,t)| : (x,t) \in \partial\mathcal{D} \times [0,T]\} \le 1.$$

Then

$$\| \boldsymbol{u}_\varepsilon|_{\overline{\Gamma}_T} \|_{[\mathrm{C}(\overline{\Gamma}_T)]^m} = 1$$

and consequently

$$(11.3.8)\qquad \mathcal{K}(\mathcal{D},T) \ge \sup_{0<t\le T} |\boldsymbol{u}_\varepsilon(y,t)|.$$

In the case $\mathcal{D}=\mathbb{R}^n$ the last inequality is obvious.

We take the inner product of the vector

$$\frac{\partial \boldsymbol{u}_\varepsilon}{\partial t}-\sum_{j,k=1}^n \mathcal{A}_{jk}(x,t)\frac{\partial^2 \boldsymbol{u}_\varepsilon}{\partial x_j \partial x_k}+\sum_{j=1}^n \mathcal{A}_j(x,t)\frac{\partial \boldsymbol{u}_\varepsilon}{\partial x_j}+\mathcal{A}_0(x,t)\boldsymbol{u}_\varepsilon=\boldsymbol{0}$$

with $\nabla|\boldsymbol{\zeta}|$, $\boldsymbol{\zeta}\neq\boldsymbol{0}$, and transform the equality

$$\left(\frac{\partial \boldsymbol{u}_\varepsilon}{\partial t},\nabla|\boldsymbol{\zeta}|\right)-\sum_{j,k=1}^n a_{jk}(x)\left(\frac{\partial^2 \boldsymbol{u}_\varepsilon}{\partial x_j \partial x_k},\ \nabla|\boldsymbol{\zeta}|\right)+\sum_{j=1}^n\left(\mathcal{A}_j(x,t)\frac{\partial \boldsymbol{u}_\varepsilon}{\partial x_j},\ \nabla|\boldsymbol{\zeta}|\right)$$
$$+(\mathcal{A}_0(x)\boldsymbol{u}_\varepsilon,\ \nabla|\boldsymbol{\zeta}|)-\sum_{j,k=1}^n\left([\mathcal{A}_{jk}(x,t)-a_{jk}(x)I]\frac{\partial^2 \boldsymbol{u}_\varepsilon}{\partial x_j \partial x_k},\ \nabla|\boldsymbol{\zeta}|\right)=0,$$

where $a_{jk}(x)I=\mathcal{A}_{jk}(x,0)$. As a result we find

$$\frac{\partial}{\partial t}(\boldsymbol{u}_\varepsilon,\ \nabla|\boldsymbol{\zeta}|)=\sum_{j,k=1}^n a_{jk}(x)\left(\frac{\partial^2 \boldsymbol{u}_\varepsilon}{\partial x_j \partial x_k},\ \nabla|\boldsymbol{\zeta}|\right)-\sum_{j=1}^n\left(\mathcal{A}_j(x,t)\frac{\partial \boldsymbol{u}_\varepsilon}{\partial x_j},\ \nabla|\boldsymbol{\zeta}|\right)$$
$$-(\mathcal{A}_0(x)\boldsymbol{u}_\varepsilon,\ \nabla|\boldsymbol{\zeta}|)+\sum_{j,k=1}^n\left([\mathcal{A}_{jk}(x,t)-a_{jk}(x)I]\frac{\partial^2 \boldsymbol{u}_\varepsilon}{\partial x_j \partial x_k},\ \nabla|\boldsymbol{\zeta}|\right).$$

Since the coefficients of system (11.1.1) belong to the class $[\mathrm{C}^{\alpha,\alpha/2}(\overline{\mathbb{R}^{n+1}_T})]^m$, and since $\boldsymbol{v}_\varepsilon\in[\mathrm{C}^{2,\alpha}(\mathbb{R}^n)]^m$ we have, according to Eidel'man [**Ei1**], that $\boldsymbol{u}_\varepsilon\in[\mathrm{C}^{2+\alpha,\alpha/2}(\overline{\mathbb{R}^{n+1}_T})]^m$. Passing to the limit in the last equality as $(x,t)\to(y,+0)$, we obtain

$$\lim_{(x,t)\to(y,+0)}\frac{\partial}{\partial t}(\boldsymbol{u}_\varepsilon,\ \nabla|\boldsymbol{\zeta}|)=\left\{\sum_{j,k=1}^n a_{jk}(y)\left(\frac{\partial^2 \boldsymbol{v}_\varepsilon}{\partial x_j \partial x_k},\ \nabla|\boldsymbol{\zeta}|\right)\right.$$
$$\left.-\sum_{j=1}^n\left(\mathcal{A}_j(y,0)\frac{\partial \boldsymbol{v}_\varepsilon}{\partial x_j},\ \nabla|\boldsymbol{\zeta}|\right)-(\mathcal{A}_0(y)\boldsymbol{v}_\varepsilon,\ \nabla|\boldsymbol{\zeta}|)\right\}\Bigg|_{x=y}.$$

Hence, using the equalities

$$\left.\frac{\partial^2 \boldsymbol{v}_\varepsilon}{\partial x_j \partial x_k}\right|_{x=y}=-\frac{\boldsymbol{\zeta}}{|\boldsymbol{\zeta}|^2}[\,(\mathcal{H}(|\boldsymbol{\zeta}|)\boldsymbol{\xi}_j,\boldsymbol{\xi}_k)+\vartheta\delta_{jk}\,],\qquad \left.\frac{\partial \boldsymbol{v}_\varepsilon}{\partial x_j}\right|_{x=y}=\frac{\boldsymbol{\xi}_j}{|\boldsymbol{\zeta}|},\quad \boldsymbol{v}_\varepsilon(y)=\frac{\boldsymbol{\zeta}}{|\boldsymbol{\zeta}|},$$

where $j,k=1,\dots,n$, and using the identity $(\boldsymbol{\zeta},\nabla|\boldsymbol{\zeta}|)=|\boldsymbol{\zeta}|$, we find

$$\lim_{t\to+0}\frac{\partial}{\partial t}(\boldsymbol{u}_\varepsilon(y,t),\ \nabla|\boldsymbol{\zeta}|)\;=\;-\frac{1}{|\boldsymbol{\zeta}|}\left\{\sum_{j,k=1}^n a_{jk}(y)[\,(\mathcal{H}(|\boldsymbol{\zeta}|)\boldsymbol{\xi}_j,\boldsymbol{\xi}_k)+\vartheta\delta_{jk}\,]\right.$$
$$\left.+\ \sum_{j=1}^n(\mathcal{A}_j(y,0)\boldsymbol{\xi}_j,\nabla|\boldsymbol{\zeta}|)+(\mathcal{A}_0(y)\boldsymbol{\zeta},\nabla|\boldsymbol{\zeta}|)\right\}. \tag{11.3.9}$$

The function $(\boldsymbol{u}_\varepsilon(y,t),\nabla|\boldsymbol{\zeta}|)$ is continuous on $[0,T]$ and differentiable at all points of the interval $(0,T)$, its derivative

$$\frac{\partial\,(\boldsymbol{u}_\varepsilon(y,t),\nabla|\boldsymbol{\zeta}|)}{\partial t}$$

tends to a finite limit as $t \to +0$. Therefore, the function $(\boldsymbol{u}_\varepsilon(y,t), \nabla|\boldsymbol{\zeta}|)$ has the right-hand derivative

$$\frac{\partial_+ (\boldsymbol{u}_\varepsilon(y,t), \nabla|\boldsymbol{\zeta}|)}{\partial t}$$

at $t=0$, and by virtue of (11.3.9) we have

$$\frac{\partial_+ (\boldsymbol{u}_\varepsilon(y,t), \nabla|\boldsymbol{\zeta}|)}{\partial t}\Big|_{t=0} = -\frac{1}{|\boldsymbol{\zeta}|}\Bigg\{\sum_{j,k=1}^{n} a_{jk}(y)[\,(\mathcal{H}(|\boldsymbol{\zeta}|)\boldsymbol{\xi}_j, \boldsymbol{\xi}_k) + \vartheta\delta_{jk}\,]$$
$$+ \sum_{j=1}^{n}(\mathcal{A}_j(y,0)\boldsymbol{\xi}_j, \nabla|\boldsymbol{\zeta}|) + (\mathcal{A}_0(y)\boldsymbol{\zeta}, \nabla|\boldsymbol{\zeta}|)\Bigg\}.$$

From the last equality combined with (11.3.3), (11.3.4) it follows that

$$\frac{\partial_+ (\boldsymbol{u}_\varepsilon(y,t), \nabla|\boldsymbol{\zeta}|)}{\partial t}\Big|_{t=0} > 0. \tag{11.3.10}$$

Since

$$(\boldsymbol{u}_\varepsilon(y,0), \nabla|\boldsymbol{\zeta}|) = (\boldsymbol{v}_\varepsilon(y), \nabla|\boldsymbol{\zeta}|) = (\boldsymbol{\psi}_\vartheta(0), \nabla|\boldsymbol{\zeta}|) = \left(\frac{\boldsymbol{\zeta}}{|\boldsymbol{\zeta}|}, \nabla|\boldsymbol{\zeta}|\right) = 1,$$

(11.3.10) implies the existence of a constant $\delta > 0$ such that

$$(\boldsymbol{u}_\varepsilon(y,t), \nabla|\boldsymbol{\zeta}|) > 1 \tag{11.3.11}$$

for $0 < t \le \delta$. Using $|\, \nabla|\boldsymbol{\zeta}|\, |_* = 1$ and inequalities (10.1.4), (11.3.11), we obtain

$$|\boldsymbol{u}_\varepsilon(y,t)| > 1,$$

where $0 < t \le \delta$.

From this and (11.3.8) it follows that $\mathcal{K}(\mathcal{D},T) > 1$ which contradicts our assumption of validity of the maximum norm principle. Thus, we see that if $\mathcal{K}(\mathcal{D},T) = 1$, then equality (11.3.3) with $\gamma < 0$ is not valid.

Finally, if $\mathcal{K}(\mathcal{D},T) = 1$, then for all $x \in \mathcal{D}$ and for all vectors $\boldsymbol{\xi}_j \in \mathbb{R}^m, \boldsymbol{\zeta} \in \mathbb{R}^m\backslash\{\boldsymbol{0}\}$, with $(\boldsymbol{\xi}_j, \nabla|\boldsymbol{\zeta}|) = 0$, $j = 1, \dots, n$, the inequality

$$\sum_{j,k=1}^{n} a_{jk}(x)(\mathcal{H}(|\boldsymbol{\zeta}|)\boldsymbol{\xi}_j, \boldsymbol{\xi}_k) + \sum_{j=1}^{n}(\mathcal{A}_j(x,0)\boldsymbol{\xi}_j, \nabla|\boldsymbol{\zeta}|) + (\mathcal{A}_0(x,0)\boldsymbol{\zeta}, \nabla|\boldsymbol{\zeta}|) \ge 0$$

holds. Therefore, condition (ii) is necessary for validity of the maximum norm principle for solutions of system (11.1.1). □

11.4. Sufficient condition for validity of the maximum norm principle for the system $\partial \boldsymbol{u}/\partial t - \mathfrak{A}(x,t,D_x)\boldsymbol{u} = \boldsymbol{0}$ with scalar principal part

Now we give a sufficient condition for validity of the maximum norm principle for second order systems with a structural restriction.

PROPOSITION 11.3. *Let the coefficients of the system*

$$\frac{\partial \boldsymbol{u}}{\partial t} - \sum_{j,k=1}^{n} a_{jk}(x,t)\frac{\partial^2 \boldsymbol{u}}{\partial x_j \partial x_k} + \sum_{j=1}^{n}\mathcal{A}_j(x,t)\frac{\partial \boldsymbol{u}}{\partial x_j} + \mathcal{A}_0(x,t)\boldsymbol{u} = \boldsymbol{0} \tag{11.4.1}$$

satisfy the condition:

(i) *for all* $(x,t) \in \Pi_T$ *and for all vectors* $\boldsymbol{\xi}_j \in \mathbb{R}^m, \boldsymbol{\zeta} \in \mathbb{R}^m \backslash \{\mathbf{0}\}$, *with* $(\boldsymbol{\xi}_j, \nabla|\boldsymbol{\zeta}|) = 0,\ j = 1, \dots, n$, *the inequality*

$$(11.4.2)\quad \sum_{j,k=1}^{n} a_{jk}(x,t)(\mathcal{H}(|\boldsymbol{\zeta}|)\boldsymbol{\xi}_j, \boldsymbol{\xi}_k) + \sum_{j=1}^{n}(\mathcal{A}_j(x,t)\boldsymbol{\xi}_j, \nabla|\boldsymbol{\zeta}|) + (\mathcal{A}_0(x,t)\boldsymbol{\zeta}, \nabla|\boldsymbol{\zeta}|) \geq 0$$

holds. Then $\mathcal{K}(\mathcal{D}, T) = 1$.

PROOF. *1. The case* $\mathcal{D} = \Omega$. Consider the auxiliary system

$$(11.4.3)\qquad \frac{\partial \boldsymbol{v}}{\partial t} - \sum_{j,k=1}^{n} a_{jk}(x,t)\frac{\partial^2 \boldsymbol{v}}{\partial x_j \partial x_k} + \sum_{j=1}^{n} \mathcal{A}_j(x,t)\frac{\partial \boldsymbol{v}}{\partial x_j} + \mathcal{A}_0(x,t)\boldsymbol{v} + \varepsilon \boldsymbol{v} = \mathbf{0},$$

where $\varepsilon = \text{const} > 0$. We show that for any non-trivial solution

$$\boldsymbol{v} \in [\mathrm{C}^{(2,1)}(Q_T)]^m \cap [\mathrm{C}(\overline{Q_T})]^m$$

of system (11.4.3), the function $|\boldsymbol{v}(x,t)|$ cannot attain its global maximum at a point $(x,t) \in Q_T$.

We take the scalar product of the vectors (11.4.3) and $\nabla|v|$, and obtain

$$\left(\frac{\partial \boldsymbol{v}}{\partial t}, \nabla|\boldsymbol{v}|\right) - \sum_{j,k=1}^{n} a_{jk}(x,t)\left(\frac{\partial^2 \boldsymbol{v}}{\partial x_j \partial x_k}, \nabla|\boldsymbol{v}|\right) + \sum_{j=1}^{n}\left(\mathcal{A}_j(x,t)\frac{\partial \boldsymbol{v}}{\partial x_j}, \nabla|\boldsymbol{v}|\right)$$
$$(11.4.4)\qquad + (\mathcal{A}_0(x,t)\boldsymbol{v}, \nabla|\boldsymbol{v}|) + \varepsilon(\boldsymbol{v}, \nabla|\boldsymbol{v}|) = 0.$$

Taking into account the identities

$$(\boldsymbol{v}, \nabla|\boldsymbol{v}|) = |\boldsymbol{v}|,$$
$$\frac{\partial|\boldsymbol{v}|}{\partial t} = \left(\frac{\partial \boldsymbol{v}}{\partial t}, \nabla|\boldsymbol{v}|\right),$$

and

$$\frac{\partial^2|\boldsymbol{v}|}{\partial x_j \partial x_k} = \left(\mathcal{H}(|\boldsymbol{v}|)\frac{\partial \boldsymbol{v}}{\partial x_j}, \frac{\partial \boldsymbol{v}}{\partial x_k}\right) + \left(\frac{\partial^2 \boldsymbol{v}}{\partial x_j \partial x_k}, \nabla|\boldsymbol{v}|\right),$$

we transform equality (11.4.4). As a result we find

$$\frac{\partial|\boldsymbol{v}|}{\partial t} - \sum_{j,k=1}^{n} a_{jk}(x,t)\frac{\partial^2|\boldsymbol{v}|}{\partial x_j \partial x_k} = -\left\{\sum_{j,k=1}^{n} a_{jk}(x,t)\left(\mathcal{H}(|\boldsymbol{v}|)\frac{\partial \boldsymbol{v}}{\partial x_j}, \frac{\partial \boldsymbol{v}}{\partial x_k}\right)\right.$$
$$(11.4.5)\qquad \left. + \sum_{j=1}^{n}\left(\mathcal{A}_j(x,t)\frac{\partial \boldsymbol{v}}{\partial x_j}, \nabla|\boldsymbol{v}|\right) + (\mathcal{A}_0(x,t)\boldsymbol{v}, \nabla|\boldsymbol{v}|) + \varepsilon|\boldsymbol{v}|\right\}.$$

Suppose the non-trivial function $|\boldsymbol{v}(x,t)|$ takes its global maximum at a point $(x_0, t_0) \in Q_T$. Then

$$(11.4.6)\qquad \frac{\partial|\boldsymbol{v}|}{\partial t}\bigg|_{(x_0,t_0)} \geq 0,$$

$$(11.4.7)\qquad \frac{\partial|\boldsymbol{v}|}{\partial x_j}\bigg|_{(x_0,t_0)} = \left(\nabla|\boldsymbol{v}|, \frac{\partial \boldsymbol{v}}{\partial x_j}\right)\bigg|_{(x_0,t_0)} = 0,$$

and

$$(11.4.8)\qquad \sum_{j,k=1}^{n} a_{jk}(x_0,t_0)\frac{\partial^2|\boldsymbol{v}|}{\partial x_j \partial x_k}\bigg|_{(x_0,t_0)} \leq 0.$$

By (11.4.5)-(11.4.8) we have

$$\Big[\sum_{j,k=1}^{n} a_{jk}(x_0,t_0)\left(\mathcal{H}(|\boldsymbol{v}|)\frac{\partial \boldsymbol{v}}{\partial x_j},\frac{\partial \boldsymbol{v}}{\partial x_k}\right)+\sum_{j=1}^{n}\left(\mathcal{A}_j(x_0,t_0)\frac{\partial \boldsymbol{v}}{\partial x_j},\ \nabla|\boldsymbol{v}|\right)$$

$$+(\ \mathcal{A}_0(x_0,t_0)\boldsymbol{v},\ \nabla|\boldsymbol{v}|\ )\Big]\ \Big|_{(x_0,t_0)}+\ \varepsilon|\boldsymbol{v}(x_0,t_0)|\leq 0,$$

and

$$\left(\nabla|\boldsymbol{v}|\ ,\frac{\partial \boldsymbol{v}}{\partial x_j}\right)\Big|_{(x_0,t_0)}=0,$$

which contradicts condition (i) and hence $|\boldsymbol{v}(x,t)|$ cannot attain its global maximum at a point $(x_0,t_0)\in Q_T$. This implies

$$\|\boldsymbol{v}\|_{[\mathrm{C}(\overline{Q}_T)]^m}=\|\boldsymbol{v}|_{\overline{\Gamma}_T}\|_{[\mathrm{C}(\overline{\Gamma}_T)]^m}. \tag{11.4.9}$$

The vector-valued function $\boldsymbol{u}(x,t)=\boldsymbol{v}(x,t)\exp(\varepsilon t)$ solves system (11.4.1). Hence, by (11.4.9) we have

$$\max_{\overline{Q}_T}|e^{-\varepsilon t}\boldsymbol{u}(x,t)|=\max_{\overline{\Gamma}_T}|e^{-\varepsilon t}\boldsymbol{u}(x,t)|.$$

Consequently,

$$\|\boldsymbol{u}\|_{[\mathrm{C}(\overline{Q}_T)]^m}\leq e^{\varepsilon T}\|\boldsymbol{u}|_{\overline{\Gamma}_T}\|_{[\mathrm{C}(\overline{\Gamma}_T)]^m}.$$

Since ε is arbitrary, the best constant in the last inequality is equal to one.

2. The case $\mathcal{D}=\mathbb{R}^n$. Contrary to the case of the bounded domain, one cannot immediately conclude that $\mathcal{K}(\mathbb{R}^n,T)=1$ because of absence of the global maximum of the function $|\boldsymbol{u}(x,t)|$ in the layer $\mathbb{R}^{n+1}_T$.

Suppose inequality (11.4.2) holds for all points $(x,t)\in\mathbb{R}^{n+1}_T$ and for all vectors $\boldsymbol{\xi}_j\in\mathbb{R}^m$, $\boldsymbol{\zeta}\in\mathbb{R}^m\backslash\{\boldsymbol{0}\}$, with $(\boldsymbol{\xi}_j,\nabla|\boldsymbol{\zeta}|)=0$, $j=1,\dots,n$, and that the maximum norm principle is not valid for solutions of system (11.4.1) in $\mathbb{R}^{n+1}_T$. Then there exists a point $(x_0,t_0)\in\mathbb{R}^{n+1}_T$ and a vector-valued function $\boldsymbol{\Phi}\in[\mathrm{C_b}(\mathbb{R}^n)]^m$ with $|\boldsymbol{\Phi}(x)|\leq 1$, such that

$$|\boldsymbol{u}(x_0,t_0)|=\left|\int_{\mathbb{R}^n}G(t_0,0,x_0,\eta)\boldsymbol{\Phi}(\eta)d\eta\right|>1. \tag{11.4.10}$$

By (11.2.14) one can assume that supp $\boldsymbol{\Phi}\subset\mathbb{B}_\rho(x_0)$. If $|x-x_0|\geq R>\rho$ then (11.2.14) implies

$$\left|\int_{\mathbb{R}^n}G(t,0,x,\eta)\Phi(\eta)d\eta\right|\leq c_1\exp\left[-c_2\frac{|R-\rho|^2}{T}\right],$$

where $0<t\leq T$, $x\in\mathbb{R}^n\backslash\mathbb{B}_R(x_0)$. Applying the assertion of the present theorem for a cylinder with a bounded base, we get

$$|\boldsymbol{u}(x,t)|=\left|\int_{\mathbb{R}^n}G(t,0,x,\eta)\boldsymbol{\Phi}(\eta)d\eta\right|\leq 1,$$

where $(x,t)\in\overline{\mathbb{B}_R(x_0)}\times[0,T]$ and R is sufficiently large. The last inequality contradicts (11.4.10) which proves the validity of the maximum norm principle in $\mathbb{R}^{n+1}_T$. □

11.5. Criteria for validity of the maximum norm principle for the system $\partial \boldsymbol{u}/\partial t - \mathfrak{A}(x, D_x)\boldsymbol{u} = \mathbf{0}$. Certain particular cases

Propositions 11.2 and 11.3 immediately imply the following assertion.

THEOREM 11.2. *The maximum norm principle is valid for solutions of system* (11.1.2) *in the class* $[\mathrm{C}^{(2,1)}(\Pi_T)]^m \cap [\mathrm{C_b}(\overline{\Pi}_T)]^m$ *if and only if:*

(i) *for all* $x \in \mathcal{D}$

$$\mathcal{A}_{jk}(x) = a_{jk}(x)I, \quad 1 \le j, k \le n,$$

where $((a_{jk}(x)))$ *is a positive-definite* $(n \times n)$*-matrix-valued function in* $\overline{\mathcal{D}}$*;*

(ii) *for all* $x \in \mathcal{D}$ *and for all vectors* $\boldsymbol{\xi}_j \in \mathbb{R}^m, \boldsymbol{\zeta} \in \mathbb{R}^m \backslash \{\mathbf{0}\}$, *with* $(\boldsymbol{\xi}_j, \nabla|\boldsymbol{\zeta}|) = 0,\ j = 1, \dots, n$, *the inequality*

$$\sum_{j,k=1}^{n} a_{jk}(x)(\mathcal{H}(|\boldsymbol{\zeta}|)\boldsymbol{\xi}_j, \boldsymbol{\xi}_k) + \sum_{j=1}^{n}(\mathcal{A}_j(x)\boldsymbol{\xi}_j, \nabla|\boldsymbol{\zeta}|) + (\mathcal{A}_0(x)\boldsymbol{\zeta}, \nabla|\boldsymbol{\zeta}|) \ge 0 \tag{11.5.1}$$

is valid.

Now, we consider particular cases of Theorem 11.2 on necessary and sufficient condition for validity of the maximum norm principle for solutions of system (11.1.2).

COROLLARY 11.1. *The maximum norm principle is valid for solutions of the system*

$$\frac{\partial \boldsymbol{u}}{\partial t} - \sum_{j,k=1}^{n} \mathcal{A}_{jk}(x)\frac{\partial^2 \boldsymbol{u}}{\partial x_j \partial x_k} + \sum_{j=1}^{n} \mathcal{A}_j(x)\frac{\partial \boldsymbol{u}}{\partial x_j} = \mathbf{0}$$

in the class $[\mathrm{C}^{(2,1)}(\Pi_T)]^m \cap [\mathrm{C_b}(\overline{\Pi}_T)]^m$ *if and only if, for all* $x \in \mathcal{D}$,

$$\mathcal{A}_{jk}(x) = a_{jk}(x)I, \quad \mathcal{A}_j(x) = a_j(x)I, \quad 1 \le j, k \le n,$$

where $((a_{jk}(x)))$ *is a positive-definite* $(n \times n)$*-matrix-valued function and* $a_1(x), \dots, a_n(x)$ *are scalar functions in* $\overline{\mathcal{D}}$.

PROOF. Putting $\mathcal{A}_0 = 0$ in (11.5.1) we get

$$\sum_{j,k=1}^{n} a_{jk}(x)(\mathcal{H}(|\boldsymbol{\zeta}|)\boldsymbol{\xi}_j, \boldsymbol{\xi}_k) + \sum_{j=1}^{n}(\mathcal{A}_j(x)\boldsymbol{\xi}_j, \nabla|\boldsymbol{\zeta}|) \ge 0,$$

which can be valid for all $x \in \mathcal{D}$ and for all vectors $\boldsymbol{\xi}_j \in \mathbb{R}^m, \boldsymbol{\zeta} \in \mathbb{R}^m\backslash\{\mathbf{0}\}$, with $(\boldsymbol{\xi}_j, \nabla|\boldsymbol{\zeta}|) = 0,\ j = 1, \dots, n$, only if $(\mathcal{A}_j(x)\boldsymbol{\xi}_j, \nabla|\boldsymbol{\zeta}|) = 0$, that is

$$(\boldsymbol{\xi}_j, \mathcal{A}_j^*(x)\nabla|\boldsymbol{\zeta}|) = 0.$$

Since $\boldsymbol{\xi}_j$ is an arbitrary vector, orthogonal to $\nabla|\boldsymbol{\zeta}|$, we arrive at the equalities

$$\mathcal{A}_j^*(x)\nabla|\boldsymbol{\zeta}| = \lambda_j(x, \zeta)\nabla|\boldsymbol{\zeta}|, \quad 1 \le j \le n, \tag{11.5.2}$$

where $\lambda_1(x, \zeta), \dots, \lambda_n(x, \zeta)$ are scalar functions. The scalarity of the matrices $\mathcal{A}_j^*(x)$, $1 \le j \le n$, can be obtained from (11.5.2) in the same way as the scalarity of $\mathcal{A}_{jk}^*(x, 0)$, $1 \le j, k \le n$, was deduced from (10.4.13) in Proposition 11.1. □

To formulate the next corollaries of Theorem 11.2 we introduce further notation.

Let $\mathcal{T}_\zeta = \{\boldsymbol{\xi} \in \mathbb{R}^m \ :\ (\nabla|\boldsymbol{\zeta}|,\ \boldsymbol{\xi}) = 0\}$ be the space of vectors lying in the tangent hyperplane to the sphere of the norm at the point $\zeta \in \mathbb{R}^m\backslash\{0\}$. Let further $\mathcal{Z} = \{\zeta \in \mathbb{R}^m \ :\ \mathcal{H}(|\boldsymbol{\zeta}|)\mathcal{T}_\zeta = 0\ \}$. By $\boldsymbol{\nu}(\zeta)$ we denote the unit outward normal to the sphere $\mathcal{S}^{m-1} = \{\zeta \in \mathbb{R}^m : |\boldsymbol{\zeta}| = 1\}$ at the point $\zeta \in \mathcal{S}^{m-1}$. Let $\mathcal{N} = \{\boldsymbol{\nu}(\zeta) : \zeta \in \mathcal{S}^{m-1} \bigcap \mathcal{Z}\}$. By $[\boldsymbol{v}_1, \dots, \boldsymbol{v}_m]$ we mean the $(m \times m)$-matrix whose columns are m-component vectors $\boldsymbol{v}_1, \dots, \boldsymbol{v}_m$.

PROPOSITION 11.4. *Let $\mathcal{Z} \neq \emptyset$. Then condition* (ii) *of Theorem* 11.2 *for the set $\mathcal{Z}$ has the following form:*

for all $x \in \mathcal{D}$ and for any $\boldsymbol{\zeta} \in \mathcal{Z}$ the equalities

$$\mathcal{A}_j^*(x)\nabla|\boldsymbol{\zeta}| = \mu_j(x, \zeta)\nabla|\boldsymbol{\zeta}|, \tag{11.5.3}$$

$1 \leq j \leq n$, hold, where $\mu_1(x, \zeta), \dots, \mu_n(x, \zeta)$ are scalar functions.

PROOF. Let x be an arbitrary fixed point of $\mathcal{D}$. Further, let $\boldsymbol{\xi} \in \mathcal{T}_{\boldsymbol{\zeta}}$, $\boldsymbol{\xi}_j = \delta_{jk}\boldsymbol{\xi}$, where $1 \leq j \leq n$ and k is a fixed element of the set $\{1, \dots, n\}$. Then, putting $\zeta \in \mathcal{Z}$ in (11.5.1), we get

$$(\mathcal{A}_k(x)\boldsymbol{\xi}, \nabla|\boldsymbol{\zeta}|) + (\mathcal{A}_0(x)\boldsymbol{\zeta}, \nabla|\boldsymbol{\zeta}|) \geq 0,$$

which can be valid for all $\boldsymbol{\xi} \in \mathcal{T}_{\boldsymbol{\zeta}}$ only provided $(\mathcal{A}_k(x)\boldsymbol{\xi}, \nabla|\boldsymbol{\zeta}|) = 0$, that is, $(\boldsymbol{\xi}, \mathcal{A}_k^*(x)\nabla|\boldsymbol{\zeta}|) = 0$.

Since the vector $\boldsymbol{\xi} \in \mathcal{T}_{\boldsymbol{\zeta}}$ is arbitrary, the last equality yields

$$\mathcal{A}_k^*(x)\nabla|\boldsymbol{\zeta}| = \mu_k(x, \zeta)\nabla|\boldsymbol{\zeta}|.$$

Taking into account that $x \in \mathcal{D}$, $\zeta \in \mathcal{Z}$ and k are arbitrary, we complete the proof. □

The following statement results from Proposition 11.4.

COROLLARY 11.2. *Let $\mathcal{N} \supset \{\boldsymbol{\nu}_1, \dots, \boldsymbol{\nu}_m\}$ and suppose the set $\{\boldsymbol{\nu}_1, \dots, \boldsymbol{\nu}_m\}$ forms a linear independent system. Then the maximum norm principle is valid for solutions of system* (11.1.2) *in the class $[\mathrm{C}^{(2,1)}(\Pi_T)]^m \cap [\mathrm{C_b}(\overline{\Pi}_T)]^m$ if and only if:*

(i) *for all $x \in \mathcal{D}$ the equalities*

$$\mathcal{A}_{jk}(x) = a_{jk}(x)I,$$

$$\mathcal{A}_j(x) = \left([\boldsymbol{\nu}_1, \dots, \boldsymbol{\nu}_m]^*\right)^{-1} \mathrm{diag}\,\{a_{1,j}(x), \dots, a_{m,j}(x)\}[\boldsymbol{\nu}_1, \dots, \boldsymbol{\nu}_m]^*, \tag{11.5.4}$$

$1 \leq j, k \leq n$, hold, where $((a_{jk}(x)))$ is a positive-definite $(n \times n)$-matrix-valued function and $a_{1,j}(x), \dots, a_{m,j}(x)$ are scalar functions in $\overline{\mathcal{D}}$;

(ii) *for all $x \in \mathcal{D}$ and for all vectors $\boldsymbol{\xi}_j \in \mathbb{R}^m, \boldsymbol{\zeta} \in \mathbb{R}^m \backslash \{\boldsymbol{0}\}$, with $(\boldsymbol{\xi}_j, \nabla|\boldsymbol{\zeta}|) = 0$, $j = 1, \dots, n$, the inequality*

$$\sum_{j,k=1}^{n} a_{jk}(x)(\mathcal{H}(|\boldsymbol{\zeta}|)\boldsymbol{\xi}_j, \boldsymbol{\xi}_k) + \sum_{j=1}^{n}(\mathcal{A}_j(x)\boldsymbol{\xi}_j, \nabla|\boldsymbol{\zeta}|) + (\mathcal{A}_0(x)\boldsymbol{\zeta}, \nabla|\boldsymbol{\zeta}|) \geq 0$$

is valid.

PROOF. Let $\zeta_i \in \mathcal{S}^{m-1} \bigcap \mathcal{Z}$, $1 \leq i \leq m$, such that $\boldsymbol{\nu}_1 = \boldsymbol{\nu}(\zeta_1), \dots, \boldsymbol{\nu}_m = \boldsymbol{\nu}(\zeta_m)$. Since the vectors $\nabla|\boldsymbol{\zeta}|$ and $\boldsymbol{\nu}(\zeta)$, $\zeta \in \mathcal{S}^{m-1}$, are linear dependent, from (11.5.3) it follows that

$$\mathcal{A}_j^*(x)\boldsymbol{\nu}_i = \mu_j(x, \zeta_i)\boldsymbol{\nu}_i,$$

where $1 \leq j \leq n$, $1 \leq i \leq m$. Consequently,

$$[\boldsymbol{\nu}_1, \dots, \boldsymbol{\nu}_m]^{-1}\mathcal{A}_j^*(x)[\boldsymbol{\nu}_1, \dots, \boldsymbol{\nu}_m] = \mathrm{diag}\,\{\mu_j(x, \zeta_1), \dots, \mu_j(x, \zeta_m)\}.$$

The last equality implies

$$\mathcal{A}_j(x) = ([\boldsymbol{\nu}_1, \dots, \boldsymbol{\nu}_m]^*)^{-1} \operatorname{diag}\{\mu_j(x, \zeta_1), \dots, \mu_j(x, \zeta_m)\}[\boldsymbol{\nu}_1, \dots, \boldsymbol{\nu}_m]^*.$$

Then, using the notation $a_{1,j}(x) = \mu_j(x, \zeta_1), \dots, a_{m,j}(x) = \mu_j(x, \zeta_m)$, we arrive at (11.5.4) which completes the proof. □

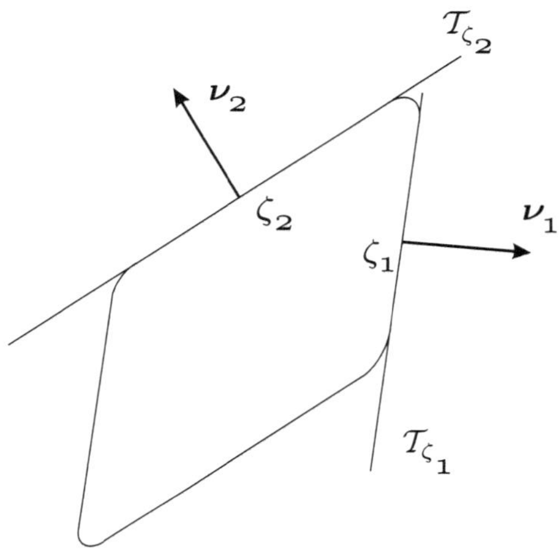

FIGURE 11.1. The unit ball of a norm in Corollary 11.2, $\mathcal{N} \supset \{\boldsymbol{\nu}_1, \boldsymbol{\nu}_2\}$

COROLLARY 11.3. *Let* $\mathcal{N} \supset \{\boldsymbol{\nu}_1, \dots, \boldsymbol{\nu}_{m+1}\}$ *and suppose any proper subset of* $\{\boldsymbol{\nu}_1, \dots, \boldsymbol{\nu}_{m+1}\}$ *forms a linear independent system. Then the maximum norm principle is valid for solutions of system* (11.1.2) *in the class* $[\mathrm{C}^{(2,1)}(\Pi_T)]^m \cap [\mathrm{C_b}(\overline{\Pi}_T)]^m$ *if and only if:*

(i) *for all* $x \in \mathcal{D}$

$$\mathcal{A}_{jk}(x) = a_{jk}(x)I, \;\; \mathcal{A}_j(x) = a_j(x)I, \;\; 1 \le j, k \le n,$$

where $((a_{jk}(x)))$ *is a positive-definite* $(n \times n)$*-matrix-valued function and* $a_1(x), \dots, a_m(x)$ *are scalar functions in* $\overline{\mathcal{D}}$*;*

(ii) *for all* $x \in \mathcal{D}$ *and for any* $\boldsymbol{\zeta} \in \mathbb{R}^m \backslash \{\mathbf{0}\}$ *the inequality*

$$(\mathcal{A}_0(x)\boldsymbol{\zeta}, \nabla|\boldsymbol{\zeta}|) \ge 0$$

is valid.

PROOF. Assume that $x \in \mathcal{D}$ and j, $1 \le j \le n$ are fixed. By A^*, for brevity, we denote the matrix $\mathcal{A}_j^*(x)$. Further, let $M = \{\, \boldsymbol{\nu}_1, \dots, \boldsymbol{\nu}_m \,\}$. According to Proposition 11.4, the elements of M are eigenvectors of the $(m \times m)$-matrix A^*. Therefore, there exists a decomposition of the set M into subsets $M_1, \dots, M_s$, $1 \le s \le m$, such that $M_1 \subset L_1, \dots, M_s \subset L_s$, where $L_1, \dots, L_s$ is the family of eigenspaces of the matrix A^*.

For the $(m \times m)$-matrix A^* with m linear independent eigenvectors we have

$$\sum_{i=1}^{s} \dim L_i = m \tag{11.5.5}$$

(see, e.g. Lankaster [**Lank**]).

Let m_i be the number of elements of the set M_i. Suppose that $s > 1$. Then $m_i < m$ for all $1 \leq i \leq s$. Since $\boldsymbol{\nu}_{m+1}$ is also eigenvector of the matrix A^*, there exists k, $1 \leq k \leq s$ such that $\boldsymbol{\nu}_{m+1} \in L_k$. By condition of the corollary, the vectors of the set $\{M_k,\ \boldsymbol{\nu}_{m+1}\}$ are linear independent. Hence, $\dim L_k \geq m_k + 1$. From this it follows that

$$\sum_{i=1}^{s} \dim L_i \geq m + 1$$

which contradicts equality (11.5.5). This means that $s = 1$, i.e. that $\mathbb{R}^m$ is the eigenspace of the matrix A^*. Since geometric and algebraic multiplicities of the eigenvalues of the matrix (11.5.4) coincide, then $a_{1,j}(x) = \cdots = a_{m,j}(x)$. Putting $a_j(x) = a_{1,j}(x) = \cdots = a_{m,j}(x)$ in (11.5.4), we arrive at the equalities $\mathcal{A}_j(x) = a_j(x)I$. From this and condition (ii) of Corollary 11.2 we obtain the inequality

$$\sum_{j,k=1}^{n} a_{jk}(x)(\mathcal{H}(|\boldsymbol{\zeta}|)\boldsymbol{\xi}_j, \boldsymbol{\xi}_k) + (\mathcal{A}_0(x)\boldsymbol{\zeta}, \nabla|\boldsymbol{\zeta}|) \geq 0,$$

which can be valid for all $\boldsymbol{\xi}_j \in \mathbb{R}^m$, $\boldsymbol{\zeta} \in \mathbb{R}^m\backslash\{\mathbf{0}\}$, $j = 1, \ldots, n$, with $(\boldsymbol{\xi}_j, \nabla|\boldsymbol{\zeta}|) = 0$, only provided $(\mathcal{A}_0(x)\boldsymbol{\zeta}, \nabla|\boldsymbol{\zeta}|) \geq 0$.

Taking into account that $x \in \mathcal{D}$ and the index $j \in \{1, \ldots, n\}$ for the matrix $A_j^*(x)$ are arbitrary, we complete the proof. □

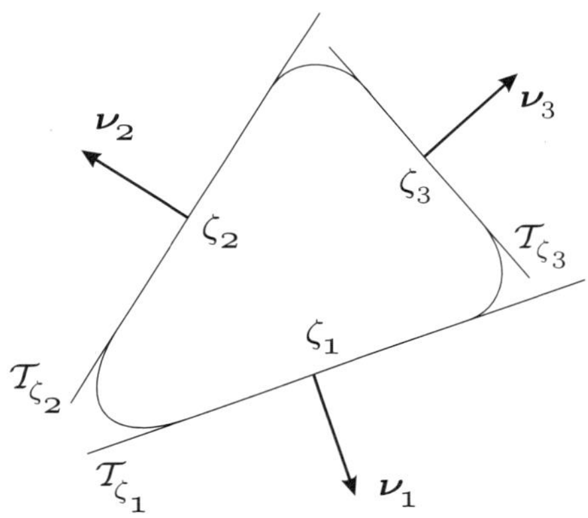

FIGURE 11.2. The unit ball of a norm subject to the conditions of Corollary 11.3, $\mathcal{N} \supset \{\boldsymbol{\nu}_1, \boldsymbol{\nu}_2, \boldsymbol{\nu}_3\}$

11.6. Example: criterion for validity of the maximum p-norm principle, $2 < p < \infty$

We now give an application of Corollary 11.2 to the p-norm in $\mathbb{R}^m$, that is to the norm

$$\|\boldsymbol{\zeta}\|_p = \left(\sum_{j=1}^{m} |\zeta_j|^p \right)^{1/p},$$

where $\boldsymbol{\zeta} = (\zeta_1, \ldots, \zeta_m) \in \mathbb{R}^m$.

In the case $p \in [2, \infty)$ the p-norm is twice continuously differentiable, which is verified by calculation of the Hessian of $\|\boldsymbol{\zeta}\|_p$:

$$(11.6.1)\quad \mathcal{H}_{jk}(\|\boldsymbol{\zeta}\|_p) = (p-1)\left(\delta_{jk}\|\boldsymbol{\zeta}\|_p^{1-p}|\zeta_j|^{p-2} - \|\boldsymbol{\zeta}\|_p^{-1}\frac{\zeta_j|\zeta_j|^{p-2}}{\|\boldsymbol{\zeta}\|_p^{p-1}}\cdot\frac{\zeta_k|\zeta_k|^{p-2}}{\|\boldsymbol{\zeta}\|_p^{p-1}}\right).$$

COROLLARY 11.4. *The maximum p-norm principle with $p \in (2, \infty)$ is valid for solutions of system* (11.1.2) *in the class* $[\mathrm{C}^{(2,1)}(\Pi_T)]^m \cap [\mathrm{C_b}(\overline{\Pi}_T)]^m$ *if and only if:*

(i) *for all $x \in \mathcal{D}$*

$$\mathcal{A}_{jk}(x) = a_{jk}(x)I, \quad \mathcal{A}_j(x) = \mathrm{diag}\,\{a_{1,j}(x), \dots, a_{m,j}(x)\}, \quad 1 \le j, k \le n,$$

where $((a_{jk}(x)))$ is a positive-definite $(n\times n)$-matrix-valued function and $a_{1,j}(x), \dots, a_{m,j}(x)$ are scalar functions in $\overline{\mathcal{D}}$;

(ii) *for all $x \in \mathcal{D}$ and for all $\boldsymbol{\xi}_j,\ \boldsymbol{\zeta} \in \mathbb{R}^m$, with $(\boldsymbol{\xi}_j, \Lambda_p\boldsymbol{\zeta}) = 0,\ j = 1, \dots, n$, the inequality*

$$(p-1)\sum_{j,k=1}^{n} a_{jk}(x)\left(\Lambda_p\boldsymbol{\xi}_j, \boldsymbol{\xi}_k\right) + \sum_{j=1}^{n}\left(\mathcal{A}_j(x)\boldsymbol{\xi}_j, \Lambda_p\boldsymbol{\zeta}\right) + (\mathcal{A}_0(x)\boldsymbol{\zeta}, \Lambda_p\boldsymbol{\zeta}) \ge 0$$

is valid, where $\Lambda_p = \mathrm{diag}\,\{\ |\zeta_1|^{p-2}, \dots, |\zeta_m|^{p-2}\ \}$.

PROOF. We have

$$\nabla\|\boldsymbol{\zeta}\|_p = \|\boldsymbol{\zeta}\|_p^{1-p}\left(\zeta_1|\zeta_1|^{p-2}\boldsymbol{e}_1 + \dots + \zeta_m|\zeta_m|^{p-2}\boldsymbol{e}_m\right),$$

where $1 \le j, k \le m$. Then, by (11.6.1),

$$\mathcal{H}(\|\boldsymbol{\zeta}\|_p)\boldsymbol{\xi} = (p-1)\|\boldsymbol{\zeta}\|_p^{1-p}\{\xi_1|\zeta_1|^{p-2}, \dots, \xi_m|\zeta_m|^{p-2}\}$$

for $(\boldsymbol{\xi}, \nabla|\boldsymbol{\zeta}|_p) = 0$. Since $\mathcal{N} \supset \{\boldsymbol{e}_1, \dots, \boldsymbol{e}_m,\}$, we put $\boldsymbol{\nu}_1 = \boldsymbol{e}_1, \dots, \boldsymbol{\nu}_m = \boldsymbol{e}_m$. Therefore, $[\boldsymbol{\nu}_1, \dots, \boldsymbol{\nu}_m] = I$. Applying Corollary 11.2, we complete the proof. □

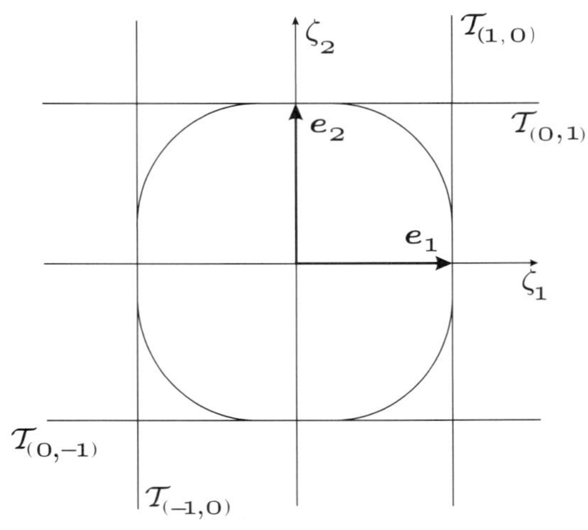

FIGURE 11.3. The unit ball of the p-norm for $m = 2$, $2 < p < \infty$, $\mathcal{N} \supset \{\boldsymbol{e}_1, \boldsymbol{e}_2\}$

11.7. Comments to Chapter 11

It is well-known that solutions of the second order uniformly parabolic equation

$$\frac{\partial u}{\partial t} - \sum_{j,k=1}^{n} a_{jk}(x,t)\frac{\partial^2 u}{\partial x_j \partial x_k} + \sum_{j=1}^{n} a_j(x,t)\frac{\partial u}{\partial x_j} + a_0(x,t)u = 0, \tag{11.7.1}$$

$a_0(x,t) \geq 0$, with real coefficients in the cylinder

$$Q_T = \{(x,t) : x \in \Omega,\ 0 < t \leq T\}, \quad \Omega \subset \mathbb{R}^n,$$

satisfy the maximum principle. In other words, for any solution of (11.7.1), where $a_0(x,t) \geq 0$, which is continuous in $\overline{Q}_T$ together with the first derivative with respect to t and derivatives with respect to x up to order two in Q_T, the inequalities

$$\min\left\{0, \min_{(y,\tau)\in\overline{\Gamma}_T} u(y,\tau)\right\} \leq u(x,t) \leq \max\left\{0, \max_{(y,\tau)\in\overline{\Gamma}_T} u(y,\tau)\right\}$$

are valid. Here $(x,t) \in \overline{Q}_T$ and $\Gamma_T = \{(y,\tau) \in \partial Q_T : 0 \leq \tau < T\}$. This means that if the initial-boundary conditions belong to the interval $[\alpha, \beta]$ containing the origin then the same holds for the solution of equation (11.7.1), i.e. any closed interval containing the origin is an invariant set of the equation.

Bibliography

[Ag1] S. Agmon, *Multiple layer potentials and the Dirichlet problem for higher order elliptic equations in the plane, I*, Comm. Pure Appl. Math., **10** (1957), 179–239.

[Ag2] S. Agmon, *Maximum theorems for solutions of higher order elliptic equations*, Bull. Amer. Math. Soc., **66** (1960), 77–80.

[ADN1] S. Agmon, A. Douglis, and L. Nirenberg, *Estimates near the boundary for solutions of elliptic partial differential equations satisfying general boundary conditions, I.*, Comm. Pure Appl. Math., **12** (1959), 623–727.

[ADN2] S. Agmon, A. Douglis, and L. Nirenberg, *Estimates near the boundary for solutions of elliptic partial differential equations satisfying general boundary conditions, II.*, Comm. Pure Appl. Math., **17** (1964), 35–92.

[AAD] L. Aizenberg, A. Aytuna, and P. Djakov, *An abstract approach to Bohr's phenomenon*, Proc. Amer. Math. Soc., **128**:9 (2000), 2611–2619.

[Alb1] G. Albinus, *Uniform estimates in maximum norms and uniqueness of solutions of the Dirichlet problem*, Math. Nachr., **89** (1979), 9–16.

[Alb2] G. Albinus, *Estimates of Miranda-Agmon type in plane domains with corners and existence theorem*, in: Constructive Function Theory '81, Sofia, 197–204.

[Ale1] A.D. Alexandrov, *Uniqueness conditions and estimates for the solution of the Dirichlet problem*, Vestnik LGU, **18** (1963), 5–29.

[Ale2] A.D. Alexandrov, *Majorization of second order linear equations*, Vestnik LGU, **21** (1966), 5–25.

[Ali1] N. Alikakos, *Remarks on invariance in reaction-diffusion equations*, Nonlinear Analysis. Theory, Methods & Applications, **5**:6 (1981), 593–614.

[Ali2] N. Alikakos, *Quantitative maximum principle and strongly coupled gradient-like reaction-diffusion systems*, Proc. Royal Soc. Edinburg, Sect. A, **94**:3-4 (1983), 265–286.

[AB] C.D. Aliprantis and K.C. Border, *Infinite Dimensional Analysis*, 3rd Ed., Springer, New York etc., 2006.

[Am] H. Amann, *Invariant sets and existence theorems for semilinear parabolic and elliptic systems*, J. Math. Anal. Appl., **65** (1978), 432–467.

[AKK] T.S. Angell, R.E. Kleinman, and J. Král, *Layer potentials on boundaries with corners and edges*, Čas. Pěst. Mat., **113** (1988), 387–402.

[At] F.V. Atkinson, *Normal solvability of linear equations in normed spaces*, Mat. Sb., Nov. Ser., **28**(70) (1951), 3–14.

[ABBO] P. Auscher, L. Barthélemy, Ph. Bénilan, and E. M. Ouhabaz, *Absence de la L^∞-contractivité pour les semi-groupes associés aux opérateurs elliptiques complexes sous forme divergence*, Potential Analysis, **12** (2000), 169–189.

[Bak1] I.J. Bakelman, *The Dirichlet problem for equations of Monge-Ampere type and their n-dimensional analogues*, Dokl. Akad. Nauk SSSR, **126** (1959), 923–926.

[Bak2] I.J. Bakelman, *The first boundary value problem for nonlinear elliptic equations*, The Second Doctor Dissertation, Leningrad State Pedagogical Institute (University), Leningrad, 1959.

[Bak3] I.J. Bakelman, *Convex Analysis and Nonlinear Geometric Elliptic Equations*, Springer-Verlag, 1994.

[Bat] P.W. Bates, *Containment for weakly coupled parabolic systems*, Houston J. Math., **11**:2 (1985), 151–158.

[BaSh] B.V. Bazaliĭ and V.Yu. Shelepov, *On the spectrum of the double layer potential on a curve of bounded rotation*, In: Boundary Value Problems for Differential Equations, Collect. Sci. Works, Naukova Dumka, Kiev, 1980, 13–30 (Russian).

[BCM] J.W. Bebernes, K.N. Chueh, and W. Fulks, *Some applications of invariance for parabolic systems*, Indiana Univ. Math. J., **28**:2 (1979), 269–277.

[BeSch] J.W. Bebernes and K. Schmitt, *Invariant sets and the Hukuhara-Kneser property for systems of parabolic partial differential equations*, Rocky Mountain J. Math., **7** (1977), 557–567.

[BB] E.F. Beckenbach and R. Bellman, *Inequalities*, Springer-Verlag, Berlin-Göttingen-Heidelberg, 1961.

[Bi1] A.V. Bitsadze, *On elliptic systems of second order partial differential equations*, Dokl. Akad. Nauk SSSR, **112** (1957), 983–986 (Russian).

[Bi2] A.V. Bitsadze, *Boundary value problems for second order elliptic equations*, Moskva, Nauka, 1966; English translation: North-Holland Publ. Company, Amsterdam, 1968.

[Bo1] E. Borel, *Démonstration élémentaire d'un théorème de M. Picard sur les fonctions entières*, C.R. Acad. Sci., **122** (1896), 1045–1048.

[Bo2] E. Borel, *Méthodes et Problèmes de Théorie des Fonctions*, Gauthier-Villars, Paris, 1922.

[BuMaz] Yu.D. Burago and V.G. Maz'ya, *Potential Theory and the Function Theory for Irregular Regions*, Zap. Nauchn. Sem. LOMI, **3** (1967); English translation: Sem. in Mathematics, V.A. Steklov Math. Inst., Leningrad, **3**, Consultants Bureau, New York, 1969.

[Burc] R.B. Burckel, *An Introduction to Classical Complex Analysis, V. 1*, Academic Press, New York-San Francisco, 1979.

[Burg] B. Burgeth, *A Schwarz lemma for harmonic and hyperbolic-harmonic functions in higher dimensions*, Manuscripta Math., **77** (1992), 283–296.

[Bus] H. Busemann, *Convex Surfaces*, Interscience, New York, 1958.

[Can] P. Cannarsa, *On a maximum principle for elliptic systems with constant coefficients*, Rend. Sem. Mat. Univ. Padova, **64** (1981), 77–84.

[Car] T. Carleman, *Über das Neumann-Poincaresche Problem für ein Gebiet mit Ecken, Inaugural Dissertation*, Tr. Almqvist and Wiksell, Uppsala, 1916.

[Ca] M.L. Cartwright, *Integral Functions*, Oxford University Press, Oxford, 1964.

[CG] R. Chichinadze and T. Gegelia, *Boundary value problems of mechanics of continuum media for a sphere*, Memoirs on Dif. Eq. and Math. Physics, **7** (1996), 1–222.

[CCS] K.N. Chueh, C.C. Conley, and J.A. Smoller, *Positively invariant regions for systems of nonlinear diffusion equations*, Indiana Univ. Math. J., **26** (1977), 373–391.

[Ci] A. Cialdea, *Criteria for the L^p-dissipativity of partial differential operators*, Operator Theory: Advances and Applications, Birkhäuser, **193** (2009), 41–56.

[CM1] A. Cialdea and V. Maz'ya, *Criterion for the L^p-dissipativity of second-order differential operators with complex coefficients*, J. Math. Pures Appl., **84** (2005), 1067–1100.

[CM2] A. Cialdea and V. Maz'ya, *Criteria for the L^p-dissipativity of systems of second order differential equations*, Ricerche Math., **55** (2006), 233–265.

[Co] D.L. Cohn, *Measure Theory*, Birkhäuser, Boston-Basel-Berlin, 1997.

[CHS] E. Conway, D. Hoff, and J. Smoller, *Large time behavior of solutions of systems of nonlinear reaction-diffusion equations*, SIAM J. Appl. Math., **35** (1) (1978), 1–16.

[CS] C. Cosner and P.W. Schaefer, *On the development of functionals which satisfy a maximum principle*, Appl. Analysis, **26** (1987), 45–60.

[Cos] M. Costabel, *Boundary integral operators on Lipschitz domains: Elementary results*, SIAM J. Math. Anal., **19** (1988), 613-626.

[DeG] E. De Giorgi, *Nouvi teoremi relativi alle misure $(r-1)$-dimensionali in uno spazio ad r dimensioni*, Richerchie Mat., **4** (1955), 95–113.

[DU] J. Diestel and J.J. Uhl, Jr., *Vector Measures*, Mathematical Surveys and Monographs, **15**, Amer. Math. Soc., Providence, RI, 1977.

[DN] A. Douglis and L. Nirenberg, *Interior estimates for elliptic systems of partial differential equations*, Comm. Pure Appl. Math., **8** (1955), 503–538.

[DS] N. Dunford and J.T. Schwartz, *Linear Operators. Part I: General Theory*, John Wiley and Sons, New York-Chichester-Brisbane-Toronto-Singapore, 1988.

[Ea] S. Earnshaw, *On the nature of the molecular forces which regulate the constitution of the luminiferous ether*, Cambridge Phil. Soc. Trans., **7** (1839), 97–112.

[Ed] R.E. Edwards, *Functional Analysis. Theory and Applications*, Holt, Rinehart and Winston, New York-Chikago-San Francisco-Toronto-London, 1965.

[Ei2] S.D. Eidel'man, *On a connection between the fundamental matrices of solutions od parabolic and elliptic systems*, Mat. sb., **35**(77) (1954), 57–72.

[Ei1] S.D. Eidel'man, *Parabolic Systems*, North-Holland and Noordhoff, Amsterdam, 1969.

[El] J. Elschner, *The double layer potential operator over polyhedral domains. I: Solvability in weighted Sobolev spaces*, Appl. Analysis, **45** (1992), 117–134.

[ERSD] K.-H. Elster, R. Reinhardt, M. Schäuble, and G. Donath, *Einfürung in die Nichtlineare Optimierung*, BSB B.G. Teubner Verlagsgesellschaft, Leipzig, 1977.

[Fa] E.B. Fabes, *Layer potential methods for boundary value problems on Lipschitz domains*, Potential theory. Survey and problems, Proceedings, Prague, 1987, J. Král, J. Lukeš, I. Netuka and J. Veselý (eds.), Lecture Notes in Mathematics, **1344**, Springer, Berlin-Heidelberg-New York, 1988.

[FJL] E.B. Fabes, M. Jodeit, and J.E. Lewis, *Double layer potentials for domains with cirners and edges*, Indiana Univ. Math. J., **26** (1977), 95–114.

[FSS] E.B. Fabes, M. Sand, and J.K. Seo, *The spectral radius of the classical layer potentials on convex domain*, Proc. IMA Conference Chicago 1990; IMA Vol. Math. Appl. **42**, Springer-Verlag, New York, 1992, pp. 129–137.

[Fe] H. Federer, *A note on the Gauss-Green theorem*, Proc. Amer. Math. Soc., **9** (1958), 447–451.

[Fi] G. Fichera, *Il teorema del massimo modulo per l'equazione dell'elastostatica tridimensionale*, Arch. Rat. Mech. Anal., **7** (1961), 373–387.

[FM] D.G. de Figueiredo and E. Mitidieri, *Maximum principles for linear elliptic systems*, Rend. Istit. Mat. Univ. Trieste, **22**:1-2 (1992), 36–66.

[Fr] L. E. Fraenkel, *An introduction to maximum principles and symmetry in elliptic problems*, Cambridge Tracts in Mathematics, **128** (2000), Cambridge Univ. Press, Cambridge.

[Fri] A. Friedman, *Partial Differential Equations of Parabolic Type*, Prentice-Hall Inc., Englewood Cliffs, N.J., 1964.

[Ga] C.F. Gauss, *Allgemeine Theorie des Erdmagnetismus*, Beobachten des magnetishen Vereins im Jahre 1838, Leipzig, 1839; Also in Werke (Collected Works), **5**, pp. 129, 373–387.

[Ge] P. Germain, *Course de Mécanique des Milieux Continus, t. I. Théorie Générale*, Masson et C^{je}, Éditeurs, Paris, 1973.

[GT] D. Gilbarg and N.S. Trudinger, *Elliptic partial differential equations of second order*, Springer Verlag, Berlin-Heidelberg-New York-Tokyo, 1983.

[GK1] I.Ts. Gohberg and M.G. Krein, *The fundamentals on defect numbers, and indices of linear operators*, Usp. Mat. Nauk, **12**:2(74) (1957), 43–118; English transl.: Transl., II Ser., Amer. Math. Soc., **10** (1960), 185–264.

[GK2] I.Ts. Gohberg and M.G. Krein, *Systems of integral equations on a half-line with kernels depending on the difference of arguments*, Usp. Mat. Nauk, **13**:2(80) (1958), 3–72; English transl.: Transl., II Ser., Amer. Math. Soc., **14** (1960), 217–287.

[GM] I.Ts. Gohberg and A.S. Marcus, *Some remarks on topologically equivalent norms*, Izv. Mold. Fil. Akad. Nauk SSSR, **10**:76 (1960), 91–94 (Russian).

[Gr] N.V. Grachev, *Representation and estimates for inverse operators of the potential theory integral equations in a polyhedron*, Proc Int. Conf. on Potential Theory, Nagoya, Masanori Kishi, ed., 1990, pp. 201–206.

[GM1] N.V. Grachev and V.G. Maz'ya, *On the Fredholm radius for operators of the double layer potential type on piecewise smooth boundaries*, Vestnik Leningr. Univ., **19**:4 (1986), 60–64 (Russian).

[GM2] N.V. Grachev and V.G. Maz'ya, *Solvability of a boundary integral equation on a polyhedron*, Report LiTH-MAT-R-91-50, Linköping Univ., Sweden, 1991.

[GR] I.S. Gradshtein and I.M. Ryzhik; A. Jeffrey, editor, *Table of Integrals, Series and Products*, Fifth edition, Academic Press, New York, 1994.

[Had] J. Hadamard, *Sur les fonctions entières de la forme* $e^{G(X)}$, C.R. Acad. Sci., **114** (1892), 1053–1055.

[Han] O. Hansen, *On the essential norm of the double layer potential on polyherdral domains and the stability of the collocation method*, J. Integral Eq. and Applications, **13** (2001), 2007–235.

[HP] G. Hile and M.H. Protter, *Maximum principles for a class of first-order elliptic systems*, J. Diff. Equations, **24** (1977), 136–151.

[HS] G. Hile and A. Stanoyevitch, *Gradient bounds for harmonic functions Lipschitz on the boundary*, Appl. Anal., **73**:1-2 (1999), 101–113.

[Hon] C.W. Hong, *Necessary and sufficient condition for a class of generalized maximum principles*, Acta Math. Sinica, **26**:3 (1983), 307–321. (chinese).

[Hop] E. Hopf, *Elementare Bemerkungen über die Lösungen partieller Differrentialgleichungen zweiter Ordnung vom elliptischen Typus*, Sitzungsber. der Preuss. Akad. der Wissenschaften, Berlin, **19** (1927), 147–152.

[Hör] L. Hörmander, *Linear Partial Differential Operators*, Springer-Verlag, Berlin-Göttingen-Heidelberg, 1963.

[Ing] A. E. Ingham, *The Distribution of Prime Numbers*, Cambridge Univ. Press, 1932.

[Je] J.L.W.V. Jensen, *Investigation of a class of fundamental inequalities in the theory of analytic functions*, Ann. of Math., **21** (1919-1920), 1–29.

[Ka] S. Kakutani, *Concrete representations of abstract (M)-spaces. A characterization of the spaces of continuous functions*, Ann. Math. (2), **42** (1942), 994–1034.

[KK1] L.I. Kamynin and B.N. Khimchenko, *On the weak maximum principle for a second order elliptic system*, Izv. Ross. Akad. Nauk, Ser. Mat., **59**:5 (1995), 73–84 (Russian); English. transl. in Izv. Math., **59**:5 (1995), 949–961.

[KK2] L.I. Kamynin and B.N. Khimchenko, *Necessary and sufficient conditions for satisfying the weak extremum principle for second order elliptic systems*, Sibirsk. Mat. Zh., **37**:6 (1996), 1314–1334 (Russian); English. transl. in Siberian Math. J., **37**:6 (1996), 1153–1170.

[KK3] L.I. Kamynin and B.N. Khimchenko, *On a weak (algebraic) extremum principle for a second order parabolic system*, Izv. Ross. Akad. Nauk, Ser. Mat., **61**:5 (1997), 35–62 (Russian); English. transl. in Izv. Math., **61**:5 (1997), 933-959.

[KA] L.V. Kantorovich and G.P. Akilov, *Functional Analysis*, 2nd ed., Pergamon, 1982.

[Kh] D. Khavinson, *An extremal problem for harmonic functions in the ball*, Canad. Math. Bull., **35**:2 (1992), 218–220.

[KO] M. Kimura and K. Otsuka, *Some remarks on the positivity of fundamental solutions for certain parabolic equations with constant coefficients*, J. Math. Kioto Univ., **28** (1988), 87–90.

[KMR1] V.A. Kozlov, V.G. Maz'ya, and J. Rossmann, *Elliptic Boundary Value Problems in Domains with Point Singularities*, Mathematical Surveys and Monographs, **52**, Amer. Math. Soc., Providence, RI, 1997.

[KMR2] V.A. Kozlov, V.G. Maz'ya, and J. Rossmann, *Spectral Problems Associated with Corner Singularities of Solutions to Elliptic Equations*, Mathematical Surveys and Monographs, **85**, Amer. Math. Soc., Providence, RI, 2000.

[Kra1] J. Král, *The Fredholm method in potential theory*, Trans. Amer. Math. Soc., **125**(3) (1966), 511-547.

[Kra2] J. Král, *Integral Operators in Potential Theory,* Lect. Notes in Math., **823**, Springer-Verlag, Berlin-Heidelberg-New York, 1980.

[KraMe1] J. Král and D. Medková, *Essential norms of a potential theoretic boundary integral operator in L^1*, Acad. Sciences of the Czech Republic, Math. Inst., preprint **123** (1997), 1–21.

[KraMe2] J. Král and D. Medková, *Essential norms of the integral operator corresponding to the Neumann problem for the Laplace equation*, Mathematical Aspects of Boundary Element Methods (M. Bonet et al. eds.), Chapman & Hall/CRC Res. Notes Math., **414** (2000), pp. 215–226.

[KraMe3] J. Král and D. Medková, *Essential norms of the Neumann operator of the arithmetic mean*, Mathematica Bohemica, **126** (2001), 669–690.

[KW1] J. Král and W.L. Wendland, *Some examples concerning applicability of the Fredholm-Radon method in potential theory,* Aplikace Matematiky, **31** (1986), 293–298.

[KW2] J. Král and W.L. Wendland, *On the applicability of the Fredholm-Radon method in potential theory and the panel method,* Panel methods in fluid mechanics with emphasis in aero-dynamics, (J. Ballmann, R. Appler, W. Hackbush, eds.), Notes on Numerical Fluid Mechanics, **21** (1988), Vieweg, Braunschweig, 120–136.

[Kras1] Ju.P. Krasovskiĭ, *Isolation of singularities of the Green's function*, Izv. AN SSSR, **31** (1967), 977–1010; English transl.: Math. USSR Izv., **1**, 935–966.

[Kras2] Ju.P. Krasovskiĭ, *Growth estimates for derivatives of solutions of homogeneous elliptic equations near the boundary*, Dokl. AN SSSR, **184** (1969), 534–537; English transl.: Soviet Math. Dokl., **10**, 120-124.

[Krat1] W. Kratz, *On the maximum modulus theorem for Stokes functions*, Appl. Anal., **58** (1995), 293–302.

[Krat2] W. Kratz, *An extremal problem for Stokes functions in the plane*, Analysis, **17** (1997), 219–225.

[Krat3] W. Kratz, *The maximum modulus theorem for the Stokes system in a ball*, Math. Zeitschrift, **226** (1997), 389–403.

[Krat4] W. Kratz, *An extremal problem related to the maximum modulus theorem for Stokes functions* J. Anal. and its Appl., **17**:3 (1998), 559–613.

[Kr1] G.I. Kresin, *On the exact constant in the inequality for a norm of solutions of the Lamé system in a half-space*, Functional Differential Equations, **1** (1993), 132–135.

[Kr2] G. Kresin, *Some sharp estimates for the gradient of bounded or semibounded harmonic functions*, Functional Differential Equations, **16**:2-3-4 (2009), 435–446.

[KM1] G.I. Kresin and V.G. Maz'ya, *On the maximum of displacements in the visco-elastic half-space (three parametrical model)*, Vestnik Leningr. Univ., 22 (1985), 47–51 (Russian).

[KM2] G.I. Kresin and V.G. Maz'ya, *On the exact constant in the inequality of Miranda-Agmon type for solutions of elliptic equations*, Izvestia Vys. Uch. Zaved., Matem. Ser., **5** (1988), 41-50 (Russian); English transl.: Izvestia VUZ. Matematika, **32**:5 (1988), 49–59.

[KM3] G.I. Kresin and V.G. Maz'ya, *Criteria for validity of the maximum modulus principle for solutions of linear strongly elliptic systems*, Potential Analysis, **2** (1993), 73–99.

[KM4] G.I. Kresin and V.G. Maz'ya, *Criteria for validity of the maximum modulus principle for solutions of linear parabolic systems*, Ark. Math., **32** (1994), 121–155.

[KM5] G.I. Kresin and V.G. Maz'ya, *The norm and the essential norm of the double layer elastic and hydrodynamics potentials in the space of continuous functions*, Math. Meth. in Appl. Sc., **18** (1995), 1095–1131.

[KM6] G.I. Kresin and V.G. Maz'ya, *On the maximum modulus principle for linear parabolic systems with zero boundary data*, Functional Differential Equations, **5**:1-2 (1998), 165–181.

[KM7] G.I. Kresin and V.G. Maz'ya, *On the maximum principle with respect to smooth norms for linear strongly coupled parabolic systems*, Functional Differential Equations, **5**:3-4 (1998), 349–376.

[KM8] G.I. Kresin and V.G. Maz'ya, *Criteria for validity of the maximum norm principle for parabolic systems*, Poten. Anal., **10** (1999), 243–272.

[KM9] G.I. Kresin and V.G. Maz'ya, *Best constants in the Miranda-Agmon inequalities for solutions of elliptic systems and the classical maximum modulus principle for fluid and ellastic half-spaces*, Appl. Anal., **82**:2 (2003), 157–185.

[KM10] G. Kresin and V. Maz'ya, *Sharp pointwise estimates for solutions of strongly elliptic second order systems with boundary data from L^p*, Appl. Anal., **86**:7 (2007), 783–805.

[KM11] G. Kresin and V. Maz'ya, *Sharp Real-Part Theorems. A Unified Approach*, Lect. Notes in Math., **1903**, Springer-Verlag, Berlin-Heidelberg-New York, 2007.

[KM12] G. Kresin and V. Maz'ya, *Optimal estimates for the gradient of harmonic functions in the multidimensional half-space*, Discrete and Continuous Dynamical Systems, series B, **28**:2 (2010), 425–440.

[KM13] G. Kresin and V. Maz'ya, *Sharp pointwise estimates for directional derivatives of harmonic functions in a multidimensional ball*, J. Math. Sc. (New York), **169**:2 (2010), 167–187.

[KM14] G. Kresin and V. Maz'ya, *Sharp real-part theorems in the upper half-plane and similar estimates for harmonic functions*, J. Math. Sc. (New York), **179**:1 (2011), 144–163.

[KM15] G. Kresin and V. Maz'ya, *Sharp real-part theorems for high order derivatives*, J. Math. Sc. (New York), **181**:2 (2012), 107–125.

[Ku] H.J. Kuiper, *Invariant sets for nonlinear elliptic and parabolic systems*, SIAM J. Math. Anal., **11**:6 (1980), 1075–1103.

[KGBB] V.D. Kupradze, T.G. Gegelia, M.O. Basheleishvili, and T.V. Burchuladze, *Three-dimensional Problems of the Mathematical Theory of Elasticity and Thermoelasticity,* North-Holland, Amsterdam, 1979.

[Lad] O.A. Ladyzhenskaya, *The Mathematical Theory of Viscous Incompressible Flow*, Gordon and Breach, New York, 1963.

[LSU] O.A. Ladyzhenskaya, V.A. Solonnikov, and N.N. Uraltseva, *Linear and Quasilinear Equations of Parabolic Type*, Transl. of Math. Monographs, **23**, Amer. Math. Soc., Providence, RI, 1968.

[Land1] E. Landau, *Über den Picardschen Satz,* Vierteljahrschr. Naturforsch. Gesell. Zürich, **51** (1906), 252–318.

[Land2] E. Landau, *Beiträge zur analytischen Zahlentheorie*, Rend. Circ. Mat. Palermo, **26** (1908), 169–302.

[LL1] L.D. Landau and E.M. Lifshitz, *Fluid Mechanics, Sec. ed., Course of Theoretical Physics*, **6**, Butterworth-Heinemann, 1987.

[LL2] L.D. Landau and E.M. Lifshitz, *Theory of Elasticity, Third. ed., Course of Theoretical Physics*, **7**, Butterworth-Heinemann, 1986.

[Lan] M. Langer, *L^p-contractivity of semigroups generated by linear partial differential operators*, In: The Maz'ya Anniversary Collection, Vol.1: On Maz'ya's work in functional analysis, partial differential equations and applications, Birkhäuser (1999), 307–330.

[LM] M. Langer and V. Maz'ya, *On L^p-contractivity of semigroups generated by parabolic matrix differential operators*, J. Funct. Anal., **164** (1999), 73–109.

[Lank] P. Lankaster, *Theory of Matrices*, Academic Press, New York-London, 1969.

[Land] N.S. Landkof, *Foundations of Modern Potential Theory*, Springer-Verlag, 1972.

[Lem] R. Lemmert, *Über die Invarianz konvexer Teilmengen eines normierten Raumes in bezug auf elliptische Differentialgleichungen*, Comm. Partial Diff. Eq., **3**:4 (1978), 297–318.

[LS] S. Lenhart and P. Schaefer *On Comparison results for classical and viscosity solutions in elliptic systems*, in: "Evolution Equations", G. Ferreyra, G. R. Goldstein, F. Neubrander ed., Marcel Dekker, 1995, pp. 269–275.

[Lev1] E.E. Levi, *Sull'equazione del calore*, Reale Accad. dei Lincei, Roma, Rendiconti (5), **162** (1907), 450–456.

[Lev2] E.E. Levi, *Sull'equazione del calore*, Annali di Mat. Pura et Appl., **14** (1908), 187–264.

[Li] E. Lindelöf, *Mémoire sur certaines inégalités dans la théorie des fonctions monogènes et sur quelques propriétés nouvelles de ces fonctions dans le voisinage d'un point singulier essentiel*, Acta Soc. Sci. Fennicae, **35**:7 (1908), 1–35.

[Lo1] Ya.B. Lopatinskiĭ, *On a method of reducing boundary value problems for systems of differential equations of elliptic type to regular integral equations*, Ukrain. Mat. Žurnal, **5**:2 (1953), 123–151 (Russian).

[Lo2] Ya.B. Lopatinskiĭ, *On a type of singular integral equations*, Teor. Prikl. Mat., No. 2 (1963), 53-57 (Ukrainian).

[LG] J. López-Gómez, *The Strong Maximum Principle*, 385 p., to appear.

[LoMo] J. López-Gómez and M. Molina-Meyer, *The maximum principle for cooperative weakly coupled elliptic systems and some applications*, Diff. Int. Eq., **7**:2 (1994), 383–398.

[MaRo] A.J. Makintyre and W.W. Rogosinski, *Extremum problems in the theory of analytic functions*, Acta Math., **82** (1950), 275–325.

[MaRu1] P. Maremonti and R. Russo, *On the maximum modulus theorem for the Stokes system*, Ann. Scuola Norm. Sup. Pisa (IV), **30** (1994), 630–643.

[MaRu2] P. Maremonti and R. Russo, *On existence and uniqueness of classical solutions of the stationary Navier-Stokes equations and to the traction troblem of linear elastostatics*, Quaderni di Matematica. Classical Problems in Mechanics, **1** (1997), 171–251.

[Ma] A.A. Markov, *On mean values and exterior densities*, Mat. Sb., **4(46)**:1 (1938), 165–191.

[Ma1] V.G. Maz'ya, *Boundary integral equations of elasticity in domains with piecewise smooth boundaries*, EQUADIFF 6 Proc., Brno Univ., (1985), 235–242.

[Ma2] V.G. Maz'ya, *Boundary Integral Equations*, Encyclopedia of Math. Sc., **13**, Springer, Berlin-Heidelberg-New York, 1991.

[MK] V.G. Maz'ya and G.I. Kresin, *On the maximum principle for strongly elliptic and parabolic second order systems with constant coefficients*, Mat. Sb., **125(167)** (1984), 458–480 (Russian); English transl.: Math. USSR Sb., **53** (1986) ,457–479.

[MazMik] V.G. Maz'ya and S.G. Mikhlin, *The Cosserat spectrum of the equations of elasticity theory*, Vestnik Leningrad. Univ., **22**:13 (1967), 58-63 (Russian).

[MPl] V. Maz'ya and B.A. Plamenevskiĭ, *Estimates in L_p and Hölder classes and the Miranda-Agmon maximum principle for solutions of elliptic boundary value problems in domains with singular points on the boundary*, Math. Nachr., **81** (1978), 25–82, Engl. transl. in: Amer. Math. Soc. Transl., **123** (1984), 1–56.

[MR2] V. Maz'ya and J. Rossmann, *On the Agmon-Miranda maximum principle for solutions of elliptic equations in polyhedral and polygonal domains*, Ann. Global Anal. Geom., **9** (1991), 253–303.

[MR3] V. Maz'ya and J. Rossmann, *On the Agmon-Miranda maximum principle for solutions of strongly elliptic equations in domains of $\mathbb{R}^n$ with conical points*, Ann. Global Anal. Geom., **10** (1992), 125–150.

[MR] V. Maz'ya and J. Rossmann, *A maximum modulus estimate for solutions of the Navier-Stokes system in domains of polyhedral type*, Math. Nachr., **282**:3 (2009), 459–469.

[MR1] V. Maz'ya and J. Rossmann, *A maximum modulus estimate for solutions of the Navier-Stokes system in domains of polyhedral type*, Math. Nachr., **282**:3 (2009), 459-469.

[MR4] V. Maz'ya and J. Rossmann, *Elliptic Equations in Polyhedral Domains*, Mathematical Surveys and Monographs, **162**, Amer. Math. Soc., RI, 2010.

[MSh] V. Maz'ya and T. Shaposhnikova, *Jacques Hadamard, a Universal Mathematician*, Amer. Math. Soc. and London Math. Soc., 1998.

[MS] V. Maz'ya and P. Sobolevskiĭ, *On the generating operators of semigroups*, Uspekhi Mat. Nauk, **17** (1962), 151-154 (Russian).

[McL] W. McLean, *Strongly Elliptic Systems and Boundary Integral Equations*, Cambridge, Cambridge University Press, 2000.

[McMS] P. McMullen and G.C. Sheppard, *Convex Polytopes and the Upper Bound Conjecture*, London Math. Soc. Lect. Notes Series, **3**, Cambridge University Press, 1971.

[Me1] D. Medková, *Invariance of the Fredholm radius of the Neumann operator*, Čas. Pěst. Mat., **115** (1990), 147–164.

[Me2] D. Medková, *On essential norm of the Neumann operator*, Mathematica Bohemica, **117** (1992), 393–408.

[Me3] D. Medková, *The boundary value problems for Laplace equation and domains with non-smooth boundary*, Arch. Mathematicum (BRNO), **34** (1998), 173–181.

[MI] H. Minkowski, *Theorie der konvexen Körper, insbesondere Begründung ihres Oberflächenbegriffs*, Posthum. Gesammelte Abhandlungen von Herman Minkowski (ed. D. Hilbert), v. 2, Teubner, Leipzig, 1911.

[Mir] C. Miranda, *Partial Differential Equations of Elliptic Type*, Springer-Verlag, 1970.

[Mi1] C. Miranda, *Formule di maggiorazione e teorema di esistenza per le funzioni biarmoniche di due variabili*, Giorn. Mat. Bataglini, **78** (1948-1949), 97–118.

[Mi2] C. Miranda, *Teorema del massimo modulo e teorema di esistenza e di unicità per il problema di Dirichlet relativo alle equazioni ellitiche in due variabili*, Ann. Mat. Pura Appl., ser. 4, **46** (1958), 265–311.

[Mi3] C. Miranda, *Sul teorema del massimo modulo per una classe di sistemi ellitici di equazioni del secondo ordine e per le equazioni a coefficienti complessi*, Istit. Lombardo Accad. Sci. Lett. Rend. A, **104** (1970), 736–745.

[MS] E. Mitidieri and G. Sweers, *Weakly coupled elliptic systems and positivity*, Math. Nachr., **173**, (1995), 259–286.

[Mi1] I. Mitrea *Spectral radius properties for layer potentials associated with the elastostatics and hydrostatics equations in nonsmooth domains*, J. Fourier Anal. Appl., **5** (1999), 385–408.

[Mi2] I. Mitrea, *On the spectra of elastostatic and hydrostatic layer potentials on curvilinear polygons*, J. Fourier Anal. Appl., **8** (2002), 443–487.

[IM3] I. Mitrea, *On the traction problem for the Lamé system on curvilinear polygons*, J. Integral Equations Appl., **16** (2004), 175–218.

[MT] I. Mitrea and W. Tucker, *Some counterexamples for the spectral radius conjecture*, Differ. Integr. Eq., **16** (2003), 1409–1439.

[Mo] T. Moutard, *Notes sur les équations aux dérivées partielles*, J. de l'École Polytechnique, **64** (1894), 55–69.

[Mu] N.I. Muskhelishvili, *Some Basic Problems of the Mathematical Theory of Elasticity*, Noordhoff, Groningen-Holland, 1953.

[Na] D.G. Natroshvili, *Effective solution of the basic boundary value problems for homogeneous ball*, Trudy Inst. Prikl. Matem. Tbil. yn-ta, **3** (1972), 127–140.

[Ne1] I. Netuka, *Fredholm radius of a potential theoretic operator for convex sets*, Čas. Pěst. Mat., **100** (1975), 374–383.

[Nikd] O.M. Nikodým, *Contribution à la théorie des fonctionelles linéaires en connexion avec la théorie de la mesure des ensembles abstraits*, Mathematica, Cluj, **5** (1931), 131–141.

[Nikl] S.M. Nikol'skii, *Linear equations in linear normed spaces*, Izv. Akad. Nauk SSSR, Ser. Mat., **7**:3 (1943), 147–166.

[Nir1] L. Nirenberg, *A strong maximum principle for parabolic equations*, Comm. Pure Appl. Math., **6** (1953), 167–177.

[Nir2] L. Nirenberg, *The maximum principle and principal eigenvalue for second order elliptic equations in general bounded domains*, Problemi attuali del'analisi e della fisica matematica: dedicato a Gaetano Fichera nel suo 70^0 compleanno: atti del simposio internazionale, Taormina, 15-17 Ottobre, 1992; a cura di P.E. Ricci, Dip. di Mat., Universitá di Roma "La Sapinza" (1993), 189–194.

[Ny] K. Nyström, *Boundary value problems for parabolic Lamé systems in time-varying domains*, Indiana Univ. Math. J., **48**:4 (1999), 1285–1355.

[Ot] K. Otsuka, *On the positivity of the fundamental solutions for parabolic systems*, J. Math. Kioto Univ., **28** (1988), 119–132.

[Pa] A. Paraf, *Sur le problème de Dirichlet et son extension au cas de l'équation linéaire générale du second ordre*, Annales de la Faculté des Sciences de Toulouse (I), **6** (1892), 1-75.

[PP1] V.Z. Parton and P.I. Perlin, *Integral Equation Methods in Elasticity*, MIR, Moscow, 1982.

[PP2] Parton, V.Z. and Perlin P.I. (1981). *Methods in the Mathematical Theory of Elasticity*, Nauka, Moscow (Russian).

[Ph] R.R. Phelps, *Convex Functions, Monotone Operators and Differentiability*, Lect. Notes in Math., **1364**, Springer-Verlag, Berlin-Heidelberg-New York, 1989.

[Pic1] M. Picone, *Maggiorazione degli integrali delle equazioni totalmente alle derivate parziali del secondo ordine*, Annali di Mat. Pura et Applic. (4), **7** (1927), 145–192.

[Pic2] M. Picone, *Maggiorazione degli integrali di equazioni lineari ellitico-paraboliche alle derivate parziali del secondo ordine*, Accad. Naz. dei Lincei. Atti Rendiconti (6), **5** (1929), 138–143.

[Pic3] M. Picone, *Sul problema della propagazione del calore in un mezzo privo di frontiera, conduttore, isotropo, e omogeneo*, Mat. Ann., **101** (1929), 701–712.

[Pini] B. Pini, *Sui sistemi equazioni lineari a derivate parziali del secondo ordine dei tipi ellittico e parabolico*, Rend. del Sem. Mat. della Univ. di Padova, **22** (1953), 265-280.

[Po] G. Polya, *Liegt di Stelle der größten Beanspruchungen an der Oberfläche?*, Z. Angew. Math. Mech., **10** (1930), 353–360.

[PW] M.H. Protter and H.F. Weinberger, *Maximum Principles in Differential Equations*, Prentice-Hall, Inc., Englewood Cliffs, N.J., 1967; Springer-Verlag, New York Inc., 1984.

[Pr] M.H. Protter, *Maximum principles*, in "Maximum principles and eigenvalue problems in partial differential equations", P.W. Schaefer ed., Pittman Research Notes, Math. ser., **175** (1988), pp. 1–14.

[PBM1] A.P. Prudnikov, Yu.A. Brychkov and O.I. Marichev, *"Integrals and Series, Vol. 1, Elementary Functions"*, Gordon and Breach Sci. Publ., New York, 1986.

[PBM2] A. P. Prudnikov, Yu. A. Brychkov and O. I. Marichev, *Integrals and Series, Vol. 3: More Special Functions*, Gordon and Breach Sci. Publ., New York, 1989.

[PS] P. Pucci and J. Serrin, *The Maximum Principle*, Progress in Nonlinear Differential Equations and their Applications, **73**, Birkhäuser-Verlag, Basel, 2007.

[Ra] Yu.N. Rabotnov, *Elements of Hereditary Solid Mechanics*, Mir Publishers, Moscow, 1980.

[Rad1] J. Radon, *Über lineare Functionaltransformationen und Functionalgleichungen*, Sitzungsber. Akad. Wiss. Wien, Abt. 2a, **128**:7 (1919), 1083–1121.

[Rad2] J. Radon, *Über die Randwertaufgaben beim logarithmischen Potential*, Sitzungsber. Akad. Wiss. Wien, Abt. 2a, **128**:7 (1919), 1123–1167.

[Raj] C.T. Rajagopal, *Carathéodory's inequality and allied results*, Math. Student, **9** (1941), 73–77.

[Rat1] A. Rathsfeld, *The invertibility of the double layer potential operator in the space of continuous functions defined on a polyhedron: The panel method*, Appl. Analysis, **45** (1992), 135–177.

[Rat2] A. Rathsfeld, *The invertibility of the double layer potential operator in the space of continuous functions defined on a polyhedron, The panel method. Erratum*, Appl. Analysis, **56** (1995), 109–115.

[RW1] R. Redheffer and W. Walter, *Invariant sets for systems of partial differential equations. I. Parabolic equations*, Arch. Rat. Mech. Anal., **67** (1978), 41–52.

[RW2] R. Redheffer and W. Walter, *Invariant sets for systems of partial differential equations. II. First-order and elliptic equations*, Arch. Rat. Mech. Anal., **73** (1980), 19–29.

[RS] S. Rempel and G. Schmidt, *Eigenvalues for spherical domains with corners via boundary integral equations*, Int. Equat. Oper. Theory, **14** (1991), 229–250.

[Ri1] F. Riesz, *Sur les opérations fonctionnelles linéaires*, C. R. Acad. Sci. Paris, **149** (1909), 974–977.

[Ri2] F. Riesz, *Untersuchungen über Systeme integrierbarer Funktionen*, Math. Ann., **69** (1910), 449–497.

[Ro] R.T. Rockafellar, *Convex Analysis*, Princeton Univ. Press, Princton, N.J., 1970.

[RPKAS] A.E. Rodkina , A.S. Potapov, M.I. Kamenskiĭ, R.R. Akhmerov, B.N. Sadowsky. *Measures of noncompactness and condensing operators*, Birkhäuser Verlag, Basel Boston Berlin, 1992.

[Rus] I.A. Rus, *Un principe du maximum pour les solutions d'un systeme fortement elliptique*, Glasnik Mat., **4(24)**:1 (1969), 75-77.

[Rusc] St. Ruscheweyh, *Über einige Klassen im Einheitskreis holomorpher Funktionen*, Berichte der Mathematisch-statistischen Sektion im Forschungszentrum Graz, Bericht **no. 7** (1974), 12 S.

[Russ1] R. Russo, *On the maximum modulus theorem for the steady-state Navier-Stokes equations in Lipschitz bounded domains*, Appl. Anal., **90**:1-2 (2011), 193–200.

[Sa1] K.B. Sabitov, *Maximum modulus principle for some classes of second order elliptic and hyperbolic systems*, Differentsial'nye uravnenia, **27**:2 (1991), 272–278 (Russian); English. transl. in Differential Equations, **27**:2, (1991), 194–199.

[Shae] C. Schaefer, *Invariant sets and contractions for weakly coupled systems of parabolic differential equations*, Rend. Mat., **13** (1980), 337–357.

[Sch] B.-W. Schulze, *On a priori estimates in uniform norms for strongly elliptic systems*, Sibirsk. Mat. Zh., **16**:2 (1975), 384–394 (Russian); English. transl. in Siberian Math. J., **16**:2 (1975), 297–305.

[SW] B.-W. Schulze and G. Wildenhain, *Methoden der Potentialtheorie für elliptische Differentialgleichungen beliebiger Ordnung*, Akademie-Verlag, Berlin 1977; Birkhauser Verlag, Basel, 1977.

[Shap] Z.Ya. Shapiro, *The first boundary value problem for an elliptic system of differential equations*, Mat. Sb., **28(70)**:1 (1951), 55-78 (Russian).

[Shel1] V.Yu. Shelepov, *On the index of an integral operator of the potential type in the space L_p*, Dokl. Akad. Nauk SSSR, **186** (1969), 1266–1268; English transl.: Sov. Math Dokl., **10/A** (1969), 754–757.

[Shel2] V.Yu. Shelepov, *Investigations of matrix integral equations in the space of continuous functions by Ya. B. Lopatinskiĭ's method*, General Theory of Boundary Value Problems, Collect. Sci. Works, Naukova Dumka, Kiev, 1983, pp. 220-226 (Russian).

[Shel3] V.Yu. Shelepov, *On the index and spectrum of integral operators of potential type along Radon curves (Russian)*, Mat. Sb., **181**:6 (1990), 751–778; English transl.: Math. USSR Sb., **70**:1 (1991), 175–203.

[Shen] Z. Shen, *Boundary value problems for parabolic Lamé systems and a nonstationary linearized system of Navier-Stokes equations in Lipschitz cylinders*, Amer. J. of Mathematics, **113** (1991), 293–373.

[Shi] G.E. Shilov, *Mathematical Analysis: Second Special Course*, Nauka, Moscow, 1965; English transl.: *Generalizes Functions and Partial Differential Equations*, Gordon and Breach, New York, 1968.

[Si] B. Sirakov, *Some estimates and maximum principles for weakly coupled systems of elliptic PDE* , Nonlinear Anal., **70**:8 (2009), 3039–3046.

[Sm] J. Smoller, *Shock Waves and Reaction-Diffusion Equations*, Springer-Verlag, Berlin-Heidelberg-New York, 1983.

[So1] V.A. Solonnikov, *On general boundary value problems for systems elliptic in the Douglis-Nirenberg sense*, Izv. Akad. Nauk SSSR, ser. Mat., **28**:3 (1964), 665–706 ; English transl. Amer. Math. Soc. Transl. (2), **56** (1966), 193–232.

[So2] V.A. Solonnikov, *On boundary value problems for linear parabolic systems of differential equations of general form*, Proc. Steklov Inst. Math., **83** (1965), 3-163; Proc. Steklov Inst. Math., **83** (1965), 1-184.

[So3] V.A. Solonnikov, *On Green's matrices for elliptic boundary value problems, I*, Trudy Mat. Inst. Steklov, **110** (1970), 107–145; English transl.: Proc. Steklov Inst. Math., **110** (1970), 123–170.

[So4] V.A. Solonnikov, *On Green's matrices for elliptic boundary value problems, II*, Trudy Mat. Inst. Steklov, **116** (1971), 181–216; English transl.: Proc. Steklov Inst. Math., **116** (1971), 187–226.

[So5] V.A. Solonnikov, *On a maximum modulus estimate of the solution of stationary problem for the Navier-Stokes equations*, Zap. Nauchn. Sem. S.-Petersburg Otdel. Mat. Inst. Steklov (POMI), **249** (1997), 294–302; Transl. in J. Math. Sci. (New York), **101**:5 (2000), 3563–3569.

[Sp] R.P. Sperb, *Maximum principles and their applications*, Academic Press, New York-London-Toronto-Sydney-San Francisco, 1981.

[SW] O. Steinbach and W.L. Wendland, *On C. Neumann's method for second order elliptic systems in domains with non-smooth boundaries*, J. Math. Anal. Appl., **262** (2001), 733–748.

[Ste] H. Steinhaus *Additive und stetige Funktionaloperationen*, Math. Zeitschr., **5** (1919), 186–221.

[SW] J. Stoer and C. Witzgall, *Convexity and Optimization in Finite Dimensions I*, Springer, Berlin-Heidelberg-New York, 1970.

[Str] R.S. Strichartz, *Boundary values of solutions of elliptic equations satisfying H^p conditions*, Trans. Amer. Math. Soc., **176** (1973), 445–462.

[Sty1] T. Stys, *Aprioristic estimations of solutions of a certain elliptic system of partial differntial second order equations*, Bull. Acad. Pol. Sc., **13** (1965), 639–640.

[Sty2] T. Stys, *On the unique solvability of the first Fourier problem for a parabolic system of second order linear differential equations*, Prace Mat., **9** (1965), 283–289 (Russian).

[Szá] O. Szász, *Ungleichheitsbeziehungen für die Ableitungen einer Potenzreihe, die eine im Einheitskreise beschränkte Funktion darstellt*, Math. Z., **8** (1920), 303–309.

[Sze] P. Szeptycki, *Existence theorem for the first boundary value problem for a quasilinear elliptic system*, Bulletin Acad. Polon. des Sciences, **7** (1959), 419–424.

[Ta] A. Tartaglione, *On existence, uniqueness and the maximum modulus theorem in plane linear elastostatics for exterior domains*, Ann. Univ. Ferrara - Sez. VII - Sc. Mat., **47** (2001), 89–106.

[Ti] E.G. Titchmarsh, *The Theory of Functions*, Oxford University Press, 2nd ed., Oxford, 1949.

[Ve] G.C. Verchota, *Layer potentials and regularity for the Dirichlet problem for Laplace's equation in Lipschitz domains*, J. Funct. Anal., **59** (1984), 419–424.

[VV] G.C. Verchota and A.L. Vogel, *Nonsymmetric systems on nonsmooth planar domains*, Trans. Amer. Math. Soc., **349**:11 (1997), 4501–4535.

[VH] A.I. Volpert and S.I. Hudyaev, *Analyses in Classes of Discontinuous Functions and Equations of Mathematical Physics*, Martinus Nijhoff, Dordrecht, 1985.

[Wal] W. Walter, *Differential and Integral Inequalities*, Springer, Berlin-Heidelberg-New York, 1970.

[Was] J. Wasowski, *Maximum principles for a certain strongly elliptic system of linear equations of second order*, Bull. Acad. Polon. Sci., **18** (1970), 741–745.

[Wh] L.T. Wheeler, *Maximum principles in classical elasticity*, in: "Mathematical problems in elasticity", R. Russo ed., Singapore World Sc., (1996), pp. 157–185.

[We1] H.F. Weinberger, *Invariant sets for weakly coupled parabolic and elliptic systems*, Rend. Mat., **8** (1975), 295–310.

[We2] H.F. Weinberger, *Some remarks on invariant sets for systewms*, in "Maximum principles and eigenvalue problems in partial differential equations", P.W. Schaefer ed., Pittman Research Notes, Math. ser., **175** (1988), pp. 189–207.

[Zaa] A.C. Zaanen, *Integration*, North-Holland, Amsterdam, 1967.

[Za] L. Zalcman, Picard's theorem without tears, *Amer. Math. Monthly*, **85**:4 (1978), 265–268.

[Zh1] C. Zhou, Maximum principles for elliptic systems, *J. Math. Anal. and Appl.*, **159** (1991), 418–439.

[Zh2] C. Zhou, Maximum principles for parabolic systems coupled in both first-order and zero-order terms, *Intern. J. Math. & Math. Sci.*, **17**:4 (1994), 813–816.

List of Symbols

Point Sets

Ω	domain in $\mathbb{R}^n$ with closure $\overline{\Omega}$ and boundary $\partial\Omega$	1
Q_T	cylinder $\Omega \times (0,T]$ in the space $\mathbb{R}^{n+1}$	1
$\overline{\Gamma}_T$	parabolic boundary of the cylinder Q_T, where $\Gamma_T = \{(x,t) \in \partial Q_T : 0 \le t < T\}$	1
$\mathbb{R}^n_+$	upper half-space $\{x = (x_1, \ldots, x_n) : x_n > 0\}$ in $\mathbb{R}^n$	3
$\mathbb{R}^{n+1}_T$	layer $\mathbb{R}^n \times (0,T]$ in $\mathbb{R}^{n+1}$	4
$\boldsymbol{S}^{n-1}$	unit sphere in $\mathbb{C}^n$ centered at 0	16
$\mathbb{S}^{n-1}_-$	lower hemisphere $\{x \in \mathbb{R}^n : \lvert x\rvert = 1, x_n < 0\}$	24
$\mathbb{B}_r$	ball in $\mathbb{R}^n$ with radius r centered at 0	27
$\mathbb{R}^n_+(\boldsymbol{\nu})$	half-space $\{x \in \mathbb{R}^n : (x, \boldsymbol{\nu}) > 0\}$, where $\boldsymbol{\nu}$ is a unit n-dimensional vector	27
$\mathbb{S}^{n-1}$	unit sphere in $\mathbb{R}^n$ centered at 0	27
$\mathbb{B}$	unit ball in $\mathbb{R}^n$ centered at 0	34
$\mathbb{B}_r(y)$	ball in $\mathbb{R}^n$ with radius r centered at y	44
$\mathbb{R}^n_+(\mathcal{O})$	tangent space to $\partial\Omega$ at a point $\mathcal{O} \in \partial\Omega$	77
$\mathbb{D}$	disk $\lvert z\rvert < 1$ in the complex plane $\mathbb{C}$	148
$\mathbb{C}_+$	upper half-plane of the complex plane $\mathbb{C}$	149
$\mathcal{B}$	Borel set in $\mathbb{R}^n$ with interior int $\mathcal{B}$, closure $\overline{\mathcal{B}}$, boundary $\partial\mathcal{B}$ and complement $\mathbf{C}\mathcal{B} = \mathbb{R}^n \backslash \mathcal{B}$	154
$\partial^* E$	reduced boundary of a set $E \subset \mathbb{R}^n$	157
Π_T	set $\mathcal{D} \times (0,T]$, where $\mathcal{D}$ is either a bounded domain in $\mathbb{R}^n$ or $\mathcal{D} = \mathbb{R}^n$	205
$\overline{S}_T$	cylindrical surface, where $S_T = \partial\Omega \times (0,T]$	237
$\boldsymbol{B}$	unit ball of a generalized norm	252
$\mathcal{S}^{m-1}$	unit sphere of the generalized norm	253

Vectors

$\boldsymbol{e}_\sigma$	n-dimensional unit vector joining the origin to $\sigma \in \mathbb{S}^{n-1}$	56
$\boldsymbol{e}_{xy}$	n-dimensional unit vector joining the point x to point y	56
$\boldsymbol{e}_i$	unit vector of the i-th coordinate axis	60
$\boldsymbol{r}_{xy}$	n-dimensional vector joining the point x to point y with the length r_{xy}	154
Conv S	convex hull of a vector set S	255
Span S	linear span of a vector set S	256

Set Functions

$\omega(x, \mathcal{B})$	solid angle at which a set $\mathcal{B} \subset \partial\mathbb{R}^n_+$ is seen from $x \in \mathbb{R}^n_+$	24
H_k	k-dimensional Hausdorff measure in $\mathbb{R}^n$	154
mes_n	Lebesque measure in $\mathbb{R}^n$	154
$P(\mathcal{B})$	perimeter of the set $\mathcal{B}$ in the sense of Caccioppoli and De Giorgi	155
$\omega_{\mathcal{D}}(p, \mathcal{B})$	solid angle at which the set $\mathcal{B} \cap \partial\mathcal{D}$ is seen from $p \in \mathbb{R}^n$	156
$\Psi_{\mathcal{D}}(p, \mathcal{B})$	matrix-valued set function	161

Functions

Spaces

$[\mathbf{C}^k(\Omega)]^m$	space of complex m-component vector-valued functions with continuous derivatives up to order k in Ω	$33, 49$
$[\mathbf{C}^k(\Omega)]^m$	space of complex m-component vector-valued functions with continuous derivatives up to order k in Ω	$33, 49$
$[\mathrm{C}^{k,\alpha}(\overline{\Omega})]^{m\times m}$	space of real $(m \times m)$-matrix-valued functions whose elements have continuous derivatives up to order k and satisfy the Hölder condition with exponent α on $\overline{\Omega}$	33
$[\mathbf{C}^{k,\alpha}(\overline{\Omega})]^{m\times m}$	space of complex $(m \times m)$-matrix-valued functions whose elements have continuous derivatives up to order k and satisfy the Hölder condition with exponent α on $\overline{\Omega}$	$33, 49$
$\mathrm{C}_0^k(G)$	space of real functions with continuous derivatives up to order k with compact support in G	33
$\mathrm{C}_0(G)$	space of real continuous functions with compact support in G	33
$[W_p^l(\Omega)]^m$	Sobolev space of m-component vector-valued functions on Ω with each component in $W_p^l(\Omega)$	33
$[\mathring{W}_p^l(\Omega)]^m$	Sobolev space of m-component vector-valued functions on Ω with each component in $\mathring{W}_p^l(\Omega)$	33
$[L^p(G)]^m$	space of real vector-valued functions $\boldsymbol{u} = (u_1, \dots, u_m)$ for which $\lvert\boldsymbol{u}\rvert^p$ is integrable on $G \subset \mathbb{R}^n$	$79, 131$
$h^p(\mathbb{R}^n_+)$	Hardy space of harmonic functions on $\mathbb{R}^n_+$ which can be represented as the Poisson integral	117
$h^p(\mathbb{B})$	Hardy space of harmonic functions on $\mathbb{B}$ which can be represented as the Poisson integral	132
$BV(\mathbb{R}^n)$	space of locally integrable functions on $\mathbb{R}^n$ whose gradients (in the distributional sense) are finite vector-valued charges on $\mathbb{R}^n$	155
$[\mathrm{C}^{(k,1)}(\Pi_T)]^m$	space of real m-component vector-valued functions on Π_T whose derivatives with respect to x up to order k and first derivative with respect to t are continuous	205
$[\mathrm{C}_{\mathrm{b}}^{k,\alpha}(\mathbb{R}^n)]^m$	space of real m-component vector-valued functions on $\mathbb{R}^n$with continuous and bounded derivatives up to order k which satisfy the uniform Hölder condition with exponent α	205
$[\mathrm{C}_b^{k+\alpha,\alpha/2\ell}(\overline{\mathbb{R}^{n+1}_T})]^m$	space of real m-component vector-valued functions with derivatives up to order k with respect to x which are bounded in $\mathbb{R}^{n+1}_T$ and satisfy the uniform Hölder condition with exponent α with respect to the parabolic distance	205

$[\mathring{\mathrm{C}}(\overline{Q}_T)]^m$	space of m-component vector-valued functions from $[\mathrm{C}(\overline{Q}_T)]^m$ vanishing on $\overline{S}_T$	237

Operators

D_x	$(\partial/\partial x_1, \dots, \partial/\partial x_n)$	23
$\mathfrak{A}_0(D_x)$	principal homogeneous part of the operator $\mathfrak{A}(D_x)$	23
$F[\,\cdot\,]$	Fourier transform	26
Δ	Laplace operator	29
grad	gradient	29
div	divergence	29
$\mathfrak{C}_0(D_x)$	principal homogeneous part of the operator $\mathfrak{C}(D_x)$	30
$\mathfrak{A}(x, D_x)$	linear differential operator of the second order with real $(m \times m)$-matrix-valued coefficients defined on $\overline{\Omega}$	33
$\mathfrak{A}_0(x, D_x)$	principal homogeneous part of the operator $\mathfrak{A}(x, D_x)$	33
$\mathfrak{A}(D_x)$	linear differential operator of the second order whose coefficients are real constant $(m \times m)$-matrices	33
$[A, B]$	commutator of operators A and B	34
$\mathfrak{C}(x, D_x)$	linear differential operator of the second order with complex $(m \times m)$-matrix-valued coefficients defined on $\overline{\Omega}$	49
$\mathfrak{C}(D_x)$	linear differential operator of the second order whose coefficients are complex constant $(m \times m)$-matrices	50
D_x^β	$\partial^{\lvert\beta\rvert}/\partial x_1^{\beta_1} \dots \partial x_n^{\beta_n}$, where $\beta = (\beta_1, \cdots, \beta_n)$ is a multi-index of the order $\lvert\beta\rvert = \beta_1 + \cdots + \beta_n$	93
$P(D_x)$	elliptic operator of order 2ℓ with constant complex coefficients	93
$P_0(D_x)$	principal homogeneous part of $P(D_x)$	93
Δ^2	biharmonic operator	98
$F^{-1}[\,\cdot\,]$	inverse Fourier transform	100
$T_\kappa^{(n)}$	matrix-valued integral operator generated by the vector-valued elastic/hydrodynamic double layer potential $\boldsymbol{W}_\kappa^{(n)}(q)$	153
$\mathfrak{P}(x, t, D_x)$	linear differential operator of order 2ℓ with real $(m \times m)$-matrix-valued coefficients defined on Π_T	205
$\mathfrak{P}_0(x, t, D_x)$	principal homogeneous part of the operator $\mathfrak{P}(x, t, D_x)$	205
$\mathfrak{L}(x, t, D_x)$	linear differential operator of order 2ℓ with complex $(m \times m)$-matrix-valued coefficients defined on Π_T	215

$\mathfrak{L}_0(x,t,D_x)$	principal homogeneous part of the operator $\mathfrak{L}(x,t,D_x)$	215
$\mathfrak{A}_0(x,t,D_x)$	principal homogeneous part of the operator $\mathfrak{A}(x,t,D_x)$	256
$\mathfrak{A}(x,t,D_x)$	linear differential operator of the second order with real $(m\times m)$-matrix-valued coefficients defined on Π_T	280

Other Symbols

$\nabla_k u$	gradient of order k of a function u	$1, 93$
∇u	gradient of u	3
ω_n	area of the unit sphere in $\mathbb{R}^n$	3
$\mathcal{B}_X$	σ-algebra of Borel subsets of X	9
$\lvert\cdot\rvert$	the length of a vector in Euclidean or unitary space	10
$(\cdot,\cdot)$	inner product of vectors in Euclidean or unitary space	10
$\boldsymbol{\lvert}\cdot\boldsymbol{\rvert}$	variation of a scalar or vector-valued measure	$10, 12$
$\Re$	real part	$16, 148$
$\Im$	imaginary part	$16, 149$
δ_{ij}	Kronecker delta	64
$\lVert\cdot\rVert_p$	the norm in L^p-space	$79, 131$
$\partial u/\partial\boldsymbol{\ell}$	derivative of u in the direction of a unit vector $\boldsymbol{\ell}$	105
$\Lambda_p(u)$	best approximation of a function u on $\mathbb{S}^{n-1}$ by a constant in the norm of $L_p(\mathbb{S}^{n-1})$	108
d_x	distance from the point x to the boundary of domain	116
$\mathcal{Q}_{\mathcal{D}}(u;x)$	characteristics of bounded or semibounded functions in a domain $\mathcal{D}\subset\mathbb{R}^n$	116
$\operatorname{osc}_{\mathcal{D}}(u)$	oscillation of a function u on a domain $\mathcal{D}\subset\mathbb{R}^n$	148
$\operatorname{ess}\lVert L\rVert$	essential norm of a linear bounded operator L acting on a Banach space $\mathfrak{B}$	151
v_n	volume of the unit ball in $\mathbb{R}^n$	154
$\operatorname{dist}(p,\mathcal{B})$	distance from a point p to a set $\mathcal{B}$	154
$R(L)$	Fredholm radius of a linear bounded operator L acting on a Banach space $\mathfrak{B}$	197
C_L	continuity degree of a linear bounded operator L acting on a Banach space $\mathfrak{B}$	198
$d[(x,t),(x',t')]$	parabolic distance between the points (x,t) and (x',t') in $\mathbb{R}^{n+1}$	203
$\boldsymbol{\lvert}\cdot\boldsymbol{\rvert}$	norm in the Minkowski sense (generalized norm)	252
$\boldsymbol{\lvert}\cdot\boldsymbol{\rvert}_*$	dual generalized norm	252
$[\boldsymbol{v}_1,\dots,\boldsymbol{v}_m]$	$(m\times m)$-matrix whose columns are m-component vectors $\boldsymbol{v}_1,\dots,\boldsymbol{v}_m$	256
$\emptyset$	empty set	292

Index

Selected Published Titles in This Series

183 **Gershon Kresin and Vladimir Maz′ya,** Maximum Principles and Sharp Constants for Solutions of Elliptic and Parabolic Systems, 2012

182 **Neil A. Watson,** Introduction to Heat Potential Theory, 2012

181 **Graham J. Leuschke and Roger Wiegand,** Cohen-Macaulay Representations, 2012

180 **Martin W. Liebeck and Gary M. Seitz,** Unipotent and Nilpotent Classes in Simple Algebraic Groups and Lie Algebras, 2012

179 **Stephen D Smith,** Subgroup Complexes, 2011

178 **Helmut Brass and Knut Petras,** Quadrature Theory, 2011

177 **Alexei Myasnikov, Vladimir Shpilrain, and Alexander Ushakov,** Non-commutative Cryptography and Complexity of Group-theoretic Problems, 2011

176 **Peter E. Kloeden and Martin Rasmussen,** Nonautonomous Dynamical Systems, 2011

175 **Warwick de Launey and Dane Flannery,** Algebraic Design Theory, 2011

174 **Lawrence S. Levy and J. Chris Robson,** Hereditary Noetherian Prime Rings and Idealizers, 2011

173 **Sariel Har-Peled,** Geometric Approximation Algorithms, 2011

172 **Michael Aschbacher, Richard Lyons, Stephen D. Smith, and Ronald Solomon,** The Classification of Finite Simple Groups, 2011

171 **Leonid Pastur and Mariya Shcherbina,** Eigenvalue Distribution of Large Random Matrices, 2011

170 **Kevin Costello,** Renormalization and Effective Field Theory, 2011

169 **Robert R. Bruner and J. P. C. Greenlees,** Connective Real K-Theory of Finite Groups, 2010

168 **Michiel Hazewinkel, Nadiya Gubareni, and V. V. Kirichenko,** Algebras, Rings and Modules, 2010

167 **Michael Gekhtman, Michael Shapiro, and Alek Vainshtein,** Cluster Algebras and Poisson Geometry, 2010

166 **Kyung Bai Lee and Frank Raymond,** Seifert Fiberings, 2010

165 **Fuensanta Andreu-Vaillo, José M. Mazón, Julio D. Rossi, and J. Julián Toledo-Melero,** Nonlocal Diffusion Problems, 2010

164 **Vladimir I. Bogachev,** Differentiable Measures and the Malliavin Calculus, 2010

163 **Bennett Chow, Sun-Chin Chu, David Glickenstein, Christine Guenther, James Isenberg, Tom Ivey, Dan Knopf, Peng Lu, Feng Luo, and Lei Ni,** The Ricci Flow: Techniques and Applications, 2010

162 **Vladimir Maz′ya and Jürgen Rossmann,** Elliptic Equations in Polyhedral Domains, 2010

161 **Kanishka Perera, Ravi P. Agarwal, and Donal O'Regan,** Morse Theoretic Aspects of p-Laplacian Type Operators, 2010

160 **Alexander S. Kechris,** Global Aspects of Ergodic Group Actions, 2010

159 **Matthew Baker and Robert Rumely,** Potential Theory and Dynamics on the Berkovich Projective Line, 2010

158 **D. R. Yafaev,** Mathematical Scattering Theory, 2010

157 **Xia Chen,** Random Walk Intersections, 2010

156 **Jaime Angulo Pava,** Nonlinear Dispersive Equations, 2009

155 **Yiannis N. Moschovakis,** Descriptive Set Theory, Second Edition, 2009

154 **Andreas Čap and Jan Slovák,** Parabolic Geometries I, 2009

153 **Habib Ammari, Hyeonbae Kang, and Hyundae Lee,** Layer Potential Techniques in Spectral Analysis, 2009

152 **János Pach and Micha Sharir,** Combinatorial Geometry and Its Algorithmic Applications, 2009

151 **Ernst Binz and Sonja Pods,** The Geometry of Heisenberg Groups, 2008

150 **Bangming Deng, Jie Du, Brian Parshall, and Jianpan Wang,** Finite Dimensional Algebras and Quantum Groups, 2008

149 **Gerald B. Folland,** Quantum Field Theory, 2008

148 **Patrick Dehornoy, Ivan Dynnikov, Dale Rolfsen, and Bert Wiest,** Ordering Braids, 2008

147 **David J. Benson and Stephen D. Smith,** Classifying Spaces of Sporadic Groups, 2008

146 **Murray Marshall,** Positive Polynomials and Sums of Squares, 2008

145 **Tuna Altınel, Alexandre V. Borovik, and Gregory Cherlin,** Simple Groups of Finite Morley Rank, 2008

144 **Bennett Chow, Sun-Chin Chu, David Glickenstein, Christine Guenther, James Isenberg, Tom Ivey, Dan Knopf, Peng Lu, Feng Luo, and Lei Ni,** The Ricci Flow: Techniques and Applications, 2008

143 **Alexander Molev,** Yangians and Classical Lie Algebras, 2007

142 **Joseph A. Wolf,** Harmonic Analysis on Commutative Spaces, 2007

141 **Vladimir Maz′ya and Gunther Schmidt,** Approximate Approximations, 2007

140 **Elisabetta Barletta, Sorin Dragomir, and Krishan L. Duggal,** Foliations in Cauchy-Riemann Geometry, 2007

139 **Michael Tsfasman, Serge Vlăduţ, and Dmitry Nogin,** Algebraic Geometric Codes: Basic Notions, 2007

138 **Kehe Zhu,** Operator Theory in Function Spaces, Second Edition, 2007

137 **Mikhail G. Katz,** Systolic Geometry and Topology, 2007

136 **Jean-Michel Coron,** Control and Nonlinearity, 2007

135 **Bennett Chow, Sun-Chin Chu, David Glickenstein, Christine Guenther, James Isenberg, Tom Ivey, Dan Knopf, Peng Lu, Feng Luo, and Lei Ni,** The Ricci Flow: Techniques and Applications, 2007

134 **Dana P. Williams,** Crossed Products of C^*-Algebras, 2007

133 **Andrew Knightly and Charles Li,** Traces of Hecke Operators, 2006

132 **J. P. May and J. Sigurdsson,** Parametrized Homotopy Theory, 2006

131 **Jin Feng and Thomas G. Kurtz,** Large Deviations for Stochastic Processes, 2006

130 **Qing Han and Jia-Xing Hong,** Isometric Embedding of Riemannian Manifolds in Euclidean Spaces, 2006

129 **William M. Singer,** Steenrod Squares in Spectral Sequences, 2006

128 **Athanassios S. Fokas, Alexander R. Its, Andrei A. Kapaev, and Victor Yu. Novokshenov,** Painlevé Transcendents, 2006

127 **Nikolai Chernov and Roberto Markarian,** Chaotic Billiards, 2006

126 **Sen-Zhong Huang,** Gradient Inequalities, 2006

125 **Joseph A. Cima, Alec L. Matheson, and William T. Ross,** The Cauchy Transform, 2006

124 **Ido Efrat,** Valuations, Orderings, and Milnor K-Theory, 2006

123 **Barbara Fantechi, Lothar Göttsche, Luc Illusie, Steven L. Kleiman, Nitin Nitsure, and Angelo Vistoli,** Fundamental Algebraic Geometry, 2005

122 **Antonio Giambruno and Mikhail Zaicev,** Polynomial Identities and Asymptotic Methods, 2005

121 **Anton Zettl,** Sturm-Liouville Theory, 2005

120 **Barry Simon,** Trace Ideals and Their Applications, Second Edition, 2005

119 **Tian Ma and Shouhong Wang,** Geometric Theory of Incompressible Flows with Applications to Fluid Dynamics, 2005

118 **Alexandru Buium,** Arithmetic Differential Equations, 2005

117 **Volodymyr Nekrashevych,** Self-Similar Groups, 2005